GW00493812

The Cable and Telecommunications Professionals' Reference

PSTN, IP and Cellular Networks, and Mathematical Techniques

Edited by Goff Hill

ELSEVIER

AMSTERDAM • BOSTON • HEIDELBERG • LONDON
NEW YORK • OXFORD • PARIS • SAN DIEGO
SAN FRANCISCO • SINGAPORE • SYDNEY • TOKYO

Focal Press is an imprint of Elsevier

Focal Press

Acquisitions Editor: Angelina Ward
Assistant Editor: Doug Shults
Publishing Services Manager: George Morrison
Project Manager: Marilyn E. Rash
Marketing Manager: Christine Degon Veroulis
Typesetting and Illustrations: SNP Best-set Typesetter Ltd., Hong Kong
Cover Design: Eric DeCicco
Cover Image: Courtesy of istockphoto.com, © Martin Hendriks
Interior Printer: Sheridan Books
Cover Printer: Phoenix Color Corp.

Focal Press is an imprint of Elsevier
30 Corporate Drive, Suite 400, Burlington, MA 01803, USA
Linacre House, Jordan Hill, Oxford OX2 8DP, UK

Copyright © 2007, Elsevier Inc. All rights reserved.

No part of this publication may be reproduced, stored in a retrieval system, or transmitted in any form or by any means, electronic, mechanical, photocopying, recording, or otherwise, without the prior written permission of the publisher.

Permissions may be sought directly from Elsevier's Science & Technology Rights Department in Oxford, UK: phone: (+44) 1865 843830, fax: (+44) 1865 853333, e-mail: permissions@elsevier.com. You may also complete your request online via the Elsevier homepage (*http://elsevier.com*), by selecting "Support & Contact," then "Copyright and Permission," and then "Obtaining Permissions."

Recognizing the importance of preserving what has been written, Elsevier prints its books on acid-free paper whenever possible.

Library of Congress Cataloging-in-Publication Data

The cable and telecommunications professionals' reference / edited by Goff Hill. — 3rd ed.
    p.   cm.
  Rev. ed. of: Telecommunications engineer's reference book / edited by Fraidoon Mazda.
    2nd ed. 1998.
  Includes bibliographical references and index.
  ISBN-13: 978-0-240-80747-8 (alk. paper)
  ISBN-10: 0-240-80747-2 (alk. paper)
 1. Telecommunication.   I. Hill, Goff.   II. Telecommunications engineer's reference book.
  TK5101.M37 2007
  621.382—dc22   2007003442

British Library Cataloguing-in-Publication Data

A catalogue record for this book is available from the British Library.

For information on all Focal Press publications,
visit our website at *www.books.elsevier.com*.

07 08 09 10 11   10 9 8 7 6 5 4 3 2 1

Working together to grow
libraries in developing countries

www.elsevier.com | www.bookaid.org | www.sabre.org

ELSEVIER   BOOK AID International   Sabre Foundation

To Mary, Joanna, Stephen, Suzanne, and Eleanor

Contents

Preface

The past decade has seen massive change in the telecom industry. The industry structure, regulation, technology, and services have all developed progressively. New technologies and services have appeared, displacing long-established ones, and the convergence of different networks, technologies, and services compounds the change.

The predecessor to this reference was called the *Telecommunications Engineer's Reference Book*. The recent trends toward convergence of voice, data and video technologies, and services has suggested that the title, scope, and content of the book should reflect these changes; for this reason, this new reference is titled *The Cable and Telecommunications Professionals' Reference*. Some of the content in the earlier edition has become dated and a substantial amount of new material is now included, so this text remains an authoritative reference book. Since some of the earlier content remains fundamentally relevant, it is retained in this volume.

Inevitably this means that some of the earlier material relating to the displaced technologies should, in some cases, be downsized and, in other cases, removed. Even so, the huge amount of new material entering the scene means that covering the full scope would lead to a significant increase in the length of the book. The current edition is therefore divided into three volumes and I have attempted to associate related material in each. Later volumes will address broadband network technologies and the ways in which they support new services, and core transport network technologies and the changes that are enabling them to deal with variable types of traffic.

Because of the many relationships among different topics, it is not possible to split them into three completely independent, separate volumes. The following are the topics covered in each:

- Volume 1: PSTN, IP and cellular networks, and mathematical techniques
- Volume 2: Core transport systems
- Volume 3: Broadband access

The topics in Volume 1 seem to form a logical grouping because of the historical influence of telephone networks on today's communications systems. Traditionally, telephone networks have been provided primarily, but not entirely, to carry voice traffic. Because of their ubiquity and versatility, they have also often been used to carry data services. Early analogue networks have been replaced by digital networks. Cellular radio and intelligent control processes have led to huge progress in mobile data services. More recently, the availability of the Internet and the Voice over Internet Protocol has provided an entirely new paradigm for data and multimedia services. Mobile networks themselves are developing into ubiquitous networks, able to offer a wide range of data and video, as well as voice services. The Internet, which began as a data network, is now able to support voice and other real-time services.

These three networking approaches (circuit-switched, Internet, and cellular mobile) therefore provide the basis for this first volume, which is divided into six parts. Part One provides an overview of the industry structure and discusses standards and regulation. The underlying rationale for the OSI 7-layer reference model is also introduced in this part, as it underpins modern network architectural design.

Part Two deals with traditional circuit-switched, landline telephone networks (often referred to as Public Switched Telephone Networks or PSTN), and switching systems. While this existing infrastructure is increasingly regarded as being in decline, it is likely to be with us for some time yet and this volume would be incomplete without its inclusion.

Part Three describes basic principles of the Internet and Internet telephony. This includes the TCP/IP protocols, the carriage of real-time voice signals over the Internet, and developments leading to Next Generation Networks (NGN). The principles of Local Area Networks (LANs) are also covered here.

Part Four explains the principles of cellular mobile networks and the wide range of developments that have led to 2G, 2.5G, and 3G networks. Other wireless topics will be covered in later volumes.

Network design and optimization depend very heavily on mathematical techniques. The mathematical methods used in the design of cable and telecommunications networks have progressed with the technology. The basic methods, as well as more recent developments, are therefore discussed in Part Five, together with coverage of statistical methods, traffic simulation, queuing theory, and teletraffic.

In addition to the specialist mathematical methods used in network design, the more basic mathematical techniques associated with cable and telecommunications networks are covered in Part Six. These are reproduced from the earlier *Telecommunication Engineer's Reference Book* on the grounds that the basic techniques are "time invariant."

I am greatly indebted to the authors who have contributed to this volume of *The Cable and Telecommunications Professionals' Reference*. They are all experts in their fields and work at the forefront of technology. I should like to express my thanks to them and to their respective organizations for all the support they have given.

Contributors

J. Barron, BA, MA (Cantab)

J Barron (1926–1998) read physics at Cambridge University and served as a lecturer at Cambridge University in both physics and engineering.

Yael Baruch
Yael Consultancy

Yael Baruch studied Chemistry at Ben-Gurion University, Beer-Sheva (BS) and at the Hebrew University of Jerusalem (MS), after which she changed to computer studies at Sivan College, Tel Aviv. Yael has a Director's qualification from Tel Aviv University in the Faculty of Management and qualified as a Mediator at the Israel Centre for Negotiation and Mediation. She has 23 years of experience in cellular systems, including 1G to 3.5G, pagers, and military radios. She gained this experience in various assignments around the world, including in Europe, the United States, and Israel. Yael worked in leading companies, such as Motorola, Alcatel, Orange, Qualcomm, and has accomplished experience in many diverse areas (research and development, management, marketing, and more). In addition, she has participated in communications standard committees as 3GPP, ETSI, and ITU for 14 years. Working as a consultant for the past five years, Yael has served numerous public and private sectors and works as a Project Evaluator for the European Commission in Brussels.

John Buckley, MA, MSc, DIC, CEng, MIEE, MCMI, Chartered Engineer
Telecoms and Management Consultant

John Buckley is the author of *Telecommunications Regulation*, published by IEE Books in 2003. Now an independent consultant, Buckley has more than 35 years of experience in the telecommunications industry. He gained his first degree in physics from Oxford University and his Master's degree in computing science from Imperial College, London. He worked with BT on telecommunications switching systems, real-time software development, and the methods of achieving ultrareliable software. He went on to pursue project management and product development for digital PBX systems before returning to advanced research in network futures and network architectures, on which subjects he published a number of papers. Working as a consultant for the past 10 years, he has served numerous public and private sector clients from Britain, Europe, Russia, and the United States on technical and market research-oriented projects, including work for the European Commission. He is a corporate member of the IEE, the IEEE, and the Chartered Management Institute.

Ian J. Dilworth, BSc, PhD
University of Essex

Ian Dilworth is an active researcher, consultant, and teacher in the Radio and RF areas. Above all, he is an expert *practical* RF engineer in these fields from antenna to RF hardware. He has more than 30 years experience as a consultant. Before becoming a consultant, he acquired wide experience working for EMI Ltd. in the United Kingdom, COMSAT labs in the United States, CRC in Canada, and SIRIM in Malaysia. He consults on RF issues to national and syndicated publications (e.g., the U.K.'s *Times* newspaper) and actively designs and delivers RF/Radio courses for academia and industry in Europe, the United States, and Asia. Ian Dilworth is an RF expert to the European Commission (EC) with involvement in multidisciplinary EC projects. He has authored more than 80 academic research papers and more than 40 other publications, including patents. In his university role, he has managed research teams and has supervised many MSc/PhD projects as well as acting as an external PhD examiner worldwide. Currently, he holds Knowledge Transfer Partnerships (KTPs), which aim to pass expertise from academia to U.K. industry in the EMC design and radio propagation areas.

Arnaud Doucet, MEng, PhD
University of British Columbia

Arnaud Doucet received the MEng from Telecom INT in 1993, then a PhD from the University Paris XI, Orsay in 1997. Over the years he has been a Research Associate at Cambridge University (1998–2001); a Senior Lecturer at Melbourne University (2001–2002); a University Lecturer (U.K. Assistant Professor) at Cambridge University (2002–2005); and an Associate Professor, Department of Computer Science and Department of Statistics, at the University of British Columbia (2005–present).

J. E. Flood, OBE, DSc, FCGI, FInstP, CEng, FIEE
Aston University

John Flood is Emeritus Professor of Electrical Engineering at Aston University. He joined the university after 20 years in the telecommunications industry. Professor Flood has served on many bodies concerned with the development of telecommunications and has been chairman of the IEE Professional Group on Telecommunication Networks and chairman of the British Telecommunications Standards Committee.

Harold C. Folts, BSEE, MSSM

William J. Fitzgerald, MSc, PhD
University of Cambridge

William J. Fitzgerald received the BSc, MSc, and PhD in physics from the University of Birmingham, England. Following seven years at the Institut Laue Langevin in Grenoble, France, he was a Professor of Physics at the ETH in Zurich for five years. He then joined the Engineering Department in Cambridge where he is the Professor of Applied Statistics and Signal Processing and a Fellow of Christ's College.

R. J. Gibbens, MA, Dip Math Stat, PhD
University of Cambridge

Richard Gibbens is a Fellow and Senior Lecturer in the Computer Laboratory at the University of Cambridge, U.K. Richard's research interests are in mathematical modeling of networks. He teaches in both computer science and mathematics. Richard is coauthor of *Dynamic Alternative Routing* published by Prentice Hall International (UK) Ltd.

Goff Hill, BSc, MIET
GTel Consultancy Ltd.

Goff Hill is managing director of GTel Consultancy, which provides advice on broadband and core transport networks and is active in supporting Europe's Framework Program research. Globally recognized as a pioneer of "optical network" technology, Goff has more than 40 years of experience in telecommunications. This includes 17 years in optical telecommunications as a Technical Group Leader and Project Manager with BT, and two years as a Chief Network Architect with ilotron and later with Altamar Networks. He has acted as a Special Editor for *IEEE Communications Magazine* and for the *Journal of Lightwave Technology*. He has published more than 70 technical papers and has delivered lectures for university and professional short courses. Goff has an honors degree in electrical engineering from the University of Newcastle-upon-Tyne and is a Member of the Institution of Engineering and Technology.

David K. Hunter, BEng, PhD, SMIEE
University of Essex

David Hunter is a Reader in the Department of Electronic Systems Engineering at the University of Essex, U.K. In 1987, he obtained a first-class honors BEng in electronics and microprocessor engineering from the University of Strathclyde, U.K., and a PhD from the same university in 1991 for research on optical TDM switch architectures. After that, he remained at Strathclyde to research optical networking and optical packet switching, holding an EPSRC Advanced Fellowship from 1995 to 2000. After spending a year as a Senior Researcher in Marconi Labs in Cambridge, he moved to the University of Essex in August 2002, where his teaching has concentrated on TCP/IP, computer networks, and network performance evaluation. David has authored or coauthored more than 100 publications and has acted as an external PhD examiner for the Universities of Cambridge, London, and Essex. From 1999 until 2003, he was an Associate Editor for the *IEEE Transactions on Communications*, and he has been an Associate Editor for the *IEEE/OSA Journal of Lightwave Technology* since 2001. He participated in editing an Optical Networks Special Issue of that journal, published in December 2000. David is a Senior Member of the IEEE and a Professional Member of the ACM.

Daniel Mallah
Clikely Consulting

Daniel Mallah is the managing director of Clikely Consulting. He has been a consultant for telecom manufacturers in Israel and abroad for the past six years. For the past 27 years, he has worked with several telecommunications manufacturers in Israel and has more than 15 years of experience in the Access Network (the "last mile"). Daniel holds an MS in electronics from the Université de Paris and is presently completing an MBA at Bar Ilan University. He is also the Chairman of the "Groupe Telecom," a networking group of Israeli-French speakers active in high tech and telecom domains.

Fraidoon Mazda, MPhil DFH DMS MBIM, CEng FIEE

Fraidoon Mazda was the previous editor of this book, which was published as the *Telecommunications Engineer's Reference Book*. He has held various senior technical positions within the electronics and telecommunications industry and was previously a technical manager for Nortel (northern Telecom). He has edited a series of eight pocket books for Focal Press, which were derived from earlier editions of this reference book.

Elena Punskaya, BS, MSci, MPhil., PhD
University of Cambridge

Elena Punskaya received the BS and MSci degrees in mechanical engineering from the Bauman Moscow State Technical University, Moscow, in 1997, and the MPhil and PhD degrees in information engineering from the University of Cambridge, U.K., in 1999 and 2003, respectively. She is currently a Fellow of Homerton College, Cambridge, U.K., where she is a Director of Studies in Engineering. Her research interests focus on Bayesian inference, simulation-based methods, in particular, Markov chain Monte Carlo and particle filtering, and their application to digital communications.

Martin Reed, BEng, PhD, MIET, MIEE
University of Essex

Martin Reed is a Lecturer at the University of Essex, U.K. He is an active researcher in the fields of transmitting media over packet networks and traffic engineering in IP networks; having published more than 25 papers in the fields of network control and audio signal processing. He was an early experimenter in spatialized (multichannel) voice for multiparty conferencing transmitted over ATM and IP networks with a 1998 project funded in the United Kingdom by BT. This earlier work has inspired projects investigating a number of issues with transmitting the next generation of voice applications over packet networks, that range from network level traffic control to signal processing requirements such as echo cancellation.

Tony Richardson, MSc, CEng, MIEE
System Technology Solutions Ltd.

Within the TeleManagement Forum, Tony is technical director of Strategic Liaisons and NGN Management, managing technical and business relationships between TMF and other industry groups; he also manages TMF's program of work on Next Generation Network (NGN) management. In addition, he is managing director of System Technology Solutions (STS) Ltd., a small company specializing in the provision of telecommunications and information technology (IT) consultancy services. Prior to forming his own company, Tony had more than 30 years of experience in telecommunications developments working for BT. When he left BT, he was the senior engineering advisor on Management System Platforms in the Ipswich Systems Engineering Centre, which had approximately 800 development staff. Tony is a Chartered Electronics Engineer and holds a Master's degree in computer science (University of Hertfordshire).

Eur Ing Sydney F. Smith, BSc (Eng), CEng, FIET
Telecommunications Consultant

Sydney Smith has worked for Siemens, ITT, and Nortel on the design and development of all types of telephone switching systems from upgrades of Strowger and manual equipment through to the hardware and software for modern digital systems. He represented ITT on the CCITT/ITU Study Group, which was responsible for defining Signaling System No 7. Sydney also spent some years as a lecturer in telecommunications.

Phil Whiting, BA, MSc, PhD
Bell Labs Research

Phil Whiting received his BA degree from the University of Oxford, his MSc from the University of London, and his PhD in queuing theory from the University of Strathclyde. After a post-doctorate at the University of Cambridge, Phil's interests have centered on wireless. In 1993, he participated in the Telstra trial of Qualcomm CDMA in South Eastern Australia. He then joined the Mobile Research Centre at the University of South Australia, Adelaide. Since 1997, he has been with Bell Labs. His main interests are the mathematics of wireless networks, particularly stochastic models for resource allocation, information and coding theory. Phil's current research includes scheduling in wireless MIMO networks, the ensemble analysis of LDPC codes, and schemes for performing hybrid ARQ.

Part One
Standards and
Regulation

The economic well-being of a country has long been dependent on a good, reliable telecommunications service. The telecommunications sector underpins all forms of commercial, industrial, and domestic life, and its smooth running is vital. As the information society develops, and new technologies bring more advanced means of communications, this dependence continues to grow. Governments at both national and continental levels therefore take a keen interest in establishing and maintaining the health of their telecommunications industry sectors, and they set up regulatory bodies to oversee the market and ensure its proper functioning. Regulatory infrastructures at national and regional levels are therefore created to ensure that the interests of consumers are properly served, that the market operates smoothly, and that competition is open and fair.

In itself, good regulation does not guarantee a healthy industry sector. That calls also for strong industrial players with dynamic interactions. Standards bodies and related organizations play a vital role in promoting constructive cooperation as the industry develops and deploys the latest technologies within the market. The standards bodies, fora, and related trade associations strive to ensure full advantage is taken of rapidly developing technological capabilities, and they provide vital links in the value chain to deliver solutions to end users.

The shape of the regulatory environment is also important to the standards groups, as a healthy sector is essential for the successful deployment of standardized products. Regulatory bodies monitor and analyze the market and its operation and invite opinions from industry on a wide range of regulatory matters. Individual entities are invited to make submissions to regulatory bodies, but in addition to this, lobbying groups such as the European Telecommunications Network Operators Association (ETNO) and European Competitive Telecommunications Association (ECTA) were formed to ensure a stronger, united voice representing the interests of their particular groups.

In this part of *The Cable and Telecommunications Professionals' Reference*, an introduction is provided to both standards (Chapter 1) and regulation (Chapter 2). Further reference to specific standards and standards bodies is made throughout the volume. Also in this part, the open systems interconnection reference model, a particular standard that has fundamental relevance to virtually all telecommunications systems, is summarized in Chapter 3.

1 Standards

Yael Baruch

Introduction

A standard is like a cookbook for experts about a specific kind of food, for example sushi, that was written by a group of chefs. The book contains functional rules about making sushi, but without specific recipes. It is left to the chef's creativity to get the best sushi results following the restricting rules (rules like taste, texture, quality, and so on).

The cookbook describes:

- How you can use it (eat it in a sushi bar or at a romantic dinner—the service)
- Its functional parts (different types of sushi: rice part, fish part—the network architecture)
- How you can combine this food with other foods (salads, tea—the interfaces)
- For each type of sushi, a description of required, related parameters (such as quality of the food, possible colors, taste, texture—the technical parameters and measurements)

Eating sushi or making it according to the chef's recommendations is up to the readers to decide. It is not necessary that people who read the book make or eat sushi.

After publication, the book can be located in library or in shops, or given as a present to anyone who wishes it (for sushi promotion). These locations can contain several cookbooks about sushi. Different chefs (vendors) following the same book (standard) usually produce different sushi (different products). Different chefs following different books produce different sushi (different products). This is because different books (standards) exist on the same subject (sushi).

The same applies to standards. Standards are written by multiple writers from different market sectors (vendors, service providers, content providers, users, consultancies, and so on). A standard is written for a specific subject that can be wide or narrow. For example, the GSM technology standard is very wide, but a standard for printer page size (like A4 or A3) is very narrow. A standard can be written, but the market forces determine whether it will be implemented or not. For example, a very good digital pager standard by the name of ERMES was published around the same time that the GSM mobile standard was published. The GSM standard has become very common and successful in the world, while ERMES is almost unknown. One of the reasons for that was that for the same price, GSM had all the features ERMES had, and much more. The people preferred to buy GSM phones rather than ERMES pagers. Therefore, although ERMES was published and the vendors produced a good product, no wide market demand was generated for it.

Sometimes the authorities play an essential role in the standard definition. Standards concerned with the use of radio spectrum are a good example of the need for authority involvement. It is not enough that the standard defines a set of frequencies for the use of the vendors and service providers who want to implement the standard. This set of frequencies must be approved by the country authorities, taking account of the competing demands and special interests, and depending on the frequencies that are available.

Standards Body Types

An "official standards body" usually makes the standards. Nevertheless, sometimes in parallel with the official standard body, other bodies contribute or compete with it. The standards body types are given below. A list of major standards bodies and related organizations, together with their Web sites, is given in the references.

Official Standards Bodies

An official standards body can be from the following types:

- Global (worldwide) such as ITU, IEC, ISO, 3GPP, 3GPP2, and IETF
- Regional such as ETSI, ETNO, and ECTA for Europe
- National such as ANSI and TIA for North America, ARIB in Japan, the TTA in Korea, the CCSA in China, and the NPL in the United Kingdom

In practice, the extensive coverage by ANSI and TIA across North America extends their scope to that more akin to a regional entity. These are a few of the well-known organizations, but there are many others (see a list at the end of this chapter), often working together toward shared objectives.

The names *global*, *regional*, and *national* express the type of group members (that also write the standard) rather than the place where the standard is implemented. As explained before, the implementation of a standard depends strongly on market forces. For example, the GSM standard was created in ETSI (a European body), but implemented all over the world. Usually, the standards bodies aim first to address the related area market needs (i.e,. ETSI for European market needs, TIA for American needs, and so on).

Forum Bodies

In addition to the standards bodies, several types of forum exist mainly to debate and support the work of the standards bodies. These may be of various types including open, closed, international, and worldwide fora. There may be many reasons why fora are created, including the following:

- To answer problems raised in a standards body during the definition phase of a specific standard. For example, if the standard's process is proceeding slower than required, and market forces push for early results, then it is possible that a group of companies with common interests will open a forum to accelerate the work.

- To lobby and influence processes in the official standard. Where a specific group of active companies in the official standards body wants to push particular interests, unite forces for that, and lead it in their direction, they might open a forum. Usually these kinds of actions are performed in background discussions during the official standard process. The group will choose to open a forum in case of ambitious goals that are hard to achieve and that require specific work and plans. This type of forum will probably be a closed forum in which only authorized members can participate.

- To define and standardize an untreated subject for which no official standards body has taken responsibility. The interested group will open a forum to define this subject and normally open it for public comment in an open forum.

De Facto Standards

Although a de facto standard is not a standards body but a standard, it is relevant to explain this term here. If the market commonly has adopted the implementation of a specific company, or a standard specified by an unofficial standards body, then this standard is called a de facto standard. When this kind of standard exists, it can be very difficult for competitors to enter the market with different directions, including those companies that implement an official standard on the subject.

Qualcomm is a special example of a company that has greatly succeeded in the standards arena as its development came to be a de facto standard. Starting as a small company, Qualcomm succeeded in introducing a new standard into the market that included unique company patents (cellular CDMA). After the standard was accepted by the market, Qualcomm experienced exponential growth. This of course is not the customary example and very few companies can achieve this level of success. Microsoft is another example of a company whose products are de facto standards.

Benefits Arising from Standards

The standard brings benefits to all involved in the value chain (the industry, network operators, content providers, service providers, vendors and end users, and so on). Each group with a different interest may need a standard for different reasons. However, it is not questionable that standards can bring benefits to all.

The most important benefits of standards are:

- They ensure that equipment from different suppliers can interwork. This benefits the user by enabling competitive procurement.

- They enable equipment supplied by a manufacturer to be used for different applications and in different regions. This increases market size and reduces costs due to economies of scale. This benefits both suppliers and consumers.

- They provide the means to deal with a changeable environment. Today businesses must to do more with fewer resources in changeable conditions and environments. It is difficult to keep up to date, to understand the complex technologies, to be a specialist in all system parts, and to respond effectively to changes, because of the speed at which change takes place. There are several challenges that business and industry need to deal with to maximize profit, improve performance, and soften risks. The standardization process influences all of these. Standards help to maximize return by lowering the cost of change, by increasing cooperation, and by enabling improved response to change.

A company that plans its products and services while being involved in the standards process increases its chances of producing a product that is suitable for the market. A company that does not operate in this way can still develop and offer a wonderful service. If the standards have not been embraced as part of the company's market research and strategy, the final product will most likely encounter stiff competition. This competition will come from cheaper standard products and strong unified market forces in other directions.

Standards bring the following benefits to network operators, content providers, and service providers:

- Lower price and better quality may be achieved by interworking between different equipment from different vendors. This encourages competition between vendors.
- High market demand, due to cheaper products and easy availability.
- Reduce time-to-market, through interworking and layered block network architecture that make implementation simpler and integration quicker.
- Softer risks:
  - The risk is spread out among vendors instead of on one proprietary vendor (vendor independence).
  - Standards are generally able to last longer than specific vendor solutions.
  - Cheaper investment.

The benefits for equipment manufacturers are

- Easier development process, because they can use standard components to build their product and do not need to reinvent the wheel. The standard "building blocks" approach applied to system elements, applications, and infrastructure products makes changes easier to implement, and brings the possibility of reusable components, improved productivity, and common terminology.
- Reduced cost and time-to-market as a result of easy integration, and cheaper and shorter development processes.
- Creation of a variety of products, solutions, and sales options. Because cooperation and interworking are improved (even in the presence of competition), a variety of products can be developed for the same time and cost as a single nonstandard product. Moreover, complex collaborative projects can be constructed that would otherwise be outside the scope of a single company (for example, a satellite communication network like Iridium).
- A bigger and more extensive market due to interworking, ability to install the equipment in different environments, and more market demand (cheaper products).

For users, standardized products

- Are cheaper.
- Are more varied.

- Have a common set of attributes.
- Are easy to manage.
- Are user friendly.
- Have multiple resources of vendors and services providers.

Participation in Standards

After understanding why standards are useful and worthwhile to use, the next question is: What is it that drives companies to participate in the standard creation process? Why not let others do the job and just use the results? The risk of course is that no one will do the job, unless the existence of the standard is of common interest and the benefits brought by participating in the standards process are much higher than the cost of participation. Companies prefer to invest in the cooperative work rather than operate without standards (i.e., large companies such as service providers and vendors such as Orange, IBM, and Motorola).

It is these large companies that tend to take the lead in the standards creation process and have both the skills and resources to drive the standards forward. For example, for the 3GPP standards:

- Two to four people from each "large company" participated in a relatively small technical group that determined the telephone services.
- Three to six people and more from each "large company" participated and led the group that determined the network architecture.

Medium and small companies are more likely to come and go depending on their specific needs and interests in particular subjects.

Participation in a standards body is more important where:

- The company's market is changing rapidly and its product area is regulated by standards.
- The company is innovative and the standards do not fully support its innovation.
- The company development and strategy are planned for the long term.
- The company is adapting to larger market segments.
- The company is aiming to lead its field or wanting to closely follow the market.

Since participation in the process takes time, manpower, and significant investment, a company needs to consider carefully whether it should participate, and in what parts and groups of the standards-making process. So, for example, companies with the following characteristics may not need to participate:

- Their products are not in an area governed by standards or should not be in a standardized area (e.g., military products).
- Their products are in a unique area (e.g., design).
- Their products are in a stable area.
- Their product belongs to a niche market.
- They do not want to lead the market but to follow from a distance.

By-product Benefits

Although the primary aim of participating in the standards process is to define the technical standard, a number of additional benefits can accrue to the participating companies as a by-product of the main standards process. These benefits help a company to ensure that its product designs will meet the standards requirements and therefore will be well positioned in the market. Conversely they can ensure that the standards will cater to the product.

Work on the standard is divided into work groups and layers that vary according to the subject depth. Therefore, results vary according to the different work groups and their type of activity (i.e., a technical group may bring technical benefits).

Participation in the standards processes can assist the formulation of company strategy, marketing plans, finance, and development. Although marketing benefits may be accumulated, it should be emphasized that direct marketing and sales during standard meetings are not considered to be an acceptable part of the process, and these matters need to be handled separately.

Participation in the standards process goes together with:

- Long-term company strategy
- Preconceptual and methodical thinking about development issues
- Thorough understanding of technical issues

All of these reduce the number of ad hoc changes and promote superior and efficient long-term development planning.

Today's systems are sophisticated, interworking with many other elements and composed of products that offer a variety of features. This requires expertise in multiple and varying areas that are not always present in the company. Participation in technical groups gives the company additional background. For example, a participant in a standards technical group can learn additional technical information and gain advice from the world's foremost experts.

The following are marketing and strategic by-product benefits:

- An understanding of the market forces
- Awareness of new products that are to be introduced into the market
- Hearing the viewpoint and various needs of the market players (operator, customers, and manufacturers)
- An understanding of the reasons behind the needs
- Learning about insights into the future market, technological trends, market forces, and new products
- Product alignment that may lead to an enlarged market and bring financial success

The preceding points could be taken from the high-level groups of the standard body. Examples of these include the following groups of the 3GPP standard body:

- Project Coordination Group (PCG) that is responsible for overall time frame and management of technical work to ensure that the 3GPP specifications are produced in a timely manner
- Service and System Aspects (SA) that are responsible for the overall architecture and service capabilities of systems based on 3GPP specifications
- SA1 (Services) that works on the services and features for 3G
- SA2 (Architecture) that is in charge of developing the architecture and high-level definitions of 3GPP network entities and features

Challenges and Major Considerations

Companies participating in standards face a number of challenges. First, they need to consider how a particular activity could affect the company and how the results of the activity could affect the company. This would logically include consideration of the company's direction and future products in formulating inputs into the standards process so that objectives could be met.

During standards meetings there are likely to be detailed technical discussions with experts of competitors and careful consideration should be given to the scope of discussions and the nature of detailed technical information that can be disclosed. Sometimes political factors may hide the real technical problems and these need to be recognized to distinguish them from the technical explanations. Moreover, the timing of a particular standards activity (usually very long) may not suit the company strategy and initiative may need to be taken to accelerate a particular aspect.

Standards development is time consuming and there may be a balance to strike between allocating the time of scarce resources, or a particular expert, to standards formulation or to product development. At the same time, the cost of membership of relevant bodies, manpower, and expenses must be factored in. It can be difficult to balance expenditure on participation in a particular activity with the longer-term benefits because it is difficult to quantify the likely long-term results.

Major Considerations for a Participating Company

The decision to participate and adopt an active stance in standardization is a strategic decision that should be made at the senior corporate level, with full, direct, and vigorous collaboration with the marketing and development departments. The company must recognize that most of the results are not immediate and that standards are part of a long-term work process.

Analysis of future trends based on results from standards groups, the integration of standards viewpoints into corporate practice, and indirect marketing all require time and effort. One year is the minimal timeframe for results.

Standards work can consume considerable expenditure. This is particularly important for small and medium-sized companies. It would not be unusual for annual expenses to reach tens of thousands of dollars.

An experienced corporate delegate should represent the company in active operations. The experience of the delegate is important to obtain proper, and relatively rapid, results, and avoid causing harm to the company by misrepresentation. The delegate needs to have a background in technical issues, business issues, standard group structure, standardization processes, and rules. The delegate should be an expert in the topic under study (company and standard topics). He or she should also have good information search skills and be able to maintain good working contacts with individuals in the standards groups. Ideally, he or she should be a persuasive presenter and have good interpersonal skills, the ability to lead topics, and the talent in forging and maintaining alliances.

In large companies, entire departments may be dedicated to this issue. Only after undergoing several months or more than a year of training for this work do corporate representatives make presentations and operate actively, after an extended period of passive participation and in-depth study. Outside consultants and experts can be an excellent solution for small-sized firms who are unable to finance such an intensive process by themselves.

On the other hand, the work in the range of standard bodies and work groups is similar in principle. With sufficient experience in a number of standards organizations, it becomes possible (with a limited investment) to lead an issue in a standards group that is new and unfamiliar to the company's expert in standards. Therefore several companies can cooperate by hiring a consultant to participate on their behalf and hold down their costs.

Intellectual Property Rights

Inserting intellectual property rights (IPR) into a standard can have a significant impact on the standardization process, and it is very difficult to achieve such a goal. Usually, only the big companies succeed, and normally, only few IPRs are entered to a standard (if any).

Inserting IPR into standards will influence the way the standard is eventually written and this will slow the process. Generally an owner of an IPR that is not in a standard is fully entitled to hold and benefit from any rights that he or she may own, including the right to refuse the granting of licenses.

However, if the IPR is to be used within a standard, it shall make fair, reasonable, and nondiscriminatory licensing terms available to all or no fee at all. If this is not the case, then it is against the rules and cannot be combined with the standard. If no agreement is reached, the standard could be stacked forever. This could be a situation where participants open a forum to explore alternative approaches.

Important Links

ANSI: American National Standards Institute—*http://www.ansi.org/*
ETNO: European Telecommunications Network Operators Association—*http://www.etno.be/*
ECTA: European Competitive Telecommunications Association—*http://www.ectaportal.com/en/*
ETSI: European Telecommunications Standards Institute—*http://www.etsi.org/*
FCC: Federal Communications Commission—*http://www.fcc.gov/*
GSM Europe: GSM Europe—*http://www.gsmworld.com/gsmeurope/index.shtml*
IEEE SA: Institute of Electronic and Electrical Engineers Standards Association—
    http://standards.ieee.org/
IETF: The Internet Engineering Task Force—*http://www.ietf.org/*
INTUG: The International Telecommunications Users Group—*http://www.intug.net/*
IPv6 Forum: IPv6 Forum—*http://www.ipv6forum.com/index.php*
IRG: The Independent Regulators Group—*http://irgis.anacom.pt/site/en/*
ISO: International Organization for Standardisation—*http://www.iso.org/iso/en/*
    ISOOnline.frontpage
ITU: International Telecommunications Union—*http://www.itu.int/home/index.html*
ITU-T: ITU Telecommunication Standardisation Sector—*http://www.itu.int/ITU-T/*
ITU-R: ITU Radiocommunication Sector—*http://www.itu.int/ITU-R/*
SCC: Standards Council of Canada—*http://www.scc.ca/*
3GPP: Third Generation Partnership Project—*http://www.3gpp.org/*
3GPP2: Third Generation Partnership Project 2—*http://www.3gpp2.org/*

Acknowledgment   The author wishes to acknowledge the valuable help and advice given by Goff Hill in preparing this chapter.

2 Telecommunications Regulation

John Buckley

Introduction

Telecommunications regulation is the setting and enforcement of the rules of participation in the telecommunications services market, and is a major feature of the marketplace of most countries in the modern era. Regulation has a long history in the United States, where the 1934 Telecommunications Act placed obligations of various sorts on AT&T, the private company then providing many places in the United States with long-distance and local communications on a monopoly basis.

In many other countries where telecommunications services had been previously provided under government control, regulation arrived in the 1980s and 1990s when these countries sought to introduce private provision into their telecommunications services markets. Paradoxically, then, modern regulation exists to support a process often known as deregulation, that is, the replacement of monopoly services by diversity and competition. A more detailed treatment of this and other topics in this chapter may be found in *Telecommunications Regulation*, published by IEE books.[1]

Goals of Regulation

The principal goal of modern regulation is the creation, maintenance, and nourishment of a competitive market for telecommunications services. Regulation has to combat the power of monopolies and powerful incumbent undertakings to act in ways that disadvantage, and

---

[1] *Telecommunications Regulation*, John Buckley, IEE Books (2003).

even exclude, competitors. The focus on competition aims at higher-order objectives of social and public good under the assumption that open, competitive markets generally produce the best outcomes in terms of maximum well-being for the greatest number of people. A typical statement of objectives for telecommunications regulation may include the following elements.

- A competitive market
- Telecommunications services meeting all national and international needs
- Promotion of the interests of consumers
- Consumers enjoying best or near-best deals compared with similar economies
- A basic set of universally available, affordable services

In understanding regulation, it is helpful to be very clear that good regulation is not market management. The theory that competition produces the best outcomes implies that we will not, and should not try to, know in advance what those outcomes will be. Regulation will certainly influence a market, but it cannot predetermine it. Attempts to do this, perhaps when the regulatory authority has been captured by a political or industry interest, are a hazard of regulation. Inappropriate regulation can do harm. The best regulator of economic activity is, according to the theory, competition itself. The wise regulatory authority will, therefore, nourish competition to the point of being able to lessen its activity and ultimately withdraw altogether from that market.

Regulation and Competition

Commonly accepted economic theory states that competitive markets produce an optimal allocation of resources. The optimum, known as *allocative efficiency*, has no potential reallocation that could make someone better off while at the same time leaving no one worse off. Customers buy from suppliers offering the most desirable products at the lowest prices, while suppliers use resources efficiently. Less efficient suppliers will lose customers to competitors with lower prices or, having reduced profits, will eventually have to retire from the market. Nonetheless, a market may fail to generate this optimum, and when this happens there exists *market failure*. The most important cause of market failure in the telecommunications services market is *dominance* or *market power*. An undertaking may be said to be in a dominant position or in a position of significant market power[2] (SMP) whenever it can act over a sustained period to exclude competitors from the market and can act in some measure independently of the interests of customers, competitors, and suppliers.

The market power of large undertakings in telecommunications services may arise from any or all of a variety of factors. These include monopoly supply, entrenched market share, established large-scale operations having horizontal and vertical integration, ownership of key resources such as distribution channels, technology, or patents, and the control of infrastructure not readily duplicated. Before any regulation can sensibly be applied to a market, the regulator[3] must carry out a *market analysis* that identifies undertakings having significant market power and understands the reasons for the market power. If a market is poten-

---

[2] The terms *dominance* and *significant market power* are used more or less synonymously in this chapter. Dominance is the more commonly used term in the United States, and significant market power in Europe.

[3] The word *regulator* is used in this chapter synonymously with *regulatory authority*, referring in both cases to the body corporate or official charged with responsibility for setting and enforcing the rules.

tially competitive in principle but happens to have a dominant undertaking, it needs rules preventing the dominant undertaking from excluding competitors, in the eventual hope that a competitive market might emerge.

An example of this is the long-distance communications market, now competitive in many countries. Another market might have a *natural monopoly*, perhaps where one undertaking controls a not readily duplicated infrastructure. In that case, the rules need to recognize an ongoing monopoly and impose price controls to simulate the missing discipline that a competitive market would have exerted. This is typical of "last mile" access networks in many countries. Very large undertakings playing in different markets may possess dominance in some though not all of the markets in which they operate. In this case, rather than having one set of rules for a given undertaking, market segmentation will be necessary with different rules of play applying to its separate operations in each market. Additional provisions may ensure that large undertakings do not unfairly lever advantages gained in a dominated market into other markets.

Excluding Competition

The ways in which undertakings with market power can exclude competition are well known, and regulators must set rules preventing undertakings with significant market power from using these practices. These rules may be specifically framed or generally framed in terms of a requirement not to indulge in anticompetitive practice or in restriction of competition.

Predatory Pricing
This is the practice of setting prices below their true cost, with the intention of driving competitors from the market, then subsequently raising prices to recoup the initial losses. Here as in other areas, regulation will seek to ensure that prices are cost related.

Margin Squeeze
This can occur when an undertaking provides an intermediate input (say access or transit switching) both for competitors on a wholesale basis and on an internal basis for its own retail service. The undertaking can weaken or destroy its competitors' businesses by raising the wholesale price, or reducing its retail price, or both together. The remedy calls for cost-related pricing and various rules and tests for margin squeezes.

Discriminatory Pricing
Charging different prices for the same product with different customers constitutes discriminatory pricing. Undertakings with significant market power are normally prohibited from doing this, as it would otherwise be easy for them to offer advantageous prices to the most profitable accounts, or to those customers with highest propensity to switch suppliers.

Bundling and Linkage of Sales
These take place when products and services are sold in combination. While some combinations are natural and unexceptional, for example an exchange line and its maintenance, others present an opportunity for abuse of market power should they force a customer to take an unwanted product in order to obtain a desired product or service. Some kinds of pricing packages may have the same effect as a linked sale. Large undertakings usually have

considerable purchasing power and may misuse their market power by conducting *exclusive trading arrangements* with their suppliers. This would be especially iniquitous if an undertaking used its strength to persuade a supplier not to supply a competitor, and this would normally be forbidden.

Cross-subsidy

This is the transfer of costs between the integrated operations of a large undertaking, providing opportunities for predatory pricing, margin squeeze, and exclusion of competition. Cross-subsidy is extremely complex to detect and measure; undertakings with significant market power are frequently required to show separate accounts for their different businesses and to demonstrate cost orientation in the transfer prices applied. The uses that undertakings make of intellectual property (such as patents, designs, and interface specifications) and general information resources may encourage or obstruct competition, so regulation may lay down rules about this.

Although no one should ask an undertaking to give away valuable information rights free of charge, regulation may typically require it to be offered at a fair price related to the cost of creating the information. Some information resources, such as a telephone directory, may constitute an infrastructure resource not readily duplicated, so the powerful telecommunications undertaking may be required to make its information available to others, or to publish the telephone numbers of its competitors' customers, all on fair, cost-related, and nondiscriminatory terms.

The Rules of Regulation: Principles

Regulation of a market such as telecommunications services is achieved by means of the rules, which may be known as regulations, conditions, provisions, or by other names. Undertakings in the regulated market are authorized to participate in the market only so long as they obey these rules. If they break the rules, they are subject to enforcement actions such as fines and, ultimately, expulsion from the regulated market. The rules are published by a regulatory authority, such as the Office of Communications (OFCOM) in the United Kingdom or the Federal Communications Commission (FCC) in the United States.

The job of the regulatory authority is to make the rules, to enforce them, to review their operation, and to change them as necessary following the set procedures. The rules have the force of law, the regulatory authorities having in the first place received their powers to regulate through primary legislation passed by national governing bodies. Within Europe, the European Commission issues directives having strong influence on regulation and regulatory frameworks; member governments must transpose these into national legislation under European Union treaty obligations. In some countries, the regulatory function is distributed among various bodies, for example between state and federal regulators in the United States, or between national regulatory authorities, government departments, and other bodies.

Administrative Law

Regulatory rules have to follow general provisions of *administrative law* laid down in primary legislation. Firstly, the rules must be objective and transparent. This means that the purpose of a rule and the problem it seeks to solve must be stated, as also must be the logic by which it addresses the problem. Rules may not introduce additional elements. They cannot, for example, be the product of political bargaining processes. Second, rules must be appropriate, reasonable, and proportionate to the problem being solved. Finally, rules must be nondiscriminatory. They cannot discriminate between different undertakings except where there is a just and revealed reason. An obvious case where discrimination will most definitely be required is that of undertakings designated as having significant market power, for which more stringent conditions must apply. The different regulation of undertakings having significant market power compared with the others is known as *asymmetric regulation*.

Regulatory Regimes

There are two fundamentally different types of regulatory regime: *specific authorization* regimes and *general authorization* regimes. The former were common in Europe in the early days, although the latest (2003) European Regulatory Framework has moved over to a general authorization regime. Under a specific authorization regime, each undertaking receives an operating license containing privileges and conditions that apply to that under-taking only. Asymmetric regulation is achieved by the variation of conditions, and in prac-tice it was the licenses of the former monopolists that formed the core of European regulation before the current regime. General authorization regimes have a published set of rules, invit-ing any undertaking that so wishes to participate in the market subject to the rules. Most or many of the rules will be conditional, varying according to the market power designation of the undertaking. General authorization regimes are in principle the more open regimes, as they leave it to the market rather than the regulator or the government to decide how many undertakings there will be. The United States has an essentially general authorization regime, although individual undertakings require operating licenses.

Amount of Regulation

For the sake of healthy, competitive markets, regulation should be no more than necessary. The European Framework insists that regulators *must not* intervene in any market classified as competitive. The reasons are plain. Competitive discipline is the best form of regulation. Regulation is expensive to operate and enforce. Finally, there is regulatory hazard since reg-ulators can accidentally or otherwise cause harm by intervening inappropriately, or by par-alyzing a market with uncertainty and inaction.

The next sections examine the principles of regulation in the European Union and in the United States. The descriptions are also useful, however, for readers outside these coun-tries and regions, partly because these models are followed in many other countries and also because the principles are generally applicable. For the same reason, readers *within* either of these areas will find it useful to read both sections.

Regulation in the European Union

Telecommunications regulation in the European member states is strongly governed by the directives of the European Commission. These directives (under Article 249 of the Treaty of Rome) must be transposed into national laws in the manner best fitting the legal frame-works and institutions of each member state. The highly simplified description that follows can at best give only the flavor and principles of the regulations. Any more serious study will require access to the original documents and the publications of the *National Regulatory Authorities* of the member states.

Under the directives, each member state must set up a National Regulatory Authority (NRA), which must be independent of any telecommunications undertakings and be impartial and transparent. Examples of NRAs in various countries include:

- The United Kingdom: OFCOM[4]
- Ireland: ComReg[5]
- France: ARCEP[6]
- Germany: Federal Network Agency[7]

The quartet of European directives is as follows. These directives and other legislative documents may be obtained from the European Commission Web site.[8]

- The Framework Directive:[9] about methods, principles, and policy
- The Authorization Directive:[10] about authorization, rights, and obligations
- The Access Directive:[11] about interoperability and interconnection
- The Universal Service Directive:[12] about universal service obligations and consumer rights and protection

Framework Directive

The European Framework Directive brings together the regulation of transmission networks and services (though not content and consumer equipment), because of the convergence of telecommunications, media, and information technology. Each member state must appoint a competent, independent, impartial, and transparent National Regulatory Authority (NRA). Each country must also have an effective, independent, and competent body to which any organization can appeal against a regulatory decision affecting it. The NRA must consult interested parties about any decision affecting them using published consultation procedures, and must make consultations public subject to normal commercial confidentiality.

The policy objectives of European regulation are laid down in Article 8 of the Framework Directive, very briefly summarized as follows:

- Promoting competition
- Contributing to the development of a Europe-wide market

---

[4] Office of Communications, *www.ofcom.org.uk.*
[5] Commission for Communications Regulation, *www.odtr.ie.*
[6] L'autorité de régulation des communications électroniques et des postes, *www.art-telecom.fr.*
[7] Die Bundesnetzagentur für Elektrizität, Gas, Telekommunikation, Post und Eisenbahnen; information available at *www.bundesnetzagentur.de.*
[8] *http://europa.eu.int/eur-lex.*
[9] Directive 2002/21/EC of the European Parliament and of the Council of 7 March 2002 on a common regulatory framework for electronic communications networks and services.
[10] Directive 2002/20/EC of the European Parliament and of the Council of 7 March 2002 on the authorization of electronic communications networks and services.
[11] Directive 2002/19/EC of the European Parliament and of the Council of 7 March 2002 on access to, and interconnection of, electronic communications networks and associated facilities.
[12] Directive 2002/22/EC of the European Parliament and of the Council of 7 March 2002 on universal service and users' rights relating to electronic communications networks and services.

- Promoting the interest of all citizens of the European Union
- Ensuring that users, including disabled users, derive maximum benefit in terms of choice, price, and quality
- Consumer protection
- Efficient infrastructure investment and innovation
- Provision of a universal service

NRAs must arrange the effective management of radio spectrum, numbering and addressing resources, and rights to construct infrastructure on principles of transparency and nondiscrimination.

NRAs must conduct an analysis of the market, showing the segmentation of the market into separate parts and designating undertakings deemed to have significant market power. Specific regulatory requirements may be placed on undertakings designated as having significant market power, while regulatory intervention may not be applied to an effectively competitive market.

Authorization Directive

The European Authorization Directive prescribes a general authorization regime without limitation on the number of undertakings. Undertakings may commence operations and are granted the facility to apply for rights to construct infrastructure, for radio spectrum and numbering resources, and to negotiate and obtain interconnection and access with other undertakings. Where limitation is placed on the grant of any right such as radio spectrum, the reason for the limitation must be made known and is subject to consultation. Any principles governing the allocation of scarce resources must be objective, transparent, nondiscriminatory, and proportionate. Fees charged for the allocation of scarce resources must be objectively justified, transparent, nondiscriminatory, and proportionate in relation to their intended purpose.

An NRA may place obligations on all undertakings in a market, for example, for the following:

- A requirement to notify and register its activity
- A duty to contribute to a universal service obligation
- A fee to pay for the regulation of the market
- A duty to provide information for defined regulatory, operational, and statistical purposes

Any fees must be objective, transparent, and proportionate. They also must be imposed in a manner that minimizes additional administrative costs and attendant charges. NRAs may impose financial or other penalties, up to withdrawal of authorization, on undertakings that fail to comply with their obligations or fail to supply information that demonstrates compliance. Special procedures are available to deal swiftly with any noncompliance that threatens public health or safety.

Access Directive

The European Access Directive governs the manner in which telecommunications undertakings interconnect with one another, with the aim of creating and nourishing competitive markets and ensuring the end-to-end interoperability of services. "Access" refers here to internetwork access rather than end-user access networks. Some obligations bear on all undertakings within a market, while others are imposed only on those with significant market power. All undertakings:

- Have the right to request interconnection with another's network.
- Must negotiate with another requesting interconnection.

Where an undertaking has control of end users, then, regardless of market power, the NRA may insist that it interconnect with other undertakings if it does not do so already. The NRA may lay down technical or operational conditions on the provision of that access.

Undertakings with significant market power may face much stronger regulation of their arrangements for interconnect with other undertakings. These include:

- Obligation to interconnect
- Transparency, with open publication of technical specifications, network characteristics, terms and conditions for supply and use, and prices
- Nondiscrimination, both with other undertakings and with respect to internal use of its services and use by associated undertakings
- Accounting separation, where there is a requirement for nondiscrimination or to prevent unfair cross-subsidy
- Obligation to give access to "specific network facilities"
- Price controls

The term *specific network facilities* refers to the potentially many, various parts of a dominant undertaking's network. These include the unbundled local loop, collocation, building and facility sharing arrangements, intelligent network services, mobile roaming services, and other services sold on a wholesale basis. The Directive does not go so far as to insist on access being given to *all* of these, but leaves NRAs the discretion to decide on the basis of technical feasibility and the long-term safeguarding of competition. Local loop unbundling is, however, separately and specifically mandated in Europe.[13] Price control is appropriate wherever the undertaking is in a position to maintain excessive prices or exert a price squeeze. However, the determined prices should be compatible with efficiency and sustainable competition, and should allow the undertaking a reasonable rate of return on capital employed, recognizing the risk of the investment.

Universal Service Directive

The European Universal Service Directive provides for the universal provision of a minimum set of services, lays down other requirements for consumer protection, and deals with the requirements laid on undertakings that provide universal service. Should universal services be provided on a noncommercial basis, then arrangements for the financing of the service are described.

A minimum set of services must be made available to all citizens regardless of geographic location and on an affordable basis, bearing in mind national economic conditions. Member states may insist on price caps, common tariffs, and geographical price averaging. The provision of the services must minimize market distortions, in particular the provision

---

[13] Regulation (EC) No 2887/2000 of the European Parliament and of the Council of 18 December 2000 on unbundled access to the local loop.

of services at prices that depart from normal commercial conditions, while safeguarding the public interest. Member states may require tariff options that depart from normal commercial practice to support people identified as having low incomes or special social needs. The minimum service set need not include *all* services offered in the market, but must include telephone service capable of supporting local, long-distance, and international voice calls as well as fax and data transmission at rates compatible with functional Internet access. Other universally provided services include directory services, directory inquiries, public pay telephones, and access to the emergency number (112 in Europe) without needing a means of payment at pay telephones.

Certain undertakings may be designated as providers of particular universal services in certain parts of the territory. The designation mechanism should be efficient, objective, transparent, and nondiscriminatory, whereby no undertaking is a priori excluded from being designated. Designated undertakings must meet service standards specified by the NRA, and may not bundle products in ways that compel customers to purchase unwanted services in order to obtain universal service. Where the burden of providing universal service is considered unfair for a designated undertaking to bear, the NRA may calculate the costs. These may then be met from public funds or by a levy on other undertakings in the market. In the latter case, the NRA must apply principles of transparency, least market distortion, nondiscrimination, and proportionality in determining the contributions.

NRAs may apply price controls to services provided by undertakings with significant market power in markets that are not deemed competitive. Price controls may include:

- Price capping
- Control of individual prices
- Cost-oriented pricing
- Pricing in line with comparable markets

All undertakings must provide customer contracts and itemized bills, ensure the integrity of their networks, provide information about their service quality, support access to operator assistance and directory inquiries, and support the European access codes "00" for international dialing and "112" for emergency calls. All network operators must provide number portability, while undertakings having significant market power must also provide carrier selection and carrier preselection.[14]

Regulation in the United States

Telecommunications has a long and varied history in the United States. The late nineteenth century pioneering companies of the telegraph and telephone industry consolidated in the early twentieth century into the private AT&T monopoly. In contrast, the corresponding consolidation in Europe was usually into government-controlled monopolies. The AT&T monopoly, recognized then as being a natural monopoly, was regulated as a *common carrier*, a position confirmed under the 1934 Communications Act. Under a public interest remit, the Act required AT&T to provide services for all on a nondiscriminatory basis and at

---

[14] Carrier selection is the facility for the user of a telecommunications terminal to choose on a call-by-call basis the undertaking to carry the call. Carrier preselection is the facility for the renter of a telecommunications service to nominate a chosen carrier for all calls of a given class, such selection then to apply by default to all calls without the need to dial an access code. Carrier preselection is known as *dialing parity* in the United States.

regulated tariffs, and placed a universal service obligation on AT&T. The same act established the national regulatory authority, the Federal Communications Commission (FCC).

The development of competition took place by a series of steps. In 1969, the FCC licensed MCI to provide long-distance, point-to-point services. In 1977, MCI won the right to provide publicly available long-distance services, though legal wrangling followed over tariffs and interconnection with AT&T. The Modification of Final Judgement (MFJ) of 1984, the culmination of much litigation, brought about the deregulation of the U.S. telecommunications services industry. It dictated the divestiture of AT&T into a long-distance operation, and seven separate local undertakings, the Regional Bell Operating Companies (RBOCs). The long-distance market was fully opened to competition, while the RBOCs continued as regulated natural monopolies providing local services. The RBOCs:

- Provided local exchange service
- Handled interexchange traffic but only between exchanges within the same Local Access and Transport Area (LATA)
- Had regulated tariffs
- Had to provide equal access (i.e., on an equal basis) to any of the long-distance undertakings
- Could not discriminate in favor of AT&T in certain areas

Analyzing the long-distance market into dominant and nondominant undertakings, the FCC granted forbearance of the 1934 Act requirements on nondominant undertakings. Other requirements included the formation by dominant undertakings of separate subsidiaries for their value-added service operations, as a defense against unfair cross-subsidy. Though local competition was not mandated by the MFJ, there was some local competition where state regulatory authorities encouraged it. AT&T was reclassified as a nondominant, long-distance undertaking in 1995, showing the success of U.S. deregulation in the long-distance market.

The 1996 Telecommunications Act replaced Title II, the part of the 1934 Act dealing with telecommunications, so creating the current regulatory framework of the United States.

Major Authorities

The United States divides the regulatory function among a number of distinct bodies. The major rule-setting regulatory authorities are the:

- Federal Communications Commission (FCC),[15] which sets, administers, and enforces the rules at the national level
- Public Utility Commissions (PUC) in each state,[16] which are responsible for tariff and interconnection matters relating to intrastate communications

Additionally, other bodies have specified roles. The National Association of Regulatory Utility Commissioners (NARUC)[17] coordinates action between PUCs, and between the PUCs and the FCC. At the federal level:

---

[15] *www.fcc.gov.*

[16] The links are too numerous to list here, and some states have more than one PUC. Links may be obtained via NARUC's Web site—*www.naruc.org.*

[17] *www.naruc.org.*

- The National Telecommunications and Information Administration (NTIA)[18] deals with government spectrum allocations and advises on policy
- The Bureau of Competition of the Federal Trade Commission[19] deals with antitrust and competition matters
- The Antitrust division of the Department of Justice (DoJ)[20] deals with antitrust and competition matters
- Congress sets policies, which the FCC is committed to implement
- The Courts interpret the antitrust, competition, and regulatory laws and in practice generate some of the key decisions

The FCC is subject to administrative law, having authority to act within its competence and under guidelines. The Administrative Procedure Act requires that the FCC publish its proposed actions with reasons, conduct consultations, and review its decisions when so petitioned. Its actions can be, and often are, subject to challenge in the courts. A significant issue is the division of roles between the PUCs and the FCC, bearing in mind the Tenth Amendment to the U.S. Constitution, which reserves to states all powers not expressly given to the federal government. The 1996 Act has reduced the powers of the state regulatory authorities under the *preemption doctrine* that states may not use their powers in ways that contradict or frustrate the purposes of federal laws.

Regulation Details

The U.S. regulatory regime is based on individual operating licenses and a common set of regulatory rules. To obtain operating licenses from the FCC and/or PUCs as appropriate, undertakings may have to provide information about technical and financial ability, as well as satisfy various tests, for example concerning foreign ownership. Interested parties may raise public interest objections. The licenses do not, however, contain company-specific regulatory rules in the style of the operating licenses seen in Europe before the present (2003) regime.

Local Liberalization

The 1996 Act aimed to introduce more competition into the U.S. telecommunications services market, and in particular to liberalize the local markets. It declared illegal *any* attempts to restrict competition in the intrastate markets, or to prohibit any undertaking from providing service. It envisaged three modes of competition for the incumbent local exchange carriers (ILECs)[21]:

- Full facilities competition, by undertakings constructing access to customers
- Competition by undertakings using unbundled elements of the ILEC networks
- Competition by resale of the services of the ILECs

Obligations placed on the ILECs include:

- Interconnection at any feasible point in the network on fair and nondiscriminatory terms

---

[18] *www.ntia.doc.gov/.*
[19] *www.ftc.gov/.*
[20] *www.usdoj.gov/atr/.*
[21] The ILECs are the dominant, former monopoly local service undertakings.

- Nondiscriminatory access to unbundled network elements at any technically feasible point on just, reasonable, and nondiscriminatory terms
- Resale on a wholesale basis of any retail service provided, again on reasonable and nondiscriminatory terms
- Provision of information affecting interoperability
- Provision of physical collocation for equipment where necessary for interconnection or unbundling, again on reasonable and nondiscriminatory terms

In return, ILECs were to be permitted to enter the long-distance (that is, inter-LATA, interstate, and international) communications markets, though subject to satisfying a further checklist of measures to encourage competition.

All undertakings in the local exchange market, ILECs and new entrants alike, must:

- Provide number portability.[22]
- Provide dialing parity,[23] access to numbers, directory inquiries, and operator assistance.

Interconnection and Unbundling

The requirements of interconnection at any point in the ILECs' networks and for access to any feasible unbundled network elements have, in practice, proved a problem, as they furnish little limitation based on technical difficulty or economic desirability. A great deal of litigation has taken place, therefore, to develop the criteria on which it is reasonable to have unbundled access to ILEC networks.

Universal Service

The 1996 Act clarified and broadened the universal service obligation and the arrangements for funding it. The general cornerstone of the universal service remains that of basic services for the public at just, reasonable, and affordable rates. Additionally, new provisions address high-cost locations, low-income households, and key education and health care institutions. All undertakings must provide discounted services to schools, libraries, and rural and nonprofit health care facilities.

Telecommunications Regulation in Practice

The modern era of regulated, competitive telecommunications services has existed for more than two decades. While it is still early for us to be completely sure that this has been of overall benefit to humankind, the present signs are positive and hopeful. The former monopolies have shed much of their inefficiency. At the same time, there are more varied services than ever that are cheaper in real and often in nominal terms than twenty years ago. Much of this cost reduction, however, stems from unprecedented technological progress. Many countries now have fully competitive long-distance and value-added service markets. Nonetheless, the local access, or "last mile," part of telecommunications remains uncom-

---

[22] Number portability is the facility for a user of a telecommunications service to retain his or her number when changing the supplier of that service. Greater detail appears in a later section.

[23] Dialing parity provides consumers with equal access to telecommunications service providers, and allows them to have calls routed via a chosen undertaking, different from that providing basic telephone service, without the need to dial additional digits or access codes. This facility is known as "carrier preselection" in Europe.

petitive (outside the downtown business districts) in many markets, dampening optimistic hopes of an early withdrawal of explicit regulation from the sector.

Local Access Market

The status of the local access market remains an ongoing concern. Regulation has to be based on a market analysis, which determines whether the market in question is competitive, prospectively competitive or noncompetitive, and which undertakings possess significant market power and if so what is the reason for that market power. The regulatory authority must take action to prevent restriction of competition, and to insert the missing discipline that a competitive market would have exerted. The question is whether competition is possible in local access, or, conversely, whether this market is a natural monopoly requiring ongoing sector-specific regulation.

The United States and European countries, such as the United Kingdom, took different views on this in the 1980s. The 1984 MFJ acknowledged a natural access monopoly, and partitioned the industry into an access sector with the seven regulated RBOCs each having a territorial monopoly, and a long-distance sector with AT&T, MCI, and later entrants such as U.S. Sprint. The United Kingdom, while acknowledging dominance in its access sector, thought that the monopoly might be contestable by alternative means of access, notably cable television systems and wireless local loops. While Europe, the United States, and other countries have found that some competition is possible, the dominance of the incumbent[24] usually remains.

Price Control

Markets with a dominant undertaking generally require some form of price control. Determining the exact form and level of price control is one of the most difficult and sensitive tasks a regulator must perform. If a regulator sets very tight price controls, then there will be early popularity with residential and business customers. If tight price controls are set for interconnection and the wholesaling of unbundled services to other telecommunications undertakings, then this will encourage competitors to enter the market. However, the mere number of competitors is not a success measure. The correct regulatory goal is a properly functioning competitive market whose undertakings have viable businesses. If some of them owed their existence only to unrealistically low input prices or to the operation of regulation in general, there would not be an economically efficient market.

Forcing a dominant undertaking to subsidize competitors in this way would have two effects. First, it would weaken the ability and desire of the undertaking to invest in its network. Second, it would discourage competition and entrench the monopoly, since why would new entrants challenge a monopoly and invest for themselves when they can get its services cheaply?

Price control for monopoly services is usually based on *long-run incremental costs*, these being the reasonable economic costs of providing the service, including the long-run costs of capital investment, using the best available technical solution and providing for a reasonable return on capital. Other types of price control, such as direct price control or price control relative to a consumer prices index, are based on a reasoned estimate or judgment

---

[24] The term *incumbent* refers to the former, monopoly undertaking. Incumbents normally possess dominance or significant market power.

of the future costs of providing the service. Direct price control provides *incentive regulation*, since the undertaking can help itself and its profitability by being more efficient within the price control envelope.

Further types of price control, for example *cost-plus* and *retail-minus* pricing, are appropriate for various types of wholesale and retail services. Retail-minus price control might be particularly apposite for an innovative service that had involved significant risk investment to bring it into being. When one undertaking buys a wholesale service from another, it should not only bear the direct cost of the input service, but also bear some of the cost of risk, if the risk is significant.

Secondary Competition

It is possible to reduce the dependence of a market on regulation by means of secondary competition. Two possible approaches in the access market are as follows. The first is by the wholesale provision of unbundled loop elements, such as local loops. The second is by full-service wholesaling, where the whole service is still provided by the dominant undertaking but is resold by a different undertaking under its own branding as a retail offering to the end customer.

The regulator normally insists that the pricing of wholesale services and unbundled network elements be transparent and nondiscriminatory. In particular, the regulator will require accounting separation to show that the prices are nondiscriminatory with respect to the internal price the undertaking charges its own retail businesses. Although the regulatory authority will naturally keep an eye on the cost orientation of any charges, it may be able in practice to relax price control if satisfied that the wholesale product provides a genuine opportunity for competition and that the opportunity is in fact being taken.

Evolution of Regulation

Regulators need to be well aware that markets and key issues change over time. Regulation is not a simple process of merely taming the incumbent. New forms of dominance can and will occur. Mergers, acquisitions, and reconsolidation may create old and new monopolies in different guises and under new names. Oligopoly, where there is competition between a small number of large undertakings, needs careful attention as joint or individual significant market power can arise in this situation. Call termination is a monopoly, since only one network has access to a particular customer or line, and regulators have needed to act on unreasonable call termination pricing notably by mobile operators in Europe. Mobile roaming charges are a similarly captive cost by which undertakings can and do exploit their customers. Consumer concerns have increased for many regulators in recent years, as they have faced the need to act on malpractices such as slamming,[25] premium rate number scams,[26] unwanted telemarketing calls, and silent calls from telemarketing call centers.

---

[25] *Slamming* is the imposition of a charge on a customer's bill, or the change of the nomination of chosen network undertaking under a carrier preselection or dialing parity facility, without the customer's authority. Slamming is illegal in most jurisdictions.

[26] Premium rate number scams take various forms, but usually a telecommunications user is forced, persuaded, or influenced to make a call with a high charge rate while being unaware of the true purpose of the call and the rate of charge.

Interconnection

Interconnection is an obvious requirement for a fully functioning, competitive telecommunications services market. In this way, any caller may call any destination or value-added service, regardless of who provides the calling party's network, the called party's network, and any transit networks employed. Likewise, an undertaking may then construct dedicated private circuits using paths across other networks. All undertakings are, therefore, normally obliged to negotiate for interconnection when approached by another. Nondominant undertakings will normally need no encouragement to interconnect with some, but not necessarily all, other networks. Dominant undertakings, which alone could maintain a viable business while refusing to interconnect, are obliged not only to negotiate but also actually to interconnect.

In some countries, regulators always determine the prices. Elsewhere, regulators determine prices when bilateral negotiations have failed. In many regimes, dominant undertakings are required to publicize reference interconnection terms, which they must apply in a nondiscriminatory way. Changes may be subject to prepublication and consultation requirements with the regulator and other interested parties. For practical purposes, undertakings may set up working parties and coordinating committees, with regulatory oversight where appropriate, to facilitate interface changes and to resolve issues that arise.

Interoperability

Besides offering interconnection, networks must offer functional interoperability. That is, services must actually work in an end-to-end fashion across two or more networks. A distinction can be drawn between *noncooperative network services*, such as information services and value-added services (which may need no more than working transmission of voice or data), and *cooperative network services*. These need consistent interoperating functionalities in calling and called networks and any intervening transit networks. Examples include voice telephony, calling line identification, and number portability.

The choice of points of interconnection may be a source of contention. One network undertaking may wish to interconnect with another at the closest point, while the other may prefer to interconnect only at dedicated access points, such as trunk-switching nodes. There is clear scope for undertakings to inflict costs on one another, and regulators normally determine that any restriction of interconnect points by a dominant undertaking must be objectively justified and should follow the manner in which it uses it own network. Sufficient information must be disclosed about physical locations and technical interfaces to permit interconnection to happen.

Interfaces

Technical interfaces are very important both for interoperability and for the possible exercise of market power. All undertakings, and particularly those with dominant positions, may restrict competition and inflict costs on other networks both through the use of proprietary technical interfaces and through the manner and timing of changes to technical interfaces. In many countries, the incumbent may use or have used proprietary technical interfaces, or national variations on standard frameworks (for example, with the CCITT No 7 common channel interexchange signaling system) dating from monopoly days.

Often, competitors will adopt the legacy interface, obtaining it by way of their equipment suppliers; if the dominant undertaking imposes charges for the use of intellectual property, these must be on genuinely cost-oriented and nondiscriminatory terms. Regulators may need to act to prevent dominant undertakings from using proprietary interfaces and technical information in a discriminatory way with equipment suppliers. It is wise for regulators to set a general objective of moving toward international standards. This will in the long run lower the overall costs of the industry, although it may be unrealistic to achieve this goal quickly. Standard interfaces may also lower barriers to competition, as smaller undertakings may not be able to apply or re-engineer proprietary technical interfaces as cheaply or as efficiently as larger competitors.

Access and Transit

Besides an obligation to interconnect, dominant undertakings normally have an obligation to provide *transit interconnection* on regulated terms, allowing their networks to be used to complete switched or permanent connections between other networks. This follows because it would be unreasonable to expect all undertakings in a competitive market to interconnect with every other undertaking. In many markets, however, the market for long-distance transit may attract other players. Should use of the incumbent's network be avoidable, then the market might be competitive with less or even no need for price regulation.

Permanent and Indirect Access

A very important subclass of interconnection is interconnection for access. This is where an undertaking relies on the use of another's network, usually the incumbent's network, as the medium by which its customers connect to it. This may take two forms: *permanent access* via private circuits and *indirect access*. The first of these is normally appropriate only to high-traffic, business customers because it has a high resource cost. Indirect access is common for business and residential users. Under indirect access, the calling customer's access connection is with the access network provider (usually the incumbent), who charges a line rental and who provides a switched connection for every call to the chosen network and forwards dialed digits to it.

Basic Indirect Access
There are three forms of indirect access. *Basic indirect access* relies on the caller dialing an access code to reach the chosen network, then further digits to identify himself or herself to the chosen network, and finally the digits to signify the destination. Notwithstanding that some of this might be relieved through preprogrammed dialer units or software in private exchange systems, the procedure is burdensome and therefore anticompetitive in its effect. It is therefore mainly of only historical interest. In Europe the caller would have a separate billing relationship with the chosen network undertaking, although, in the United States, the charges for calls were billed in bulk through the RBOC.

Easy Access
A simplification of basic indirect access is *easy access*. This circumvents the need for customer identification, as the access network passes the *calling line identity* on to the chosen network. The need for a short access code remains, however.

Carrier Preselection or Dialing Parity

Indirect access without the need to dial an access code is achieved through *carrier preselection* (in European terminology) or *dialing parity* (in U.S. terminology). This facility, now mandated by regulators in many countries, allows a customer to nominate a chosen network to receive all his or her calls, and to register the choice with the access network undertaking. This data is stored in the customer's class-of-service data, and the access network routes calls by default to the chosen network, with calling line identity. There is no need to dial an access code, and most customers are unaware of the network selection process. Variations on the service may allow customers to make different nominations for different classes of call. The network choice may be overridden on a call-by-call basis using an access code (as with easy access). The access network may have a code to signify its own network, allowing the caller to select this from other possibilities on an equal-access basis.

Under carrier preselection, the customer remains a customer of the access network undertaking, and pays it a basic line rental. This is different from arrangements where a retail telecommunications service provider manages the entire service by buying the access network undertaking's service on a wholesale basis. In this case, the customer has no direct billing relationship with the access network undertaking and is not its customer. This arrangement may, or may not, make use of a carrier preselection facility.

Telecommunications Numbering

Regulators normally nowadays oversee the strategic management of national telecommunications numbering and addressing schemes, a role that incumbents typically fulfilled in monopoly days. This is partly because there is scope for the obstruction of competition in numbering, but mainly because a well-functioning numbering scheme is an essential ingredient of a competitive telecommunications services market.

The regulator must ensure that there are sufficient telephone numbers to allow all undertakings that so wish to enter the market. The supply of numbers must be commercially neutral: Numbers must not distort competition in the form of shorter, more easily memorized or otherwise preferable numbers for particular undertakings. The regulator must protect consumers by preventing unnecessary volatility (that is, frequent number changes), and should ensure that the numbering scheme is user friendly (that is, good, pleasing, coherent, and easy to use). While regulators are responsible for the strategic development of numbering, the day-to-day management of allocations may be subcontracted. Regulators have discovered that numbering issues can generate a lot of public and media attention that is not necessarily helpful.

Numbering Demands

Pressure on numbering schemes has been a fact of life for the last thirty years in most countries. This arises from expanding demand. Even though main exchange connection growth—the principal driver throughout most of the twentieth century—no longer dominates, the demands of new entrants and new services show little abatement. Mobile stations now exceed fixed stations in many markets, all of them requiring numbers. The recent arrival of *Voice over Internet Protocol* (VoIP) services presages additional demand. The response to these pressures has been much activity in number changing, while the simple coherence of the national numbering systems developed from the 1950s for long-distance dialing is being eroded.

Management of Numbering Systems

Different countries have responded to numbering pressure in different ways, often depending on the legacy of earlier numbering schemes, the capabilities of switching systems, and simple national taste. Many towns and cities have had number changes with an increase of number length, so that Paris, Tokyo, London, and other places have wholly eight-digit local numbering. Various countries, for example Australia and the United Kingdom, have increased the overall number length of their national schemes. Some countries have imposed national number dialing in certain localities or across the entire country, where the whole number with area code must be dialed for *all* calls, even local calls. This is usually to facilitate the pathway to other reorganizations.

Area splitting reorganizations, where second, third, and more area codes are deployed in an area previously served by one code, have been common in countries unable for one reason or another to contemplate longer numbers. This has notably been the method of reorganization in the United States, where an established fixed format presents a high barrier to number length change. Often the area is actually split geographically, with the result that some users suffer a change of area code though not of local number.

A way of avoiding the need for number change is the overlay method, where new codes serve precisely the same area as the old and next-door neighbors might find themselves with different area codes. Usually a solution of desperation, overlay numbering tends to be anti-competitive in its effect, since new entrants' customers are more likely to have the new area code, while existing undertakings retain the more widely known and recognized old code. The technique of *number pooling* is widely employed in the United States and elsewhere. This reduces the granularity by which numbers are allocated to specific undertakings and their individual exchanges.

Historically, the minimum allocation has been 10,000 numbers (four subscriber digits), while allocating smaller blocks lessens the potential wastage inherent in the block size. This solution carries a penalty in the amount of data each exchange must carry to route calls, and of course it only postpones rather than removes the need for more fundamental reorganization later.

Number Portability

Number portability is the ability of the user of a particular telephone number to conserve his or her number while changing some aspect of the service. The change might be a change of physical location, of service type, or of the undertaking providing the service. Service provider portability has been seen as an important feature of a competitive market since, otherwise, the need to change number and inform contacts is a barrier for a consumer to change supplier. Most countries' regulators mandate service provider portability in one form or another. In practice, the exact requirements and technical solutions may vary between markets, for example for fixed numbers, mobile numbers, and nongeographic added-value service numbers such as 800[27] for toll-free calls.

The provision of service provider portability, which we will now call *number portability* in line with many countries' terminology, is complex and expensive. However, it is widely

---

[27] In the United States, 800 is the historic area code for toll-free numbers. Other countries may use different prefixes for these services. These numbers are known as "nongeographic" or "generic" in that they denote a service, and do not have the locality inherent in a normal area code.

believed that the overall gains to the industry of assisting telecommunications users to migrate to the most efficient provision of service outweigh its costs. Number portability cuts across the normal manner by which networks analyze dialed digits to obtain the destination of a call. Normally, switching systems work in a hierarchical manner examining firstly the area code to determine the main location, then the leading local digits to identify the serving network and local exchange, and finally the remaining digits to denote a particular line or group of lines. Portability intervenes in this process by requiring identification and exceptional handling of ported numbers.

Number portability is a cooperative network service requiring interoperable functionality in calling, called, and transit networks. Therefore, while most regulators try to be technologically neutral, leaving markets to find the best solutions, regulators usually cannot in this case escape some commitment in consultation with interested parties to a general technical solution. There are two basic types of portability solution, though each class permits a range of variations.

Modified call forwarding solutions involve some dependence on the "donor" network, that is the network that would have naturally owned the number but for its being a ported number. It is this network that provides knowledge that the number in question is ported, and supplies information about the real destination for that number. In a common form of this solution, calls to ported numbers must be routed via the donor network (or even the donor exchange) of the number, incurring extra carriage costs whenever this is an inefficient routing of the call.

Intelligent network solutions rely on an external database that tracks ported numbers. Calling and transit exchanges have to screen numbers against the database while processing each call. Different countries have made differing choices. The United States uses the intelligent network method, also using the technology to support number pooling. Other countries chose modified call forwarding methods mainly because these are quicker and cheaper to set up at the start, although they tend to entail higher running costs. Some countries chose modified call forwarding methods in conjunction with good intentions of migrating to an intelligent network solution in the fullness of time.

Local Loop Unbundling

Local loop unbundling (LLU) occurs when a telecommunications undertaking makes its local access connections available, that is, its "copper loops," on a wholesale basis to other network operators or service providers, who use them to create a retail customer service. The second undertaking is sometimes known as an LLU operator. Loop unbundling has become a prominent issue for regulation partly because it is a potential method of injecting competition in markets where the copper loop owner is dominant, and partly because of its importance in the provision of broadband Internet access using *Asymmetric Digital Subscriber Loop* (ADSL) and similar *Digital Subscriber Loop* (DSL) technologies. With few exceptions, most undertakings possessing copper loops have tended to resist loop unbundling, and so regulatory intervention has been necessary to make it happen. Local loop unbundling is mandatory for dominant loop-owning undertakings in Europe and the United States.

Commercial and Operational Arrangements

Both commercial and operational arrangements for loop unbundling are complex. The regulated price of unbundled loops is normally cost oriented and based on long-run incremental costs. From an operational point of view, there must be means for the LLU operator to obtain physical connection to the unbundled loops, and this usually implies a requirement for *collocation*, where the LLU operator is given space within the loop-owning undertaking's buildings to house its equipment. The physical and commercial arrangements for collocation have encountered difficulties, in some cases with the loop-owning undertakings having insufficient space for collocation. Fortunately, the introduction of digital switching in the 1980s and 1990s led to many exchange buildings having spare space vacated by the older mechanical switches, and this has eased the problems of collocation at many locations.

No matter how good the eventual arrangements, LLU operators face an unavoidable disadvantage relative to loop-owning undertakings, of needing to provide *back-haul* connection to their own networks. For the loop-owning undertaking, it is more than likely that its buildings already coincide with network nodes and access points. For these reasons, true loop unbundling, although available, has not been used in some countries as much as regulators might have hoped.

Loop unbundling is a topical subject in the provision of broadband Internet access, where the LLU operator provides the DSL multiplex equipment and offers a branded retail *Internet service provider* (ISP) product. Most loop-owning undertakings have served this market by providing DSL services on their own account, in many cases offering a wholesale "bitstream DSL" service where they lease the data path over their ADSL loops to ISPs. This allows ISPs to enter the DSL Internet broadband access market without becoming full LLU operators, and many use this arrangement.

Variations of Loop Unbundling

Variations of local loop unbundling are *part-loop unbundling* and *shared loop unbundling*. Both are usually mandatory offerings, and both are designed to increase the opportunity for effective competitive exploitation of a dominant undertaking's copper loops. Part-loop unbundling takes place when the unbundled part of the loop is not the whole length of the loop from exchange to customer premises, but a part, normally between the nearest access network flexibility point, typically a street cabinet, and the customer premises. This is especially useful to an LLU operator hoping to use *very high bit rate DSL* (VDSL) for video services. As these have only a short reach, over perhaps 500 meters of loop, it is usual to deploy VDSL in conjunction with a *fiber-to-the-cabinet* distribution system.

Shared loop unbundling takes place when the LLU operator takes use of a loop for the DSL service but leaves the loop-owning undertaking still providing the *plain ordinary telephone service* (POTS) over the loop. This is normally mandated, because refusal by the loop-owning undertaking to do this would make it harder for a DSL service provider with neither desire to provide nor expertise in voice telephony to enter the market.

DSL Technical Standards

The high-frequency signals used for DSL have the potential to impair communication, causing interference and disrupting both the DSL and normal telephony uses of copper pairs. The exact effects are unpredictable as they depend on the physical characteristics and quality of local loops and are likely to be at their worst with antiquated line plant. Regulators, therefore, need to set technical standards for DSL transmission. The simplest approach is to nominate approved products and systems, although this is a restraint on innovation.

More complex prescriptions, such as an *access network frequency plan*, permit any system provided it emits no greater power in any frequency band than that provided for in the plan.

Universal Service Obligations

Universal nondiscriminatory service to every citizen regardless of location was an obligation placed on European government-run telecommunications monopolies in Europe and on the common carrier AT&T in the United States. The *universal service obligation* (USO) continues in today's competitive markets. European regulation lists the services that have to be provided on a universal basis. The services have to be affordable, bearing in mind national economic conditions. They include:

- Local, long-distance, and international voice calls
- Fax
- Data transmission at rates compatible with functional Internet access
- Directory services
- Directory inquiries
- Public pay telephones
- Access to the emergency number without needing a means of payment at pay telephones

Additional features that may be required include support for physically and cognitively impaired users, support for low-income users (perhaps in the form of a low-user scheme), and geographical price averaging (where charges do not vary with location) or support for high-cost locations. Both European and U.S. regulations recognize that with technological development it will be necessary to review an evolving definition of the universal service set from time to time.

The universal service obligation may involve service provision on a noncommercial basis, that is, cost transference, so regulators must ensure that any costs involved are efficient, proportionate, and fair. European and U.S. regulation allows for the twin possibilities of funding a USO through public (taxpayer) funding and by a levy on the undertakings in the industry. In the latter case, the allocation formula must be fair, transparent, and nondiscriminatory. European regulation prevents a levy being placed on foreign undertakings that do not actually provide service within a member state, thus preventing it being loaded onto interconnection charges or linked with rights to interconnect.

When deciding the correct level of subsidy for a universal service undertaking, it is worth bearing in mind that universal service brings benefits as well as costs to its provider. It gives it a brand position as "the" provider: This has economic value as this will cause it to capture customers who might otherwise have considered other providers. Arguments in favor of taxpayer funding are that transfers and subsidies should be explicit, while industry levy funding may be convenient to implement and can be justified in that added destinations and sources of traffic benefit all users and all providers of service.

Although dominant undertakings are most likely to be universal service providers in their countries, it is important that means should be provided for others having the ability and desire to provide universal service to tender such service in a nonexclusive market. Means of selection should then be fair, transparent, and nondiscriminatory.

The 1996 Telecommunications Act in the United States places additional universal service obligations in the form of Internet connections at preferential rates to schools, libraries, and rural and nonprofit healthcare facilities. Elsewhere, it is unusual for such provisions to be enacted as part of telecommunications regulation, although similar goals are addressed through different frameworks. Many countries have social policy initiatives for education, healthcare, government, rural broadband, and rural development, where in these cases the government enters the market as an explicit customer or joint customer for the desired provision.

New Technologies, Radio Spectrum, and Voice over IP

Regulation and technology possess a complex and interesting relationship between them. On the one hand, regulation should be technologically neutral, allowing markets to find the best solutions to problems. It follows that regulators should not prescribe technological implementation, or employ technology as a basis for regulatory rules. Regulators should not stifle innovation. On the other hand, regulators have to take an active interest in technology, since technology changes markets and may provide the means of solution of difficulties in market and consumer affairs. New technology may change the cost characteristics of a market, so challenging existing market power but also creating fresh opportunities for market power to develop. Technology may change the prevailing service quality expectations in a market, which may have repercussions in the areas of consumer protection and of universal service obligations. When new technology appears, regulators must make careful study and impact analysis, taking care on the one hand not to be too prescriptive and on the other to reduce uncertainty.

Radio technology is advancing very rapidly. It is widely applied to mobile communications and to the new generations of mobile data and broadband services. Ubiquitous mobile telephony and data access is a vital adjunct to modern life, while new technology is making possible new applications on the scales of the wide-area, metropolitan area, local-area, home-area, and personal-area networks. Diverse applications will enable communication not only between people but also between autonomously communicating devices. This phenomenon has been described as the *Internet of Things*.

Present and predicted growth in these services is creating expanding demand for spectrum, lack of availability of which will stifle development. At the same time, new technologies are creating dynamic ways of using spectrum much more efficiently and effectively than in the past. Radio spectrum has traditionally been allocated by methods of central control that prescribe not only the spectrum allocated to each license holder, but also detailed definition of the equipment to be used and the transmitter locations and powers. Recent work is considering the development of spectrum as a property right that can be freely used and traded so long as detriment to others is not thereby caused. The technical, commercial, and regulatory frameworks are under active progress in many countries and regions.

Voice over Internet Protocol (VoIP) has emerged as a new medium for the making of telephone calls, usually over broadband data links, either from computers equipped with suitable handsets and software, or from normal telephones backed by an adapter. This technology reduces the cost of voice telephony and with increasing take-up is bringing about changes in the marketplace. At its first appearance, most regulators were content to regard it as another computer application that, while having niche and specialist application, did not significantly impact the voice telephony market (although regulators in some countries may have viewed it as illegal bypass). On this basis, regulators left it alone, granting forbearance from the general conditions applying to publicly provided telephone services. With significant and growing numbers of users and even greater public awareness in many countries, regulators have to review this relaxed attitude.

The regulation of voice over IP telephony services applies to publicly offered services. It does not affect network operating undertakings using IP transmission primarily as an underlying technology in their national and international networks, or implementing so-called "next generation" networks. In these cases, the publicly presented products do not change and therefore remain under the same regulation as before. The principal problems with consumer VoIP service are to ensure that consumers are well informed about the benefits and limitations of what they are buying, and to decide how many of the traditional obligations on telephony service providers should be laid on the new service, including:

- Access to an emergency service
- Provision of location information to the emergency services
- Availability of directories, directory inquiries, and operator assistance
- Number portability
- Obligations of interconnection, interoperability, and service quality

The types of information that consumers should have before buying a service include most of the preceding, together with service reliability, service quality, whether they have to provide power and whether there is backup in the event of power cuts or problems with the broadband connection. The issue of emergency access is very important in the United States and elsewhere; there is valid debate whether provision of a reliable emergency access should be obligatory for reasons of social welfare, even if some well-informed customers might willingly choose services that do not offer it. Relevant to the debate is the extent to which it might be assumed that VoIP telephony users also have access to a traditional voice line. This might be a reasonable presumption in markets where DSL services are sold only in addition to ordinary voice service over the copper loop.

3 Open Systems Interconnection Reference Model

Harold C. Folts

Introduction

Effective and meaningful interchange of information is an essential element in the operation of enterprises and the conduct of business activities. The dramatic advances in computer/communications technology are now providing comprehensive capabilities for the establishment and the evolution of distributed information systems fully interconnected and integrated to serve as the foundation of business around the world.

These distributed information systems have diverse operational requirements and will be supported by continually advancing technologies through a vast, worldwide, multivendor marketplace of products. The question that immediately arises is: How can compatibility among the variety of available systems, designs, technologies, and manufactures be realized without constraining innovation, performance, and ongoing evolution?

In 1978, the International Organization for Standardization (ISO) established a massive standards development program called Open Systems Interconnection to establish an architecture and family of standards that would serve as the generic basis for compatibility among systems for information interchange. These standards remain fundamentally relevant to today's systems. This chapter presents the perspective, concepts, and functions of the Reference Model for Open Systems Interconnection (OSI), defined in International Standard ISO 7498, and the structure of the comprehensive family of International Standards that have been established for distributed information systems.

OSI Environment

The OSI architecture and family of standards, which specify the services and protocols for interchange of information between systems, have been defined to provide an operating environment for the implementation of distributed information systems from a multivendor marketplace. Interchange of information is in digital form and can convey data, voice, and image communications. The interconnected telecommunication resources can be dedicated transmission paths or switched services on a demand basis. Switched paths interconnecting communicating users can be on a fully reserved basis or on a demand basis using various switching technologies. An illustrative distributed information system is shown in Figure 3.1. The many different types of systems shown contain Application Processes (APs) that may need to communicate. APs can be manual, computerized, or physical. For example:

1. A manual process could be a person operating a "point of sale" terminal entering data or receiving an output.
2. A computerized process could be an operational program in a "host" computer performing its task, such as an accounting program processing a payroll.
3. A physical process could be the operation of a "robot" in a manufacturing plant.

Communications are often required in order for APs to perform their designated tasks. The data that an AP requires may need to be retrieved from a remotely located database. The operations that an AP requires may not be supported by its local processing resources, and therefore it must share additional resources in a remotely located system to accomplish

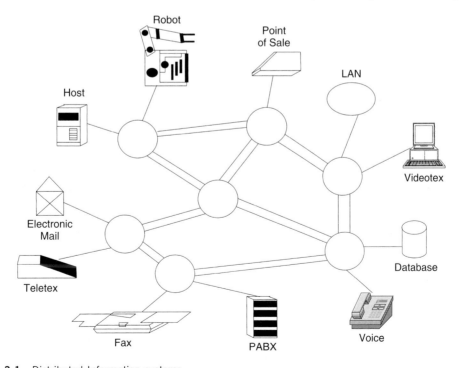

Fig. 3.1   Distributed Information systems.

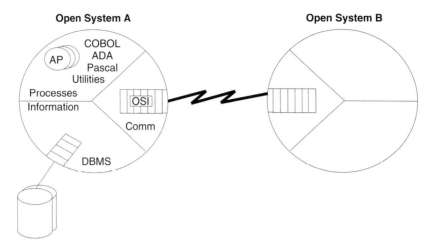

Fig. 3.2 Functional components of a computer-based system.

its task. Finally, on completion of its operations, the AP may need to deliver the resulting data to a remotely located database for further processing or analysis.

The interconnecting network shown in Figure 3.1 provides the telecommunication paths for information interchange between the systems. The OSI architecture defines an orderly structure of functionality to facilitate successful communications between APs. There are two basic components to ensure successful communications. The first is the transparent movement of the data between systems. The second is ensuring that the data arriving at the destination AP is in a meaningful form that can be immediately recognized and processed.

OSI standards define only those functions that are necessary to facilitate communications between systems. They do not describe the specific implementation, design, or technology that is used. These are left to the innovation of the system developer.

The functional components of a computer-based system are shown in Figure 3.2. The processing part, information and memory part, and internal communications (comm) part are not constrained by the standards. The OSI elements are additional functions in the communications part that are invoked only when communications with other systems are required. Systems that implement the OSI functions are therefore called Open Systems and can participate in information interchange using the OSI environment.

Layered Architecture

Basic Principles

There are many ways to describe and characterize systems for information technology applications. Many arbitrary alternatives could be just as effective, but it is imperative to have as much inherent flexibility as possible in the structure and to have a solution that is globally agreeable. Only then can an architecture be established as a generic basis that will continue to evolve to accommodate advancing technology and expanding operational requirements.

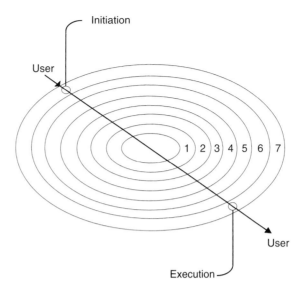

Fig. 3.3 Onion skin structure.

The concept of layering has been widely accepted. Only a minimum number of layers have been defined, thereby keeping the structure basically simple, while no individual layer should be so functionally complex that it is unwieldy.

The modular structure of layering provides a great deal of flexibility by

- Enabling implementations to change over time with new technology.
- Allowing tailoring of the invoked functions to optimize a specific operating configuration.
- Satisfying the broadest range of applications.

A layered architecture, as shown in Figure 3.3, can be referred to as an "onion skin structure" to describe a communications environment. The users on either side represent the corresponding APs that are interchanging information. In the communication process, the functions of each of the layers (layer 7 through 1) are invoked in the originating system and conveyed to the destination system where they are executed from layer 1 through 7 in the destination system. Very simply, in a point-to-point configuration, each layer is traversed twice, first for initiation and second for execution of the appropriate functions to ensure a successful communication. As interconnecting configurations become more complex with diverse paths through switched telecommunication resources, some of the layers may be traversed multiple times in performing relaying functions to support the information flow between systems.

An overlay of the layered architecture to real systems is shown in Figure 3.4. Each open system has its local system environment (LSE), which includes the basic processing, information, and communication components shown in Figure 3.2. Part of the OSI environment (OSIE) is also shown in each system. The AP resides in the LSE of a system and binds to the OSIE for communicating with another AP in a remotely located system. Each system has a component of each of the layers that is referred to as a layer entity. An instance of a layer entity represents the set of functions of the specific layer that are active in a system to

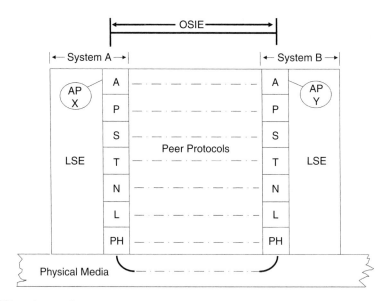

Fig. 3.4 OSI environment.

support the instance of communication. The functions are invoked in the layer entities in the originating system and executed by the respective layer entity in the destination system. Specifically, the functions invoked by the entity in layer 7 of system A are executed by the entity in layer 7 of system B.

These functions are logically conveyed between systems using peer protocols, which are defined for each of the layers to perform particular tasks. Each descending layer in the originating system adds its functions to the layer above as the communication proceeds layer by layer. The physical media, which are not part of the OSIE, provide the interconnecting transmission paths between the systems. At the destination system, each layer in turn executes its functions and passes the action to the next layer, in ascending order.

Layer Concept

The concept of a layer is illustrated in Figure 3.5. OSI characterizes an (N)-layer as a subdivision of the OSI architecture that interacts with elements in the next layer above, the (N + 1)-layer, and the next layer below, the (N − 1)-layer. The (N)-entity is an active element within the (N)-layer in a real system, and peer-entities are (N)-entities within the same layer, but in different systems.

The (N)-service relates the collective capability of the (N)-layer and underlying layers provided to support the (N + 1)-entity. Associated with each (N)-service are primitive interactions that pass various parameters. The primitives will be described later in this chapter. The semantics of the services and primitive interactions are defined in the family of OSI standards to facilitate an understanding of the dynamics of the OSIE but are not critical to OSI conformance and interoperability between systems.

Peer-entities communicate with each other using an (N)-protocol, which is a set of rules and procedures to facilitate an orderly communication in the performance of the associated

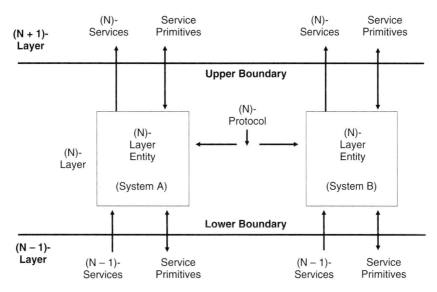

Fig. 3.5 Concept of a layer.

functions. The peer protocols are the critical specifications of the OSI family of standards that facilitate interoperability between systems and are the basis for determining OSI conformance.

In the case of the uppermost layer, there is no (N + 1)-layer, but this is the point at which the AP binds to the OSIE during an instance of a communication. There is also no (N − 1)-layer associated with the lowest layer because that is the point of interface between the OSIE and interconnecting physical transmission media.

Layer Services

While each layer performs its functions independently of other layers, interaction between layers is essential to ensure full cooperation in support of the communication. The services are provided to the (N + 1)-layer (see Figure 3.5), which is characterized as the (N)-service-user in each system by the companion technical report, ISO TR 8509.

This concept is illustrated in Figure 3.6, which applies to any of the layer boundaries. The (N + 1)-entity in each system is the service user of the functions of the (N)-layer and any lower layers, which are characterized collectively in Figure 3.6 as the (N)-service-provider. Therefore, each of the layer services can be defined from specific vantage points using this model.

Four interaction primitives are defined to further characterize the semantics of the dynamics of operation within a system as follows:

1. **REQUEST** is provided by the originating service user to activate a particular service by the service provider.
2. **INDICATION** is provided by the service provider to the destination service user to advise that a particular service is activated.

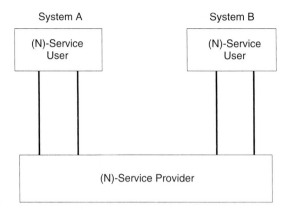

Fig. 3.6 Layer service model.

3. **RESPONSE** is provided by the service user back to the service provider in response to an incoming indication primitive.
4. **CONFIRM** is returned by the service provider to the originating service user to acknowledge completion of a requested service.

The primitives carry important parameter information between the layers. The nature of the parameters may vary for different services, but they can include address information, data to be transferred, and invocation of service options. The specifics of the services and associated primitives for each layer are covered by the specific layer service definitions, which are part of the OSI family of standards.

Communication between Peer Entities

Two modes of communication are defined in the OSI Reference Model: connection-mode and connectionless-mode. Connection-mode operation was the original basis for OSI. Later, the connectionless-mode was defined in ISO 7498 ADI.

Connection-mode Operation

The connection-mode functions through three phases of operation in support of an instance of communication. These are establishment, data transfer, and release. The initial definition of the OSI Reference Model included only the connection-mode for all the layers.

An (N)-connection represents an association that has been established between (N + 1)-entities and establishes the path for information flow that is maintained for the duration of the instance of communication, as shown in Figure 3.7. It is a result of the (N)-layer and the underlying layers collectively. Each layer operating in the connection-mode follows the procedure to establish the association, or connection, with its peer entity. After termination of the communication, the connection of each layer is released.

The establishment of a connection allows negotiation of parameter values and options between corresponding entities. It has a distinguishable lifetime that can support interchange of multiple data units over a period of time.

In contrast, the connectionless-mode does not establish and maintain a relationship between corresponding entities. Each transfer of data in the connectionless-mode is a separate, independent action for that layer. It has often been called the "message mode" or

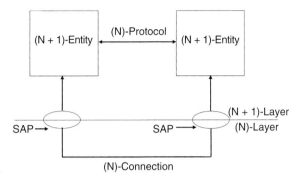

Fig. 3.7 (N)-connection.

"datagram mode," where each data unit is routed to the destination on an individual basis with no regard to data units transferred before or after it.

Connectionless-mode Operation

The connectionless-mode simplifies operations where the overhead of error control and recovery is not needed. It also eliminates the overhead of establishment and release of connections for a faster reaction communication. Typically, the connectionless-mode has been applied to routing data to destination systems through telecommunication resources. In this case, some layers are operating in the connectionless-mode, while others are operating in the connection-mode. A discussion regarding the use at different layers is included later in this chapter.

In the future, the upper layers directly supporting applications may also use the connectionless-mode for certain operations, such as broadcast to multiple APs.

Relaying

A layer entity can also provide a relaying function to facilitate the routing of data along the path to the destination system. An (N)-relay-entity does not provide any service to an (N + 1)-entity but uses only the services from a receiving (N − 1)-entity. It processes the incoming data unit and forwards it to the (N − 1)-entity associated with the next path for the data to flow through. This concept is illustrated in Figure 3.8. The relay system is identified between the two end systems, which are the source and destination systems for the instance of communication.

A relay system may have a number of possible paths to select from for routing data to the addressed destination, as illustrated by the lines radiating from the ellipse in Figure 3.8. Relay systems apply to both packet and circuit switching and can function in either the connection-mode or the connectionless-mode depending on the specific operational configuration. The circles in Figure 3.1 along the paths interconnecting the end systems are examples of intermediate relay systems that participate in OSIE to support the communications.

Identifiers

The principles of naming and addressing have been included in the OSI structure to facilitate the mapping of the information flow through the layers within a system and to find the

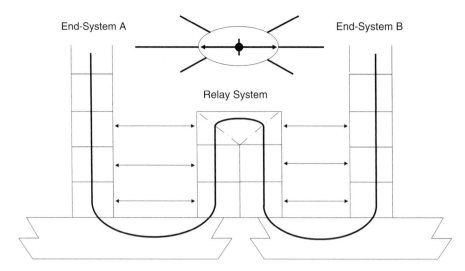

Fig. 3.8   Relay function.

path between systems for transmission of the data to the destination system. The conceptual port between an (N)-layer and an (N + 1)-layer is called a Service Access Point (SAP), which is shown in Figure 3.7. The location of the SAP is identified by an (N)-address, which may be globally unique or may only be specific to a particular instance of communication, depending on the layer involved.

A single SAP can support multiple instances of connections, which are identified by (N)-connection-endpoint-identifiers. A layer boundary may support more than one SAP between the same or different entities of the (N)-layer and (N + 1)-layer.

The provision of (N)-titles is used to uniquely identify individual (N)-entities to facilitate the routing and relaying functions. An (N)-directory is the function that is used to translate a global title into an (N − 1)-address to which the (N)-entity is attached. The complex and abstract subject of Naming and Addressing is dealt with in detail in Part 3 of ISO 7498, and the provisions for global Network SAP addressing are defined in the OSI Network Layer Service Definition ISO 8348 AD2. The addressing structure of the upper layers is defined in the Application Layer Structure of ISO 9545.

Data Units

As discussed earlier in this chapter, a communication activity from an AP in a system progresses from layer to layer as the required functions are initiated to support the communication. In turn, at the destination system, the incoming communication ascends through the layers as the functions are executed to provide a successful communication to the destination AP.

This transfer of information between the systems involves the construction and forwarding of control information and data units that convey the actions and substance of the communication. Figure 3.9 presents the basic data unit structure that passes layer by layer within a system. A service data unit (SDU) is the data passed to the (N)-layer from the

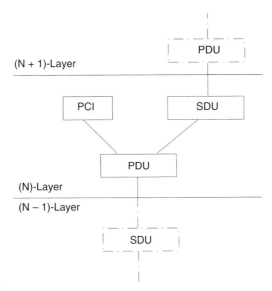

Fig. 3.9   Data units.

(N + 1)-layer for forwarding to the peer (N + 1)-entity in the destination system. Therefore, the (N)-layer is responsible for transferring the SDU transparently, so that it is delivered to the destination (N + 1)-entity exactly as it was received by the (N)-entity in the source system.

The SDU is received by the (N)-entity as a parameter of a request primitive issued by the (N + 1)-entity. The (N)-entity then produces the protocol control information (PCI), which initiates the required functions and provides the appropriate parameter information for execution by the peer (N)-entity in the destination system.

The combination of the SDU and PCI becomes the protocol data unit (PDU) for transfer to the peer (N)-entity via the (N − 1)-connection or connectionless path, depending on the mode of operation. The (N)-PDU becomes the (N − 1)-SDU, which is then processed in a similar manner with the addition of an (N − 1)-PCI to produce the (N − 1)-PDU for the next layer below. Recursively, this process occurs in each descending layer until the data leaves the OSIE in transmission on the physical media.

At the destination system, the PDU enters a layer as a parameter of an indication primitive. The layer entity processes the PCI part, executes the functions, and discards the PCI, leaving the SDU to pass to the next ascending layer via an indication primitive. Again, each layer in turn recursively processes the data units until the final data is delivered to the destination AP in a form that is meaningful for processing in the local environment of the destination system.

The layer protocols in the family of OSI standards support the many functions in the different layers. The protocols define the peer entity interactions, along with the syntax and information content of the PCIs that convey the actions between communicating systems. These protocol standards are the specifications that provide the compatibility and interoperability between systems in the worldwide, multivendor marketplace of products for information technology applications.

Additional Layer Elements

The OSI Reference Model defines many additional layer elements that further characterize the OSIE. The various elements defined can apply at some or all layers, depending on the specific service requirements for the particular layer.

Multiplexing and Splitting

The layer entities have been shown as mapping on a one-to-one basis from layer to layer. However, additional flexibility is provided through the mapping of multiple entities to provide for more effective use of resources as needed.

Multiplexing involves the mapping of multiple (N)-connections to a single (N − 1)-connection. This configuration enables bandwidth sharing of the connection to support multiple communications. The demultiplexing is done in the respective peer entity in the destination system.

Splitting involves the mapping of a single (N)-connection to multiple (N − 1)-connections. This configuration enables expansion of the bandwidth to support a communication that requires a higher throughput of data than can be supported by the bandwidth of a single (N − 1)-connection. Splitting can also apply to high survivability situations where the loss of a sole (N − 1)-connection would be seriously disruptive to an operation, whereas the loss of one of many connections in a split configuration would only reduce the bandwidth.

Data Transfer

There are two basic types of data transfer: normal and expedited. The unit of transfer by a layer is its protocol data unit (PDU). Every layer has the function for transfer of data in support of the instance of communication.

Connection-mode

In the connection-mode, normal data transfer is accomplished by a layer through an (N − 1)-connection that has been established with the destination (N)-entity. Some of the layers will also employ the following additional functions:

1. Flow control regulates the rate of data transfer, so that PDUs do not arrive at the destination (N)-entity faster than they can be processed. Peer-to-peer protocol flow control mechanisms are generally employed to coordinate the rate of transfer between the peer (N)-entities. Flow control may also be applied across the layer interface between an (N + 1)-entity and an (N)-entity.

2. Segmenting enables a single (N)-SDU to be broken up into smaller units for processing into multiple (N)-PDUs. At the destination (N)-entity, the SDU segments are reassembled into a single (N)-SDU before passing to the (N + 1)-layer. An (N)-protocol mechanism is employed to keep track of the segments during transfer so they can be recombined at the destination.

3. Blocking enables multiple (N)-SDUs to be mapped into a single (N)-PDU. Although this function is specified in the OSI Reference Model, it has not been employed in any of the OSI protocols or configurations.

4. Concatenation enables multiple (N)-PDUs to be mapped to a single (N)-SDU in the (N − 1)-layer.

5. Sequencing returns data units to the same order at the destination (N)-entity that they were sent in from the originating (N)-entity. This situation could particularly occur in connectionless-mode data transfer where each data unit could take a different route to the destination system. The (N)-protocol PCI contains a sequence number that is used for the resequencing function.

6. Acknowledgment provides confirmation for delivery or receipt of PDUs by a destination (N)-entity. Generally, the (N)-protocol PCI provides the acknowledgment mechanism and sequence numbers to facilitate this function, which generally provides for recovery of lost data units.

7. Error detection and notification identifies corruption of data during the transfer process between (N)-entities. Recovery from such errors can be accomplished by the same mechanism used for lost data units found by the acknowledgment function.

8. Reset provides for reinitialization of an (N)-connection without the loss of associations and connections of the upper layers.

9. Routing provides for relaying PDUs across multiple (N)-entities for establishing a connection and transferring data, or for connectionless-mode data transfer.

The connection-mode data transfer also provides for expedited data, which is not intended to be used on a regular basis. Generally, expedited data is limited in the amount of data transferred and in its frequency of transfer. It is not subject to the same flow-control constraints, if any, as normal data transfer. The basic rule is that an expedited data unit should not arrive at the destination any later than a normal data unit sent after the expedited data unit. In other words, it should not be any slower than normal data but may be faster. Expedited data service is invoked as required by the upper layers. Expedited data may also be acknowledged by the destination (N)-entity.

Connectionless-mode

Data transfer in the connectionless-mode involves minimal functionality. Data units are sent, routed, and delivered with no knowledge being maintained by the (N)-layer of their status along the way. Some have referred to this mode as "send and pray" because there is no assurance that delivery has been successful. On the other hand, the shortcomings of this simple approach can be compensated for by use of appropriate functions at higher layers, which are operating in the connection-mode.

Specific OSI Layers

As explained earlier in this chapter, the number and functional configuration of the specific layers for the OSIE are somewhat arbitrary. A number of arrangements to satisfy the requirements are possible. However, the agreement has been made internationally for the OSI configuration, which provides a solid generic basis to characterize distributed information systems while establishing a structure for accommodating advances in technology and expanding operational requirements through continuing evolution.

The required functions to support successful communications between APs are distributed among seven architectural layers. The layers, as shown in Figure 3.10, are defined in the following sections.

In summary, the upper three layers provide the functions for a meaningful communication so that information reaching the destination AP can be processed and is compatible

| Application |
| Presentation |
| Session |
| Transport |
| Network |
| Data Link |
| Physical |

Fig. 3.10 Layers of the OSI model.

with the local systems environment. The lower four layers provide the connectivity for inter-change of data between systems and collectively represent the "bit pipe" for transparent data transfer.

Application Layer

The structure of the application layer has changed from that defined in the OSI Reference Model of ISO 7498. The current concept defines the application layer entity as consisting of two types of service elements. The Association Control Service Element (ACSE) of ISO 8649 and ISO 8650 establishes the association between communicating APs and sets the context of the communication. The context refers to the other operational elements that are needed to support the specific nature of the communication. These are defined as the Application Service Elements (ASEs), which may be used individually or in multiples to satisfy specific operational contexts. Example ASEs include File Transfer, Access, and Management (ISO 8571); Virtual Terminal (ISO 9040 and ISO 9041); Transaction Processing (ISO 10025); Commitment, Concurrency, and Recovery (ISO 9804 and ISO 9805); Remote Operations (ISO 9072); Reliable Transfer (ISO 9066); Remote Database Access (ISO 9579); and others. There is a richness of functionality to select from in the application layer.

Presentation Layer

The presentation layer is defined in terms of its semantics and described in terms of an abstract syntax of the data types involved. The Abstract Syntax Notation No. 1 (ASN.1) of ISO 8824 has been widely accepted for use in OSI specifications. When data is transferred through the OSIE, an encoding agreed on by the peer presentation entities is needed. The common presentation layer service and protocol of ISO 8822 and ISO 8823 are used to establish the transfer syntax that will be used to support the communication. The Basic Encoding Rules of ISO 8825 are now widely used in OSI applications, but other transfer syntaxes can also be selected.

Session Layer

The session layer facilitates orderliness in a communication. A session connection is established, and the specific functions that are to be used are selected. Among the choices in typical use are use of either two-way simultaneous or two-way alternate data transfer, expedited data transfer, synchronization and recovery of data, activity management, and exception reporting. The session service and protocol are specified in ISO 8326 and ISO 8327, respectively. The connectionless-mode in the upper three layers is under development.

Transport Layer

The transport layer provides for the transparent transfer of data between peer session layer entities and ensures that the appropriate quality of service, which was requested by the originating session entity, is maintained. In this regard, the transport layer is required to select the network layer service that most closely matches the requested requirement. In addition, it may also have to invoke additional functionality to ensure the proper quality of service. Five classes of connection-mode protocol have been defined in ISO 8073 to accommodate the variety of operational conditions. The transport service is defined in ISO 8072. There is also a connectionless-mode transport protocol specified in ISO 8602.

Network Layer

The network layer provides the functions for the telecommunication resources that provide the paths for transfer of data between systems. Real telecommunication services, such as may be provided by a public data network, are defined by OSI as subnetworks. Paths between systems may involve switched services and interconnection of multiple subnetworks en route. Both connection-mode and connectionless-mode operations have been widely applied in the network layer. The Network Service Definition is provided in ISO 8343 and a number of standards are available for the various protocols and configurations that can be employed.

Data Link Layer

The data link layer provides the means for synchronizing the bit stream flowing to and from the physical layer and for detection of errors due to transmission problems. Typically, local area networks (LANs) operate the data link in connectionless-mode, while wide area network (WAN) applications operate in the connection-mode. In the connection-mode, the additional functions of flow control and error recovery are included. The Data Link Layer Service Definition is specified in ISO 8886.

Physical Layer

The physical layer provides the functions for transparent transmission of the bit stream between data link entities. When the physical layer is active, the bits flow through to the transmission medium. The electrical characteristics and physical connector serve as the interface between the OSIE and the external transmission media. The Physical Layer Service Definition is specified in ISO 10022. A number of physical interfaces are available for a variety of transmission environments.

Additional Provisions

Three additional parts of the OSI Reference Model have been published that cover some other very important aspects of distributed information systems. These are summarized as follows:

ISO 7498-2: Security Architecture provides the framework for development of various security provisions to protect system integrity and unauthorized access to user data.

ISO 7498-3: Naming and Addressing provides the concepts and principles for identifying objects through naming and location of objects through addressing mechanisms.

ISO 7498-4: Management Framework provides the concepts and principles for management of the resources of the OSIE to ensure continuing and effective operation.

Conclusions

The OSI architecture establishes a generic foundation for implementation and evolution of distributed information systems to support a diversity of operational requirements. While the richness of functionality that is available can lead to significant complexity, not all the capabilities are needed for every application. It is important to select only those functions that are needed for particular operational configurations, and no more. Therefore, the complexity will vary as needed to fulfill specific requirements.

OSI is a living architecture that will facilitate evolution to new technologies and operational requirements, but the basic structure must be preserved to ensure an orderly transition on a common basis. New functions may be added to layers in the future, while some of the existing functions will become disused. Even some layers may be inactive in certain applications. Nevertheless, keeping the basic OSI structure intact will ensure a consistent evolution in the future.

Appendix: International Standards

The following references used in this chapter are International Standards published by the International Organization for Standardization, 1 rue Varembe, Geneva, Switzerland. They are also widely available online or from national standards bodies such as ANSI in the United States.

For brevity, the following abbreviations have been used in the titles:

IPS – Information Processing Systems
ISO 7498, IPS–OSI – Basic Reference Model
ISO 7498 ADI, IPS–OSI – Basic Reference Model, Addendum 1: Connectionless-Mode Transmission
ISO 7498-2, IPS–OSI – Basic Reference Model—Part 2: Security Architecture
ISO 7498-4, IPS–OSI – Basic Reference Model—Part 4: Management Framework
ISO 7498-3, IPS–OSI – Basic Reference Model—Part 3: Naming and Addressing

ISO 8072, IPS–OSI – Transport Service Definition

ISO 8073, IPS–OSI – Connection Oriented Transport Protocol Specification

ISO 8326, IPS–OSI – Basic Connection Oriented Session Service Definition

ISO 8327, IPS–OSI – Basic Connection Oriented Session Protocol Specification

ISO 8348, IPS–DC – Network Service Definition

ISO 8348 AD2, IPS–DC – Network Service Definition, Addendum 2: Network Layer Addressing

ISO TR 8509, IPS–OSI – Service Conventions

ISO 8571, IPS–OSI – File Transfer, Access and Management (5 parts)

ISO 8602, IPS–OSI – Protocol for Providing the Connectionless-Mode Transport Service

ISO 8649, IPS–OSI – Service Definition for the Association Control Service Element

ISO 8650, IPS–OSI – Protocol Specification for the Association Control Service Element

ISO 8822, IPS–OSI – Connection Oriented Presentation Service Definition

ISO 8823, IPS–OSI – Connection Oriented Presentation Protocol Specification

ISO 8824, IT–OSI – Specification of Abstract Syntax Notation One (ASN. 1)

ISO 8825, IT–OSI – Specification of Basic Encoding Rules for Abstract Syntax Notation One (ASN. 1)

ISO 8886, IPS–Data Communication – Data Link Service Definition for OSI

ISO 9040, IT–OSI – Virtual Terminal Basic Class Service

ISO 9041, IT–OSI – Virtual Terminal Basic Class Protocol (2 parts)

ISO 9066, IPS–TC – Reliable Transfer (2 parts)

ISO 9072, IPS–TC – Remote Operations (2 parts)

ISO 9545, IT–OSI – Application Layer Structure

ISO 9579, IT Database Languages – Remote Database Access (2 parts)

ISO 9804, IT–OSI – Service Definition for the Commitment, Concurrency and Recovery Service Element

ISO 9805, IT–OSI – Protocol Specification for the Commitment, Concurrency and Recovery Service Element

ISO 10022, IT–OSI – Physical Service Definition

ISO 10026, IT–OSI – Distributed Transaction Processing (5 parts)

IT – Information Technology

OSI – Open Systems Interconnection

TC – Text Communication

Part Two

Circuit-switched Networks and Digital Switching Systems

Circuit-switched telephone networks have provided voice communications for more than a century. Over much of the period, their development has been relatively slow, with major changes taking place only rarely. For example, in the 1950s subscriber trunk dialing was introduced, in the 1970s mobile phone networks emerged, and in the early 1980s new digital and switching transmission caused a transition from analogue to digital technology, which was later accelerated by the appearance of optical fiber transmission. Since then, new calling features enabled by digital operation and voice processing have progressively improved the service over the years and, more recently, Internet telephony has taken hold. So although major change has been rare, development has nevertheless been continuous.

But telephone networks have long been used to carry more than just voice traffic. The analogue local bandwidth and 64 kb/s long-distance capability of conventional circuit-switched networks have offered connections for a variety of non-voice services, including, for example, facsimile, data, and surveillance. In more recent years, the existence of the telephone network has supported narrowband access to the Internet at speeds up to 56 kb/s. Without the telephone network it is fair to say that the Internet would not have achieved the widespread and massive growth seen over the past decade. However, Internet growth has been so massive that the telephone network architecture soon had to be modified to support the huge and growing volume of data traffic being carried over it.

In particular, flat-rate charging led to telephone circuits being held open for long periods of time when Internet data was being sent only infrequently between customer and Internet Service Provider (ISP). This led to inefficient use of telephone circuits and switches and had an adverse effect on the telephone service. The adopted solution was to provide private circuits between local exchanges and ISPs and, so, bypass many of the telephone switching stages.

Once the Internet took hold, the demand for higher speeds outstripped telephone circuit capacity and new broadband access technologies coupled directly to the underlying transport infrastructure emerged. Before long, the new broadband infrastructure began to demonstrate that it could not only support the faster data connections that were in demand, but could also efficiently support Internet telephony using Voice over Internet Protocol (VoIP) and even video over IP. VoIP is a technology that is disrupting the long-established "circuit-switched," narrowband telephone network. Internet basics and Internet telephony are therefore covered in Part 3 while broadband technologies will be covered in more detail in a later volume of *The Cable and Telecommunications Professionals' Reference*.

The public-switched telephone network has been described as the largest machine in the world. It relies on faithful transmission of signals from one telephone to another, whether they are in the same town or on opposite sides of the world. It relies on signaling systems that allow any telephone user to send instructions into the network. These can set up links to a wanted destination or make reliable connections via a switching system. Chapter 4 provides a broad discussion of the principles of circuit-switched telephone networks and general architecture while Chapter 5 explains the operation of digital switching systems and the associated signaling and control methods.

4 Circuit-switched Telephone Networks

J. E. Flood

Introduction

A telecommunication network is required to transmit messages between any of its users (who are usually called customers or subscribers), and the messages may be conveyed by signals that are either digital or analogue. A digital signal can have only discrete values. The simplest digital signal is a binary signal having only two values. Binary-coded data from a digital computer is an obvious example. However, multilevel signals are also used. An analogue signal is a continuous function of time and at any instant it can have any value between limits set by the maximum power that can be transmitted. A speech signal is an obvious example. A television signal is a mixture of analogue and digital, since it transmits both picture contents and synchronizing pulses.

Channels used to transmit signals also may be either digital or analogue, but it does not follow that a digital signal always requires a digital channel and an analogue signal always requires an analogue channel. A digital signal may be transmitted over an analogue channel by modulating a carrier wave. An example is the use of a modem (modulator/demodulator) to transmit digital data over a telephone line. An analogue signal can be transmitted over a digital channel by using analogue-to-digital conversion. An example is the use of pulse-code modulation (PCM) to send speech as a digital signal.

The earliest telecommunication networks were digital, since they were used for telegraphy. The invention of the telephone required the provision of analogue networks. The amount of traffic resulting from telephony soon began to exceed the traffic from telegraphy. It then became economic for telephone networks to be used also for telegraphy, in order to avoid the need for separate line plant.

Analogue signals are affected by attenuation, distortion, noise, and interference. However, a digital signal is unaffected by noise or interference unless it is so great that decoding errors are caused (e.g., if "0" is replaced by "1" or "1" by "0"in a binary code). An analogue signal is thus progressively degraded as it is transmitted over a long distance, whereas a digital signal can be almost immune from crosstalk, noise, interference, and variation of attenuation. Pulse-code modulation enables these advantages to be obtained for telephone transmission and the invention of the transistor made it economical. Consequently, PCM was widely adopted for telephone transmission during the second half of the twentieth century.

If a network is to serve a large number of customers, it is not economical to provide a line from each customer to every other, particularly as some pairs may seldom or never converse. Instead, customers are provided with lines to a smaller number of switching centers that provide connections from customers who are calling to those who are being called by them. For example, if a network has 10,000 customers, it would require 99.99 million lines to interconnect them all directly. If each customer has a line to a central switching center, only 10,000 lines are needed. Such switching centers are called central offices in North America and telephone exchanges in the United Kingdom.

The earliest form of switching used in telecommunication networks was called message switching. If a customer in town A wished to send a telegram to a customer in town B and there was no telegraph line between A and B, but there was between A and C and between C and B, the operator at A would send the message to C. The operator at C would receive the message and retransmit it to B. Later, the messages were transmitted and received on paper tape and eventually the whole process was made automatic by using digital computers.

Simple message switching is not very satisfactory for data traffic because of the large variation in message durations. Messages can vary in length from a single keyboard character to very large streams of data. The user of a terminal who is interrogating a distant computer needs a quick response and will not obtain it if the call has to wait in a queue while a large file is exchanged between two mainframe computers. For this reason, data networks use a modified form of message switching called packet switching. Long messages are split into a number of small ones called packets. Thus, the single packet from the terminal user is sent between packets of the large computer file instead of waiting until its transmission is finished, so delay is minimized. Packet switching is used on the Internet and in local area networks, which are discussed later.

The invention of the telephone brought the requirement for concurrent communication in both directions in real time and this requirement could not be met by message switching. It was necessary for the lines of a calling and a called customer to be connected together in a switching center for the duration of their call. This was called circuit switching and has formed the basis of all telephone networks.

The earliest telephone exchanges were manually operated. Later, automatic exchanges were introduced, in which the connections were made by electromechanical switches responding to numbers dialed from the subscribers' telephones. Eventually, advances in electronics enabled digital electronic switching systems to replace electromechanical telephone exchanges and to make connections between digital circuits (Flood, 1995). Electronic control methods for these systems were developed, based on central digital processors using stored-program control (SPC). This introduced great flexibility, since software changes can cater for differing exchange sizes and configurations and the services provided to customers (Redmill and Valdar, 1995). These advances enabled integrated digital networks (IDN) to be introduced in which all switching is digital and all transmission between exchanges is also digital. By the end of the twentieth century, the world's principal telephone networks had become fully digital.

In addition to its transmission and switching functions, a telecommunication network must perform signaling functions. Signals must be sent between customers and their exchanges and between exchanges in order to instruct the switches when to set up and release connections and how to route the calls. A telecommunication network may therefore be considered as a system consisting of three interacting subsystems as follows:

- Transmission
- Switching
- Signaling

These are described more fully in the following sections.

In addition to providing "dial up" connections, a public telecommunications operator (PTO) leases circuits in its network to customers who generate sufficient traffic to justify this. Examples are companies' private telephone networks linking their premises at different sites and private data networks. A large public switched telephone network (PSTN) thus becomes a bearer for different kinds of traffic carried for many different organizations. These include mobile network operators, whose radio base stations are connected by fixed links to mobile-service switching centers that also provide interconnection between the mobile network and the PSTN. Other customers are competing telephone companies that are smaller than the PTO and so find it uneconomical to build their own networks.

Until recently, most of the traffic carried by PSTNs was telephone, and data traffic was a small proportion of the total. However, a striking feature of the last few years has been a very large growth of data traffic as a result of rapidly increasing Internet use. In many major networks, the amount of traffic generated by data communication now exceeds that for telephony. It used to make economic sense to provide data services by packet-switched circuits in a network designed primarily for circuit-switched speech. Now, it would appear to make sense to design a packet-switched network and to use this also for speech.

The Voice over Internet Protocol (VoIP) has been developed to enable packet switches to route voice calls in the form of packets as if they were data. Major telecommunication operators have already begun the process of conversion (Nunn, 2005). British Telecom, for example, plans to change over completely to a packet-switched network and expects this to be serving most of its customers by the year 2008 (Beal, 2004).

Internet and local area network operations, which are based on packet-switching principles, will be described in later chapters. This chapter is concerned with circuit-switched networks. These do provide a reference model for newer networks, such as mobile telephony and IP telephony.

Network Services

The customers of a PTO require many different services (Van Duuren et al., 1992). These can be divided into two categories:

1. Teleservices, in which the provision of the service depends on particular terminal equipment (e.g., a telephone or teleprinter)
2. Bearer services, which present the customer with transmission capacity that can be used for any desired function (e.g., private circuits)

Some services consist of providing a connection from one customer to another (e.g., for a telephone call or a transfer of data). Other services require connecting the caller to a service. This may be provided by the PTO, say for operator assistance or directory inquiries, or it may be provided by a separate company as a "value-added" service, such as a message service. Other services require a call to be connected from one network to another, for example to the PSTN of a competitor or to a cellular-radio mobile telephone network. Different services appear to require different networks. However, these all use the same transmission bearer network and a subscriber obtains access to them through the same customer access network.

Transmission

Attenuation

The complete path of a telephone connection includes the air path from the talker's mouth and that to the listener's ear, in addition to their telephones and the switched connection between them. The overall attenuation of this path is expressed in terms of its overall loudness rating (OLR) in decibels (dB) and it is measured by comparing the perceived loudness of the received sound with that from a standard speech path defined by the International Telecommunications Union, Telecommunications Sector (ITU-T) (Richards, 1973). Users dislike connections that are too loud as much as those that are too quiet. Subjective tests carried out to determine users' opinions on a variety of telephone connections showed that there is a preferred range of overall loudness ratings from about 5 dB to 15 dB.

A national transmission plan should allocate the permissible overall attenuation so as to obtain the most economical design. Subscribers' lines are the ones that are the most numerous and contribute significantly to the total cost of the network. To minimize this cost, subscribers' lines use the minimum size of cable conductors. Consequently, the transmission plan should allocate as much loss as possible to the subscribers' lines at the two ends of a connection.

In an analogue PSTN, the economic necessity of allocating a large limiting attenuation to subscribers' lines, together with the loss variations of the intervening circuits, results in some connections whose losses are excessive. For example, the British Telecom analogue network had a value of 59 dB for its limiting OLR. Fortunately, most calls did not follow the adverse routings that gave this result. By means of statistical studies of routings and their transmission impairments, it was possible to prove that only a very small proportion of connections were unsatisfactory. It was unfortunate that the locations of some users caused them to make such calls frequently!

In an integrated digital network, the transmission links have zero loss with zero tolerance, so the subscribers' lines at the two ends of the connection are the only cause of variation of loss between different connections. This eliminates the wide variation of losses encountered in analogue networks and enables every connection to have a loudness rating within the preferred range. If digital transmission is extended over the access network to the customer's premises, even this variation is eliminated.

Delay

Propagation delay is inevitable with any form of transmission over distance; however, delays do not exceed a few tens of milliseconds for connections within a country. For international connections, the ITU-T recommends that the maximum one-way delay should not exceed 400 ms. Since a link via a geostationary satellite has a one-way delay of 260 ms, two such links may not be used in tandem.

When circuit switching is replaced by packet switching, delays increase considerably. Packetizing and depacketizing cause delays. Queues in switches cause delays that may vary randomly during a call, causing jitter. On the Internet, the protocol allows successive packets to take different routes, so they can encounter different delays and can even arrive in the wrong order. These long and varying delays are acceptable for data transmission, since they do not cause errors, but they can have a great effect on the quality of speech transmission. It is therefore not surprising that early attempts to send speech over the Internet resulted in very poor performance. Subsequently, complex methods of quality control have been developed that are installed at the periphery of the network and enable telephone connections to be made over the Internet with a performance approaching that obtained with circuit switching (Rudkin, 2004).

Four-Wire Circuits

Long-distance circuits use amplifiers to overcome their high attenuation. Amplifiers are usually unidirectional devices. An amplified circuit therefore has a separate channel for transmission in each direction, as shown in Figure 4.1. Such a circuit is called a four-wire circuit, although the channels may be provided in high-capacity multiplex systems over optical fibers or microwave radio instead of wires. A connection made up of such circuits will be joined at each end to two-wire circuits leading to the subscribers' stations. A four-wire circuit is connected to its two-wire terminations by means of a four-wire to two-wire

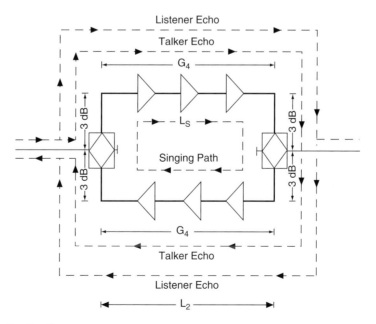

Fig. 4.1   Four-wire circuit.

terminating set containing a hybrid transformer and a line-balance impedance (Flood and Cochrane, 1991).

If the four-wire circuit were to be directly connected to the two-wire circuit at each end, this would complete a loop around the channels in the two directions which could oscillate or "sing." The terminating set prevents this. If the balance impedance exactly matches the impedance of the two-wire line, the hybrid transformer provides transmission between the two-wire lines and the four-wire circuit (with 3 dB loss), but there is infinite loss across it between the two channels of the four-wire circuit.

Thus, it is theoretically possible for a four-wire circuit to provide zero loss, or even gain, between its two-wire ends. However, in practice, line balances are not perfect, so the losses across the hybrid transformers are not infinite. If the trans-hybrid loss is too low, or the gain (G_4) in either direction is too high, the net gain around the loop will be greater than zero and the circuit can still sing. In practice, there is a tolerance on G_4 and line balances cannot exactly match the two-wire lines. Consequently, the gains G_4 must be set to provide an adequate stability margin to cater for these variations. As a result, there is usually an overall loss (L_2), typically 3 dB, between the two-wire ends (Hills and Evans, 1973).

Even if a four-wire circuit is stable, it still impairs transmission if its trans-hybrid losses are non-infinite. As shown in Figure 4.1, a signal from one two-wire circuit can return to it (attenuated and delayed) after traversing the four-wire circuit in each direction. This is called talker echo. This signal can then be transmitted again over the four-wire circuit to produce listener echo. In each case, the echo follows the direct signal after a delay equal to twice the one-way propagation time of the four-wire circuit. This is known as the round-trip delay. For the listener, echo reduces the intelligibility of the conversation. For the speaker, it can actually interrupt speech and is thus even more troublesome. It is therefore necessary to ensure that echoes are attenuated sufficiently to make these effects unobjectionable.

The annoying effect of echo increases with its delay and so the greater echo attenuation required. This can be achieved by making the overall loss increase with distance. However, there is a limit to the extent to which this can be done. This is usually reached when the round-trip delay is about 40 ms. This delay is exceeded on long transcontinental and intercontinental circuits particularly if these use satellite links. It is then impossible to obtain both an adequately low transmission loss and an adequately high echo attenuation; on such circuits; it is necessary to control echo by fitting echo suppressors or echo cancellers.

Echo cancelers operate in the four-wire digital circuit and usually cancellation takes place in two stages. First, a digital adaptive filter generates a model of the voice signal passing through the echo canceller. In the returning direction the voice path passes back through the cancellation system, and the echo canceller compares the return signal and the model to cancel any echo. This process removes more than 80 to 90 percent of the echo across the network. In the second stage of the process a nonlinear processor eliminates any remaining residual echo by attenuating the signal below the noise floor.

Digital Transmission

PSTNs use a range of digital transmission systems, operating over different transmission media: copper, microwave radio, and optical fiber (Flood and Cochrane, 1991). The basic building block of these systems is the primary multiplex. This uses PCM to modulate the

Table 4.1   Error Performance Objectives for an International Hypothetical Reference
Connection at 64 kbit/s

Performance Classification	Objective
(a) Degraded minutes	<10% of 1-min intervals to have a bit error ratio $>1 \times 10^{-6}$
(b) Severely errored seconds	<0.2% of 1-s intervals to have a bit error ratio $>1 \times 10^{-3}$
(c) Errored seconds	<8% of 1-s intervals to have any errors (i.e., >92% to be error free)

Note: A total averaging period of any one month is suggested as a reference.

signal of each telephone channel onto a 64 kbit/s pulse train and the signals are combined by time-division multiplexing to form an assembly of 30 or 24 channels occupying approximately 2 Mbit/s or 1.5 Mbit/s. (The latter is used in North America.) These assemblies can form tributaries that are combined to form higher-order multiplexes.

The synchronous digital hierarchy (SDH) is sent over optical fibers. In the United States, it is known as SONET (synchronous optical network) and the lower-order multiplexing stages (up to 155 Mb/s) differ from the European standards. It assembles signals from modules called synchronous transport modules at level 1 (STM-1). This module has a bit rate of approximately 155 Mbit/s and can accommodate 63 2 Mbit/s bit streams (i.e., 1,890 telephone channels). The signals of four STM-1 modules can be multiplexed together to provide a synchronous transport module, STM-4, at 622 Mbit/s and four of these can be combined to form a synchronous transport module, STM-16, at 2.48 Gbit/s. Thus, STM-16 can accommodate up to 30,240 telephone channels.

Because the number of telephone channels that can now be transported over a single optical fiber is so enormous, the cost per kilometer of providing each channel is extremely small. Consequently, the cost of providing a long-distance telephone connection is almost independent of the distance involved.

Even in a digital network (where attenuation and its variation are not affected by the number of links in tandem, as in analogue networks), there are impairments that affect the quality of the received signals. Quantization distortion occurs as a result of the analogue signal being converted to digital format with a limited number of discrete signal levels represented. This occurs only once in each telephone connection when the signal is carried digitally from end to end across the network and it is held down to acceptable levels through design choice. However, because of noise buildup and distortions that accumulate in the transmission media, the digital error rate increases with the number of links in tandem. Expressing error performance as a long-term mean error rate is not very meaningful, however, because errors tend to occur in bursts.

The ITU-T has defined (in Recommendation G821) three parameters for measurement of error performance in order to take account of the distribution as well as the number of errors and has set objectives for them for a 27,500 km hypothetical international connection, as shown in Table 4.1. Parameters (a) and (b) are aimed at the needs of telephony and (c) is more relevant for data transmission services.

Switching

The earliest telephone exchanges used switchboards worked by operators. The manual exchange was unique among switching systems, since a single switch (i.e., the operator)

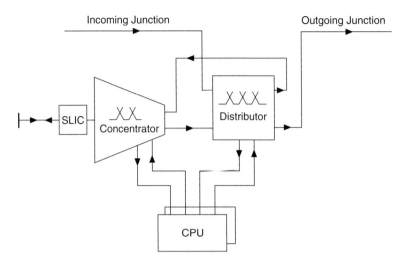

CPU = central processor unit, SLIC = subscriber's line interface card

Fig. 4.2   Trunking of digital telephone exchange with stored-program control.

could make a connection to any of several thousand lines by inserting a plug in a jack within arm's reach. When automatic systems were introduced, they used switches of much smaller capacity (Flood, 1995). For example, the Strowger two-motion switch could select only one out of 100 or 200 lines. A typical crossbar switch could connect any one of ten incoming trunks to any one of 20 outgoing trunks. To enable any subscriber to obtain a connection to any other, it became necessary to route the call by means of a network of switches having several stages, as described in Chapter 5.

Subscribers' lines are often not very busy. Trunks between switches inside an exchange and between exchanges can thus be more heavily loaded than subscribers' lines and fewer are needed. Subscribers' lines can therefore be terminated at the outside of an exchange on a switching network having fewer trunks looking into the network than on the outside looking to the subscribers. These initial switching stages are called concentrators. A large exchange will have many concentrators, so a further switching network, having several switching stages, is needed to make a connection from the calling subscriber's concentrator to that of the subscriber being called, as shown in Figure 4.2. These stages are called the distributor. Junction circuits to and from other exchanges are more heavily loaded than customers' lines, so they terminate on switches in the distributor network, as shown in the figure.

The switching networks shown in Figure 4.2 can make the connections required for the following types of calls:

- Between two subscribers on the same exchange
- Between a calling subscriber and an outgoing junction to another exchange
- From an incoming junction to a called subscriber on the exchange
- For a transit call between two other exchanges using this exchange as a tandem switching center

If an exchange is to serve only as a tandem switching center, it will have no customers' lines and thus no concentrators.

In an integrated digital network, all the junction and trunk circuits between exchanges are digital (using PCM) and the switches in the exchanges make digital connections between them. The concentrators and distributor shown in Figure 4.2 are then digital switches. However, the subscribers normally have analogue telephones and exchange lines. Analogue-to-digital and digital-to-analogue conversions are carried out by a PCM codec (i.e., coder/decoder) in the subscriber's line interface card (SLIC). Since the distributor and concentrator are both digital, the PCM link between them can extend beyond the exchange building, thus enabling the concentrator to be at a location remote from the exchange. This has enabled digital exchanges to serve a larger number of subscribers spread over a wider area than was possible with electromechanical exchanges. If digital transmission is extended also over the subscriber's line, it can be used for sending data at 64 kbit/s or even 2 Mbit/s to provide access to an integrated services digital network (ISDN).

The operators of manual exchanges provided many different services using the same basic apparatus by exchanging information verbally with the subscribers and performing different sequences of actions stored in their memories. The introduction of electromechanical switches controlled by telephone relays made exchanges much less flexible in the services they could offer, since they had lost the intelligence provided by the operator. A modern electronic exchange performs basic actions very rapidly, so it can be controlled by a single central digital processor, as shown in Figure 4.2. (However, this is replicated to provide continuity of service in the presence of faults.) The processor is controlled by stored programs, so the actions it performs can be changed by modifications to the software. The use of stored-program control (SPC) has enabled some of the versatility of the manual exchange to be regained and new services can be offered to the subscribers (Redmill and Valdar, 1990; Flood, 1995).

The ways in which calls are processed differ for different kinds of customers. They are determined by the class of service (COS) of a customer's line. This is in two parts, concerned with originating and terminating calls respectively. Examples of different originating COS are lines with ordinary telephones and those with payphones, to which different call-charging arrangements apply. Some lines may be barred from making long-distance calls. Others may be barred from making any outgoing calls because of unpaid bills. Examples of different terminating COSs are a customer having a single line and one having a group of lines to a private branch exchange. In the latter case, a busy tone is not sent to a caller until every line has been tested and found to be busy.

In electromechanical systems, lines having different COS had to be segregated into different groups connected to switches having different control circuits. It was therefore only practicable to cater for a small number of different COS. In a stored-program controlled exchange, the COS of a customer forms part of the data stored for the customer and is used by the central processor when processing the customer's outgoing and incoming calls. Many more COS can be provided, any COS can be associated with any customer and it can readily be changed.

Signaling

Customer Line Signaling

A customer's line is normally fed with current by a central battery in the exchange. If the telephone handset is in the "on-hook" position, the line loop is open-circuited and no current flows. If the handset is lifted, the switch hook completes the circuit; current flows in the line and is detected by the exchange. The following conditions are thus applied to the line by the customer:

1. Idle: no current
2. Busy: line current
3. Calling or answering: change from idle to busy
4. Clearing (i.e., disconnecting at the end of a call): change from busy to idle

In the reverse direction, when the telephone is on-hook, the exchange calls the customer by sending low-frequency alternating current to ring the bell. When the telephone is off-hook, the exchange sends back audio-tone signals (e.g., dial tone, ring tone, or busy tone) or recorded announcements.

With the introduction of automatic exchanges, the customer was no longer able to tell an operator the number of the customer to be called. Telephones were therefore fitted with dials and the customer signaled each digit of a called number by using the dial to break the line loop, with the corresponding number of disconnections at about 10 pulses per second. The introduction of subscriber trunk dialing and then international dialing resulted in long strings of digits and the dial was considered to be too slow. Push-button telephones were therefore introduced with multifrequency signaling. To receive such signals, the exchange requires a device called a register that is more complicated than the simple relay circuit needed to receive loop/disconnect dialing. However, it is used for only a short time during the setting up of a connection, so a small number can be shared in common by a large number of customers' lines. Dual-tone multifrequency (DTMF) signaling is used (Flood, 1995). Each digit is sent as a combination of two frequencies, one from each of two groups of four frequencies. In addition to the digits "1" to "0," the telephone keypad has two additional buttons with the symbols "*" and "#." These can be used by an SPC exchange for services which are under the control of customers (e.g., to instruct the exchange to route incoming calls to another telephone).

In an ideal signaling system, every signal should be acknowledged before the next is sent. In customer line signaling, the initial off-hook "seize" signal is acknowledged by a "proceed to send" signal in the form of a dial tone. The "address" signal in the form of the dialed number is acknowledged by the exchange sending back a ring tone, busy tone, or "number unobtainable." At the end of the call, the backward clear signal is answered by a forward clear, or vice versa.

Channel-associated Signaling

If a customer whose line is on one exchange wishes to make a call to a customer on another exchange and there is a direct route between the exchanges, signals must be sent between them in order to establish the connection. If there is no direct route, the connection may have to be established through tandem switching centers and signals will need to be sent to them. The traditional way of exchanging these signals is to send them over the circuits being taken into use for the connection. This is called channel-associated signaling. Signals sent in the direction away from the caller and toward the called customer are said to be in the forward direction and those in the reverse direction are called backward signals. The basic signals required between exchanges for a simple telephone call are as follows:

- Call request or seize (forward)
- Address signal (forward)
- Answer (backward)
- Clear (forward and backward)

For a call between two exchanges connected by a physical circuit, DC loop/disconnect signaling can be used as on customers' lines for the seize, address, and clear signals. When the answer signal is required, there is already line current flowing and this is necessary until the clear signal is given by its termination. It was therefore arranged for the answer signal to be given by reversing the direction of the line current and for the reversal to be detected by a polarized relay.

With the introduction of frequency-division multichannel carrier systems for long-distance transmission, there was no longer a physical pair of wires for each channel that could be used for signaling. Voice-frequency (VF) systems were therefore developed for sending signals within the 4 kHz bandwidth allocated to each channel (Welch, 1979). These systems were of two types: outband and inband. In an outband system, a small part of the frequency band was separated by a narrowband filter and used as a separate channel for a signaling frequency above the 300 Hz to 3.4 kHz speech band.

To use outband signaling in a network, all its routes must use carrier systems equipped with outband signaling in order to interwork successfully. Inband signaling uses a frequency within the 300 Hz to 3.4 kHz speech band. It has the advantage that it can be used for any connection that will provide satisfactory speech transmission. However, it cannot be used to send signals over a connection while it is being used for speech.

PCM time-division multiplex systems were designed from the beginning to provide the equivalent of outband signaling (Flood and Cochrane, 1991). The 2 Mbit/s primary multiplex has 32 time slots, but it provides only 30 channels. Time slot zero is used for frame alignment and time slot 16 is used for signaling. The 8 bits of channel 16 are shared between the 30 speech channels by a process of multiframing. This gives each channel, in each direction, up to four independent signaling channels at 500 bit/s. It therefore enables a much larger number of different signals to be exchanged than is possible with DC signaling. The 1.5 Mbit/s PCM system used in North America has 24 channels, with eight bits per time slot all used for speech. In every sixth frame, the least-significant bit is used for signaling instead of speech. It has been found that this "bit stealing" has a negligible effect on the quantization distortion. The process gives every speech channel, in each direction, an associated signaling channel at 1.3 kbit/s or two at 650 bit/s.

Inter-register Signaling

For multilink connections in a large network of exchanges, the need for rapid signaling of address information is obvious; so exchanges were equipped with registers, and multi-frequency address signals were sent between them. Each digit is sent by a single pulse, consisting of two frequencies out of six (2/6 MF), instead of the train of pulses needed to send a digit by loop/disconnect dialing. This gives 15 possible digit values, so five extra signals are available besides the digits "1" to "0."

The Bell R1 system (Welch, 1981) uses combinations of two frequencies from six spaced at intervals of 200 Hz between 700 Hz and 1700 Hz for forward signals. There are 12 forward signals as follows: start of pulsing (KP), digits 1 to 0, end of pulsing (ST). No signals are acknowledged, so no frequencies are provided for backward signaling. The ITU-T R2 system provides both forward and backward signals, using two groups of six frequencies spaced at 120 Hz in the range 540 Hz to 1980 Hz. More than 15 signals can be sent in each direction because one frequency combination is used as a "shift" signal, enabling each of the others to have two meanings. The ITU-T defined the system as a regional one. Thus, PTOs were able to choose how to use the additional signals and different versions appeared.

Networks may use either link-by-link or end-to-end signaling. In link-by-link signaling, information is exchanged between only adjacent registers in a multilink connection. This has two advantages: Signals suffer only the transmission impairments (e.g., attenuation and noise) of a single link and different links may use different signaling systems. The latter means that, if a network is being modernized, not all registers need to be converted simultaneously. However, every register involved in a call must receive, store and retransmit the complete address information for the call. Link-by-link signaling is therefore slow and post-dialing delays are long, particularly if backward acknowledgments are used in addition to forward signals.

In end-to-end signaling, the register in the originating exchange controls the set-up of the entire connection. Each intermediate register receives only the address information needed to select the outgoing route to the next exchange. It is then released and the originating register sends to the next exchange in the connection. This requires all registers throughout the network to be compatible and for their VF receivers to cope with the transmission impairments of several links in tandem (e.g., by having a greater dynamic range). However, end-to-end signaling reduces the register holding times and post-dialing delay. It has therefore been used in most networks. The North American R1 system used link-by-link signaling, but the ITU-T R2 system used end-to-end-signaling.

An inter-register signaling system cannot be used for seize, answer and clear signals. No register is connected when an incoming seize signal is received, since it is this signal that initiates the connection to a register. The register is released after it has set up a connection through its exchange and sent out routing digits, so it cannot receive answer and clear signals. Line signaling, using one of the methods described earlier, is therefore required in addition to inter-register signaling.

Common-channel Signaling

If a network has SPC exchanges, a call that passes through two exchanges must be processed by the central processor in each. If channel-associated signaling is used, signaling takes place over an individual speech circuit on the route between them that has been selected for the connection being made. As shown in Figure 4.3(a), the central processor in the first exchange sends forward signals to the speech circuit for transmission to the second exchange.

Here, the signals must be detected on the speech circuit and transferred to the central processor. Similarly, backward signals from the processor of the second exchange must be passed to the speech circuit, sent back to the first exchange, detected there, and transferred to its central processor. This is an unnecessarily complicated route for communicating between the two processors! If a high-speed data link is provided between the central processors of the two exchanges, as shown in Figure 4.3(b), it can provide a channel for all the signals needed between the two exchanges. This is known as common-channel signaling (CCS).

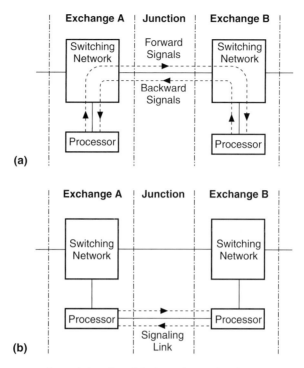

Fig. 4.3 Principle of common-channel signaling: (a) channel-associated signaling between central processors; (b) common-channel signaling between central processors. *Source:* From Flood (ed.), 1997.

Common-channel signaling has the following advantages:

1. Signals can be exchanged between the processors much more quickly than with channel-associated signaling.
2. A much wider repertoire of signals can therefore be used. As a result, customers can be offered more services.
3. Signals can be altered to provide new services by software changes, without any modifications to signaling hardware.
4. There is no longer signaling apparatus on every junction circuit, thus giving considerable cost saving.
5. Since there is no line signaling, both-way working can be used; i.e. calls on a junction circuit may originate from either end.
6. Signals may be sent while a call is in progress. This enables customers to alter connections that have already been set up (e.g., to transfer a call elsewhere or to enable an additional party to join in).
7. Signals can be exchanged between processors for purposes other than call processing (e.g., for maintenance or network management functions).

The error rate for CCS must be much lower and the reliability much higher than for channel-associated signaling. Failure of a line signaling unit, or even a register, results in loss of only a small fraction of the traffic on a route using channel-associated signaling. With CCS, failure of the data link between two processors prevents any calls being made between two exchanges. When channel-associated signaling is used, the successful exchange of signals over a circuit checks that the circuit is working. CCS does not do this, so a separate means (i.e., automatic routine testing) must be provided to monitor the circuits.

CCS systems use message-based signaling. Processors assemble messages into message units containing sequence numbers and check bits for error control. Since successive messages between two processors usually relate to different calls, each message must contain a circuit-identity code that indicates the speech circuit to which it refers. Since messages pass directly between processors, no connection has to be made before an address message is sent; therefore, no seize signal is required. In a multilink connection, the processor of the originating exchange is not involved in signaling between the processors of subsequent exchanges (i.e., link-by-link signaling is used).

The ITU-T No 7 signaling system (Manterfield, 1999) sends three types of signal units:

- The link-status signal unit: used for link initialization and flow control.
- The fill-in signal unit: used to maintain alignment when there is no signal traffic.
- The message signal unit; its information field transfers signal information for a particular application, called a user part.

The user parts defined by the ITU-T include the telephone-user part, data-user part, mobile-user part, and ISDN-user part. CCS systems are now being used for purposes other than call establishment—for example, to update a location register in a cellular mobile radio network, to interrogate a remote database in an intelligent network, for traffic management, and for operations management and administration. This led to the specification of an additional part called the transactions capability part. The message signal unit indicates the user part appropriate to its message by means of a service-information octet, which precedes its signaling-information field of up to 272 octets containing the information to be transmitted.

In an integrated digital network, the CCS links between exchanges form a signaling network separate from the speech network. However, as its bearers it uses 64 kbit/s channels in the PCM systems used for speech transmission between the exchanges. In a 30-channel system, time slot 16 is used and a 24-channel system uses time slot 24, thus reducing its capacity to 23 speech channels. A CCS link operating at 64 kbit/s can normally provide the signaling for 1,000 or 1,500 telephone circuits. Common-channel signaling is described in more detail in Chapter 5.

Network Architecture

General

A network consists of links joining different locations, known as nodes. In a public-switched telecommunication network, each customer's station is a node. Switching centers form other nodes. At some nodes, certain circuits are not switched but are joined semipermanently to form part of the same route between two other nodes. Customers also require connections to other nodes that provide services, such as recorded messages, emergency

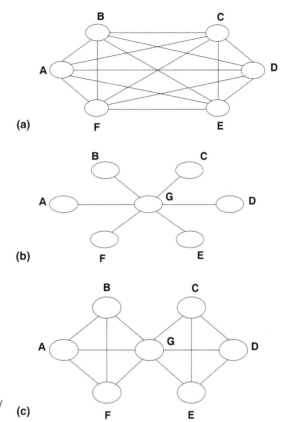

Fig. 4.4 Simple network configurations: (a) fully interconnected mesh network; (b) star network; (c) partially interconnected mesh network.

services, and operators for providing assistance in making certain calls. Consequently, a PSTN may be considered as the totality of its links and of its nodes, which are of the following types:

- Customer nodes
- Switching nodes
- Through-transmission nodes
- Service nodes

If each subscriber's station in a telecommunication system must be able to communicate with every other station, one possibility would be to provide a circuit between each pair of stations, as shown in Figure 4.4(a). If there are n stations, the number of circuits required is $n (n - 1)$. Thus, the cost is approximately proportional to n^2. If n is large, such a fully interconnected mesh network would obviously be enormously expensive. If, instead, each station is provided with a circuit to a central switching system, or exchange, as shown in Figure 4.4(b), the number of circuits required is then only n. Clearly, the star network of Figure 4.4(b) is much cheaper than the mesh network of Figure 4.4(a), unless n is very small.

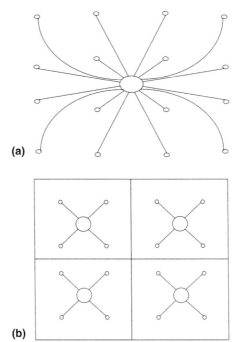

(a)

(b)

Fig. 4.5 Effect of number of exchanges on lengths of subscribers' lines: (a) single-exchange area; (b) multi-exchange area.

If an exchange area is large and contains a large number of stations, many of the subscribers' circuits will be long and therefore expensive, as shown in Figure 4.5(a). If the area is divided into several separate areas, each with its own exchange, as shown in Figure 4.5(b), subscribers' circuits are shorter and their total cost is less. However, the costs of switching equipment and its accommodation increase with the number of exchanges. Thus, as shown in Figure 4.6, there is an optimum number of exchanges for which the total cost of the network is a minimum.

In a multi-exchange area, subscribers connected to each exchange will wish to communicate with subscribers connected to other exchanges. Thus, there is a need for circuits between the exchanges. These circuits are called junctions. The number of junctions required on each route between a pair of exchanges depends on the traffic between them. If a direct junction route is provided between every pair of exchanges, these routes form a fully interconnected mesh network, as shown in Figure 4.4(a). An alternative is to provide a junction route from each exchange to a central switching center, to form a star network as shown in Figure 4.4(b). The exchanges to which subscribers' circuits are connected are called local exchanges and the central switching center is called a tandem exchange.

In a star junction network with a tandem exchange there are fewer junction routes than in a fully interconnected mesh. However, each route incurs the additional cost of its associated switching equipment in the tandem exchange. Therefore, if there is sufficient traffic between a pair of local exchanges to ensure that direct junctions are heavily loaded and they are cheap because the distance is short, it is more economical to provide direct junctions between these exchanges than to route the traffic indirectly through a tandem exchange. Consequently, a multi-exchange area will often contain direct junction routes between nearby

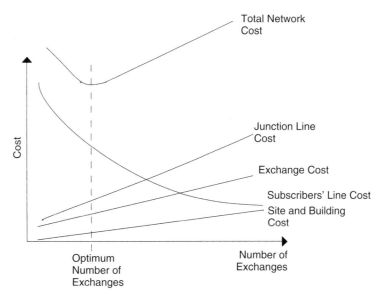

Fig. 4.6 Variation of network cost with number of exchanges.

local exchanges that have a high community of interest, and traffic between the other exchanges will be routed indirectly via a tandem exchange. The network thus consists of a mixture of a star and meshes, as shown in Figure 4.4(c).

In a multi-exchange area, only a minority of calls generated at a local exchange will be to other subscribers served by that exchange. Most traffic will be to or from other exchanges in the same area or outside it. Thus, the principal function of a local exchange is not to carry traffic between its own subscribers, but to provide an interface between these subscribers and the rest of the network.

National Networks

A national telecommunications network (Pita et al., 1993; Valdar, 2006) links many towns and cities, so it contains many multi-exchange areas. There is thus a need for a network of long-distance routes between them; this constitutes the trunk network. It is obviously uneconomical for the trunk network to interconnect all the local exchanges. However, each local exchange has junctions to a tandem exchange, so this provides a convenient interface with the trunk network. In ITU-T (formerly CCITT) terminology, it is called the primary trunk switching center. In the United Kingdom, it was called a group switching center (GSC) and in North America a Class 4 office. In the United Kingdom, a GSC usually acted as a trunk exchange, as a tandem exchange, and as a local exchange serving nearby subscribers. However, in a large city, these functions may be performed by separate exchanges.

Just as it is usually uneconomical for all local exchanges to be directly interconnected in a multi-exchange area, it may be uneconomical for all primary trunk switching centers to be directly interconnected. Thus, a primary center may have direct trunk routes to those nearest to it, but traffic to other primary centers is routed via tandem switching centers, known as secondary trunk centers, each serving a region of the country. Again, these may

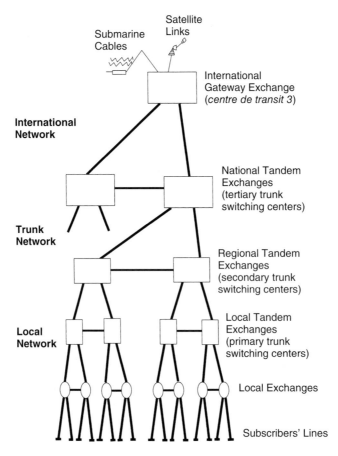

Fig. 4.7   National telecommunications network.

not be fully interconnected and traffic between some secondary centers is routed via tertiary centers.

It can be seen that a complete national network contains a hierarchy of exchanges, as shown in Figure 4.7. At the apex of the hierarchy, there will be one or more international gateway exchanges that provide the interface to the rest of the world. In ITU-T terminology, these are called CT3 exchanges (*centres du transit 3*).

A national network can be seen to consist of a hierarchy of interconnected networks, as follows:

1. A subscriber's premises network. For example, this may consist of extension lines connected to a private automatic branch exchange (PABX) for telephony or a LAN for data.
2. The subscribers' distribution network, which is sometimes called the access network or local-loop plant. This connects subscribers' telephones or PABXs to a local exchange.
3. The junction network, which interconnects a group of local exchanges and connects them to the primary trunk center.

4. The core or trunk network, which interconnects primary centers throughout the
   country.

There are variations in practice and in terminology throughout the world. In North America, junctions are called trunks and trunk circuits are called toll circuits. In ITU-T terminology, networks (2) and (3) from the preceding list together are called the local network; however, in the United Kingdom this term is usually reserved for (2) alone.

Where subscribers' distribution areas are large, it is economical to locate the switching equipment directly associated with a group of subscribers at a location close to the subscribers but remote from the exchange. This equipment is called a remote concentrator (Flood, 1997). It concentrates traffic from a number of subscribers' circuits on a smaller number of circuits leading to the exchange. A concentrator may also be used as a temporary expedient when unexpected development in one part of an exchange area uses up all the cable pairs leading to it. In this way, service can be provided to new subscribers before an additional cable can be laid. If remote concentrators are used, an additional level of switching is introduced into level (2) of the hierarchy.

A subscriber's private network may use a private automatic exchange (PAX) that, unlike a PABX, does not connect its lines to the PSTN. On the other hand, a private network may extend over many different locations using circuits in networks (2), (3), and (4) rented from the PTO. Multinational companies may have private circuits extending through the international network to link their offices in different countries.

In a national network, as shown in Figure 4.7, the minimum configuration of star-connected exchanges is usually augmented by direct routes interconnecting exchanges where a high community of interest generates sufficient traffic. There is thus a backbone route joining each switching center at the lowest level to the highest level via intermediate centers, together with transverse branches between centers at the same level. There may also be some direct routes between centers at different levels that violate this pattern, if there is sufficient traffic to justify them.

In the case of breakdown of a direct route between two exchanges at the same level in the hierarchy, it is possible for traffic between them to use their backbone routes to the switching node at the next higher level to obtain a tandem connection. In older systems this required manual intervention. However, modern networks provide automatic alternative routing, as described in a later section on routing. This contributes to the resilience of the network (i.e., its ability to cope with equipment failure and traffic overload).

The resilience of the transmission bearer network can be improved by the use of route diversity. For example, some circuits between nodes A and B may be routed directly and others may be routed via node C. Thus, some circuits on routes A–C and B–C carry traffic between A and C and between B and C (which is switched at C), and other circuits make up the route A–B (and are not switched at C). Either route between A and B can still carry traffic if the other fails. Resilience may also be improved by adding a service protection network. On busy routes requiring several high-capacity transmission systems, an extra system is provided that can be brought into service if one of the normal systems fails.

The number of levels in the hierarchy of a national network depends on the relative costs of transmission and switching. A country that is small and densely populated will have relatively short distances between its primary centers and large amounts of traffic between them. A country that is large and sparsely populated will have longer distances and less traffic between its primary centers. Consequently, the cost per circuit of providing direct routes between primary centers is much less in the former case than in the latter. As a result, a small, densely populated country will have more of its primary centers directly connected and fewer levels in its hierarchy than will a country that is large and sparsely populated.

Analogue Networks

Figure 4.7 is representative of the analogue networks developed by major countries during the mid-twentieth century. For example, the U.K. national public network had a hierarchy of four levels: local exchanges, group switching centers, district switching centers, and main switching centers. The North American network had five levels of switching centers: Class 5 (end offices), Class 4 (toll centers), Class 3 (primary centers), Class 2 (sectional centers), and Class 1 (regional centers). In a typical network, each switching center had a route to only one center at the next higher level. There was thus a single backbone route from each local exchange to the top of the hierarchy.

Digital Networks

During the latter part of the twentieth century, there was a large increase in telephone traffic. There was also a dramatic decrease in the cost of providing long-distance circuits because advances in technology caused successive increases in the number of channels that could be carried by a coaxial cable, a microwave radio link, or, eventually, an optical fiber. The use of digital switching integrated with digital transmission produced further cost reductions. These factors made it economical to provide many more direct routes between switching centers, thus requiring fewer transit switching centers and a hierarchy with fewer levels. It also made it economical for an exchange to have routes to two (or more) centers at the next-higher level instead of the single backbone route of the former analogue network. This increases the resilience of the network. If one of the two routes breaks down, traffic can still use the other.

Since the introduction of digital exchanges has reduced the per-line cost of switching, quite a high proportion of the cost of an exchange is due to its central processors. This makes it desirable for a digital local exchange to have more subscribers' lines than an analogue exchange has, and so it serves a larger area. Increasing the size of exchange areas favors the use of remote concentrators. Analogue circuits from subscribers' premises are terminated at concentrators that require PCM systems to connect them to the main exchange. In addition to reducing the per-line cost of call processing, having larger local exchanges reduces the number of exchanges in an area, so reducing the cost of junction plant and tandem switching. Having fewer local exchanges also reduces operation and maintenance costs.

The integrated digital network of British Telecom (BT) (Muir and Hart, 1987) is shown in Figure 4.8. Local areas are served by digital cell center exchanges (DCCEs). Subscribers obtain access to a DCCE to process their calls via a remote concentrator unit (RCU) unless they are situated very close to the DCCE. The concentrator is connected to the main exchange by two low-capacity digital transmission systems, usually operating at 2 Mbit/s or 8 Mbit/sec. These are diversely routed in order to protect against failures. Concentrators parented on about 400 DCCEs replaced more than 6,000 local exchanges.

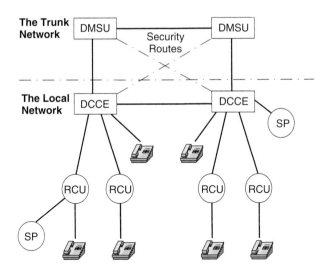

DCCE = digital cell center exchange, DMSU = digital main switching unit,
RCU = remote concentrator unit, SP = service provider

Fig. 4.8 Digital network of British Telecom.

The BT trunk network contains only one level of switching centers. These exchanges are called digital main switching units (DMSUs) and are fully interconnected. The digital trunk network contains only 60 DMSUs, whereas the analogue network contained nearly 400 GSCs. Each DCCE has junction routes to at least two DMSUs. Thus, it is still able to connect trunk calls to and from its subscribers in the event of the breakdown of one route or its DMSU.

The digital long-distance network of AT&T (Ash and Mummert, 1992) is also nonhierarchical. As shown in Figure 4.9, one class of tandem switching center replaces the Class 1 and Class 2 offices (regional and sectional centers) of the previous analogue network. However, traffic entering the network continues to be routed by a hierarchy of Class 3 and Class 4 offices (primary centers and toll centers).

In a fully interconnected, nonhierarchical network, such as the trunk networks shown in Figures 4.8 and 4.9, a direct route between two exchanges can have an alternative via any other exchange. This gives a very large number of options for routing. Each exchange has a routing table containing its permitted options and selects from these in some preferred order. The routing table can be changed from time to time depending on traffic conditions (Stacey and Songhurst, 1989; Ash, 1992). In the AT&T network, this is called real-time network routing (RTNR).

In a country where there are competing long-distance networks, customers obtain access to them via a trunk exchange. In the United Kingdom, this is the DMSU. In the United States, customers in a local-access and transport area (LATA) obtain access to one of the long-distance network operators, known as interexchange carriers (IXCs), at a switching center called a point of presence (POP), as shown in Figure 4.10. The LATAs belong to local-exchange carriers. Each carrier completes toll calls that are within its own LATA, but it does not convey calls between LATAs; this is done via an IXC. End offices may have direct routes to a POP or may be connected via an access tandem, which replaces a Class 4 office in the previous analogue network.

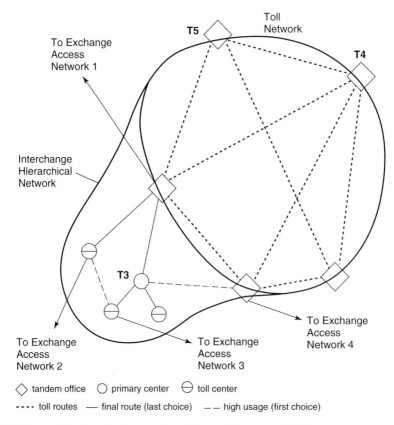

Fig. 4.9 ATT digital toll network. *Source:* From Ash and Mummert, 1992.

Modern networks have exchanges using stored-program control and common-channel signaling. The data links used for common-channel signaling between digital exchanges in the PSTN make up a separate self-contained network, as shown in Figure 4.11. However, this signaling network uses channels in the basic digital transmission bearer network, just as does the PSTN.

To maintain synchronism of the PCM frames in all the digital transmission links and exchanges, it is necessary to distribute synchronizing signals from a national reference clock source. This results in a synchronizing network linking it to all the digital exchanges. Finally, centers remote from exchanges have been introduced to collect bulk data (e.g., traffic statistics, billing, and so on) from exchanges and to provide man–machine interfaces for maintenance and making software changes (Waliji, 1980). Management centers receive real-time information on traffic flows and allow traffic to be rerouted to minimize congestion (Bealby, 1990). These require an administration data network connecting the remote centers to the nodes of the PSTN. This is discussed in more detail in a later section on network management. It is seen that a complete IDN contains four separate networks, as shown in Figure 4.11. However, all of these use channels in the basic digital transmission network, making five networks in all.

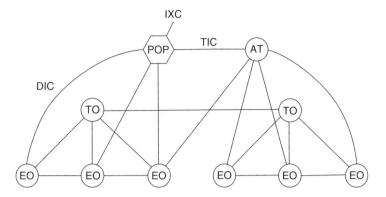

AT = access tandem, DIC = direct inter-LATA connection, EO = end office,
IXC = interexchange carrier, POP = point of presence, TIC = tandem inter-LATA
connection, TO = tandem office

Fig. 4.10 Local-access and transport area in the United States. *Source:* From Ash and
Mummert, 1992.

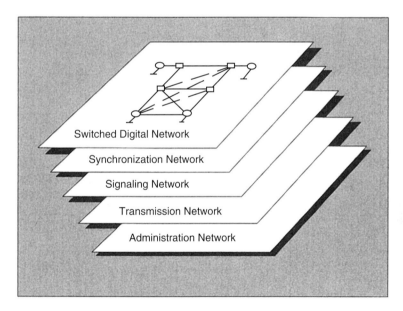

Fig. 4.11 Component networks of an integrated digital network.

Integrated Services Digital Networks

The final stage in the evolution of an integrated digital network is to extend digital transmission through the access network to customers' premises, which enables the IDN to
become an Integrated Services Digital Network (ISDN). This permits the transmission of
high-speed data in addition to speech and so can provide many additional services. It also
improves telephony performance because there is no longer any variation of attenuation
due to different lengths of customers' lines.

Two forms of access to an ISDN have been standardized by the ITU-T:

1. Basic-rate access (ITU-T Recommendation I.420). In each direction, the customer's line carries two 64 kbit/s "B" channels plus a 16 kbit/s "D" channel for a common signaling channel.

2. Primary-rate access (ITU-T Recommendation I.421). Two lines are provided to carry a complete PCM frame in each direction. In countries that use 2 Mbit/s PCM, this provides 30 "B" channels plus a 64 kbit/s "D" channel for signaling in time slot 16. In countries that use 1.5 Mbit/s PCM, it provides 23 "B" channels plus a "D" channel in time slot 24.

Intelligent Networks

New services can often be provided by modifying the call-processing programs of SPC exchanges. However, making consistent software upgrades to the many interconnected exchanges in a national network is costly and time-consuming. Consequently, such changes are made infrequently, making the introduction of new services a slow process.

A possible solution to this problem is to separate the software that controls basic functions, such as setting up and supervising connections, from the software for providing more-complex services. These include, for example, freephone services, calling-card services, and value-added services. The more-complex services can be controlled by a centralized processor with a large database, called a service control point (SCP). This can be remote from the exchanges setting up the connections. A network that has been enhanced in this way is called an intelligent network (IN) (Flood, 1997). The exchange that makes the required connection is called a service switching point (SSP). This can be at any level in the network hierarchy (e.g., a local exchange, a trunk exchange, or an exchange in a special overlay network). BT enhanced all its trunk exchanges (DMSUs) to provide an SSP capability (Wyatt et al., 1995).

The architecture of an IN is shown in Figure 4.12. The software of an SSP is modified so that various events can trigger it to suspend normal call processing and request the intervention of the SCP. These events may include the caller's class of service, the code digits dialed, or some subsequent event (e.g., no reply to ringing). As a result, the SSP sends information to the SCP and resumes call processing when it receives information from the SCP on how to proceed.

In addition to the SCP, the IN uses intelligent peripherals (IPs) to perform specialized functions. These include digit-collection units and voice-guidance systems that give instructions to users by recorded announcements. The messages are passed between the SCP and SSPs and IPs by means of common-channel signaling. A network with a number of SCPs may also have a service-management system (SMS) connected to the SCPs by data links.

Private Networks

Organizations operating in several locations often have sufficient traffic between their sites for it to be economical to join them with a private network, instead of sending the traffic through the PSTN. A private network is usually provided by leasing circuits from a national PTO or even, for a multinational company, the PTOs of several countries.

Organizations commonly have separate networks for telephony and data communication. A voice network links their private automatic branch exchanges (PABXs) and a wide

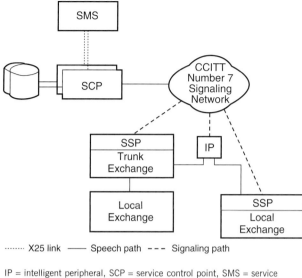

········ X25 link   ——— Speech path   – – – Signaling path

IP = intelligent peripheral, SCP = service control point, SMS = service
management system, SSP = service switching point

Fig. 4.12   Intelligent network architecture. *Source:* From Flood (ed.), 1997.

area network (WAN) links their local area networks (LANs) for data communications. ISDN
technology is applicable to private networks as well as to public networks. Use of integrated-
services PBXs (ISPBXs) enables both voice and data traffic to share the same digital circuits.
The VoIP protocol enables telephony to be provided over a packet-switched network.
Packet-switched private exchanges are now available to send both voice and data over the
same private network.

Small private networks may consist of only a few direct junctions between PABXs.
Larger ones may use some PABXs to provide tandem switching on routes between others.
Automatic alternative routing can be applied in a private network. A PABX may be programmed so that, when all private circuits to another PABX are busy, traffic to it overflows
onto the PSTN.

Some PTOs offer a Centrex service, in which users' telephones are served by switching
equipment in a public exchange instead of by a PBX on the customer's premises. Line costs
are, of course, greater, but a charge by the PTO replaces the cost of a separate PABX. Use of
Centrex is attractive if staff members work in separate locations, each of which would need
a separate PABX.

A further development is to route traffic over a public network instead of using a private
network. This is called a Virtual Private Network (VPN). Initially, VPNs were provided over
the PSTN, but subsequently the Internet has been used. The customer's staff forms a closed
user group within it. The economic justification is that less capacity is needed to carry the
private traffic if mixed with public traffic on large-capacity routes than if separate small
routes are provided. A VPN must give the same services and have the same numbering
scheme and signaling protocols as a private network. These usually differ from those of the
PSTN and so require an additional database and different software. It is preferable for these
to be in a central location rather than in every local exchange, so an intelligent network
(as described in the preceding Intelligent Networks section) is required. Introduction of a
nationwide VPN would encourage more telecommuting and teleworking (Huws et al., 1990).

The staff of an organization can work at home and communicate with colleagues as if they were using the same PBX in the same building.

Numbering

National Schemes

To establish a connection through a network, it is necessary for a caller to inform the switching center of the required address of the subscriber being called (i.e., the called subscriber's directory number). This determines both the route for the call and the charging rate. Thus, a numbering plan is required to allocate a unique number to each subscriber connected to the network (Buckley, 1994; Wittering, 1994; Bishop, 1995). At first, each numbering scheme applied only to a single exchange and exchanges were identified by the names of their towns. Later, linked numbering schemes were applied to multi-exchange areas.

In a linked numbering scheme, the "local" numbering scheme covers a number of exchanges in the area, so that a call from any exchange in the area uses the same number to reach a particular subscriber. Thus, the first part of a subscriber's directory number is an exchange code and the remainder is the subscriber's number on the exchange. For example, a 5-digit linked numbering scheme has a theoretical capacity for ten 4-digit exchanges or 100 3-digit exchanges, or a combination.

The subsequent introduction of subscriber trunk dialing (STD) for long-distance calls required the development of national numbering plans (Barron, 1959; Francis 1959; Clarke and Osborne, 1952; Nunn, 1952). Later, the introduction of international subscriber dialing (ISD) made it necessary for national numbering plans to conform to an international numbering plan.

In a country where there are several competing networks, such as in the United Kingdom and the United States, it must be possible for the customers of each network to make calls to the customers of all the other networks. It is thus necessary for the customers of the different networks to have directory numbers that are within a single national numbering plan. Consequently, it is the responsibility of the government regulator to develop and to maintain this plan.

Numbering plans may be either open or closed. An open plan does not have a fixed number of digits. A closed plan has a fixed number of digits to be dialed for all calls, regardless of the geographical locations of the calling and called customers. Modern switching systems can cater to numbers of varying lengths, so many national numbering schemes contain numbers having several different lengths, up to a given maximum number of digits.

There must always be an upper limit to the maximum number of digits in a national numbering scheme to comply with international requirements. The ITU-T recommended (ITU-T, 1989a) that the maximum number of digits for an international call be 12. The maximum number of digits in a national number is thus $12 - N$, where N is the number of digits of the country's code in the world numbering plan.

Since each subscriber has a directory number in his or her own exchange, a unique national number can be obtained by adding to this number a code identifying the exchange. These codes could be allocated in a purely arbitrary manner. This has the advantage of being able to accommodate the largest possible number of exchanges and makes it easy to handle unforeseen developments. It has the disadvantage that subscribers must always dial a complete national number, even for a local call. Also, each originating trunk exchange must be able to identify the code of every local exchange in the country to determine the routings and charges for calls. Instead, the country is usually divided into geographical areas, each having a separate code and containing a number of local exchanges. The originating trunk exchange is then required to identify only the area code to determine the routing and charging rate for each call.

In general, a national number consists of three parts:

1. An area code
2. An exchange code
3. The subscriber's local-exchange number

In addition to dividing the country into numbering areas (each with an area code) and determining a local numbering plan for each area, it is necessary to determine appropriate dialing procedures for both trunk and local calls.

The use of a single dialing procedure (i.e., the full national number is dialed for both trunk and local calls) is only justified if trunk traffic represents a high proportion of the total traffic or if the country is sufficiently small for national numbers to contain few digits. Otherwise, it is preferable to use two procedures, whereby local calls are obtained by dialing the local number and other calls by dialing the national number. The local exchange must be able to differentiate between them. This is usually done by means of a trunk prefix digit for national calls. Usually, this digit is "0." Recognition of the prefix causes the local exchange to route the call to the trunk exchange.

An additional prefix can also be used to differentiate between trunk calls and international calls. In the United Kingdom, the international prefix is "00." The first digit (0) routes the call to the trunk exchange. The next digit (0) switches the call to the international gateway exchange. This two-digit prefix is followed by the international code for the required country and the national number of the called subscriber in that country.

As an example, in the United Kingdom, the largest cities have a uniform 7-digit numbering scheme, consisting of a 3-digit exchange code and 4-digit subscriber's number. These areas are identified by 3-digit area codes. Other areas have 3-digit codes and numbering schemes with uniform 6-digit numbers or mixed schemes not exceeding six digits. The total number of dialed digits, excluding the trunk prefix, thus varies between eight and nine. This complies with the ITU-T recommendation that national numbers (excluding prefix digits) shall not exceed 12 digits minus the country code (for the United Kingdom, the country code is 44, and $12 - 2 = 10$). The long-term objective is to have a national number of 10 digits with 6-digit or 7-digit local numbers (McLeod, 1990).

A national numbering scheme must make a generous allowance for growth in the number of subscribers for up to 50 years ahead. However, growth is not uniform and the numbering schemes in some areas become exhausted before those in others. For example, the London area, which previously had a single numbering code (1), had to be split into two (Bannerjee et al., 1989): 207 for inner London and 208 for outer London. Similar changes have been made in a number of American cities. In France, the entire country has been divided into two numbering zones: L'Isle de France covering the region around Paris and La Province covering the rest of France.

The capacity of national numbering schemes is reduced by the need to avoid ambiguity by not using the trunk prefix as the initial digit of any local-exchange code. Similarly, the international prefix must not form the initial part of any area code. There is also a need to reserve codes for accessing various services. For example, in the United Kingdom, the code

Table 4.2 U.K. Dialing Prefixes as of April 1995

Prefix	Allocation
01	Current geographic PSTN
02	New geographic scheme
03	Future geographic area codes
04	Unallocated
05	Reserved for corporate numbering
06	Unallocated
07	Personal numbering, paging, mobile
08	Specially tariffed services: freephone, national call charged at local rate
09	Premium rate services

999 is used for emergency services, 100 is used to obtain operator assistance, 118xxx is used for directory inquiries, and so on. Thus, the digits 0, 1, and 9 are not available as initial digits for local numbers. This reduces the capacity of the numbering schemes of all local areas. The numbering capacity of local areas is also reduced by the need to provide direct dialing in (DDI) to PABXs. This requires each extension to have a number within the scheme for the area. The advent of ISDN also has an impact. It may be necessary for a basic-access subscriber to have up to eight numbers for receiving calls from the PSTN.

In countries where there is competition among PTOs, codes must be reserved to enable subscribers to obtain access to different networks. Different trunk prefixes are used to enable subscribers to choose different carriers for their long-distance calls. In the United States, they are known as "carrier identification codes." These may be dialed manually or the network operator may supply an automatic dialer. A subscriber on an SPC local exchange may choose a particular long-distance carrier in advance and this information is stored in the database of the office to cause all toll calls to be routed to the chosen carrier. In the United Kingdom, the national numbering scheme was changed in 1995 to accommodate several different networks by adding an additional digit after the initial "0" of the trunk prefix, as shown in Table 4.2.

In countries such as the United Kingdom, where there is competition among PTOs, it is a disincentive to competition if a subscriber has to change her directory number when changing from one network operator to another. There is thus a demand for number portability from a "donor" network to a "recipient" network. If number portability is provided, a local exchange must recognize from a database if a number that it receives has been transferred to another network. Since rival networks are usually interconnected at trunk exchanges, the call must then be rerouted to the subscriber's new local exchange as if it were a trunk call originating from the subscriber's former local exchange (Flood, 1997).

International Numbering

The introduction of international subscriber dialing made it necessary for every subscriber's station in the world to have a unique number. In the ITU-T world numbering plan, each subscriber's number consists of a country code followed by the customer's national number (ITU-T, 1989a).

For numbering purposes, the world is divided into zones, each given a single-digit code. Each country within a zone has the zone number as the first digit of its country code.

Table 4.3 World Numbering Zones

Code	Zone
1	North America (including Hawaii and Caribbean Islands, except Cuba)
2	Africa
3 and 4	Europe
5	South America and Cuba
6	South Pacific (Australiasia)
7	CIS (formerly USSR)
8	North Pacific (Eastern Asia)
9	Far East and Middle East
0	Spare code

Table 4.4 Examples of Country Codes

Zone	Country	Country Code
1	United States	1
1	Canada	1
2	Egypt	20
2	Liberia	231
3	France	33
3	Portugal	351
4	United Kingdom	44
4	Slovakia	421
5	Brazil	55
5	Ecuador	593
6	Australia	61
6	New Zealand	64
7	Russia	7
7	Kazakhstan	7
8	Japan	81
8	China	86
8	Hong Kong	852
9	India	91
9	Pakistan	92
9	Azerbaijan	994

However, the European numbering zone has been allocated two codes because of the large number of country codes required within it. The codes for the world numbering zones are listed in Table 4.3.

Within each zone, every country has been allotted a single 2- or 3-digit code number. For example, within zone 3 (Europe) Holland has the code 31 and Albania has 355. The 3-digit codes have been allocated to smaller countries, having fewer digits in their national numbering plans, to minimize the total number of digits in subscribers' international numbers. Exceptions occur where an integrated numbering plan covers an entire zone; countries in these zones require only a single-digit code. Thus, 1 is the country code for all countries in the North American numbering plan (Nunn, 1952) and the country code for the former USSR is 7. Some examples of country codes are given in Table 4.4.

The existence of a world numbering plan places restrictions on the national numbering plan of each country. This enables subscribers in other countries to make international calls into each country's national network. The restrictions are:

1. The number of digits in a subscriber's world number is limited to an absolute maximum of 12. The number of digits available for a national numbering plan is thus $12 - N$, where N is the number of digits in the country code.

2. National numbers must consist of all numbers and no letters. Countries that had used letters on telephone dials had chosen different ways of allocating the 26 letters of the alphabet among the 10 numerical characters. Thus, exchange codes could have been dialed as different numbers from outside the country.

3. National numbering plans must include international prefixes to avoid ambiguity when national and international numbers have the same initial digit.

In 1989, the ITU-T made recommendations for an international numbering plan to cover ISDNs (ITU-T, 1989b). This extended the existing numbering plans for PSTNs. Thus, a subscriber's ISDN access can normally be indistinguishable from PSTN access and can be provided by the same local exchange.

The maximum length of international numbers was extended to 15 digits. The area code became a national destination code (NDC). This arose from the present situation whereby some "area" codes have been used for access to mobile networks and special services (which are nongeographic). An international exchange may analyze six digits to determine a route. These can therefore include digits within the NDC in addition to the country code, in order to route a call to the appropriate network in the destination country.

Provision has also been made for sub-addressing. This uses additional digits following a subscriber's national number, which may be transmitted to the destination for use on the subscriber's premises. These can select the appropriate terminal for an ISDN connection or provide DDI on a PBX without using up numbers in the area numbering scheme. The sub-address field can range from four digits for simple applications up to 40 digits for use in open systems interconnection (OSI). For OSI, a global scheme for the identification of network service access points has been developed by the International Standards Organization (ISO).

The full implementation of this plan requires that stored program-controlled digital exchanges have penetrated all countries so that they can handle 15-digit international numbers and the increased number of digits to be analyzed (ITU-T, 1989c).

Routing

Routing in National Networks

National networks are hierarchical, as shown in Figure 4.7. In addition to the minimum configuration of star-connected exchanges, there are direct routes interconnecting those exchanges, between which there is heavy traffic. A backbone route joins each switching center to the highest level via intermediate centers. There are transverse branches between some centers at the same level and there may also be some other direct routes between centers at different levels that violate this pattern.

If it is assumed that a country is already divided into local-exchange areas and the locations of their exchanges have already been decided, a routing plan should be developed to determine:

- Which exchanges should be interconnected by direct junctions, and which connections should be made indirectly via tandem switching centers.
- The number and location of tandem switching centers.
- The number of levels of tandem switching to be used in the network.
- Whether automatic alternative routing is to be used and, if so, under what conditions.

The routing plan must be consistent with the plans for numbering, charging, transmission, and signaling.

Large groups of circuits are more efficient than small groups because of their higher occupancy (i.e., traffic per circuit), as previously noted. If there is a large amount of traffic between two exchanges, it is economical to provide a direct route between them. If there is little traffic between two exchanges, it is more economical to combine this with traffic to other destinations to produce a large amount of traffic over a common route to a tandem switching center. The correct solution obviously depends on the cost of the circuits as well as the amount of traffic. If circuits are cheap, it is less expensive to lightly load them than to incur the cost of switching equipment in a tandem exchange. Thus, many direct routes are provided between local exchanges in a small area with a high subscriber density but not to more distant exchanges.

In some networks, automatic alternative routing is used (Clos, 1954). Direct routes are under-provided with circuits; when all circuits on a direct route are busy, traffic overflows to a fully provided tandem route through a switching center at a higher level in the hierarchy. An under-provided direct route is called a high-usage route, and the fully provided indirect route to which its traffic finally overflows is called a final route. Only a small proportion of calls use the complete backbone of final routes, since transverse routes are used whenever these are free.

The traffic levels at which direct routes, tandem routing, and automatic alternative routing should be used depend on the relative costs of direct and tandem routes (including associated switching and signaling equipment). Low-traffic and high-cost direct routes indicate tandem working. High-traffic and low-cost direct routes indicate direct connection. Intermediate situations favor automatic alternative routing, as shown in Figure 4.13 (Rapp 1964). Automatic alternative routing is now widely employed to provide network resilience under fault conditions, rather than for traffic efficiency.

In an IDN, all trunk exchanges may be fully interconnected, as shown in Figures 4.8 and 4.9. There is then a direct route between each pair of exchanges, and an indirect route between them using any other exchange as a tandem switching center. This provides a wide choice of alternative routes. Since digital networks have SPC exchanges, their routing translations can be readily changed. Moreover, common-channel signaling enables the routing tables of exchanges to be changed remotely. The exchanges' processors can be linked by common-channel signaling to a central network-management center. Its staff can monitor traffic on all the routes at frequent intervals and change routings to bypass failures and congestion.

Variations on the basic automatic alternative routing scheme have been introduced. In dynamic alternative routing (Stacey and Songhurst, 1989), if a call is successful on a given

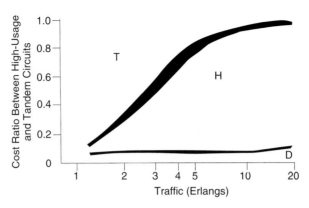

Fig. 4.13   Domains of employment of tandem (T), high usage (H), and direct (D) circuits. *Source:* From Rapp, 1964.

routing, then that current choice is retained. If, instead, the call fails on the current choice, then a new choice is selected for the next one. The introduction of common-channel signaling between exchanges, together with exchanges having routes to more than one center at the next higher level, makes complex alternative routing schemes possible. Thus, if a call encounters congestion at a subsequent exchange in the chosen route, it is possible to "drop back" or "crank back" to a preceding exchange and choose a different outgoing route there.

Automatic alternative routing will reroute calls away from a high-usage route whenever it is unable to carry traffic, which can happen if there is a breakdown instead of high traffic. This is beneficial, since some traffic can still reach its destination, whereas it could not if there were only the direct route. However, the large amount of additional traffic offered to the final route can cause severe congestion, resulting in the loss of calls to destinations served only by that route (which could be to most other exchanges in the network). A solution to this difficulty is trunk reservation (Ingham and Elvidge, 1989; Katzschner, 1973). A proportion of the circuits on the final route is reserved for calls that can take only that route. Thus, these calls still obtain a reasonable grade of service when the overflow traffic offered to the route is abnormally large.

Sophisticated automatic alternative routing can cause congestion originating at one part of the network to spread as traffic overflows to other routes. This is obviously undesirable, so network management (Macurdy, 1973; IEEE, 1988) has been introduced. A central management center can monitor traffic on different routes and, if necessary, reduce or entirely cut off traffic originating from switching centers to prevent the overload from spreading.

Call management can also provide economies if the busy hours in different parts of a network do not coincide. Calls that would normally be routed through a part of the network that is busy can be routed instead through another part of the network that is less busy. Thus, overall, fewer circuits are required to carry the total traffic. In the United States, there is a time difference of three hours between the East and West coasts, with a corresponding difference between the busy hours. Thus, it is economical to route long-distance calls between North and South either via the East or the West of the country at different times in the day.

International Routing

The world network developed for international subscriber dialing was hierarchical (Munday, 1967). It had three levels of international exchanges, called *centres du transit* and designated CT1, CT2, and CT3. The international gateways of national networks were provided by centers CT3, as shown in Figure 4.7.

The introduction of digital switching and transmission, together with common-channel signaling, has produced changes in international routing similar to those described above for national networks. The international network has become mainly nonhierarchical. There are many more direct routes between countries, and fewer calls require tandem routing. Automatic alternative routing is used extensively and common-channel signaling enables use of crank-back. In countries with more than one international gateway exchange, the proportional traffic distribution facility (PTDF) may be used. This ensures that the national network shares outgoing traffic to another country evenly among these gateways.

In the international network, it is particularly advantageous to make use of time zone differences by employing dynamic routing. For example, it is economical to route traffic between the United Kingdom and other European countries via the United States in the morning, when the time difference ensures that there is little traffic between Europe and the United States.

Charging

The cost of providing a telecommunication network consists of its capital cost and current operating expenses. All of these costs must be met by the income obtained by the PTO from its customers. It is equitable that the charges paid by each customer be related, whenever possible, to the proportion of the costs incurred in providing the services used. For this reason, the charges to the customer are levied in the following ways:

- An initial charge for installing the customer's line
- An annual rental or leasing charge
- Call charges

The customer's share in the capital cost should be covered by the connection charge and part of the rental. Part of the operating cost would be incurred even if the network carried no traffic, so this should also be recovered by rental charges. Those parts of the capital cost and operating cost that are traffic-dependent should be recovered by call charges.

Because a major part of the cost of a customer access network is independent of traffic, some operators do not have a separate charge for local calls but include them in the rental. This is known as a flat-rate tariff. Others charge for each call, which is known as message-rate charging. The quantity of switching equipment and of junction and trunk transmission plant required depends on the busy-hour traffic. Calls made at off-peak times incur virtually no capital cost, since no plant would be saved if these calls were not made. Because of these relative costs, and to discourage peak demand, it is common to have call charges vary with the time of day, being cheaper during evenings and on weekends than during working hours on weekdays.

Traditionally, charges for long-distance calls have been proportional to distance × time, the distance being that between the originating and terminating primary centers. However, with high-capacity transmission systems, such as those on optical fibers, the cost per channel-kilometer is extremely low because the transmission medium is shared by many channels. Thus, the cost of a call in a trunk network is now almost independent of distance. As a result, a typical PTO may have only two charging rates for ordinary calls: a local rate for calls within the same or a nearby area and a national rate for long-distance calls. In addition, there may be

1. Freephone calls, in which the charges for calls are made to the called customer instead of the caller.
2. Premium-rate calls for calls to various services or to mobile-telephone networks. The higher income from the call charges is then shared between the PTO and the service provider.
3. Rates for international calls to different countries.

Dimensioning

The traffic carried by a group of trunks is defined as the average number of calls in progress on the trunks. It is measured in erlangs. Thus, the traffic A in erlangs is given by Equation 4.1, where h is the mean call duration and C is the mean number of call arrivals in time T.

$$A = Ch/T \qquad \text{(4.1)}$$

For a single trunk, A is the proportion of the time interval for which the trunk is busy. It is called the occupancy of the trunk.

It is uneconomical for a telecommunication network to have sufficient circuits for all subscribers to make calls simultaneously. Instead, sufficient circuits are provided to make the probability of a subscriber finding congestion (i.e., all circuits busy) on the required route acceptably small in the busy hour. This probability is called the grade of service (GOS). Thus, a high grade of service corresponds to a poor service and vice versa. In a circuit-switched network, if a call encounters congestion, the call is lost. In a message-switching or packet-switching network, if a call encounters congestion, it enters a queue and so is delayed. The problem of dimensioning a route is thus to determine from the traffic forecast the number of circuits required to provide the grade of service specified by the network operator (Farr, 1988).

For a circuit-switched route it can be shown (Bear, 1988; Flood, 1995) that the grade of service (i.e., the probability of a lost call) is given by Erlang's equation (Equation 4.2).

$$E_{1,N}(A) = \frac{\dfrac{A^N}{N!}}{\displaystyle\sum_{k=0}^{N} \dfrac{A^k}{k!}} \qquad \text{(4.2)}$$

where A is the traffic offered (in erlangs) and N is the number of trunks on the route.

For a queuing system, provided that the probability of delay is small, the GOS is approximately given by Equation 4.3.

$$E_{2,N}(A) = \frac{N}{N-A} E_{1,N}(A) \qquad \text{(4.3)}$$

The assumptions made in deriving the above formulae include full availability, random (Poisson) call arrivals, and a negative exponential distribution of holding times. Full availability means that any call can use any trunk on the route, provided that it is free.

Equation 4.2 is well suited to calculate the GOS, given A and N. However, the dimensioning problem is this: Given A and the required GOS, determine N. Equation 4.2 is not suited for this. Fortunately, tables are available giving the traffic capacity of different numbers of trunks for various grades of service. An example (Bear, 1988) is shown in Table 4.4. By looking down the column for the specified grade of service until the required traffic is exceeded, one can read across to determine the number of trunks needed. For example, to handle 20 erlangs of traffic with one lost call in 100, the table shows that 30 trunks are required. Table 4.5 applies to a lost-call system. Tables are also published for queuing systems (Hunter et al., 1988).

Specifying a grade of service is, ultimately, a matter of judgment. It should be sufficiently large to make the provision of trunks economical, but not so large as to cause undue customer dissatisfaction. In practice, grades of service may vary between 0.001 for trunks within an exchange to 0.1 for expensive long-distance and international circuits.

Grades of service are specified for busy-hour traffic. At other times of the day, the probability of loss is much smaller. Moreover, routes are dimensioned to cater to the traffic forecast at the end of a planning period for which plant is provided. Thus, if traffic grows in accordance with the forecast, the GOS will initially be much better than the specified value and increase toward it over the provisioning period. However, if traffic grows more quickly than was forecast, the specified GOS will be reached during the provisioning period and, toward the end of it, the GOS will be much worse than that specified. This situation has occurred often because telecommunications have expanded so rapidly in recent years.

Of course, a subscriber is interested in the grade of service of a complete connection, which often consists of several routes connected in tandem, whereas Equations 4.2 and 4.3 give the grade of service for a single link in such a connection. However, if a connection contains n links with GOS given by $B_1, B_2, \ldots, B_n$, the overall loss probability is given by Equation 4.4.

$$B = B_1 + B_2(1 - B_1) + \ldots + B_n \overset{n-1}{\underset{k=1}{\pi}} (1 - B_k) \tag{4.4}$$

Thus, if the losses are small (which they should be) then B is given approximately by Equation 4.5.

$$B = B_1 + B_2 + B_3 + \ldots + B_n \tag{4.5}$$

Grades of service are specified for the busy-hour traffic. Thus, if the busy hours of the links are non-coincident, the loss probabilities of some links may be very low when those of others are high. The overall loss is then mainly determined by the links that are busiest at the time considered.

It has been noted that large groups of trunks are more efficient than small groups, since each trunk can have a higher occupancy for a given grade of service. In the example shown in Figure 4.14, trunks in a group of five each carry less than $0.2E$ of traffic, whereas those in a group of 40 each carry over $0.6E$ with the same grade of service. The price paid for the

Table 4.5   Traffic Capacity for Full-availability Groups

Number of trunks	1 lost call in 50 (0.020) E	100 (0.010) E	200 (0.005) E	1000 (0.001) E	Number of trunks	1 lost call in 50 (0.020) E	100 (0.010) E	200 (0.005) E	1000 (0.001) E
1	0.020	0.010	0.005	0.001	51	41.2	38.8	36.9	33.3
2	0.22	0.15	0.015	0.046	52	42.1	39.7	37.7	34.2
3	0.60	0.46	0.35	0.19	53	43.1	40.6	38.6	35.0
4	1.1	0.9	0.7	0.44	54	44.0	41.5	39.5	35.8
5	1.7	1.4	1.1	0.8	55	44.9	42.4	40.4	36.6
6	2.3	1.9	1.6	1.1	56	45.9	43.3	41.2	37.5
7	2.9	2.5	2.2	1.6	57	46.8	44.2	42.1	38.3
8	3.6	3.1	2.7	2.1	58	47.8	45.1	43.0	39.1
9	4.3	3.8	3.3	2.6	59	48.7	46.0	43.9	40.0
10	5.1	4.5	4.0	3.1	60	49.6	46.9	44.8	40.8
11	5.8	5.2	4.6	3.6	61	50.6	47.9	45.6	41.6
12	6.6	5.9	5.3	4.2	62	51.5	48.8	46.5	42.5
13	7.4	6.6	6.0	4.8	63	52.5	49.7	47.4	43.3
14	8.2	7.4	6.7	5.4	64	53.4	50.6	48.3	44.2
15	9.0	8.1	7.4	6.1	65	54.4	51.5	49.2	45.0
16	9.8	8.9	8.1	6.7	66	55.3	52.4	50.1	45.8
17	10.7	9.6	8.8	7.4	67	56.3	53.4	51.0	46.7
18	11.5	10.4	9.6	8.0	68	57.2	54.3	51.9	47.5
19	12.3	11.2	10.3	8.7	69	58.2	55.2	52.8	48.4
20	13.2	12.0	11.1	9.4	70	59.1	56.1	53.7	49.2
21	14.0	12.8	11.9	10.1	71	60.1	57.0	54.6	50.1
22	14.9	13.7	12.6	10.8	72	61.0	58.0	55.5	50.9
23	15.8	14.5	13.4	11.5	73	62.0	58.9	56.4	51.8
24	16.6	15.3	14.2	12.2	74	62.9	59.8	57.3	52.7
25	17.5	16.1	15.0	13.0	75	63.9	60.7	58.2	53.5
26	18.4	17.0	15.8	13.7	76	64.9	61.7	59.1	54.4
27	19.3	17.8	16.6	14.4	77	65.8	62.6	60.0	55.2
28	20.2	18.6	17.4	15.2	78	66.8	63.5	60.9	56.1
29	21.0	19.5	18.2	15.9	79	67.7	64.4	61.8	57.0
30	21.9	20.3	19.0	16.7	80	68.7	65.4	62.7	57.8
31	22.8	21.2	19.9	17.4	81	69.6	66.3	63.6	58.7
32	23.7	22.0	20.7	18.2	82	70.6	67.2	64.5	59.5
33	24.6	22.9	21.5	19.0	83	71.6	68.2	65.4	60.4
34	25.5	23.8	22.3	19.7	84	72.5	69.1	66.3	61.3
35	26.4	24.6	23.2	20.5	85	73.5	70.0	67.2	62.1
36	27.3	25.5	24.0	21.3	86	74.5	70.9	68.1	63.0
37	28.3	26.4	24.8	22.1	87	75.4	71.9	69.0	63.9
38	29.2	27.3	25.7	22.9	88	76.4	72.8	69.9	64.7
39	30.1	28.1	26.5	23.7	89	77.3	73.7	70.8	65.6
40	31.0	29.0	27.4	24.4	90	78.3	74.7	71.8	66.5
41	31.9	29.9	28.2	25.2	91	79.3	75.6	72.7	67.4
42	32.8	30.8	29.1	26.0	92	80.2	76.7	73.6	68.2
43	33.8	31.7	29.9	26.8	93	81.2	77.5	74.5	69.1
44	34.7	32.5	30.8	27.6	94	82.2	78.4	75.4	70.0
45	35.6	33.4	31.7	28.4	95	83.1	79.4	76.3	70.9
46	36.5	34.3	32.5	29.3	96	84.1	80.3	77.2	71.7
47	37.5	35.2	33.4	30.1	97	85.1	81.2	78.2	72.6
48	38.4	36.1	34.2	30.9	98	86.0	82.2	79.1	73.5
49	39.3	37.0	35.1	31.7	99	87.0	83.1	80.0	74.4
50	40.3	37.9	36.0	32.5	100	88.0	84.1	80.9	75.2

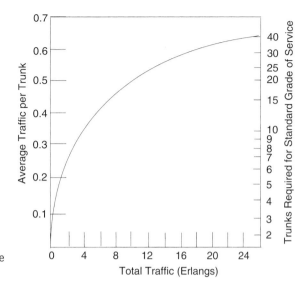

Fig. 4.14 Trunk occupancy for full-availability groups of various sizes (group sizes designed to give grade of service = 0.002).

greater efficiency of a large group of trunks is greater susceptibility to overload. For the same example, Figure 4.15 shows that an increase in traffic of 10 percent increases the loss probability for a group of 5 trunks by 45 percent, but for a group of 100 trunks it increases by 550 percent.

For this reason, it is common practice to specify two grades of service: one for the specified traffic and a higher value for a given overload. For small groups of trunks the former criterion determines the number required, whereas the latter criterion determines the number of trunks when the traffic is high and the number of trunks is large. For example, if it is specified that the grade of service shall be 0.002 but not deteriorate beyond 0.01 when the traffic increases by 10 percent, then Figure 4.15 shows that the former criterion is the more stringent when the number of trunks is less than 70 but the overload criterion applies for more than 70 trunks. Many public network operators therefore use traffic tables based on such dual criteria. BT adopted tables based on three criteria. If the permitted GOS is B when the normal traffic is offered, it shall not exceed $2.5B$ for a 10 percent overload or $6.25B$ for a 20 percent overload.

If automatic alternative routing is used, the traffic that overflows from a high-usage route to the tandem route consists of only the peaks of the traffic that is offered to the high-usage route. At times when the number of calls is less than the number of trunks on the high-usage route, the overflow traffic is zero. Consequently the overflow traffic is non-Poissonian; it is "peaky." Equation 4.2 and traffic tables based on it should not be applied. It is necessary to take into account the variance of the traffic as well as its mean value. Methods for determining the numbers of high-usage and tandem circuits required have been developed (Wilkinson, 1956; Wallstrom, 1969; Rapp, 1964).

Results obtained from analytical formulae may be inaccurate in practical cases, particularly when complex alternative-routing schemes are involved, because the assumptions have been oversimplified. The alternative is to obtain results by means of a computer simulation of the system under study (Bear, 1988; Ghanbari et al., 1997). Public network

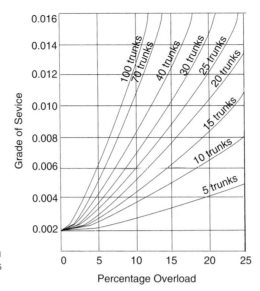

Fig. 4.15 Effect of overload on grade of service for trunk groups of different sizes.

operators have extensive computer programs for this purpose. Other programs, which may be used in planning private networks, are available commercially.

Network Planning

A national telecommunications network is large and complex, and different parts of it are planned by different groups of engineers. It is therefore essential that it adheres to common standards in order to obtain satisfactory performance. National plans are needed to govern the design of a network and its constituent local, junction, and long-distance networks. They include the following:

- The transmission plan
- The numbering plan
- The charging plan
- The routing plan
- The signaling plan
- The grade of service required (end to end)
- The capabilities of switching systems
- The economics of direct and indirect routing
- Network management requirements

These considerations are interrelated. For example, charging, routing, and numbering are closely related. Both the route for a call and its charging rate are determined by the directory number of the called station. The flexibility with which these can be handled is governed by the capabilities of the switching equipment employed. Either transmission or signaling standards may limit the size of a local-exchange area, or the number of links that may be used in tandem for a trunk connection. These place restrictions on the routings that may be employed.

The national plans for numbering, charging, transmission, and signaling thus form a set of standards that govern planning both for a national network and its constituent local networks. The specified grade of service determines the number of circuits required on each route within it (i.e., the dimensioning of the network).

Customer-access networks may account for more than one-third of the total cost of a PSTN, so economy in their provision is essential and the smallest practicable wire gauge is used. Consequently, the largest part of the allowable attenuation for a limiting connection is allocated to the customer's line and the smallest part to the trunk network. Similarly, the overall grade of service of a connection is the sum of those of its parts. It is economical to provide circuits generously where they are cheap and less generously where they are expensive, in order to obtain a specified overall grade of service at minimum cost.

The national plans for numbering, charging, transmission, and signaling thus govern planning both for a national network and its constituent local networks. Performance standards must be adequate both for connections within the network and for those that extend into other networks: for example, private networks, networks of competitors, mobile radio networks, and connections into other countries via international circuits. A PSTN must provide its customers with a service of satisfactory quality at a price they are willing to pay (Oodan et al., 1999). Network planners must therefore achieve a compromise between best performance and lowest cost.

Network planning is a continual process. It involves monitoring the present state of the network, producing plans to meet requirements for growth and enhancement, implementing the plans, and auditing the outcomes (Flood, 1997). The end product of planning is a large capital expenditure on plant often having a long economic lifetime. Since investment in the network and its running costs, together with the revenue obtained from it, dominate the finances of a PTO, network planning is inextricably linked to its business planning.

Network Management

The management of a PSTN is inevitably complex (Flood, 1997). It is necessary to consider several different aspects, as follows:

- The customers
- The network operator
- The network itself

Relationships between these aspects are shown in Figure 4.16.

Users are mainly concerned with the services provided by a network. They are affected by the quality of service (Oodan et al., 1999) and, of course, its cost. The introduction of new services must also be considered.

Network operators try to control the costs of operating their networks to maximize profits. This requires managing all operations that affect the maintenance and operation of the network at four levels, as follows:

1. *The business level:* This is the management of the business itself (i.e., sales, customer administration and billing, profit-and-loss accounting, inventory control, investment planning, and so on).

2. *The service level:* This is the management of services offered to users. Many services are provided besides basic telephone and data switching and transport. Thus, service provision is separate from network operation and needs to be managed separately.

3. *The network level:* This includes maintaining a database recording the installed system, optimization of routes and flow control (as described earlier in the section on routing), planning changes to enhance or extend the system, planning for emergencies, and management of the installed equipment.

4. *The network element level:* This includes the installation of the equipment; the prediction, detection, and repair of faults; and the management of repairs and alterations.

Computer tools are used in all these activities and, since a telecommunications network is distributed over a wide area, a data network is needed to reach the elements to be monitored and controlled. This leads to an overlay network for administration of the network, as shown in Figure 4.11. In a complex network, many different proprietary systems are involved, so standard interfaces are needed between them. International cooperation has resulted in a set of open-systems standards for this purpose (Black, V., 1992).

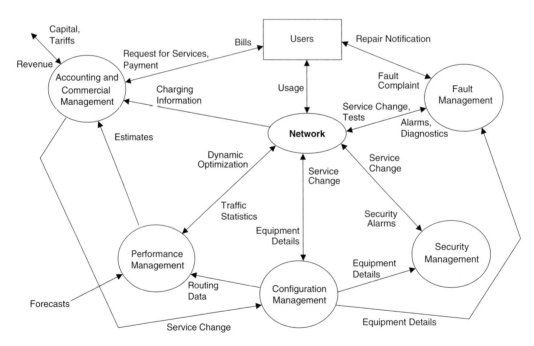

Fig. 4.16   Network management functions. *Source:* From Flood (ed.), 1997.

References

Ash, G. R. (1992) Real-Time Network Routing in the AT&T Network, *Proceedings of IEEE Global Telecommunications Conference,* pp. 802–809.

Ash, G. R., and Mummert, V. S. (1992) AT&T Carves New Routes in Its Nationwide Network, *AT&T Bell Labs Rec.* 62, pp. 18–22.

Bannerjee, V., Rabundrakumar, K., and Szczrch, B. J. (1989) London Code Change, *Brit. Telecom. Eng*, 8, p. 134.

Barron, D. A. (1959) Subscriber Trunk Dialling, *Proc. IEE* 106B, p. 341.

Beal, M. (2004) Evolving Networks for the Future—Delivering BT's 21st Century Network, *J. Coms. Net* 3, pp. 4–10.

Bealby, A. G. (1990) Network Administration Implementation Programme, *Brit. Telecom. Eng.* 8, p. 212.

Bear, D. (1988) *Principles of Telecommunication Traffic Engineering, Third Edition*, Peter Peregrinus.

Bishop, T. (1995) Freeing the Network for Competition, *Telecom. Eng. J.*, April, p. 75.

Black, V. (1992) *Network Management Standards: the OSI SSNMP and CMOI Standards*, McGraw-Hill.

Buckley, J. F. (1994) Telecommunications Numbering, *Electron. Commun. Eng.*, June, p. 119.

Clarke, A. B., and Osborne, H. S. (1952) Automatic Switching for Nationwide Telephone Service, *Bell Sys. Tech. J.* 31, p. 823.

Clos, C. (1954) Automatic Alternate Routing of Telephone Traffic, *Bell Lab. Rec.* 32, p. 51.

Farr, R. E. (1988) *Telecommunications Traffic, Tariffs and Costs*, Peter Peregrinus.

Flood, J. E., and Cochrane, P. (eds.) (1991) *Transmission Systems*, Peter Peregrinus.

Flood, J. E. (1995) *Telecommunications Switching, Traffic and Networks*, Prentice-Hall.

Flood, J. E. (ed.) (1997) *Telecommunication Networks*, 2nd edition, IEE.

Francis, H. E. (1959) The General Plan for Subscriber Trunk Dialling, *Post Office Elec. Eng.* 51(January), p. 258.

Ghanbari, M., Hughes, C. J. M., Sinclair, M. C., and Eade, J. P. (1997) *Principles of Performance Engineering for Telecommunications and Information Systems*, IEE.

Hills, M. T., and Evans, B. G. (1973) *Transmission Systems*, Allen and Unwin.

Hunter, J. M., Lawrie, N., and Peterson, N. (1988) *Tariffs, Traffic and Performance*, Com. Ed. Publishing.

Huws, U., Kurte, W. B., and Robinson, S. (1990), *Teleworking: Towards the Elusive Office*, Wiley.

IEEE (1988) *J. Selected Areas Commun.*, SA6; papers on Telecommunications Network Operations and Management.

Ingham, A. R., and Elvidge, A. M. (1989) Trunk Reservation with Automatic Alternative Routing. In *6th U.K. Teletraffic Symposium*.

ITU-T (1989a) *Recommendation E163, Numbering Plan for the International Telephone Service* (9th Plenary Assembly), ITU-T.

ITU-T (1989b) *Recommendation E164, Numbering for the ISDN Era* (9th Plenary Assembly), ITU-T.

ITU-T (1989c) *Recommendation E165, Timetable for Co-ordinated Implementation of the Full Capability of the Numbering Plan for the ISDN Era* (9th Plenary Assembly), ITU-T.

ITU-T (1989d) *Recommendation X121, International Numbering Plan for Public Data Networks* (9th Plenary Assembly), ITU-T.

Katzschner, L. (1973) Service Protection for Direct Final Traffic in DDD Networks, In *3rd International Teletraffic Congress*.

Macurdy, W. B. (1973) Network Management in the United States—Problems and Progress. In *Proceedings of 7th Teletraffic Congress, Pt. 2*, p. 621.

Manterfield, R. J. (1999) *Telecommunications Signalling*, IET.

McLeod, N. A. C. (1990) Numbering in Telecommunications, *Brit. Telecom. Eng.* 8, p. 225.

Muir, A., and Hart, G. (1987) The Conversion of a Telecommunications Network from Analogue to Digital Operation. In *IEE National Conference on Telecommunications Networks*.

Munday, S. (1967) New International Switching and Transmission Plan Recommended by the CCITT for Public Telephony, *Proc. IEE* 114, p. 619.

Nunn, A, (2005) Voice Service Trends—Past and Future, *J. Coms. Net* 4, pp. 9–12.

Nunn, W. H. (1952) Nationwide Numbering Plan, *Bell Sys. Tech. J.* 31, p. 851.

Oodan, A., Ward, K., and Mullee, T. (1999) *Quality of Service in Telecommunications*, IEE.

Pita, I., et al. (1993) Planning National Telecommunication Networks, *Electr. Commun.*, 2nd Quarter, p. 188.

Rapp, Y. (1964) Planning of a Junction Network in a Multi-exchange Area, *Ericsson Tech.* 20, p. 77.

Redmill, F. J., and Valdar, A. R. (1990) *SPC Digital Telephone Exchanges*, Peter Peregrinus.

Richards, D. L. (1973) *Telecommunication by Speech*, Butterworth.

Rudkin, S. (2004) Session-Based Quality of Service, *Brit. Telecom. Technol.* 22, pp. 86–94.

Stacey, R. R., and Songhurst, D. J. (1989) Dynamic Routing in British Telecom Network. In *International Switching Symposium*.

Tobin, W. J. E., and Stratton, J. (1960) A New Switching and Transmission Plan for the Inland Trunk Network, *Post Office Elec. Eng.* 53, p. 75.

Valdar, A. (2006) *Understanding Telecommunications Networks*, IET.

Van Duuren, J., Kastelein, P., and Schoute, F. C. (1992) *Telecommunication Networks and Services*, Addison-Wesley.

Waliji, A. A. (1980) Architecture of System X: the Local Administration Centre, *Post Office Elec. Eng.* 73(1), p. 36.

Wallstrom, B. (1969) Methods for Optimizing Alternative Routing Networks, *Ericsson Tech.* 25, p. 3.

Webster, S. (1989) The Digital Derived Services Intelligent Network, *Brit. Telecom. Eng.* 8, p. 144.

Welch, S. (1981) *Signalling in Telecommunication Networks*, Peter Peregrinus.

Wilkinson, R. I. (1956) Theories for Toll Traffic Engineering in the USA, *Bell Sys. Tech. J.* 35, p. 421.

Wittering, S. (1994) Numbers Up, *Commun. Int.*, June, p. 10.

Wyatt, T., Gozal, D., and Hancock, J. (1995) Intelligent Networks Phase 1: Intelligence in the Core Network, *Brit. Telecom. Eng.* 14, pp. 202–208.

5 Digital Switching Systems

Sydney F. Smith

Introduction

Basic Structure

The most elementary form of a switch network consists of a single square switch, where connection can be made between any one inlet and any one outlet by connecting the associated horizontal and vertical circuits at the point where they cross (i.e., the "crosspoint"). However, to provide a switching system capable of serving say 10,000 subscribers, it would not be economical either to construct a single large switch or to simply gang a large number of smaller switches to form a square matrix of $10,000 \times 10,000$ crosspoints. A more practical arrangement is to connect the switches in two or more stages.

To illustrate this, Figure 5.1 shows first a single square matrix of 81 crosspoints providing for nine paths between nine inlets and nine outlets and then how to meet the same requirement with only 54 crosspoints. The advantage of this approach is even greater with larger numbers of circuits and practical switches having typically 10×20 crosspoints or more, rather than only 3×3 as shown in the example.

This simple example provides only for the distribution of calls from the inlets to an equal number of outlets through the same number of links. In practical exchanges, it is necessary to provide only as many paths through the switching network as the number of calls that are expected to be in progress at one time. The concentration of lines onto a smaller number of paths or trunks can be achieved using the principle illustrated by another simple example in Figure 5.2.

A complete local exchange then consists of a combination of a distribution (group selection) stage and one or more line concentration units, as shown in Figure 5.3. The separate switches are interconnected through yet another switching stage that does not provide concentration or expansion (i.e., a square switch), as shown in Figure 5.1 This group switching (distribution) stage can also be used to make connections to other exchanges.

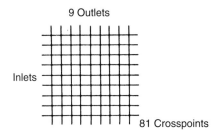

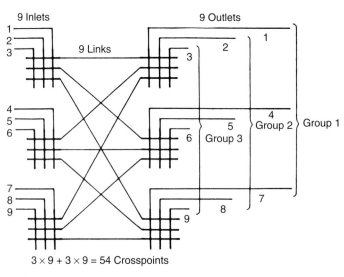

Fig. 5.1   Trunking principle of distribution stage.

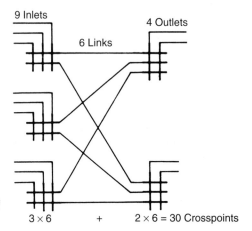

Fig. 5.2   Trunking principle of concentration stage.

To complete the exchange, two other functional blocks are needed. The call connection (and disconnection) instructions have to pass between the terminations and the exchange and also between exchanges for interexchange (junction) calls. These supervisory messages are conveyed by the exchange signaling systems, which may transmit the relevant data over

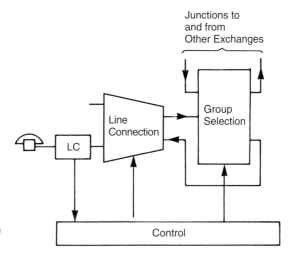

Fig. 5.3 Generalized architecture of a complete exchange.

the same paths as the calls they control or may use separate links for the purpose. Also needed is a control system for interpreting these instructions to enable appropriate paths to be set through the switches.

The operation and design of exchange switching systems can be conveniently separated into these three separate, but interdependent, functions: *switching, signaling, and control*. In practical terms, the choice of switching technology is dependent on the transmission methods used in the network concerned and this in turn influences the choice of signaling and control methods.

The early exchanges were all manually operated and some are still in service around the world. The exchange equipment consists of a switchboard at which an operator (telephonist) connects calls on demand. The term "automatic" is used to describe exchanges where the switching is carried out by machine under the remote control of the caller, who could be either a subscriber or an operator. The first public automatic exchange opened in 1892 at La Porte, Indiana (U.S.), and used rotary switches invented in 1891 by Almon B. Strowger, an undertaker in Kansas City, after whom the system is named. Since then several systems have been developed using a variety of electro-mechanical devices. One is a *crossbar* system in which the arrays of crosspoints are implemented by mechanisms in which a series of electromagnets each controls a row of contacts. Another is a *reed relay* system where a separate relay is provided at each crosspoint.

Electro-mechanical systems have now been entirely replaced in most public networks around the world by digital systems employing totally solid-state switching and stored program control (covered later in this chapter).

Types of Transmission

Transmission may be either analogue, in which the electrical signal represents directly the speech signal, or digital, in which speech or data are coded in pulses. At present, telephone instruments and subscribers' lines are generally analogue, but the trunk and junction network is almost entirely digital.

The essential feature of digital transmission, as opposed to analogue, is regeneration. The digital signals consist of a sequence of digits (or symbols) each of which has a few recognized levels (typically four or fewer) and occupies equal periods of time known as digit (or symbol) periods. Transmission systems are designed so that corruption of the digits, due to noise, interference, or distortion, does not cause the signal level in each digit period to become indeterminate. A regenerator measures the level in each digit period and generates a new digit of the original amplitude. This means that noise and corruption of the signal do not accumulate through a series of regenerators. Only two types of impairment are left: (a) digital errors that occur when excessive noise and corruption cause a regenerator to make wrong decisions about the level, and (b) jitter, which is displacement of the digit periods from their ideal positions, since time is essentially analogue even in digital transmission and so is subject to analogue types of distortion.

Switching

Types of Switching

Space Division
Analogue exchanges such as Strowger, crossbar, and reed relay systems employ space division switching in which a separate physical path is provided for each call and is held continuously for the exclusive use of that call for its duration.

Time Division
Digital and other pulse modulated systems employ time division switching in which each call is provided with a path only during its allocated time slot in a continuous cycle. Each switching element or crosspoint is shared by several simultaneous calls, each occupying a single time slot, in which a sample of the speech is transmitted. This sample may be a single pulse as in pulse amplitude modulated systems or a group of pulses as in pulse code modulated (PCM) systems.

Circuit Switching
The space and time division switching described above are both "circuit" switches in that although a call in a time-divided switch shares elements of its path with other calls, it nevertheless occupies that same path throughout the call. In most calls the nature of speech is such that the path carries information for only a relatively small proportion of the time.

Packet Switching
Strings of data or digital speech samples may be grouped together with address information to form a discrete "packet." If a complete call consists of a number of such packets, each packet is assigned an available path through the switch when it arrives. Associated packets may therefore reach a common destination over different paths through the switch.

Digital Circuit Switching

Speech itself is analogue and so too are most telephones. Analogue to digital (A/D) conversion (modulation) is, therefore, necessary at some point in each connection if digital switching is to be employed. For subscribers connected to existing analogue local exchanges this occurs when the call first encounters a PCM transmission trunk on leaving the exchange. For subscribers connected to a digital local exchange it is necessary to provide the A/D conversion on each line. The cost of this conversion constitutes a major part of the exchange cost but is amply compensated for by the overall economy of digital switching and the performance advantages in the transmission network.

The European standard is for speech encoded in 8-bit bytes at 64 kbit/s. The U.S. standard is 7-bit encoding at 56 kbit/s.

Digital Tandem (Transit) Exchange

This interconnects transmission systems that are multiplexed in groups usually of 24 or 30 speech channels on each link. The tandem exchange has to be able to connect an incoming channel in one PCM link to an outgoing channel in another PCM link. Received speech samples must, therefore, be switched in space to the appropriate link and translated in time to the required channel time slot.

A simple time switch is shown in Figure 5.4, which provides for time slot interchange between the 30 channels of an incoming PCM circuit and the 30 channels of an outgoing circuit. Each successive byte corresponding to a channel is written cyclically into the appropriate location in a 30×8 bit RAM under the control of a time slot counter synchronized with the incoming circuit clock.

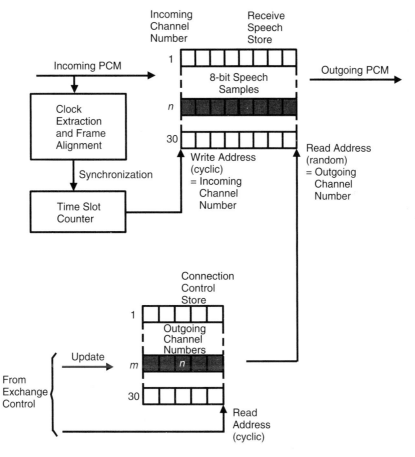

Fig. 5.4 Simple time switch.

The contents of the store are read out to the outgoing circuit under the control of a connection store. The words in the store, which is itself addressed cyclically, identify which address in the speech store is to be addressed. If the incoming time slot counter and connection store run in synchronism, the outgoing circuit carries a similar multiplex to the incoming circuit but with the time slots in a different order as determined by the contents of the connection store. In practice, the connection store may have a different speed or a different number of channels (words) from the incoming circuit.

Practical PCM systems employ additional channels for signaling and framing purposes. These can be extracted for use by the exchange control by providing additional RAM addresses from which they are read out onto a circuit connected to the exchange control processor. Each incoming circuit, therefore, requires 32×8 bit RAM addresses.

For large exchanges, each time switch serves a group of, say, 32 incoming circuits using 1024 RAM addresses. The connection store then runs at 32 times the incoming multiplex data rate to interleave all 30×32 speech channels on a single high-speed highway to the next switching stage in the exchange.

A simple TDM space switch is shown in Figure 5.5 consisting of a matrix in which the crosspoints are electronic gates controlled by a connection store. Each incoming time slot on

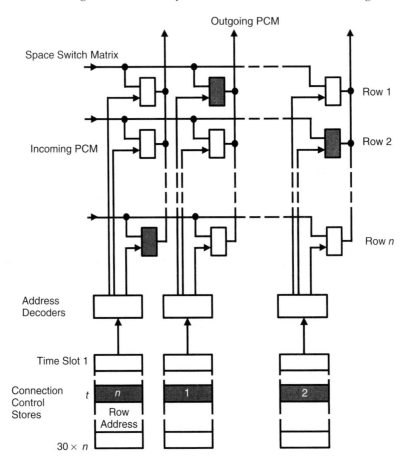

Fig. 5.5 Simple space switch.

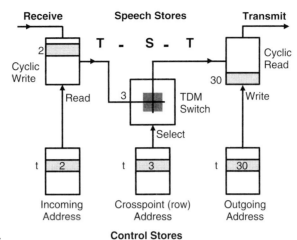

Fig. 5.6 T-S-T switch.

each incoming circuit is switched individually to the appropriate outgoing circuit, in accordance with the addresses stored in the space switch connection store.

A complete tandem exchange might use a time switch, as in Figure 5.4, to connect each incoming circuit to a row in a space switch matrix, as in Figure 5.5. The verticals of the matrix could be the outgoing circuits but are more usually highways connected to a time switch associated with each outgoing trunk group. This arrangement is shown in Figure 5.6 and is known as time-space-time (TST) switching.

If the internal exchange bit rate is high enough to interleave all the incoming channels onto a single highway, no space switch is required. Each cross-office time slot corresponds to a unique outgoing circuit/time slot combination so no outgoing time switch is required either. Allocation of highway time slots to outgoing channels is achieved by gating on a fixed pattern of strobe clock pulses. This is sometimes known as TS switching, the output gating being regarded as a "space" (S) switching stage. It is also sometimes called memory switching. The size of exchange is limited by the available technology since RAM size and speed requirements are proportional to the number of channels provided across the exchange. Larger exchanges can be constructed by interconnecting separate TS stages.

One other possible configuration is space-time-space (STS). This was used for some early designs when the only available time switching technology was to insert expensive time delay elements in the speech path.

The arrangements described above carry speech in one direction only. All "receive" channels are connected to one side of the switch, regardless of whether the circuits are incoming or outgoing in the traffic sense and all "transmit" channels are connected to the other side. Each call requires transmission in each direction and these two paths can occupy the same crosspoint in the space switch but at different time slots, usually 180 degrees out of phase. Separate time switches must be provided in each direction of transmission.

Digital Local Exchange

The switching networks discussed above provide for only the distribution of calls from the inlets to an equal number of outlets through the same number of links. In practical exchanges, it is necessary to provide only as many paths through the switching network as the number of calls that are expected to be in progress at one time.

The concentration of lines onto a smaller number of paths or trunks is provided in the form of an additional time switching stage. The number of available time slots to the main switch is less than the total number of input time slots and store locations in the concentrator speech stores they serve. This might consist typically of up to 16 groups of 256 lines, each written cyclically into a 256-word store but with only 256 time slots available for allocation to the acyclic read out from all these stores.

If these concentration stages are connected to a TST switch, we have a 5-stage network for local calls of the form T-TST-T.

Telephone subscribers' lines are generally on individual pairs employing analogue transmission. The interface to the analogue subscriber's line includes certain functions that require voltage and power levels incompatible with the digital concentrator switch and that, therefore, have to be provided at the individual line circuit (line card). The line interface functions are generally referred to as the Borscht functions:

Battery feed to line
Overvoltage protection
Ringing current injection and ring trip detection
Supervision (on-hook/off-hook detection)
Codec for analogue to digital conversion
Hybrid for two-wire to four-wire conversion
Test access to line and associated line circuit

The connection between the concentration stage and the main switching stage is a PCM multiplex and is very suitable for connection over an external transmission line. Digital exchanges are, therefore, particularly suited to provision as a single large, central switch serving a number of remote concentrators.

Teletraffic

Grade of Service

It would be uneconomical to provide switching equipment in quantities sufficient for all subscribers to be simultaneously engaged on calls. In practice, systems are engineered to provide an acceptable service under normal peak (i.e., busy hour) conditions. For this purpose, a measure called the grade of service (g.o.s.) is defined as the proportion of call attempts made in the busy hour that fail to mature because the equipment concerned is already engaged on other calls. A typical value would be 0.005 (one lost call in 200) for a single switching stage. To avoid the possibility of serious deterioration of service under sudden abnormal traffic, it is also usual to specify grades of service to be met under selected overload conditions.

Traffic Unit (the Erlang)

Traffic is measured in terms of a unit of traffic intensity called the erlang (formerly known as the traffic unit or TU), which may be defined as the number of call hours per hour (usually, but not necessarily, the busy hour), or call seconds per second, and so on. This is a dimensionless unit that expresses the rate of flow of calls and, for a group of circuits, is numerically equal to the average number of simultaneous calls. It also equates on a single circuit to the proportion of time for which that circuit is engaged and consequently to the probability of finding that circuit engaged (i.e., the grade of service on that circuit). The traffic on one circuit can, of course, never be greater than one erlang. Typically a single selector or junction circuit would carry about 0.6 erlang and subscribers' lines vary from about 0.05 or less for residential lines to 0.5 erlang or more for some business lines.

It follows from the definition that C calls of average duration t seconds occurring in a period of T seconds constitute a traffic of A erlangs given by Equation 5.1.

$$A = C\frac{t}{T} \tag{5.1}$$

An alternative unit of traffic sometimes used is the c.c.s. (call cent second) defined as hundreds of call seconds per hour (36 c.c.s. = 1 erlang).

Erlang's Full Availability Formula

Full availability means that all inlets to a switching stage have access to all the outlets of that stage. Trunking arrangements, in which some of the outlets are accessible from only some of the inlets, are said to have limited availability.

For a full availability group of N circuits offering a traffic of A erlangs, the grade of service B is given by Equation 5.2.

$$B = \frac{\dfrac{A!}{N}}{1 + A + \dfrac{A^2}{2!} + \ldots + \dfrac{A^N}{N!}} \tag{5.2}$$

This formula assumes pure chance traffic and that all calls originating when all trunks are busy are lost and have zero duration.

Busy Hour Call Attempts

An important parameter in the design of stored program control systems in particular, and common control systems in general, is the number of attempts to be processed, usually expressed in busy hour call attempts (BHCA). Typically a 10,000-line local exchange might require a processing capacity of up to 80,000 BHCA.

Blocking

If two free trunks cannot be connected together because all suitable links are already engaged, the call is said to be blocked. In Figure 5.1, if one call has already been set up from inlet 1 to outlet 1, an attempted call from inlet 2 to outlet 2 will be blocked. The effects of the blocking are reduced here by allocation of the outlets to the three routes.

Blocking can be reduced by connecting certain outlets permanently to the inlet side of the network to permit a blocked call to seek a new path. Alternatively, a third stage of switching may be provided.

It is possible to design a network having an odd number of switching stages in which blocking never occurs. Nonblocking arrays need more crosspoints than acceptable blocking networks and are uneconomical for most commercial telephone switching applications.

Control

Processors

The functions of a digital exchange switch, described above, need processor control to interpret the connections required in terms of signals received and connection store contents to be maintained.

One advantage of central control by processors is that function or service changes may be implemented by changes to a few processors instead of many registers or selectors. This advantage is further increased if the processors are based on "software" programs. This method of operation is known as "stored program control" (SPC). It is possible, in the limit, to convert into processor software almost all of the logical functions of the exchange, including signaling.

The advantages are that the variety of hardware may be minimized, so easing production and stocking problems, and that facility changes may be readily accomplished. Disadvantages are that software development can be exceedingly complex and that real-time problems may occur.

Software

The essence of SPC is that it is software driven. The functionality of the exchange control processors can be changed to suit different hardware configurations (number and type of lines, traffic, and so on) for different installations and growth during the life of the system. The features offered can also develop as requirements change over time (e.g., tariff structures, dialing codes) without major hardware replacements.

This software must, of course, meet the same standards of reliability and availability as those demanded of the switch and processor hardware. Software does not wear out like hardware. Failures can only occur due to design faults (bugs) that have not been detected during the design and development phase. The number of combinations of external inputs and system states in programs of this size makes it impossible to guarantee the complete absence of such bugs.

To provide some degree of fault tolerance, it is first necessary to design the software in discrete modules with defined interfaces between them. This enables more comprehensive testing of possible combinations within each module than would be possible on the complete system. It also allows checks to be applied to input and output conditions at the interfaces. These checks can be used either to trigger the hardware changeover mechanism or to reinitialize the software module to an acceptable state.

The modular design of software also facilitates the evolution of features during the life of the system and the implementation of distributed control.

The primary function of the software is to set up and clear down calls on demand. Calls arise concurrently in real time and in a random manner. In addition, the software is required to carry out diagnostic checks on the hardware, provide statistical data on traffic and performance, and provide call records for billing purposes. Separate modules are provided for each of the functions in the applications software. The operating system schedules the use of the hardware resources by these application programs, storing the status of all registers and such when a program is deactivated and restoring the same state when it is reactivated. This multiprogramming technique enables high-priority tasks (e.g., call setup) to be given precedence over lower-priority tasks (e.g., statistics).

Figure 5.7 shows in simplified form a typical modular breakdown of the software in an SPC exchange. The line scan program continuously updates a tabular store of the state of each line termination (e.g., free, busy, calling, and so on) and alerts the central call routing software to connection requests. The latter validates a request against a predetermined class of service for the termination and receives called number data from the line scanner or the signaling software, depending on the type of termination. The received digits are compared with the stored routing data (translation table) to enable the switch to be controlled

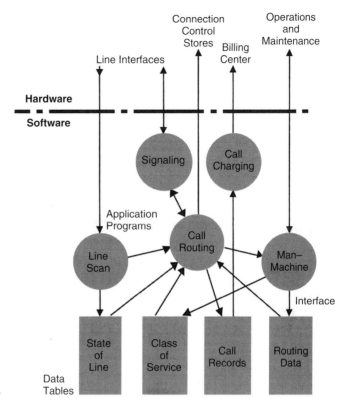

Fig. 5.7   SPC software structure.

via the connection control stores (Figures 5.4 to 5.6). Exchange staff can interrogate the status of the processor and update the routing and class of service tables through the Man–Machine Interface (MMI).

Network Aspects

The introduction of SPC has enabled electronic exchanges to provide subscriber and administration features not readily available with earlier control systems. Examples include the ability to transfer calls at will and allow calls to wait on a busy line. Where new features are self-contained within an exchange, they can be introduced by installing a new build of software. If the operation of a feature requires identical operation at all exchanges, the practical problems of modifying them all at once become immense.

One solution is to download new programs, routing data, and such, remotely via data links and to site the operations and maintenance terminals at a central location to serve a group of exchanges. Central access to the databases and processors of all the exchanges permits short-term as well as long-term changes to be made to facilities and routing. This is known as Network Management and enables plant failures to be identified and located, and traffic to be re-routed around congestion bottlenecks. A further development of the centralized control principle is the concept of the Intelligent Network in which individual exchange processors interrogate a remote database for instructions on how to process particular types of call or route such calls to a central service provider.

Operational Security

Unlike many other systems that can plan for regular down time for maintenance, telephone switching systems are required to provide virtually uninterrupted service 24 hours a day, 7 days a week, year in, year out, to tens of thousands of subscribers (and junctions). At the same time they must be adaptable to growth and operational requirements. Occasional faults are, of course, inevitable and the system design must allow for this, so that a fault does not result in loss of service to any significant number of subscribers. It is better for a fault to temporarily reduce the grade of service to all subscribers than to deny service entirely to a smaller number. As a result, various system architectures have evolved.

In electro-mechanical systems, each call follows a path wholly exclusive to itself, sufficient paths being provided to carry the predicted traffic. Failure in any path, therefore, reduces the grade of service to all terminations normally having access to it, but since other paths are available, service is not wholly cut off to any one. Electro-mechanical systems are, therefore, inherently fault-tolerant as regards path switching and, for example, Strowger-type selectors incorporate their own local control that reacts to loop-disconnect signaling over the path concerned.

In the switching area, security is normally provided by giving any termination access to multiple paths through the network. A particular example of this is the use of sectionalization where the switching network is divided into planes of switching, operating in parallel and all accessible to all terminations. Thus a fault affecting any one plane of switching will not cause loss of service to any terminations, the traffic being carried by the remaining planes, albeit at a slightly reduced grade of service.

In the control area, security is normally provided by replication. In its simplest form this would mean a single duplicated control unit. The two units can operate:

1. In parallel with a fault cutting out the defective unit
2. As worker and spare with automatic changeover in the event of a fault
3. Each in turn under the control of a time cycle

However, except in very small systems, a single duplicated control is unlikely to be adequate and a number of alternative options are available:

1. Triplication, where three identical units are used in parallel with majority decision equipment on the outputs.
2. One-in-N sparing, where a number of similar pieces of equipment are involved and one (or more) spares are provided such that, in the event of a fault, one of the spares may be switched into service in place of the faulty equipment.
3. Load sharing, where a number of similar pieces of equipment operate in parallels each carrying a proportion of the traffic. In the event of a fault, the equipment concerned is taken out of service with the traffic being carried by the remainder at a reduced grade of service.

In practice, some combination of these options is often used with the overall control area being subdivided dependent on the function to be performed. In the case of traffic dependent equipment such as registers, these are normally dimensioned to carry the predicted busy hour traffic at the required grade of service in the presence of one or more faults, depending on the predicted fault rate.

Whichever method is employed to provide the operational security, adequate check circuits and routines must be provided to enable faults to be remedied within the mean time between failures (MTBF) for the equipment concerned.

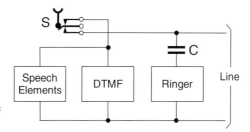

Fig. 5.8 Signaling elements of
a typical telephone instrument.

Signaling

Line Signaling

Subscriber Line Signaling

The signaling elements of a telephone instrument are shown in Figure 5.8. They comprise a dial (or a keypad for Dual Tone Multifrequency signaling, DTMF) and associated gravity switch for signaling to the exchange and a ringer (bell) or an electronic detection circuit and sound transducer for receiving calling (ringing) signals from the line.

Gravity Switch (S)   To indicate calling and clearing conditions, a contact is provided to interrupt the line current. This contact is open in the idle state and closes (makes) when a call is originated or answered. Because in many instruments it is operated by the weight of the handset, it is known as the gravity switch. But in much modern equipment, it is operated by a push button on the handset or terminal. It is also known as a "switch hook" contact and the terms "off-hook" and "on-hook" are often used to describe signaling conditions corresponding to the contact closed or open, respectively.

Dial   This is a mechanical device designed to send numerical data to the exchange as trains of up to ten short (66 ms) interruptions of the line current. This is known as loop-disconnect signaling and has now been largely superseded by DTMF signaling. The term "dialing" is still often used to refer to the process of signaling called numbers to the exchange.

Dual Tone Multifrequency   In dual tone multifrequency (DTMF), a 12-button keypad is provided with oscillators having frequencies as shown in Figure 5.9. Pressing any button causes a pair of frequencies to be generated, one from each band. For the digits 1 to 0 plus # and *, the four lower-band frequencies are paired with three of the upper-band frequencies as shown. An additional upper-band frequency of 1633 Hz is available to give four additional combinations if required for other services or functions.

Ringer   In the idle state, the gravity switch is open and no direct current flows in the line. The ringer or equivalent electronic circuit is, therefore, designed to operate on alternating current. This is connected from a common supply at the exchange typically at 75 V r.m.s. and 25 Hz connected through a 500-ohm resistance. Closing the gravity switch causes the exchange to disconnect the ringing current.

CLASS Signaling

CLASS (Customer Line Access Signaling System) is used to convey additional information to the subscriber, such as the calling line identity (CLI) at the start of a received call.

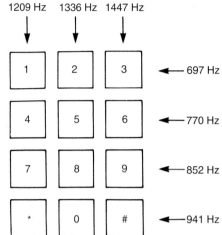

Fig. 5.9 Push-button
(multifrequency) dialing.

It uses a special telephone set to receive the CLASS signaling and display the CLI information.

Data is transferred using frequency shift keying (FSK) at 1200 baud based on V.23 signaling. The frequencies used are 1300 Hz (mark, logic 1) and 2100 Hz (space, logic 0).

Two methods are used to transmit the information:

1. Prior to ringing, the d.c. line feed is reversed. The CLASS telephone set acknowledges the reversal with a loop condition for 5 ms. This short current burst is used as wetting current. When the pulse ends, the exchange sends an in-band FSK signal with the CLI information. This is followed by the normal ringing current. This method is used by British Telecom (BT) in the United Kingdom.

2. Prior to ringing, the exchange sends a short burst of ringing current. This alerts the CLASS set to the forthcoming CLASS FSK information and is also used to wet the line. Then the exchange sends an in-band FSK signal with the CLI information. This is followed by the normal ringing current. If the subscriber reacts to the initial burst and answers the call, the CLASS phase is terminated and the call proceeds as normal. This method is used in North America and by cable operators in the United Kingdom.

There are some telephone sets available that respond to the two types of protocol.

Advanced Subscriber Features   CLASS uses in-band tones and hence cannot be used during the call. It can be used only during the call setup phase. Advanced features require signaling during the call. Several methods are used:

1. For Star Services, the subscriber can signal the requirement to obtain access to supplementary services by pressing the recall button (either earth or short break pulse recall). The exchange connects the line to an MF receiver to receive the subscriber's requests.
2. Some proprietary systems use an out-of-band channel for information transfer to and from the phone during the call.

Interexchange (Junction) Signaling

To connect calls through a network of exchanges, it is necessary to send data referring to the call between the exchanges concerned. In the simplest case, this consists of "loop/disconnect" signaling similar to that used on subscribers' lines. Each speech circuit consists of a pair of copper wires that have direct current flowing when a call is in progress, and this current is interrupted to form a train of break pulses for each digit corresponding to those produced by the telephone dial. When the called party answers, the called exchange sends back an "answer" signal to the calling exchange by reversing the direction of the current flow.

Longer circuits are unsuitable for direct current signaling (e.g., due to the use of amplifiers, multiplexing equipment, or radio links). Voice frequency (VF) signaling is then used in which pulses of alternating current are used at frequencies and levels compatible with circuits designed for handling speech currents (typically 600, 750, 2280, and 2400 Hz). This use of frequencies within the speech band (300–3400 Hz) is termed "in-band" signaling.

An alternative is the use of "out-of-band" (out-band) signaling, in which a signaling frequency (e.g., 3825 Hz) outside the speech band is separately modulated on to the carrier in a frequency division transmission system. In digital transmission systems, the signals are digitally encoded and certain time slot(s) are reserved for this purpose separate from the speech time slot(s).

These are all "channel associated" signaling systems, in which the signaling is physically and permanently associated with the individual speech channel even though no signaling information is transmitted during the greater part of the call

Common Channel Signaling

The use of stored program control has given rise to a more efficient signaling means known as "common channel" signaling. In this, one signaling channel serves a large number of speech channels and consists of a direct data link between the control processors (computers) in the exchanges concerned. Each new item of information relating to a call is sent as a data message containing an address label identifying the circuit to which it is to be applied.

Signaling System No 7

The specification for this common channel signaling system was designed by an international committee of the CCITT (now the ITU-T, see Chapter 1) and only subsequently incorporated into various manufacturers' equipment. It was originally known as CCITT No 7, often abbreviated to C7 or SS7 and has become the world standard for interexchange telephony signaling. It is used as a high-level communication link between digital exchanges on a dedicated 64 kbit/s data channel. In addition to call setup and clearing it supports other functions such as:

- ISDN calls, including call classification
- User-to-user data
- Calling Line Identification (CLI)
- Malicious Call Identification (MCI)
- Network address extension
- Distinctive ringing
- Call diversion
- Database interrogation

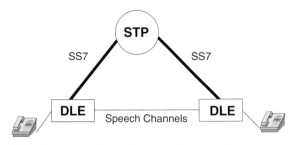

Fig. 5.10   Signal transfer point.   DLE = digital local exchange, SS7 = signaling system No. 7,
STP = signal transfer point

One SS7 64 kbit/s link can support call setup for about 2,000 telephone channels (about 60×2 Mbit/s PCM links). Thus, time slot 16 of one PCM link serves the signaling for the other PCM links. The signaling messages do not need to be carried along the same route as that carrying the related telephony traffic. Figure 5.10 shows SS7 message routing via a Signal Transfer Point (STP).

SS7 is used between exchanges for call setup and supervision and, in the SS7 signaling network, these exchanges are described as Service Switching Points (SSP). It is also used to enable SSPs to interrogate remote databases known as Service Control Points (SCP). Signals may be routed between either of these over direct connections or transit via STPs.

SS7 Functional Concept
SS7 employs a functional concept embracing processor function, user function, and message transfer function, which allows change to any one function without significant impact on the others. It involves, for example, flexible arrangements in the message transfer function to enable an error control option to be exercised without involving the exchange call processing subsystem. This implies that the error control should be associated with the signaling terminals, the processor never being involved.

This leads to a four-level structure for the elements of SS7 similar to the layers in the general seven-layer OSI model, but with the upper three layers combined into a single level 4.

Message Transfer Part
The Message Transfer Part (MTP) is the core of SS7 and comprises the three lower levels of the protocol.

MTP level 1 is the 64 kbit/s data link.
MTP level 2 inserts signaling units for transmission, each transmitted with a forward and
    backward sequence number and acknowledgment bits. The forward serial number ensures
    message transmission in the correct order. The backward serial number is used to
    acknowledge messages received from the far end.
MTP Level 3 provides message routing between signaling points in the SS7 network. It re-
    routes traffic away from failed links and signaling points and controls traffic when
    congestion occurs.

It allows user parts, which may be held on different processors, to communicate with each other by means of addressed messages. This processor subsystem addressing (type-of-service coding in the heading of a message) directs the message to the appropriate subsystem module, the subsystem addressing identifying the service for which the message is intended. Should corresponding user subsystems communicate with each other for the interchange of information not related to call handling, such messages will be conveyed by the MTP.

User Parts (Level 4)

User parts are separate subsystems or modules within the processor and its software that control the message requirements for particular services (telephony call, data call, network management, network maintenance, centralized call accounting, and so on). Such subsystems may be regarded as being the service-dependent parts of the signaling system. They define message, signaling, and call control procedures, and the function and coding of the signaling information for different services. Each user part deals with a particular service, and the concept has allowed the system to evolve to provide additional services by the addition of appropriate user parts.

ISDN User Part   The ISDN User Part (ISUP) handles signals for the setup, management, and release of circuits that carry voice and data between a calling party and a called party for both ISDN and non-ISDN calls.

Telephone User Part   The Telephone User Part (TUP) handles analogue circuits only and in some parts of the world it is used to support basic call setup and clear-down. In most countries, ISUP has replaced TUP for call management.

Signaling Connection Control Part   The Signaling Connection Control Part (SCCP) provides connectionless and connection-oriented network services and global title translation (GTT) capabilities above MTP level 3. A global title is an address (e.g., a dialed 800 number, calling card number, or mobile subscriber identification number) that is translated by SCCP into a destination point code and subsystem number. SCCP is used as the transport layer for TCAP-based services.

Transaction Capabilities Applications Part   The Transaction Capabilities Applications Part (TCAP) supports the exchange of noncircuit-related data between applications across the SS7 network using the SCCP connectionless service. Queries and responses sent between SSPs and SCPs are carried in TCAP messages. For example, an SSP sends a TCAP query to determine the routing number associated with a dialed 800/888 number and to check the personal identification number (PIN) of a calling card user.

Mobile Application Part   In mobile networks, TCAP carries Mobile Application Part (MAP) messages sent between mobile switches and databases to support user authentication, equipment identification, and roaming.

Signal Routing

The signal messages relating to a given group of speech circuits between two switching centers using a CCS system can be transferred in the following ways:

(a) *Associated mode.* The signal messages are transferred between the two switching centers over a CCS link that terminates at the same switching centers as the group of speech circuits to which the signaling link is assigned.

(b) *Nonassociated mode.* The signal messages are transferred between the two switching centers over two or more CCS links in tandem. Thus, on a different routing from the relevant speech group, the signal messages are processed and forwarded through one or more intermediate STPs. It follows from this that there may be a range of nonassociated modes that vary in the degree of rigidity imposed on the choice of path used by the signal messages. The extremes of this range can be described as the fully dissociated mode and the quasi-associated mode.

(c) *Fully dissociated mode.* The signal messages are transferred between the two switching centers via any available path in the network according to the routing principles and rules of that network. The great flexibility in the routing of signal messages demands a more comprehensive message addressing scheme than is needed for associated signaling. This is the ultimate of the principle of separate paths for signaling, and a completely separate CCS network is theoretically possible.

(d) *Quasi-associated mode.* The signal messages are transferred between two switching centers over two or more CCS links in tandem, but only over predetermined paths and through predetermined STPs. The predetermined routing permits the same method of addressing (circuit labeling) the signal messages as in the associated mode.

The constituent signaling routes of a quasi-associated signaling relationship may be associated signaling in their own right, in which case the constituent routes carry the quasi-associated signaling traffic in addition to the associated signaling traffic. Alternatively, the constituent signaling routes need not be associated signaling in their own right. When the constituent signaling routes (AC and CB) are associated signaling, the circuit labels for the speech group AB must be different from those of groups AC and CB; for the necessary discrimination the circuit labeling of signaling routes AC and CB must be increased to accommodate this. When routes AC and CB in the quasi-associated relationship are not associated, the circuit labeling concerns the speech circuit group AB only in the typical case shown.

Quasi-associated signaling may be adopted when a speech circuit group is too small to justify economic application of associated signaling. The mode may also be used for security backup for associated signaling, and for backup for another quasi-associated signaling relationship, but the circuit labeling tends to be complex in the latter case.

With quasi-associated signaling, the number of signal transfer points in the signaling path for a group of speech circuits between two switching centers should be as few as possible to minimize the signaling time of those circuits and to minimize the total signal processing load of the network. Normally one or two STPs should suffice in a quasi-associated signaling relationship.

Signal Transfer Point

The STP is a signaling center that forwards signal messages received on one CCS link for onward transmission over another CCS link, in quasi-associated or nonassociated signaling. It follows that a signal transfer point need not necessarily have any connection with a switching center. The STP analyzes the circuit label and the priority indication of every received signal message to determine the routing and offers the message to the proper signaling link for the onward transmission of the message. In doing so, it may be necessary to change (translate) the circuit label of the message according to some preset rules.

While SS7 has the potential for the various signaling modes discussed, a combination of associated and quasi-associated signaling meets the demands of most networks. The pre-

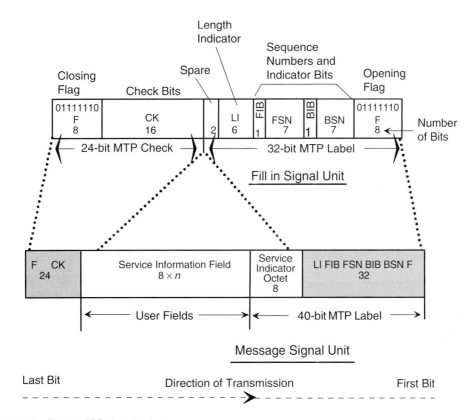

Fig. 5.11 Typical SS7 signal units.

determined signal routings possible with these modes facilitate an orderly and efficient traffic-dimensioned CCS network.

S7 Signal Unit Formatting Principles

The signaling channel sends signaling messages continuously. When there is no traffic the channel carries fill-in signaling units (FISU). A signaling message (Figure 5.11) consists of the following elements: flag (01111110) at the start and end of a message, check bits, user information (message content), service information, message length indicator, forward and backward sequence numbers, and forward and backward indicator bits. The sequence numbers and indication bits ensure correct reception of messages including message retransmission. Thus, under fault conditions messages can still be received in the correct order.

The order in which messages are received is important, for example if a telephone number is split between several messages. A simple example of message transfers to set up a telephony call is given in Figure 5.12.

In SS7, the exchange call processing subsystem is not involved in the error control function. For this reason, the system SS7 signal unit has two parts: (1) that passed to the call processor function, and (2) that concerned with error control (message transfer control— "housekeeping" part) not passed to the processor. The SS7 signal units (and the "user" and "housekeeping" component parts) always consist of an integral number of octets.

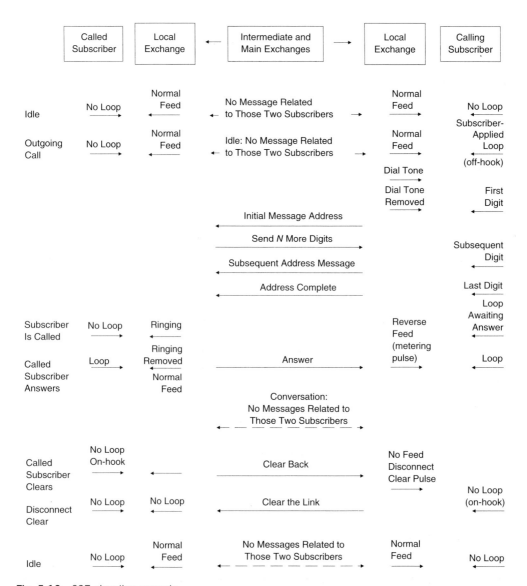

Fig. 5.12 SS7 signaling example.

Following is an explanation of various bit fields in the call processor part of the signal unit.

Flag   As the signal units are variable bit length, the receiving end must be informed by the signal unit as to the start and end of the signal unit. A flag is used to indicate the start. It must be a unique bit code pattern (01111110) and transmission means must be used to ensure that it cannot be imitated in any other part of the signal unit.

Length Indicator   This gives an indication as to the length, and thus the end, of a signal unit. Its coding indicates the number of octets that follow the BSN, BIB, FSN, FIB error control part, up to but excluding the check bit error control part. The length indicator is used to differentiate between the different types of signal unit as follows:

Length indicator = 0: fill-in (idle) signal unit
Length indicator = 1 or 2: CCS link status signal unit
Length indicator > 2: message signal unit

In international use, the length indicator is 62 octets maximum. In national use it may be up to 272 octets. For 63 to 272 octets, the length indicator is set to 63.

Service Information Octet   This is used to associate the signal information of a message with a particular user subsystem of the processor function. It is coded to indicate "international" or "national" and the service (telephony, data, and so on).

Circuit Label   The 40-bit international circuit label is divided into:

- Destination point code (14 bits) indicating the signaling point for which the signal unit is intended.
- Originating point code (14 bits) indicating the signaling point that is the immediate source of the signal unit.
- Circuit identification code (12 bits) indicating one speech circuit among those directly interconnecting the originating and destination points. This bit field may be further subdivided in the circuit number/band number philosophy to give a speech circuit time slot in a PCM system and the identity of the particular system, or an f.d.m. channel in a particular f.d.m. transmission system.

This international 40-bit circuit label may be reduced and varied in national network application of SS7 at administration's choice.

Heading, H0/H1 Octet   All telephone message signal units contain a heading octet of two parts: H0 (4 bits) and H1 (4 bits). The H0 coding identifies message groups, for example:

(a) forward address messages
(b) call supervision messages
(c) circuit supervision messages
(d) backward setup information messages

The H1 coding identifies a particular message in the relevant HO group—for example, "initial address message" in a forward address message (a), or "clear forward" in a call supervisory message (b).

This H0/H1 arrangement, giving potential for 16 H1 specific signals in each group, is more than adequate to cater to signal units presently defined. There is ample spare capacity for additional international and national message signals not presently defined.

SS7 Error Detection

SS7 adopts variable bit length messages, with a maximum message, for international operation, of 528 bits (66 octets). In national use, the message may be up to 2,176 bits (272 octets). In both cases, 16 check bits are employed. The chosen polynomial is:

$$x^{16} + x^{12} + x^5 + 1$$

If the initial state of the receive shift register is set at all zeros, and there are no transmission errors, the contents of the shift register will pass through a number of combinations dependent on the received data and will finally return to its original all-zero state. However, it is possible that the final all-zero condition could be due to other causes (e.g., the shift register itself may be faulty). To guard against such malfunctions, in SS7 the initial condition of the shift registers is set at the all-one state (instead of all-zero) and also the remainder in the transmit shift register is sent as the one's complement (i.e., all bits are inverted before transmission).

This inverted bit stream is applied to the receive shift register. Therefore, when there are no transmission errors, the final state of the receive shift register is a definite bit pattern, irrespective of the data, and any other pattern denotes a transmission error.

SS7 Error Control

For convenience, the variable bit length signals of the SS7 system will be referred to as "signal unit" in the following. It will be understood that a signal unit (SU) may comprise any type of signal and is not a fixed length standard size SU.

The main features are:

- All SUs are treated in their own right (as distinct from being assembled in blocks).
- Cyclic retransmission on error detected: When an error is detected, that message SU and all message SUs following it in the retransmission store are retransmitted. Nonmessage SUs (idle, CCS link status) are not deposited in the retransmission store and thus not retransmitted on error detected.
- All types of SU carry both a forward sequence number (FSN) and a backward sequence number (BSN).
- Nonmessage SUs transmitted between message SUs all carry the same FSN as the last previously transmitted message SU.
- All types of SUs carry a forward indicator bit (FIB) and a backward indicator bit (BIB) to control retransmission.

Following is a list of the functions of these numbers and indicator bits.

Forward Sequence Number   This is the sequence number (in cyclic order) of the SU in which it is transmitted. When a message SU is transmitted, the Forward Sequence Number (FSN) in that SU is increased by one (cyclically) over that of the previous message SU sent. The FSN enables the correct message SU sequence to be preserved during the message transfer process, and identifies message SUs when retransmission is required.

A terminal will accept only message SUs in correct sequence (discarding others) to ensure that message SUs passed by the terminal to the processor are always in correct sequential order and never duplicated.

Backward Sequence Number   The Backward Sequence Number (BSN) transmitted in an SU (of any type) transmitted from a terminal indicates the FSN of the message SU in the other direction that the terminal has just accepted. The returned BSNs (carried in all types of SUs) are not changed until that terminal accepts the next error-free and correct sequence FSN message SU in the other direction.

Backward Indicator Bit   When accompanied by an unchanged polarity of the Backward Indicator Bit (BIB), the BSN is the positive acknowledgment to clear the corresponding FSN message SU from the retransmission store.

When accompanied by a retransmission request (change in polarity of BIB—the negative acknowledgment), the BSN indicates the point in the FSN sequence (in the other direction) at which the retransmission should start.

The BIB is used by a receive terminal to signal back to the transmit terminal that a retransmission is required, the request being made by reversing the polarity of the BIBs transmitted back to the transmit terminal. Once a terminal has requested retransmission, no further reversal of BIB polarity is made until a new retransmission is required.

No action is taken by a receiving terminal on an error detected. The system waits until an error-free SU (of any type) is received before making a decision as to the action to be taken. BIB polarity is reversed to request retransmission when the FSN of this error-free SU is not greater (cyclically) than that of the last error-free (and accepted) message SU. After requesting retransmission, the terminal takes no further action until a change in polarity of received FIBs indicates the retransmission.

Forward Indicator Bit   The polarity of the Forward Indicator Bit (FIB) in SUs (of any type) transmitted from a terminal is the same as that of the BIB contained in the last SU (of any type) correctly received by that terminal. Thus, on a sequence of error-free SUs, the polarity of the BIBs received at a terminal will be the same as that of the FIBs transmitted.

Since a change in polarity of the BIB is used to request retransmission, a change in polarity of the FIB indicates that the retransmission is taking place. The combination of FIB and BIB is a "handshaking" process.

The polarity of the FIB is reversed when retransmission has started. After reversal, the polarity remains unchanged until the next retransmission has started.

Resources

Atkinson, J. (1950) *Telephony*, Pitman.
Bear, D. (1988) *Principles of Telecommunication Traffic Engineering*, Peter Peregrinus (IEE).
Briley, B. E. (1983) *Telephone Switching*, Addison-Wesley.
Hills, M. T., and S. Kano (1976) *Programming Electronic Switching Systems*, Peter Peregrinus (IEE).
ITU-T *Recommendations Q.700–Q.799 Signaling System No 7*, ITU.
ITU-T *Recommendation V.23—Data Transmission* (as used for CLASS/CLI), ITU.
Manterfield, R. J. (1999) *Telecommunications Signaling*, IEE.
Redmill, F. J., and A. R. Valdar (1990) *SPC Digital Telephone Exchanges*, Peter Peregrinus (IEE).
Smith, S. F. (1978) *Telephony and Telegraphy*, *Third Edition*, Oxford University Press.
Takamura, S., et al. (1979) *Software Design for Electronic Switching Systems*, Peter Peregrinus (IEE).

Part Three

The Internet and Data Networks

With the advent of the Voice over Internet Protocol (VoIP) and related protocols, data networks are now able successfully to carry real-time voice, and even video, services and provide a "disruptive" alternative to circuit-switched technologies. Second, although the Internet itself has very wide bandwidth, access to the Internet has often been via narrowband public telephone networks. Once Internet services via narrowband access have been generally accepted in a country, the trend has then been to migrate to broadband technologies, so that a greater range of services can be delivered with higher speed.

In this part we discuss the basic operation of the Internet and of Local Area Networks (LANs), and we describe Internet telephony and the principles behind Next Generation Networks, which are expected to eventually replace traditional circuit-switched networks. Broadband access technologies will be covered in a later volume of *The Cable and Telecommunications Professionals' Reference*.

The existence of public telephone networks and their ability to support a range of data services made it possible for the Internet to undergo rapid growth through the use of dial-up connections. The architecture of the public networks was even modified (see Part 2) to make connection to the Internet more efficient. However, the demand for bandwidth-hungry services led to the development of broadband, which, together with VoIP technology, is now positioned to provide a completely new infrastructure for voice with even greater versatility than the public telephone networks.

VoIP is now seen by many as a more versatile and economic service than telephony and is positioned to take over the role of the fixed telephone network. This will cause further major change in the industry, as telephone exchanges are replaced with giant router systems and the myriad of special services supported by the telephony network are replaced by "soft" equivalents. Already, some major telephone network operators are changing or planning to change over to the new technology and switching off their current circuit-switched networks. This change will be at least as great as other major developments that have taken place in the industry, such as the switch-over to digital working or the introduction of subscriber trunk dialing.

In this part, we provide an introduction to the Internet and the protocols that it relies on (Chapter 6), the special requirements to enable real-time voice to be carried over the Internet (Chapter 7), and the way in which the migration to Next Generation Networks is being managed (Chapter 8). Chapter 9 describes local area networks, which provide the customer premises network that completes the user connection.

6 The Internet and TCP/IP

David K. Hunter

Introduction

Networking Concepts and History of the Internet

Although the Internet was not exploited commercially until the mid-1990s, it has its origins in work by the U.S. Defense Advanced Research Projects Agency (DARPA), dating back to 1973, to develop a nonproprietary networking standard. Indeed, this work drew on pioneering work in the 1960s on packet switching. The first publication about the Internet was in 1974, and it was deployed widely by 1983. At first, the Internet was only used by the academic and military communities, but much has taken place in the last 20 years, both through exploiting new technologies and through broadening its user base. The shipping of Netscape Communicator in 1994 was an important landmark, making the World Wide Web as we now know it available to the public. Since then, the number of users and the amount of traffic on the Internet has grown at a consistently rapid pace.

This chapter describes the Internet and the suite of protocols—TCP/IP or Transmission Control Protocol/Internet Protocol—that have their origins in the DARPA work of 1973 and that make all this possible. Figure 6.1 depicts a generic data network, where process A knows that process B runs on another computer and the programmer must be aware of how the computation is split over several machines when writing the application program. This differs from a distributed computer system, where each computer is "deceived" by the operating system into thinking that all processes run on the same computer. In other words, the distributed system behaves like a single computer—the physical network topology is hidden from the applications.

The two principal devices in networks are routers and hosts. Their interaction will be discussed in detail below. A router is a specially optimized computer that is part of the network routing mechanism, which ensures that packets are forwarded in the correct way to reach the destination, whereas a host is a computer running a user application program.

Fig. 6.1 A generic data network, providing a way of getting information from one network device to another. Processes A and B are not aware of the details of network operation, but are aware that they are running on separate computers.

Internet Design Philosophy

In the early 1970s, when the Internet protocols were designed, there were a number of proprietary computer networking technologies in existence. The DARPA team wished to design a nonproprietary networking technique, for interconnection of existing networks. Their key objectives are summarized as follows:

- Use store-and-forward packet switching, since this was a natural choice for envisaged applications and the networks that were to be interconnected used packet switching.
- Continue communication even if routers or networks go down.
- Support multiple types of communications services.
- Integrate a number of separately administered networks of different types into a common entity.
- Use distributed management.

While these objectives remain relevant today, the Internet has evolved greatly since this original design phase. Moreover, it is now being used for voice and video transmission, which were not envisaged in the original data-centric concept.

Originally, the Internet was used by a relatively small group of academic users, and the provisions for addressing seemed more than ample. As the use of the Internet has increased over the years, providing addresses for all users has become an issue, and the addressing schemes described in this chapter are designed to prevent address wastage. Many envisage that a new version of IP, IPv6, will be introduced, largely for this reason.

Figure 6.2 illustrates some of the technologies deployed in the present-day Internet. The terminology, protocols, and techniques used will be explained in depth in the remainder of this chapter. The network in the diagram is composed of a variety of technologies; for example, trans-Atlantic cables, wireless networks, local area networks such as Ethernet, and connections via the telephone system via both 56 K modems and Asymmetric Digital Subscriber Line (ADSL). TCP/IP unifies these diverse technologies, making them all appear to be one network to the applications programmer. Although they will be discussed later, two further features of the diagram are relevant here:

- A Point of Presence (POP) is owned by an Internet service provider, so that subscribers can dial into it to access the Internet. Service providers usually have several POPs in many different geographic locations, so that subscribers can dial into the nearest one.
- Network Address Translation (NAT) allows each host on its local network to have its own IP address, while only requiring one address in total from the viewpoint of the external network. This is a way of easing the addressing problem described above, although IPv6 is a more satisfactory and permanent solution.

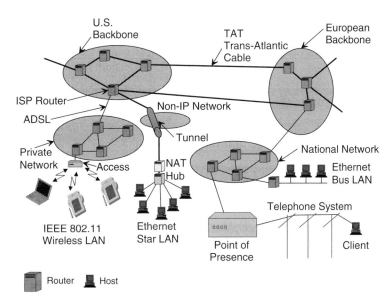

Fig. 6.2 An illustration of the wide variety of technologies that co-exist on the Internet. The TCP/IP protocol suite (layers 3–5) provides a unifying framework for the variety of networking technologies in layers 1 and 2, making them appear as one network to the application programs.

Protocol Layering and the Protocol Stack

Layering of network protocol functions is a crucial design strategy that is adopted in almost all communications networks. TCP/IP consists of very many protocols, with each protocol residing in one of five layers, facilitating protocol design, implementation, and maintenance. Each layer provides a service to the layer immediately above, while adjacent layers communicate via an interface that hides layer implementation detail; details of lower layers are hidden from higher layers.

This simplifies implementation and maintenance of the protocols, although it can cause problems—for example with TCP over wireless, where packets are often lost due to a noisy channel, but TCP interprets congestion as a sign of network congestion. The 5-layer model used in TCP/IP (Figure 6.3) is similar to the 7-layer Open System Interconnection (OSI) model, but layers 5 and 6 are missing. The five remaining layers are:

1. *Physical layer:* This interfaces to the physical transmission medium.
2. *Data-link layer:* This transfers information reliably between two nodes that are connected by a link.
3. *Network layer:* This transfers information on a best-effort basis from source to destination, perhaps over many links.
4. *Transport layer:* This transfers information reliably from source to destination, ensuring packet ordering.
5. *Application layer:* This interfaces to application programs.

A set of cooperating protocols that are interrelated in this way is called a protocol stack. The hosts at either end of the communications path implement all five layers in the protocol

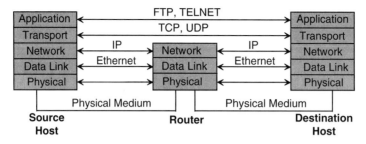

Fig. 6.3 The TCP/IP protocol stack, consisting of five layers—physical, data link, network, transport, and application.

Fig. 6.4 Two network devices communicating over a physical link, as defined by the physical layer.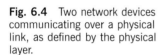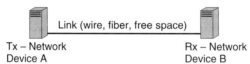

stack, whereas intermediate routers need only implement the lower three layers—physical, data link, and network. Each layer in the stack is described below.

Physical Layer—Layer 1

The physical layer deals with transmitting raw "zero" and "one" bits over a communications channel. It specifies signaling voltages, the length of bits in nanoseconds, connector pinouts, whether operation is unidirectional or bidirectional, and so on. However, the physical layer is *not* the physical transmission medium itself; this lies below the physical layer.

Information is carried from A to B via a transmission medium (Figure 6.4); in this example, it is encoded in physical form at network device A and decoded at network device B, being carried through voltage, current, infrared light, or radio waves. A link may be point-to-point or broadcast, and in both cases a device is connected to a link by a hardware network interface.

The Bit Error Rate (BER) is the ratio of the number of bits received in error (a "one" mistaken for a "zero" or vice versa) divided by the total number of bits sent. Typical BERs for wireless are better than 10^{-2}, for coaxial cable better than 10^{-6}, and for optical fiber better than 10^{-14}. The propagation speed for wireless transmission, and in free space, is approximately $3 \times 10^8 \, \text{ms}^{-1}$; for all other media it is approximately $2 \times 10^8 \, \text{ms}^{-1}$. This is important when considering the operation of higher-layer protocols, especially TCP, because the delay this introduces affects the protocol's efficiency.

Data-Link Layer—Layer 2

The data-link layer provides reliable transfer of information across a physical link, where a link is a direct physical connection between two network devices. (A connection between two users that passes through many intermediate nodes would involve many links.) It sends blocks of data (frames) with the necessary functionality:

- Synchronization, to ensure that the receiver can align successfully to bit and byte boundaries.
- Error control, to ensure that any lost packets due to errors are retransmitted.

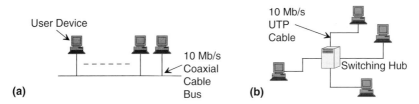

Fig. 6.5 Two common configurations of Ethernet networks: (a) a broadcast bus using CSMA/CD, and (b) a star, which uses a switched hub.

It uses the services of the physical layer to get bits over the line:

- The layer 2-to-layer 1 interface converts "zero" and "one" bits to a physical quantity on the link.
- The layer 1-to-layer 2 interface converts a physical quantity on the link into "zero" and "one" bits.

A Medium Access Control (MAC) protocol is necessary for broadcast links, to decide when each network device may access the medium.

TCP/IP does not specify layers 1 and 2, but Ethernet is commonly used, since the IEEE 802.3 Ethernet standard defines both layers. Other options instead of Ethernet include Asynchronous Transfer Mode (ATM) over Synchronous Digital Hierarchy (SDH), token ring, and Fiber Distributed Data Interface (FDDI). Common 10 Mb/s Ethernet topology configurations are shown in Figure 6.5. The bus network is passive and uses a MAC protocol called Carrier Sense Multiple Access with Collision Detection (CSMA/CD) to provide all hosts with fair access to the network. The switched Ethernet is connected in a star topology, and frames are switched to the correct host by the central switching hub; there is no broadcast shared medium. Ethernet standards exist for 10 Mb/s, 100 Mb/s, 1 Gb/s and 10 Gb/s data rates.

Network Layer—Layer 3
The network layer uses the Internet Protocol (IP) to route packets from source to destination efficiently through a network, or patchwork of networks, each using a different link layer technology. It is concerned with getting packets from the source all the way to the destination, and in order to do this, an efficient global network-interface addressing scheme is required that must be independent of layer 2 addressing, since layer 2 addressing is technology dependent.

The network layer also deals with addressing and routing, and, in the OSI model, also implements congestion control. Packets in the network layer are called datagrams in IP, an important network layer protocol. A datagram size of 1500 bytes is generally chosen when discussing Internet operation because this is the maximum permitted by the Ethernet frame size or Maximum Transmission Unit (MTU), and Ethernet networks often carry IP traffic.

Transport Layer—Layer 4
The transport layer uses the Transmission Control Protocol (TCP) to ensure that a message from a source process arrives without errors or omissions at the destination process, and also ensures that packets arrive in order. TCP also implements congestion control. In layer 2, Ethernet does not correct bit errors in the frame, but it can, however, detect them. Such errors are usually resolved in layer 4 via retransmission of erroneous packets.

Application Layer—Layer 5

Application layer protocols form an interface between the network and the application software, so they are not the application software itself. Examples of application layer protocols include:

- *File Transfer Protocol (FTP):* This is used to transfer files between a host and a server.
- *Telnet:* This allows the user to log in and use a remote machine with a text interface, Simple Mail Transfer Protocol (SMTP), which transfers e-mails between mail servers, and from a host to its mail server.
- *Post Office Protocol version 3 (POP3):* This is used by a host to read e-mails on its mail server.
- *Simple Network Management Protocol (SNMP):* This facilitates the exchange of management information between network devices, enabling network administrators to manage network performance, find and solve network problems, and plan for network growth.
- *Hypertext Transfer Protocol (HTTP):* This is the standard Web transfer protocol, used for transferring Web pages written in HTML from the server to the host that requests them.
- *Domain Name System (DNS):* This translates domain names such as *www.smithsuni. ac.uk* into IP addresses such as 110.22.67.238. This avoids the user having to think in terms of IP addresses.

Applications often have a client/server architecture, where the client application, which usually runs on a host, makes a request of the server application. The server then fulfills the request—this is a convenient way to interconnect programs distributed across different locations.

Conclusions on Protocol Layering

The protocol stack simplifies programming because, without it, programmers would need to know about:

- Creating bit-streams directly, for transmission
- Communications hardware
- Writing drivers for the hardware
- Detecting errors in byte streams and correcting them
- Routing the byte stream through the network

While a simple data link may be more efficient than a network in certain specialized applications where transmission efficiency is paramount, the additional trouble and complexity involved would be justified in only a few very special cases.

Layer Encapsulation and IP-in-IP

Because TCP/IP employs a layered protocol architecture, the interfaces between layers are of crucial importance (Figure 6.6). Between layers, a process known as *encapsulation* takes a packet from a higher layer and inserts it into the data field of the layer below, while at the destination, the data field is removed and presented to the layer above. Thus a hierarchy of packet headers is in fact transmitted by the physical layer.

Each packet header contains a field that states from which protocol in the layer above the payload it came and consequently where it should be sent at the destination. In an Ethernet packet (known as a *frame*—layer 2), this field is called the *Ethertype field*. In an IP packet (known as a *datagram*—layer 3), this is known as the *protocol field*, and in a TCP segment or UDP datagram (layer 4), it is called a *port*. The port identifies the application protocol concerned, and the combination of an IP address and a port is called a *socket*, an important concept in TCP/IP programming.

Figure 6.6 depicts three application-layer protocols:

- With *echo*, the server returns what the client has sent to it; this is used when implementing the Linux "ping" command.
- *Network Time Protocol* (NTP) synchronizes the router clocks.
- *Domain Name System* (DNS) translates domain names such as *www.techuni.ac.uk* into 32-bit IP addresses.

The data field in the data-link layer is limited in size by the protocol concerned, limiting the size of the IP datagram that may be placed within it. This size limit is called the Maximum Transmission Unit (MTU), and some common values are shown in Table 6.1. The MTU refers

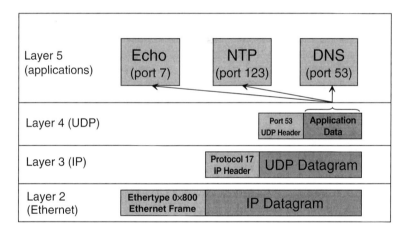

Fig. 6.6   An illustration of interlayer encapsulation. The data field in each layer holds a packet or block of data from the layer above.

Table 6.1   Default MTU Sizes for Common Link Layer Technologies

Link layer protocol	Default MTU
Point to Point Protocol (PPP)	296
X25	576
Serial Line Internet Protocol (SLIP)	1066
Ethernet	1500
Fiber Distributed Data Interface	4352
4 Mb/s token ring	4464
Token bus	8166
16 Mb/s token ring	17,914
Hyperchannel	65,535

to one link; however, the path MTU refers to an entire path, which may consist of many links, and is the minimum value of all the MTUs along the path. Hence the path MTU is the maximum size of IP datagram that can be sent over a given path.

Besides encapsulating transport layer packets, IP can encapsulate existing IP datagrams (a form of intralayer encapsulation); hence a new IP datagram is created with the existing IP datagram encapsulated within its data field. Mobile IP is an important application of this, where a mobile host is associated with both its original primary address and the secondary address of its new physical location. Datagrams that arrive destined for the primary address are intercepted by an agent, which encapsulates them via IP-in-IP, before sending them to the secondary address.

This type of encapsulation can also implement tunnel mode in IPsec (IP security), protecting traffic that must pass through an intermediate, untrusted network. The path from the point of IP-in-IP encapsulation and the corresponding point of decapsulation is called a *tunnel*, and generally exists between two firewalls in different organizations. It is called a "tunnel" because existing datagrams are placed inside new ones with different addressing information. Hence hosts on each company's LAN need not be aware of the security mechanism.

Tunnel mode may aggregate a bundle of TCP connections, which are handled as one encrypted stream. This prevents an intruder seeing who is sending how many packets to whom, since this may well be information that the senders do not wish to share. Applications of IPsec include secure Virtual Private Networks (VPNs), secure access by users to Internet Service Providers (ISPs), and electronic commerce.

IP and the Network Layer

The preceding examination of protocol layering is a necessary prerequisite for the study of Internet Protocol (IP), arguably the most important single protocol in TCP/IP. Implementation of IP addressing is an important concern, along with the addressing schemes that have been proposed since the Internet was first devised.

IPv4 Addressing

In IPv4 addressing, each host and router has a unique Internet-wide IP address that identifies the device, and the network that it is on, which could be, for example, an Ethernet. This is 32 bits long in IPv4 (2^{32} addresses), or 128 bits long in IPv6. No two devices on the Internet have the same IP address (except when NAT or Network Address Translation is involved—to be discussed later). To be exactly correct, the IP address refers to a network interface on a device, and not the device itself. For example, routers have more than one network interface, each on a different network.

Network addressing has evolved over the years, principally due to the increasing scarcity of IP addresses. It affects the efficiency and ease of implementation of Internet routing, since it is used by routing protocols when computing optimal paths through the network. It also influences how datagrams are forwarded within routers, influencing routing table organization and size. This affects the efficiency of the Internet's operation because large router tables cannot be searched quickly, slowing down network operation.

Due to expansion and increasing use, Internet addressing has developed since its first deployment both to facilitate fast and efficient forwarding of data in routers and to avoid wasting addresses. It will be explained later that if a group of physically adjacent hosts have consecutive IP addresses, then these addresses can often be summarized in just one entry in a router's table, facilitating fast searching by reducing the number of table entries. The Internet was originally developed for a relatively small group of academic users, and at that time, a 32-bit address space seemed ample. Since then, Internet use has increased dramatically, resulting in a series of addressing techniques offering increasingly efficient address usage. These are, in order of implementation:

- *Flat addressing—32-bit address:* Addresses are allocated arbitrarily, with no order or structure. Hence it is very difficult to combine multiple addresses into one routing table entry, since they are rarely consecutive.
- *Class-based addressing—classes A, B, and C:* There are a predefined number of possible customer networks in each of these three classes. In each class, a network has a predefined number of addresses assigned to it. Unfortunately, these divisions of addresses are rigid, resulting in inefficiency and address wastage.
- *Subnet addressing:* The networks from class-based addressing are subdivided into subnets. This provides a network administrator with some flexibility in dividing up the addresses allocated to a particular network. The number of subnets in each network is defined by the network administrator.
- *Classless Inter-Domain Routing (CIDR):* This is based on many of the ideas of subnet addressing, and offers the greatest efficiency in address usage. Each user network is assigned a variable-sized group of addresses, greatly reducing unnecessary wastage of addresses. This address allocation method can often allocate large groups of addresses into one routing table entry, yielding more efficient routing table organization.

IPv6 overcomes the problem of address exhaustion by allocating 128 bits for each IP address. Although IPv6 was proposed more than a decade ago, it has yet to be adopted. One reason is that Network Address Translation has delayed the need for it, since NAT is a quick fix to avoid running out of IP addresses.

Routing Table Organization

Routers contain *routing tables*, which decide the next hop for incoming IP datagrams. The router matches a datagram's IP address with a table entry, and then looks up the corresponding next hop. Both the address of the next hop router and the number of the corresponding router port are stored. Each entry in the table contains a *prefix,* which defines a block of destination addresses.

A prefix consists of a "base address" and a "mask," where the base address is indicated by "one" in the corresponding bit of the mask. The concept is described in more detail when discussing CIDR. A table entry may also define a single IP address, but this is extremely rare—the rationale behind IP routing tables is to minimize the number of entries via prefixes, making table searching faster and more efficient. When several matching prefixes exist for an IP address, the one with the longest mask is chosen. If a matching prefix cannot be found, the datagram is sent to the interface indicated by the *default* entry in the table.

Class-based IPv4 Addressing

In class-based addressing (or "classful addressing"—Figure 6.7), each IP address lies in one of four classes—A, B, C, or D. In classes A, B, and C, two fields define the network ID and the host ID (Table 6.2). Class D addresses begin 1110, and are used for multicasting, while there is also an experimental class E starting 11110.

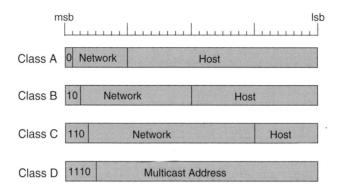

Fig. 6.7 Class definitions for class-based (or "classful") addressing. Classes A, B, and C are for point to point communication, whereas class D is used for multicasting. Classes A to C define fixed fields for identifying both the network and the host within it.

Table 6.2 Summary of Internet Address Classes A, B, and C

Class	Most significant bits	Network ID bits	Host ID bits	Maximum networks	Maximum hosts
A	0	7	24	128	16,777,214
B	10	14	16	16,384	65,534
C	110	21	8	4,194,304	254

Certain addresses are reserved for special purposes:

- 0.0.0.0 = "this host, this network"—the source address for RARP packets, discussed later
- 255.255.255.255 = "all hosts on the local network"
- 0.X.Y.Z = "host X.Y.Z, this network," class A
- 0.0.Y.Z = "host Y.Z, this network," class B
- 0.0.0.Z = "host Z, this network," class C
  - Hence a host need not know the network address of its LAN, but it must know the address class.
- X.255.255.255 = "all hosts on network X," class A
- X.Y.255.255 = "all hosts on network X.Y," class B
- X.Y.Z.255 = "all hosts on network X.Y.Z," class C
- 127.0.0.1 = "loopback address"—used for debugging and testing

Subnet Addressing

Subnet addressing (Figure 6.8) overcomes the inflexibility of classful addressing by permitting the division of a class A, B, or C address space efficiently among the individual networks (or subnetworks) within an organization. With subnet addressing, subnet masks of contiguous "one" bits followed by contiguous "zero" bits define subnetworks within a network. Without subnets, one class B address space (for example) defines one network with some 65,000 hosts, which is too big for one physical network. Here, this address space is divided among physical networks, yielding more efficient use of IP addresses.

The mask used in a router table entry's prefix depends on whether it is inside or outside the network. Inside the network, the mask contains 1 for network bits, 1 for subnet bits, and

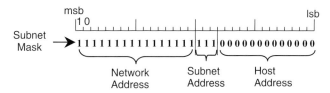

Fig. 6.8 Division of a 32-bit address into fields for the network address, subnet address, and host address. In this case, a class B address space is used, so the address begins with "10" binary.

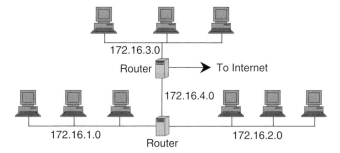

Fig. 6.9 An example of subnet addressing: the class B address space is divided into four subnets—172.16.1.0/24, 172.16.2.0/24, 172.16.3.0/24, and 172.16.4.0/24.

0 for host bits, so that the routers take account of the network address and subnet address. Outside the network, the mask contains 1 for network bits, 0 for subnet bits, and 0 for host bits, so that the routers take account of the network address, but not the subnet address.

Example of Subnet Addressing   Figure 6.9 shows network 172.16.0.0 (class B) that has been divided into four subnetworks—172.16.1.0, 172.16.2.0, 172.16.3.0, and 172.16.4.0. Inside the network, the routing table entries contain the following mask:

$$11111111\ 11111111\ 11111111\ 00000000$$

However, outside the network, the routers do not need to be aware of how the network is split up into subnetworks, so the mask there is:

$$11111111\ 11111111\ 00000000\ 00000000$$

Traffic is not broadcast over all subnetworks; each subnet only carries traffic that is necessary.

Example—Subnetting of Branch Offices   An organization has a head office with fewer than 254 branch offices (Figure 6.10), and each office has fewer than 254 hosts. The organization chooses a "hub and spokes" topology, with the head office at the center and one branch at the end of each spoke. The organization obtains the class B network addresses starting from 140.20.0.0.

In dotted decimal notation, a prefix is written as $a.b.c.d/m$, with a, b, c, and d defining the base IP address in dotted decimal as usual. The mask consists of m "one" bits followed by $32-m$ "zero" bits. The number of addresses with a mask of m bits is 2^{32-m}, where $32-m$ is the

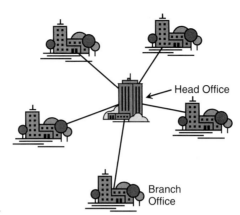

Fig. 6.10   A star network of offices, for illustrating subnetting.

Branch Office

Head Office

number of "zero" bits in the mask. In this example, subnet addresses could be allocated as follows:

- Head office is assigned the prefix 140.20.1.0/24, so the mask is 11111111 11111111 11111111 00000000, that is, 24 one bits followed by 32–24 = 8 zero bits
- Branch office 1 is assigned the prefix 140.20.2.0/24
- Branch office 2 is assigned the prefix 140.20.3.0/24 and so on

The router at the head office only needs an entry for each branch, not an entry for each host in each branch.

Forbidden Addresses

Assigning a class C network address X.Y.Z.0 to a particular host P is not acceptable, since it may result in ambiguity. To see why, suppose that another host Q, on the same network, does not know the network id. B tries sending a datagram to P, using destination address 0.0.0.0, which would be confused with "this host, this network." Also, X.Y.Z.255 would be confused with a broadcast to the network. Hence, a class C network can only support 254 hosts, not 256 as expected. Likewise, a class B network can only support 65,536 − 2 = 65,534 addresses, and a class A network can only support 16,777,216 − 2 = 16,777,214 addresses.

Assume that broadcasting over all subnets is implemented. Then having all "one" bits in the subnet id field and all "one" bits in the host id field will initiate a broadcast to all subnets. For this reason, a subnet id consisting entirely of "one" bits is forbidden because a broadcast in this subnet would be confused with a broadcast to the whole network.

Two Problems with Class-based Addressing

In classful addressing, much of the address space is wasted, which is particularly serious because, due to the growth in Internet use, addresses have become scarce. To see this, consider the three address classes:

- Class A networks are too large, with up to 16,777,214 hosts, so many host addresses are unused.
- Class B networks are just the right size, hence largely used up, with up to 65,534 hosts.
- Class C networks are too small, so not so widely used, with up to 254 hosts.

Also, with classful addressing, addresses in the same geographical region or organization tend not to be contiguous. This makes address summarization (aggregation) impossible in many cases, resulting in inefficient routing, because the routing tables become large and inefficient.

Classless Interdomain Routing

Classless Interdomain Routing (CIDR) overcomes the problems of class-based addressing. The remaining class C address space is allocated geographically, for example:

- 194.0.0.0–195.255.255.255 represents Europe
- 198.0.0.0–199.255.255.255 represents North America
- 200.0.0.0–201.255.255.255 represents Central and South America
- 202.0.0.0–203.255.255.255 represents Asia and the Pacific

Hence, an entire continent has just one routing table entry. The address space is allocated hierarchically as blocks of addresses, and *not* as class C networks. First, a prefix (a base network address plus a mask) is allocated to an ISP, which can allocate a longer prefix in its address space to each customer. Prefixes are stored as base address and mask pairs in routing tables, for example:

204.71.0.0/16 (16 msb prefix of ISPA)
204.71.1.0/24 (prefix of a customer of ISPA—1/256th of ISPA's address space)

A customer's prefix or an ISP's prefix can denote addresses covering more than one old class C address space.

In the past, an organization applied to the Network Information Center (NIC) for a network address space of the required class. Then the organization obtained Internet services from an Internet Service Provider (ISP). The NIC was responsible for Internet network addressing, so networks served by a given ISP often had unrelated addresses, and routing tables had an entry for each of the ISP's networks. It is desirable to have just one entry for each ISP, which can then use CIDR to divide up its available address space.

Organizations are now assigned blocks of IP addresses by Internet Service Providers, and in turn ISPs are allocated blocks of IP addresses by the appropriate Internet registry. Addresses are now managed by Internet Corporation for Assigned Names and Numbers (ICANN).

Example: Routing Table Size and Allocating Prefixes   Suppose all class-based network addresses were allocated randomly, and that the number of class-based networks was approximately 2^{21}. This happens to be the number of class C networks and would also be the number of rows in the routing table. Searching such a large table, and hence forwarding, would be slow. Also, routers must exchange these large tables to reflect network changes, using up network bandwidth that could otherwise be used by customer traffic.

Using CIDR, an ISP called ISPA is allocated the prefix 203.85.0.0/16, having $2^{16} = 65,536$ possible customer host addresses. ISPA is based in Asia and the Pacific; it is within the range 202.0.0.0 to 203.255.255.255. Five of its customers need 64, 64, 128, 1024, and 4096 host addresses respectively.

For 64 addresses, the least significant 6 bits of the address (to give addresses 0 to 63) are required so the prefix mask has $32 - 6 = 26$ bits. The first 64 addresses of ISPA's address space are allocated with prefix 203.85.0.0/26 so that the addresses range from 203.85.0.0 to 203.85.0.63.

The next 64 addresses are allocated with the prefix 203.85.0.64/26, where the last masked bit is a "one," rather than a "zero," as for the first allocation. This corresponds to the range 203.85.0.64 to 203.85.0.127.

The 128 addresses need 7 bits so the mask has $32 - 7 = 25$ bits, hence 203.85.0.128/25 is allocated. So far, the equivalent of one old class C network has been allocated (i.e., 203.85.0.0/24)—call this block 0.

Allocation of 1024 addresses requires 4 old class C networks, which needs 10 host bits. The mask is 22 bits and the allocation is 203.85.4.0/22 (blocks 4 to 7). The 22nd bit is a "one" to distinguish the addresses from those already allocated, for which this bit was a "zero." It is impossible to allocate addresses 203.85.1.0 to 203.85.4.255 instead, because they cannot be uniquely distinguished from other customers' addresses by one base address and one mask.

The 4096 addresses require 16 old class C blocks and 12 bits, that is, a prefix mask of $32 - 12 = 20$ bits. Blocks 16 to 31 are used, so the network prefix is 203.85.16.0/20.

Allocated also are $1 + 4 + 16 = 21$ class C networks. With random allocation, each would have had its own routing table entry, but they are now all summarized by the prefix 203.85.0.0/16, the allocation for ISPA. So the number of routing table entries decreases by 20. The fraction of ISPA's address space allocated is 21/256, about 8.2 percent.

IPv6: A New Version of the Internet Protocol

So far, ways of using IPv4's address space more efficiently—notably CIDR—have been discussed. IPv4 is the current version of IP, having a 32-bit address space, supporting up to 4.3 billion different IP addresses. While this was more than enough in the Internet's early days, when it was used by a small community of academic users, its use is now so widespread that IP addresses have become very scarce. One quick-fix solution involves the use of Network Address Translation (NAT), where a customer's network has only one IP address when accessed from outside. Although NAT is presently widely used, a more satisfactory and permanent solution is offered by IPv6, which has an address space of 128 bits ($\approx 10^{38}$ IP addresses). Progress in developing and deploying IPv6 has been slow but is accelerating.

Not only will the IPv4 address space become insufficient; its real-time delivery of audio and video is not fully satisfactory. Also, more complex addressing and routing facilities, to support collaboration between users, will be required by future applications. IPv6 retains many features of IPv4, but uses extension headers much more widely than IPv4's options, which are a similar concept. The header format has been thoroughly revised in IPv6, and along with having better support for audio and video, the protocol can easily be extended.

Instead of the dotted decimal notation used in IPv4, addresses in IPv6 are written in a type of hexadecimal notation, which is more compact. There are three main types of address. Unicast addresses send datagrams to single computers, whereas multicast addresses send them to a group of computers, which may be geographically dispersed. Finally, anycast is a new type of addressing in IPv6 that sends a datagram to the nearest computer residing at a particular location.

Domain Name System

The discussion so far has involved numeric IP addresses, which most users find uninformative and unappealing. For this reason, applications such as Web browsing and e-mail usually use domain names, for example, *www.charityappeal.org*. The top-level domain name is at the right of the domain name ("org" in this case), preceded by the second-level subdomain ("charityappeal"). The top-level domain names, and the subdomain names, are separated by dots. Application layer services resolve or translate domain names into IP addresses by making a call to a library "resolve" function. The domain name is passed as a parameter, and this function accesses the local Domain Name System (DNS) server (or name server) to carry out the translation. If this is not possible, this server accesses DNS servers at higher levels in the DNS hierarchy until the domain name is resolved.

DNS was developed in the early to mid-1980s, and is highly distributed and resilient. It uses layer 4 protocol UDP; it is a hierarchical system with more than 250 top-level domains, each representing many hosts. These top-level domains are either one of 15 generic top-level domains (e.g., *com*, *edu*, *gov*, *mil*, *net*, *org*) or one of 243 country domains, with most having a registrar administrating second-level domains.

The domain names are case-insensitive. Most U.S. organizations are under a generic domain but others are under the domain of their country. Naming follows organizational boundaries, not physical networks. For example, if the computer science and engineering departments in a university share the same Local Area Network (LAN) in the same building, they may nevertheless have different domains. Alternatively, if engineering is split over two buildings and two separate networks the hosts on both networks may belong to the same domain.

Internet Protocol

Having covered IP addressing schemes, the most important layer 3 protocol, Internet Protocol (IP) is now discussed. It provides the basic mechanism for carrying information from one network device to another. The network layer (layer 3) sends information—as independent packets called datagrams—as it becomes available. IP is entirely connectionless, with no circuits and no connection setup or teardown. It is simple and therefore easy to implement, with no overhead for circuit establishment and termination. Hence, it is economical on network resources. It handles simple "client/server" queries rapidly, with no circuit setup or pull-down.

IP has the disadvantage that information order can be lost, especially when different datagrams take different routes. Quality of service is not guaranteed since only "best-effort" transport is provided. Transmission Control Protocol (TCP, layer 4) ensures information order and provides network performance, while IP concentrates on getting packets from A to B.

IPv4 Datagram Header

Figure 6.11 shows the IPv4 datagram header—the total length is in bytes. Each row of the diagram represents four bytes or 32 bits, so that when the header is transmitted, the first row is transmitted from left to right, followed by the second row, and so on. Each row contains one or more *fields*, each of which carries out a specific function as detailed below.

The identification field increments by 1 for each datagram sent, to permit out-of-order datagrams to be detected at the receiver. The fragment offset is only 13 bits long, which has implications that will become clearer when *fragmentation* is discussed. If a datagram is too large to fit in a link layer frame, fragmentation splits it into several smaller ones. The Type of Service (ToS) field is an 8-bit field used by routers to prioritize service of arriving datagrams (Table 6.3). It permits delay to be minimized, throughput to be maximized, reliability to be maximized, or cost to be minimized, although it is not normally used in this way, as originally intended. Today's routers allow ToS functionality within organizations, but ToS

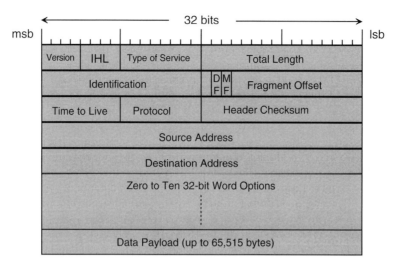

Fig. 6.11   The IP datagram header, which is between 20 and 60 bytes long.

Table 6.3   Fields in the ToS Field

Bits	Type	Description
0–2	Precedence	Levels 0–7; 0 is normal, 7 is highest
3–6	Type of service	Delay, reliability, throughput, cost
7	Reserved	For future use

has a greater role to play in IPv6. It is also re-used in DiffServ, a Quality of Service (QoS) mechanism in IP.

The version field (4 bits) is set to either 4 or 6, denoting IPv4 or IPv6. The header length (IHL, IP Header Length) is only 4 bits long, and measures the header length in groups of 4 bytes (i.e., to obtain the length of the header in bytes, multiply the IHL by 4). The minimum value of IHL is 5, corresponding to a 20-byte header, where there are no options. Options correspond to extra information in the header, which is used for added features. The maximum value of IHL is 15, corresponding to a 60-byte header with 40 bytes of options.

The total length of the IP datagram is held in a 16-bit field, and hence may in theory be up to $2^{16} - 1 = 65,535$ bytes. In practice, it is limited to a smaller value by the link-layer technology being used. However, all hosts must be able to accept datagrams up to 576 bytes in length, and longer datagrams should not be sent unless it is known that the destination host can accept them.

The source and destination IP addresses are both 32 bits long. The Time to Live (TTL) field is 8 bits long, and is decremented by one for each link (or hop) taken. The datagram is discarded if the TTL reaches zero, preventing datagrams from wandering indefinitely or going around in circles. TTL is usually initialized to a value in the range 64 to 128 at the datagram's source, depending on the implementation.

IP provides a service to a layer 4 (transport layer) protocol (i.e., sending a datagram from the source to the destination on a best-effort basis). The protocol field is one byte (8 bits long), and defines the layer 4 protocol to pass the data payload on to. As discussed previously, the protocol can be IP itself if the datagram contains an encapsulated IP

Table 6.4 Values of the IP Field for Different Transport Layer Protocols

Protocol	Code	Description
ICMP—Internet Control Message Protocol	1	Carries network error messages and network control information
IGMP—Internet Group Message Protocol	2	Supports multicast
IP—Internet Protocol	4	IP encapsulated within IP when supporting security
TCP—Transmission Control Protocol	6	Provides a reliable end-to-end connection
UDP—User Datagram Protocol	17	Implements connectionless operation with no guarantees
ISO-TP4	29	International Standards Organization (ISO) transport protocol
AH—Authentication Header	51	Packet requires security check or encryption on arrival at destination
ISO-IP	80	ISO Internet Protocol
OSPF—Open Shortest Path First	89	Carries routing information between routers

datagram. Table 6.4 shows the values of the IP protocol field for different transport layer protocols.

Fragmentation

If a datagram is too large for a link, the router fragments it. For example, Ethernet restricts IP datagrams to 1500 bytes. Each fragment is encapsulated in a smaller datagram, and fragments of the same original datagram have the same identification number.

IP reorders fragments and only reassembles them into datagrams at the destination; it has flags devoted to fragmentation in the header:

- DF = "Don't Fragment"; don't fragment the datagram
- MF = "More Fragments"; more fragments to follow
- DF = 0, MF = 1 except for last fragment where MF = 0

Each fragment carries in the fragment offset field the location of its first data byte in the original datagram; this is expressed in multiples of 8 bytes since this field is only 13 bits long. Hence, all fragments except the last one must be a multiple of 8 bytes in length.

Example of Fragmentation
A 4000-byte IP datagram (with DF = 0) from an FDDI ring is received by a router to be sent over an Ethernet LAN. The IP header is 20 bytes long (with no options), so the data is 3980 bytes. This is to be placed into IP data fields of no more than $1500 - 20 = 1480$ bytes, so the data field size in fragments is 1480 bytes + 1480 bytes + 1020 bytes, and hence the IP datagram sizes are 1500 bytes + 1500 bytes + 1040 bytes. 1480 bytes represents 185 groups of 8 bytes.

The first fragment (1500 bytes) has DF = 0 (fragmentation allowed), MF = 1 (more fragments of the datagram to follow), and offset = 0.

The second fragment (1500 bytes) has DF = 0, MF = 1, and offset = 185, since $185 \times 8 = 1480$ bytes have already been sent, and the 186th group of 8 bytes in the original datagram is the first in this fragment.

The last fragment (1040 bytes) has DF = 0, MF = 0 (no more fragments since this is the last one) and offset = $2 \times 185 = 370$ (i.e., $2 \times 185 \times 8 = 2960$ bytes already sent).

IPv4 Options

Up to 40 bytes of options are allowed in the header, to support additional features. The main IP header is 20 bytes long, and the options are less than or equal to 40 bytes; hence, the total header size is 20 to 60 bytes. These are generally used for diagnostics and troubleshooting. In strict and loose source routing, the source specifies the path to be taken, and routers send the datagram to the next address specified. The "record route" option keeps a record of (Timestamp, IP address) pairs for each router that the datagram passes through. "Router alert" implements a special service because a packet has special requirements, usually for security to route away from certain countries. Often options are not used.

IPv4 Checksum for Header Verification

Header contents are verified by each router processing the datagram. At the transmitter, the checksum field is set to zero, and one's complement addition (exclusive-OR) of all 16-bit header words is calculated. The result of the addition is placed in the checksum. Hence this procedure chooses a value for the checksum field, so that the checksum of the whole header is now equal to zero. It is important to distinguish between the checksum field and the value of the checksum for the whole header.

Each receiving router recomputes the checksum, including the checksum field in the calculation. If the result is zero, it is assumed there are no errors. The data field of the datagram is not checked, and it is left to the transport layer protocol to do this.

This scheme cannot detect two errors in the same bit position, and cannot detect swapped bytes or indeed swapped words. This IP checksum calculation was chosen to permit fast implementation in software. In Ethernet (layer 2), the Frame Check Sequence (FCS) is determined via hardware. The use of hardware allows a more sophisticated error check to take place.

IPv4 Support Protocols

The IP support protocols are necessary for IPv4 to function correctly. Internet Control Message Protocol (ICMP) reports network errors and delivers information to hosts and routers. Address Resolution Protocol (ARP) provides the layer 2 address corresponding to the layer 3 address provided. The layer 2 address is required to forward a datagram on the next link. Reverse Address Resolution Protocol (RARP), Boot Protocol (BOOTP), and Dynamic Host Configuration Protocol (DHCP) represent successive refinements of the same concept. They provide the layer 3 address corresponding to the layer 2 address provided. If a host boots and needs to know its IP address, it is provided by one of these protocols. Examples might be a workstation with no disk, which therefore does not know its IP address, or a customer connected to a service provider via a modem, who must have a new IP address allocated every time he or she powers up.

Internet Control Message Protocol

The Internet Control Message Protocol (ICMP) is the diagnostic and control protocol, informing a source of the fate of its datagrams if they do not reach their destination. It also informs hosts of the existence of local routers.

ICMP carries diagnostics about a dropped datagram to its source. For example, "Time Exceeded" indicates that the router detects that the TTL expired; "Parameter Problem" shows that the router detects via a checksum that the IP header is corrupted. With "Echo Reply," the source requests the return of an ICMP packet by the destination. Echo Reply is used to implement the "ping" command in Linux.

In addition, ICMP carries router control messages back to the source. For example, "Source Quench" indicates that the router is congested so the source must send fewer packets. However, better congestion control methods exist in TCP, so Source Quench is rarely used today. "Destination Unreachable" indicates either that the packet has its DF bit set and is too long for the network, or that the router cannot find a route. This is often not reported due to the use of "default" routing table entries. A "Redirect" means that a router knows a better route for a packet (i.e., that the packet should really have been sent by the host to another router).

ICMP also permits hosts to find their local default router, which forwards datagrams that are destined for outside the local network. With a "Router Advertisement" message, a router informs hosts on this network of its existence by sending an ICMP message (type 9) to either address 224.0.0.1 or address 255.255.255.255. The address 224.0.0.1 is a class D (multicast) address meaning "all hosts on the network." "Router Solicitation" is used by hosts to indicate a desire to receive router advertisements. Local router addresses are usually timed out after 30 minutes.

ICMP messages are transmitted in IP datagrams with the structure depending on the type of control message; however, all begin as shown in Figure 6.12. ICMP messages are held in the IP data field, where types 9 and 10 are used to supply hosts with the local router IP address. For diagnostic purposes, an ICMP message includes the IP header from the datagram causing the error and the first 8 bytes after the header.

ARP Implementation

Normally, an IP datagram is passed to Ethernet (data-link layer), and the whole IP datagram is encapsulated in the data field of an Ethernet frame (Figure 6.13). The datagram only contains the 32-bit IP address of the destination, hence the objective is to determine the 48-bit Ethernet address.

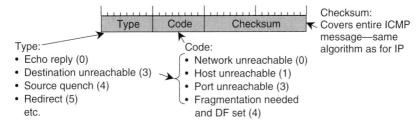

Fig. 6.12   First four bytes of an ICMP message.

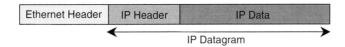

Fig. 6.13   Encapsulation of an IP datagram within an Ethernet frame.

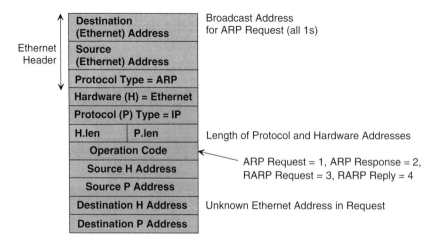

Fig. 6.14 Ethernet header and ARP packet. The packet format is flexible and supports various protocols.

Using Address Resolution Protocol (ARP), the source broadcasts the destination IP address in an Ethernet-ARP frame to all other devices on the network. The device having that IP address sends back an ARP packet with its "media" or Ethernet address, so that in this way the corresponding Ethernet address is obtained.

Now that the Ethernet address is known, the IP datagram is encapsulated in an Ethernet frame and sent to its destination. The IP address and corresponding Ethernet address are cached at the source in case further datagrams must be sent to the same destination. Hence with ARP no address translation lookup table needs to be maintained, simplifying address management.

ARP (Figure 6.14) can be used with technologies other than Ethernet (e.g., ATMARP). The same packet format is also used for RARP.

RARP, BOOTP, and DHCP

Although RARP and BOOTP have been superseded by DHCP, it is nevertheless instructive to study their operation. At boot-up, a diskless machine (nowadays known as a "thin client") does not have a record of its IP address. It can obtain it by making an "all ones" Ethernet broadcast, and a RARP server responds with its IP address. The protocol employed is Reverse Address Resolution Protocol (RARP), which uses the ARP packet format. Since the broadcast is local, each LAN must have a RARP server.

The restriction of having a local server is avoided by BOOTP. The BOOTP request is broadcast via a UDP datagram, so a router can forward it to a BOOTP server if it is on another network. BOOTP informs a booting host of necessary network information:

- Its IP address
- The network prefix
- The default local router
- Its executable code if it is "diskless"—in conjunction with Trivial File Transfer Protocol (TFTP)
- Number of seconds since midnight on 1 January 1900
- IP address of the default router and other routers that can be used
- IP address of DNS name server, printer server, time server, and so on

The Dynamic Host Configuration Protocol (DHCP) extends BOOTP and has in fact superseded it. It assigns IP addresses dynamically from a pool that is defined in the server's configuration file, hence a host does not tie up an IP address when not in use. A range of addresses is declared for each subnet, for example:

```
pool {
      range 10.0.0.10 10.0.0.99
}
```

This assigns 90 IP addresses in the range `10.0.0.10` to `10.0.0.99` to the subnet in question. DHCP is an improvement over BOOTP because BOOTP requires individual IP and Ethernet addresses to be set up in a table manually. However, DHCP may override the pool of addresses by assigning a manually assigned IP address to any Ethernet address specified by the network administrator, in the same way as BOOTP. DHCP is used by ISPs to assign IP addresses automatically to dial-up users, cable users, and ADSL users. It is also used in office networks, particularly when laptops are used extensively and are only rarely connected.

Steps Involved in a First Hop

This example explains how an IP datagram gets to its first-hop destination. Assume its source is on an Ethernet LAN, and also assume that DHCP was used to configure the datagram's source. Furthermore, assume that the source's ARP cache is empty.

From DHCP, the source knows the prefix of its network, so it examines the datagram's destination address. Via the network address field, it determines whether the destination is on its network. If this is so, the final destination is the first-hop destination. Otherwise, the source knows that the first-hop destination must be the default router on its network and, from DHCP, it has the IP address of this default router.

Either way the source now knows the IP address of the first hop, so it then sends an ARP broadcast with this IP address enclosed. This is necessary because the ARP cache is empty. A network device on the local network will recognize the IP address as its own, and it then sends an ARP response back informing the source of the link-layer (Ethernet) address of the first hop. The source then encapsulates the IP datagram in an Ethernet frame destined for that link-layer address.

Suppose the source host's address is 150.28.128.5, a class B address with 128.5 = 10000000 00000101. The source host's subnet mask is 255.255.192.0, also written as 11111111 11111111 11000000 00000000. The final destination address is 150.28.96.5, also class B with 96.5 = 01100000 00000101. The last two bits of the mask correspond to part of the subnet address, which differs between the source (10) and destination (01); hence they are on different subnets, and the first-hop destination is the local router.

Transport Protocols

IP provides a basis on which protocols in the layer above (layer 4) can build, in order to provide a useful service to application protocols in layer 5. The most important layer 4 protocol is undoubtedly Transmission Control Protocol (TCP), which, unlike IP, transfers information reliably between network devices. UDP (User Datagram Protocol) is an enhancement of IP that is used when TCP would be too complex and with too great an overhead. Transactional TCP (T/TCP) seeks to provide a compromise between TCP and UDP, whereas Real-time Transport Protocol (RTP) transports real-time data such as audio and video.

Outline of TCP Operation

Transmission Control Protocol is an important layer 4 protocol, providing a reliable duplex "byte stream pipe" between application processes on remote computers. It:

- Sets up and closes connections between the source and destination, and negotiates maximum acceptable segment length (Maximum Transmission Unit or MTU).
- Ensures the correct byte order of the message at the destination.
- Retransmits lost IP datagrams via fast retransmit and timeout.
- Manages the transmission rate of a source host via congestion control, so that the network and destination are never overloaded.

IP routes a datagram from source to destination on a best-effort basis, with no connection setup or teardown. However, TCP turns datagram exchange into a reliable process-to-process data connection, by providing circuits or "data calls." TCP delivers data reliably, in sequence and without errors, and does not overwhelm the destination or the network. TCP is implemented in the end hosts only, and not in routers. In TCP, a *port number* defines the application in layer 5 that is using it.

To set up a TCP connection, the application requests it from the TCP process. The TCP processes at the source and destination hosts then set up a connection. This is known as a "three-way handshake." It can also carry passwords if required by the application. TCP also reserves send and receive memory buffers at each end for implementation of TCP's sliding window protocol. These buffers store incoming data while the receive application is busy. TCP accesses the path MTU once the connection has been established.

After that, the application places outgoing bytes in the send buffer, and TCP slices the contents into segments (the name for packets in TCP), which are encapsulated in IPv4 datagrams. The receiving TCP process sends a TCP segment acknowledging each incoming segment taken out of its receive buffer. It is possible to "piggyback" data onto acknowledgment segments when implementing duplex communication.

If an acknowledgment segment is not received before a timeout period expires, the segment is resent, overcoming the problem of IP datagrams becoming lost in the network. The congestion control mechanism, implemented via fast retransmit and fast recovery, is triggered by packet loss.

In client/server applications, a client application makes a request of the server application (Figure 6.15), and the server fulfills the request. This is a convenient way to interconnect programs distributed across different locations. The port number on the server is "well known" or "registered," and the client must know it in advance since it is predefined for each application. The port number on the client is "ephemeral," and is defined by the client as part of the request.

TCP is also used in "peer-to-peer" applications, where the application in either peer may initiate a transaction. Table 6.5 shows some common applications that use TCP, where the port numbers refer to the server.

TCP Segment Header

As noted before, in TCP a packet is known as a segment. The header is shown in Figure 6.16. The maximum length of an IP datagram is 65,535 bytes, so the amount of data permitted in each TCP segment's data field may be obtained by subtracting the length of the IP header and the length of the TCP header from this. TCP segments are encapsulated within

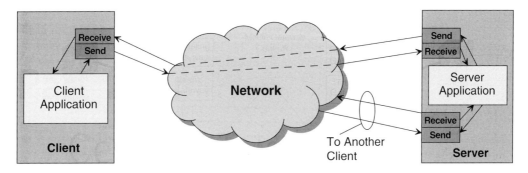

Fig. 6.15 Communications between a client and a server, often a convenient way of organizing application programs.

Table 6.5 Common Applications Using TCP

Application	Port number
File Transfer Protocol (FTP)	20
Telnet (log in at remote host)	23
Simple Mail Transport Protocol (SMTP—communicates between mail servers)	25
Hypertext Transfer Protocol (HTTP)	80
Connect user to ISP's mailbox (POP3)	110

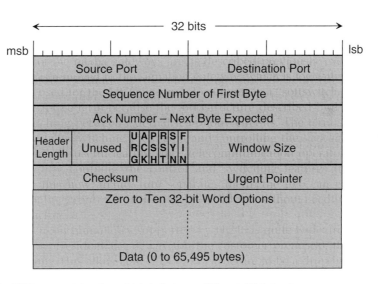

Fig. 6.16 The TCP segment header, which is between 20 and 60 bytes long.

IP datagrams with an IP protocol field of 6—this identifies TCP as the layer 4 protocol to which IP in the receiving host should send the contents of its data field. Recall that all Internet hosts must be able to accept IP datagrams of up to 576 bytes in length. Hence TCP must be able to accept segments with up to a 536-byte payload (i.e., 576 minus the lengths of both IP and TCP headers).

TCP Header Field Functions

The source and destination ports are each 2 bytes in length, and, along with the IP address of the encapsulating IP datagram, they uniquely identify the application to which the TCP must deliver the TCP segment. The combination of an IP address and a port number is often called a *socket*. Well-known port numbers exist for SMTP, FTP, and HTTP for example.

The sequence number is 32 bits long, and it is set by the source of a segment; it contains the byte position of the start of the data payload within an overall data stream. The acknowledgment number is 32 bits long and is set by the destination in an acknowledgment segment. It contains the sequence number of the next byte expected by the destination, so that the destination may transmit bytes up to a window ahead of this.

The header length is expressed in units of 32-bit words; it is 4 bits long, and indicates a header length between 20 bytes and 60 bytes. It is required because the length of the options field varies from 0 to 40 bytes.

The ACK and SYN bits are used during connection setup in the three-way handshake:

- SYN = 1, ACK = 0 is a connection request.
- SYN = 1, ACK = 1 is a connection accept and connection request.
- SYN = 0, ACK = 1 is a connection accept.

The PSH flag is set if the TCP receiver process is to send (push) both the segment payload and data already in the buffer to the application immediately. The URG flag is set to indicate that the urgent pointer points to payload data to be read by the destination application immediately. This is an application-level interrupt mechanism that is used when the Esc key is pressed in telnet and rlogin, or when aborting a file transfer in FTP.

The FIN bit is set when closing the connection from the sender; however, with a duplex connection, the other half of the connection remains open. The RST bit is set if the connection has become confused or a connection request is refused.

Options (up to ten 32-bit words) are most commonly used when announcing on setup the acceptable MTU. The window size (16 bits) is set in an acknowledgment segment, permitting the receiving TCP process to tell the sending TCP process how many bytes it is willing to accept; it is often equal to the receiver buffer space available. The sending TCP process may choose to send less if it thinks that the network is congested, and this will be discussed in detail later. Usually, this field allows a window of up to 65,535 bytes; however, an extension to TCP exists that supports larger windows.

The checksum field (16 bits—Figure 6.17) is calculated using the same algorithm as the IP header checksum; however, it includes the TCP header, TCP data, and selected data that also appears in the IP header.

Three-way Handshake

The three-way handshake sets up a TCP connection (Figure 6.18). It is convenient to name segments according to the flags set:

- SYN is the segment sent to set up a connection.
- ACK is the segment acknowledging a SYN.
- FIN is a segment declaring that no more data is to be sent.

After transmitting a SYN, the sequence number is incremented by 1, permitting unambiguous acknowledgment.

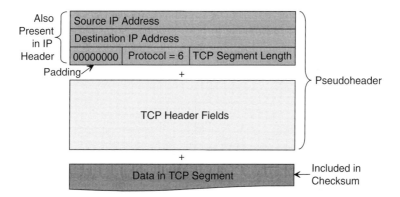

Fig. 6.17 Calculation of the TCP header checksum, which includes the segment's data field and includes fields present in the IP datagram header.

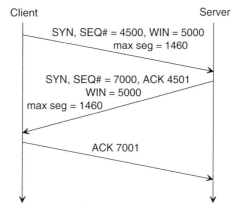

Fig. 6.18 A three-way handshake, setting up a TCP connection. A SYN sets up a connection, an ACK acknowledges a SYN, while a FIN terminates a connection.

Path MTU Discovery

Fragmentation is undesirable because it increases the likelihood of misordered datagrams, and reordering is an overhead for TCP. Also, the fragment headers represent an additional load on the network, increasing the processing load on routers. Instead, TCP uses path MTU discovery, where MTU is the Maximum Transmission Unit. TCP sends a datagram with the length of either the source network MTU or the MTU announced by destination, whichever is smaller, and DF = 1. The default destination MTU is 536 bytes.

If the datagram does not reach its destination (indicated by an ICMP message), smaller datagrams are sent until one reaches the destination. This allows the path MTU to be determined, and reduces the overhead of packet headers as much as possible without causing fragmentation.

Security Problems with TCP

Suppose there is a malicious attacker who emulates a trusted host (Figure 6.19). Each connection in TCP starts with a different Initial Sequence Number (ISN), which is not zero since this is easy to predict. If an attacker can predict the ISN, it can participate in a TCP

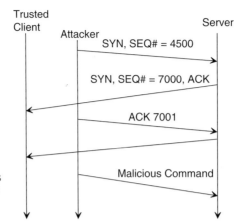

Fig. 6.19 If an attacker can guess the Initial Sequence Number, it can send malicious commands to the server, even though it does not have authorization to do so.

connection with the victim server. The attacker sets its source IP address to that of the trusted client it is impersonating, and sends a SYN segment. The attacker receives no messages from the server; they are all sent to the trusted client, including the reply from the attacker's SYN. However, the attacker does not need a segment from the server to tell it what the ISN is, and can therefore send an acknowledgment with the correct ISN to the server, followed by malicious commands.

Reducing the Number of Datagrams Transmitted

Nagle's algorithm increases the efficiency of TCP when the user is typing—for example, while using telnet to log in to another host remotely. Suppose the sending TCP process sends the first byte immediately; a 41-byte IP datagram is sent consisting of a 20-byte IP header + a 20-byte TCP header + 1 byte of data. The sending process buffers the bytes received from the application until the first segment is acknowledged, preventing a large number of wasteful 41-byte datagrams.

It is also necessary to avoid 1-byte windows (also known as the "silly window syndrome"). This occurs if the received buffer is full and the application reads one byte at a time. The receiving TCP process does not re-advertise a new window size unless it corresponds to the agreed MTU or its buffer is half empty (i.e., the receive process waits until a reasonable amount of capacity is available).

It may also be necessary to reduce the number of TCP ACK segments sent due to application actions. The receiving TCP process can block read requests by the application until there is sufficient information to read, reducing the number of calls to TCP. This increases response time, but may not be important for some applications such as FTP. It can be over-ridden using PSH.

Slow Start

Slow start is an important part of TCP's congestion control mechanism. It takes place when a TCP connection is started, and adjusts to the available capacity between the sender and the receiver. To do this, it injects new segments into the network only at the rate previous segments are acknowledged. Two variables are used throughout—*ssthresh* is used by slow start and congestion avoidance, and is initialized to 65,535. The congestion window variable (cwnd—measured in bytes) is used alongside the window advertised by the receiver. The variables ssthresh and cwnd are held at the transmitter.

The window used by the sender is either cwnd or the advertised window, whichever is smaller. The advertised window indicates flow control imposed by the receiver, and, as discussed next, cwnd indicates congestion control imposed by the sender's perception of congestion in the network.

Slow start is used only when cwnd ≤ ssthresh. It starts with cwnd being set to 1 segment and the sender transmitting 1 segment. When an ACK is received, cwnd is increased by 1 segment, resulting in an exponential increase in cwnd over time.

Example of Slow Start in TCP

Slow start is used on a line with a 10 ms round-trip time and no congestion. The receive window is fixed at 24 KByte, and the maximum segment size is 2 KByte. The time before the first full window can be sent is computed:

First	2 KByte	1 segment	$t = 0$ ms
	4 KByte	2 segments	$t = 10$ ms
	8 KByte	4 segments	$t = 20$ ms
	16 KByte	8 segments	$t = 30$ ms
	min(32, 24) KByte	12 segments	$t = 40$ ms

Therefore at time $t = 40$ ms the first full window is sent. Each time an ACK is received, the congestion window is increased by one segment.

Congestion Avoidance

Congestion avoidance is used when cwnd > ssthresh; it responds to lost segments, assuming that they are due to congestion. One indication of congestion is a timeout; however, duplicate ACKs also indicate packet loss and, hence, congestion. This is because TCP always acknowledges the last contiguous byte received; if there is a break in the bytes received due to loss of a segment, the last byte before the break is always acknowledged (Figure 6.20).

Congestion avoidance is always implemented alongside slow start, since they are both integral parts of TCP's congestion control scheme. When congestion is detected:

$$ssthresh \leftarrow 0.5 \times min\ (cwnd, advertised\ window)$$

In addition, if the congestion was indicated by a timeout, TCP goes back to slow start:

$$cwnd \leftarrow 1\ segment$$

If, on the other hand, the congestion was indicated by duplicate ACKs, fast retransmit and fast recovery are instigated, as described next. In congestion avoidance, 1/cwnd is added to cwnd whenever an ACK is received, yielding a linear increase in cwnd of 1 segment per round-trip time.

Fig. 6.20 TCP always acknowledges the last *contiguous* byte of data that it received; therefore if there is a break in the data due to a lost segment, the last byte before the break will be acknowledged repeatedly.

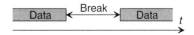

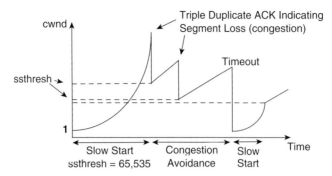

Fig. 6.21 The dynamics of evolution of the TCP congestion window size, cwnd. Duplicate ACKs result in fast retransmit followed by fast recovery, whereas a timeout implies a return to slow start.

Fast Retransmit and Fast Recovery

Before assuming segment loss, TCP waits until three duplicate ACKs have been received; these indicate a break in the data received, hence a lost segment. Three duplicate ACKs are usually too many to have been caused by misordering. TCP now enters fast retransmit phase, retransmitting lost segments without waiting for a timeout. This is appropriate when data is still flowing since ACKs have been received, so there is no need for a slow start and the reduction in transmission rate that it implies. Then TCP enters fast recovery phase by going back to congestion avoidance. The ssthresh has been set to half the window by congestion avoidance. In addition, cwnd is set to ssthresh (this is in fact a slight simplification).

It takes more than one Round-trip Time (RTT) from a data segment not being received, before the retransmitted segment arrives; hence having multiple segment losses in a window causes delay and loss of throughput in many TCP implementations. One solution to this problem is Selective ACKnowledgment TCP (SACK TCP), which uses a more sophisticated scheme for providing acknowledgments.

By employing source quench packets, ICMP sought to avoid network congestion; however, it is better to control the sending TCP process at the TCP level by using the congestion control measures described here.

Figure 6.21 shows the evolution of TCP's cwnd in the version known as Reno. Other variants, such as Tahoe and New Reno, behave slightly differently; for example, Tahoe goes into slow start after a triple ACK. The TCP sender backs off when it thinks there is congestion, and it is assumed that segment loss is due to congestion. This is not a valid assumption, for example, with radio transmission, since packet loss is very often due to the noise characteristics of the physical layer. Hence this assumption made by TCP has limited value in certain circumstances.

TCP Retransmission Timer

The correct choice of "timeout" interval for retransmission is critical to efficient transport layer operation. In TCP, this is determined dynamically. Define the round-trip time M as the time between sending the last segment and receiving its acknowledgment. The estimated *RTT* is a moving average:

$$RTT = \alpha RTT + (1 - \alpha M) \ (\alpha = 7/8)$$

The mean deviation D on the RTT is also a moving average:

$$D = \alpha D + (1 - \alpha) |M - RTT|$$

The timeout is taken to be $RTT + 4D$, where both 4 and 7/8 were chosen for computational ease. When TCP was devised, computers were not as cheap and powerful as they are now, so it was important to choose constants that could be multiplied quickly. Hence the timeout interval is dependent on the mean RTT and its deviation. If a timeout occurs it is doubled for each successive failure—this is known as "exponential backoff." TCP timeout is followed by a slow start.

Example of Send Window Size after Timeout

Suppose the TCP congestion window is set to 18 KByte, and, at this point, a timeout occurs. The maximum segment size is 1 KByte. The size of the send window is now computed, assuming the next four transmission bursts are all successful; the data transmitted in one RTT is called a "burst."

The congestion window is the amount the source thinks it can send in a window, so since a timeout has occurred it sets ssthresh to $18/2 = 9$ KByte. Then the send window size increases exponentially from 1 (slow start). When the threshold is reached, it grows linearly in time, assuming no timeouts or segment losses:

Send window = 1 KByte

Send window = 2 KByte

Send window = 4 KByte

Send window = 8 KByte

Send window = min(16, 9) KByte = 9 KByte (i.e., *ssthresh*)

The send window is 9 KByte after the fourth burst is sent.

Round-trip Time Calculation Example

Suppose the current TCP estimated round-trip time RTT_0 is 30 ms, and that the following three acknowledgments come in after 26 ms, 32 ms, and 24 ms, respectively. Assuming $\alpha = 0.9$, the new RTT estimates are calculated as follows:

$$RTT_1 = 0.9 \times RTT_0 + 0.1 \times 26 = 29.6 \text{ ms}$$
$$RTT_2 = 0.9 \times 29.6 + 0.1 \times 32 = 29.84 \text{ ms}$$
$$RTT_3 = 0.9 \times 29.84 + 0.1 \times 24 = 29.256 \text{ ms}$$

Bandwidth-delay Product and Link Utilization

The capacity of a "pipe" is the number of bits that are in transit along the transmission path at once:

bandwidth (bits/sec) × round-trip time (sec)

This is also known as the bandwidth-delay product, and is measured in bits. For long, modern, high-speed links, this can be much greater than the maximum TCP advertised window size of 65,535 bytes.

The upper bound on maximum link utilization is the advertised window size divided by the bandwidth-delay product. The propagation delay of optical fiber is approximately 5 ns per meter, so a 1000 km fibre link has a 5 ms one-way delay or 10 ms RTT. When operating at 10 Gb/s, the bandwidth-delay product is 100 Mbit or 12.5 MByte, hence the upper bound on maximum throughput is $65,535/(12.5 \times 10^6) = 0.52\%$.

To improve efficiency, TCP's receive window size should be increased, so TCP's window scale extension (nonstandard) allows windows up to 1 GByte. Hence the efficiency is increased since it is proportional to the receive window size, assuming the receive window size is not greater than the bandwidth-delay product. The window scale extension overcomes the fact that the channel is "long" (i.e., RTTs are large) and "fat" (i.e., has a large bandwidth).

If multiple segments are lost in an RTT, then it can take so long to retransmit them all that TCP times out. For example, Reno can only retransmit at most one dropped segment per round-trip time. This is much more likely with large window sizes, since there are more segments in transit that can potentially be lost. Selective Acknowledgement (SACK) is one variant of TCP that has been proposed to solve this problem.

Long Fat Network Example

In this example, two TCP/IP hosts communicate over a path of one-way length L meters, and the mean propagation velocity of segments along the path is c m/s. The hosts use a window size of W bytes, and the transmission bit rate is r bits per second. Assume that only one TCP connection (i.e., the one in question) exists along the transmission path. The objective is to derive an upper bound for the throughput in bits per second.

Using the previous quantities, the propagation time along the forward and return path is $2L/c$, and the time taken to transmit one byte is $8/r$. The number of bytes in transit along the forward and return paths is:

$$(2L/c)/(8/r) = Lr/4c \tag{6.1}$$

The TCP window size can be divided by this quantity to yield the utilization:

$$U = \min(4Wc/Lr, 1)$$

The "min" function is used because U cannot by definition be greater than 1. If W, the window size, is greater than the bandwidth-delay product, then the throughput is limited by the window size. Assuming that this is not the case, the throughput is:

$$rU = 4Wc/L \text{ bits per second}$$

Assume $c = 2 \times 10^8$ m/s and that $W = 65,535$, the maximum possible in conventional TCP. Also, the bit rate $r = 10^{10}$ bits/s = 10 Gb/s. Hence for a 1 km link, an upper bound on utilization is:

$$4Wc/Lr = (4 \times 65,535 \times 2 \times 10^8)/(1000 \times 10^{10}) = 5.25 > 1, \text{ hence } U = 1$$

Whereas for an 8800 km link, the utilization is:

$$4Wc/Lr = (4 \times 65{,}535 \times 2 \times 10^8)/(8.8 \times 10^6 \times 10^{10}) = 0.00060$$

Suppose a scientist in London has many large files of experimental data, each exceeding 100 Mbyte in length. The scientist has tried to send them using TCP to a colleague in Los Angeles but has been disappointed with the transfer rate. To improve throughput, many files could be transmitted at once; the transfer rate for each individual file is the same; however, the overall transfer rate is proportional to the number of parallel files. Another option would be to modify TCP so that it allows a window size greater than 65,535 bytes.

Assume the 8800 km link noted above is full of bits—the number of bytes in transit at any time in the forward direction is now computed. Modifying Equation 1 for one-way transmission only, this is:

$$Lr/8c = (8.8 \times 10^6 \times 10^{10})/(8 \times 2 \times 10^8) = 5.5 \times 10^7 \text{ bytes}$$

Hence the number of 1500-byte datagrams this represents is approximately:

$$5.5 \times 10^7/1500 = 36{,}667$$

In these calculations, the effect of packet header length has been neglected, including Ethernet frame headers, IP datagram headers, and TCP segment headers.

User Datagram Protocol

User Datagram Protocol (UDP) provides a lightweight datagram service between source and destination applications; it is a layer 4 (transport layer) protocol. It is simple and avoids the complexities of TCP. It is also fast, with no need for a three-way handshake before data transfer takes place. It is suited to simple client/server network applications, where the connection setup overhead of TCP is unacceptable.

On the other hand, it has the disadvantages of having no error control; hence lost or erroneous datagrams are not retransmitted because there is no mechanism for correcting transmission errors. Also, UDP does not guarantee to preserve datagram order, a problem which was solved in TCP.

The UDP header and checksum calculation is shown in Figure 6.22. The pseudo-header contains the UDP header and certain fields which are also in the IP header, where the same checksum algorithm as for IP applies. The UDP length field includes the header and the

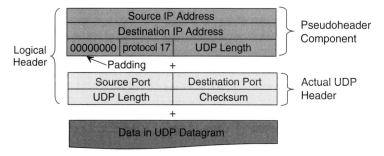

Fig. 6.22 Calculation of the UDP checksum, which includes data also present in the IP datagram, and may include the UDP data field.

Table 6.6   Some Well-known UDP Ports

Service	Port	Description
echo	7	Echo datagram back to user
daytime	13	Report time in units of a second
nameserver	53	Used by Domain Name System (DNS)
bootps	67	BOOTP server port
bootpc	68	BOOTP client port
Network Time Protocol (NTP)	123	Network time over LAN or WAN to millisecond accuracy
Simple Network Management Protocol (SNMP)	161	Supports network management

data length. To obtain the even number of bytes required by the checksum algorithm, padding with zeros takes place. Checksum calculation is optional, and the UDP header has no options. Some well-known UDP ports are shown in Table 6.6.

Transactional TCP

A transaction is a client request followed by a server response, for example, a DNS request and response. With transactions, the overhead of connection establishment and termination should be avoided or minimized; furthermore, latency should be reduced as much as possible.

Transactional TCP (T/TCP) fills the gap between UDP and TCP. UDP is too simple since it does not provide error control, and has no retransmission of lost or erroneous datagrams, whereas TCP is very sophisticated and general-purpose, and many segments must be exchanged even for a single transaction.

T/TCP is a modification to TCP, using the options field, and is intended for client server transactions. T/TCP avoids using conventional TCP unless either the client has rebooted or a first request is taking place. The client increments a 32-bit Connection Counter (CC) for each new transaction; this is initialized to zero when the client boots up. The CC is stored by the server.

During operation, the client sends client-request data, client-SYN, client-FIN, and client-CC. If the received CC is no greater than the stored CC, or the server has no stored CC for this client, it must be a re-boot or first request; hence, T/TCP must revert to a normal TCP three-way handshake to obtain state information. In this case, the round-trip time and maximum segment size are cached. (A standard window size is used throughout.)

Otherwise, just two more segments are exchanged. Firstly, the server responds with server-data, server-SYN, server-FIN, ACK of client-FIN, and echo of client-CC, and, finally, the client sends an ACK of the server-FIN.

Realtime Transport Protocol

The original Internet was designed to transport only data traffic; however, transporting all services—data, audio, and video—on one network can reduce the overall cost of managing and running a telecommunications service. Many telecommunications operators have already committed themselves to this path, or are considering doing so, due to the economies that can result.

Realtime Transport Protocol (RTP) facilitates this by transporting services such as Internet radio, Voice over IP (Internet telephony), music-on-demand, videoconferencing, and video-on-demand. Although it is, strictly speaking, a transport protocol, it is in fact implemented in the application layer, using UDP in the transport layer.

RTP itself is a thin protocol supporting real-time applications. It implements timing reconstruction at the receiver, providing buffering to compensate for jitter. This is necessary because IP offers no guarantees about delay, and each packet in a stream may experience a different delay when traveling from source to destination. It has no flow control, no error control, and no acknowledgments, since if a packet is lost, a retransmitted packet would be too late to be of use to the receiver. Each RTP payload can carry several real-time data streams of different types.

Realtime Transport Control Protocol (RTCP) is a control protocol that does not transport any user data. It provides support for real-time conferencing of groups of any size, and offers quality-of-service feedback from receivers in a multicast group. Based on this, the encoding process can adjust its rate to suit network conditions. It supports the synchronization of different media streams, and names data sources in ASCII so that they can be identified by the user.

Internet Routing

Thus far, it has been assumed that routers know where to forward incoming IP datagrams, in order to route information effectively across the network from one network device to another. Internet routing carries out this task by setting up routing tables in routers, thus defining the best route for each IP datagram from source to destination. There are a number of considerations and constraints:

- Routers receive topology information from only their immediate neighbors.
- Failure of a router or link may take place at any time, so routing must be dynamic, not static.
- Any one of several possible distance metrics may be employed, such as number of hops, reciprocal of link capacity, queue length, or propagation delay.

Routing can be implemented within (interior to) and between (exterior to) autonomous systems (AS). An AS is a set of interconnected routers and networks that is controlled by one or more network administrators working for a single entity, such as a university or all or part of a business enterprise. A distributed algorithm is used, because the Internet has no central control. An efficient routing algorithm is essential to the overall efficient operation of the Internet, because inefficient, indirect, routes imply misuse of resources, and assigning addresses inefficiently (e.g., randomly) implies very large routing tables.

Manual Routing

In manual routing, the network administrator sets up routing table entries manually. This is a viable option only if the network has a fixed topology with no router or link failures, has unique routes that are easy to determine manually, and is small, implying a reduced number of calculations. For Windows hosts and routers the command to store permanent routes is:

route – *p* **add** *base_address* **mask** *mask router_interface* **metric** *n*

where *n* is the number of hops required to reach the destination. For large, complex networks, this approach is cumbersome, error-prone, and inefficient.

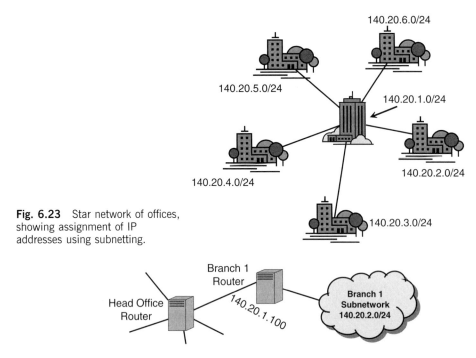

Fig. 6.23 Star network of offices, showing assignment of IP addresses using subnetting.

Fig. 6.24 Setting up the routing table in the head office, to forward datagrams to branch 1.

Subnetting of Branch Offices Revisited

An organization has a head office with fewer than 254 branch offices (Figure 6.23), where each office has fewer than 254 hosts. The organization chooses a "hub and spokes" topology, with the head office at the center and one branch at the end of each spoke. The organization obtains the class B network addresses starting at 140.20.0.0, and each office is allocated a subnet address space.

Windows NT servers are being used as routers, so consider the commands necessary to create permanent manual routes in the head office router. The route command is used at the head office for each branch office (Figure 6.24), for example, for branch 1:

route − *p* add 140.20.2.0 **mask** 255.255.255.0 140.20.1.100 **metric** 2

One interface on the branch 1 router is part of the head office subnet, so this interface has the address 140.20.1.100. The metric is equal to 2 because one hop is required to get the branch router from the head office router, and one hop is then required to get onto the destination subnet. Apart from associating each interface with its own IP address (done using the operating system of its own computer), nothing needs to be done, because the remote computers associate link layer addresses with these IP addresses via ARP.

Dynamic Routing

Dynamic routing is carried out by specialized computers that are known, not surprisingly, as routers. These run special software and may have special routing hardware. Routers use routing tables stored in memory, which are used to determine on which link to place an IP

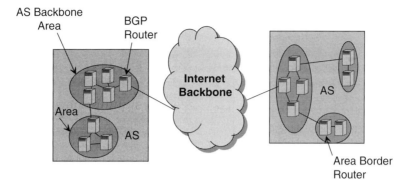

Fig. 6.25 Autonomous systems are interconnected by the Internet backbone, with each AS being divided into areas.

Table 6.7 Routing Hierarchies, Routing Protocol Types, and Example Routing Protocols

Routing hierarchy	Routing protocol type	Example protocol
Intra-area routing	Interior Gateway Protocol (IGP)	Open Shortest Path First (OSPF)
Autonomous system	Interior Gateway Protocol (IGP)	Routing Information Protocol (RIP), OSPF
Inter-AS	Exterior Gateway Protocol (EGP)	Border Gateway Protocol (BGP)

datagram so that it reaches the correct destination; the datagram destination address is used to index the table.

Routers build up routing tables by exchanging network connectivity ("topology") information with their neighbors; this information is carried in IP datagrams. The discussion below describes in detail how routers build and use routing tables to route a packet efficiently through the Internet to its destination.

Routing Hierarchy—Area, AS, Internet

In Figure 6.25, Border Gateway Protocol (BGP) routers are always on the edge of an Autonomous System (AS), and exchange routing information with BGP routers in one or more other ASs. Unlike Routing Information Protocol (RIP), Open Shortest Path First (OSPF) allows the existence of area border routers, which are on the edge of an area and connected to backbone area within the AS. Networks or groups of networks (not usually individual hosts) are referred to in routing tables.

Internet routing is carried out in a hierarchical fashion. It takes place first in an area, then in the autonomous system, then in the Internet backbone. The routing protocol used depends on the level in the hierarchy, as shown in Table 6.7. Routing Information Protocol (RIP) does not permit the use of areas. There are also proprietary routing protocols such as Interior Gateway Routing Protocol (IGRP), which are not described here. Table 6.8 describes the main features of distance vector, link state, and path-based routing—these are all important types of dynamic routing.

Distance-vector Routing

Distance-vector routing was once widely deployed, especially within RIP. Although RIP is now rarely used, it is still instructive to study this type of routing, since its shortcomings explain many of the design decisions underlying link-state routing, which is very

Table 6.8 Types of Routing Protocol

Protocol type	Description	Example
Distance vector	Distributed shortest path routing algorithm	Routing Information Protocol (RIP)
Link state	Each router builds up a "picture" of the network, through flooding	Open Shortest Path First (OSPF)
Path based	Routers exchange entire path information—consider administration and policy	Border Gateway Protocol (BGP)

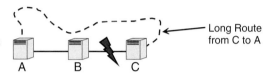

Fig. 6.26 Example to illustrate bounce and count to infinity problems. Traffic bounces between A and B until their distance metrics increase above the length of the long route from C to A.

widespread. Furthermore, distance-vector routing has much in common with path-based routing, which is widely used in Border Gateway Protocol (BGP).

In distance-vector routing, each router has an entry for every destination router in the AS, containing the distance to the destination router and the address of the first neighbor router on the route to the destination. No assumptions are made about the starting condition of each router's table. Periodically, each router sends a routing update to each of its neighbors, consisting of Distance Vectors (DV), where each DV indicates a destination and the distance to it. Hence, there is one DV for each possible destination router. Different routers may send out routing updates asynchronously, and it is not important if routing updates are occasionally dropped.

Suppose an update arrives from a neighbor. Then the router adds one hop to each distance metric in the update, to account for the hop that exists from its neighbor. It then compares each new DV with the corresponding current DV. If the new distance to any destination node is smaller than the existing value, the new route is adopted; that is, the router updates the distance and first hop in the table entry for the node in question. If the router that sent the update is the first hop on the existing route (i.e., if it is the router that the route was "learned" from previously), the new metric is used even if it is larger than the old one.

This accommodates the possibility of the initial estimate being too low. It can be shown analytically that this algorithm converges to the best result, assuming the system is static. If a failure occurs, the distance entries for paths with a broken link as first hop are set to infinity. Although the algorithm operates satisfactorily under static conditions, it is essential to consider how quickly it converges in the event of a fault.

Count to Infinity and Bounce Problems

In the "count to infinity" scenario, the DVs do not converge (Figure 6.26 and Table 6.9), and distances increase toward infinity (represented by 15 hops in RIP). As a result, datagrams pass to and fro repeatedly between two nodes (A and B in this case), known as "bouncing." Bouncing occurs here because there is a long route (shown dotted in Figure 6.26) to C from A, which is not used until d_{AC} has incremented beyond this long route's length. In the meantime, packets destined for C bounce between A and B, causing congestion.

Example—Distance-vector Routing

For simplicity, assume routers exchange DVs simultaneously every second starting from "cold" ($T = 0$); in practice, routers send out DVs to one another asynchronously. The earliest

Table 6.9   Illustration of "Bouncing" with RIP

d_{AC}	d_{BC}	RIP signaling and routing activity
2	1	All routes set up—network has converged.
2	∞	Link BC fails.
2	3	A reports DV first—B sets route via A (what happens if B reports first?). B does not realize it is on the route from A to C.
4	3	A notes B now has link to C of length 3. The new metric is used since the router that sent the update also sent the original route.
4	5	B notes A now has link to C of length 4. The new metric is used since the router that sent the update also sent the original route.
etc. . . .	etc. . . .	Both distances count toward infinity.

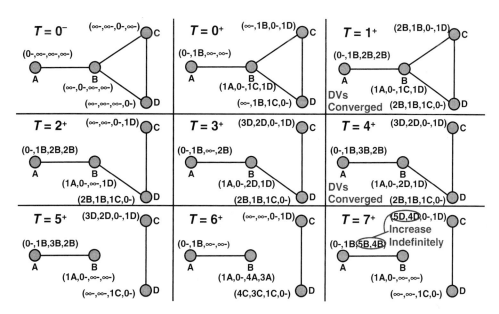

Fig. 6.27   Evolution of distance-vector routing under fault conditions. The last time a break occurs, the network does not stabilize. The inter-router distance is 1 for all adjacent routers. The notation used is dn, dn, dn, dn where *d* is the distance in hops, and *n* is the next hop. The first such pair refers to node A, next to node B, and so on.

time at which the DVs converge will be determined—to do this, each router's DVs are displayed in Figure 6.27 shortly after each integer time, immediately after DVs have been exchanged.

Just after all routers acquire the correct distance vectors, link BC breaks; the evolution of the DVs until they again converge is shown. Again, the routers advertise their distance vectors at each integer time, and routers only learn about link failures when a distance vector is not received (RIP itself is more complex as discussed shortly).

After the distance vectors converge, link DB also breaks; it will be shown that the distance vectors do not converge. The complete solution is shown in Figure 6.27.

Triggered Updates and Split Horizons
With triggered updates, routers send the routing updates on as soon as the information is available without waiting for periodic transmission of updates, so speeding up the response

to network changes. Information regarding only the changed routes is sent, and not the whole routing table. Care must be taken not to propagate out-of-date information.

In the split horizon solution to bouncing, entries are not sent along the link from which they were learned. Such entries are set to infinity ("poisoned reverse"); however, this is only a partial solution, and other problems still arise. Triggered updates and split horizon are both used in Routing Information Protocol (RIP).

RIP 1

RIP routers communicate distance vectors via port 520/udp, where a RIP message within the UDP data field contains:

- RIP version—1 in this case.
- Command field (response = 1, request = 2).
- Up to 25 DV entries, each including IP address and distance; each of these is 20 bytes, leaving plenty of room for additional fields in future versions.

Every 30 seconds, each router's entire routing table is sent to every other neighbor router. If a router receives no update about a route for 3 minutes, then that route's metric times out and is set to infinity. The entry is deleted 60 seconds later ("garbage collection"), allowing time for the invalidation to propagate. Infinity is represented by 15 hops, which is assumed to be greater than the network diameter. Triggered updates are sent when the metric for a route changes, and split horizon is also implemented.

RIP has a number of drawbacks. It does not function efficiently when implemented in large networks, because the network diameter is limited by infinity being 15 hops, and also because periodic broadcasts of routing tables consume a lot of bandwidth. Furthermore, if a router fails, distance vectors converge too slowly; it can take several minutes, and time-out and garbage collection are slow. It is especially slow and hence inefficient for large autonomous systems; during settling time, loops can occur and datagrams can be lost or misdirected.

In RIP 1, routing decisions are primarily based on hop counting, with no concept of network delays or link costs. Also, RIP 1 is oblivious to subnet addressing since it cannot identify subnet and host fields. Furthermore, it is not secure since any host sending packets from UDP port 520 is considered to be a router by its local routers. Consequently, RIP has not been recommended since 1979.

RIP 2

In spite of the drawbacks in RIP 1, RIP was still the most popular routing protocol until relatively recently, due to its simplicity. But it has now largely been superseded by more modern protocols such as OSPF, although some enhancements were introduced in RIP 2, which employs additional message fields that were left unused in RIP 1. Hence the message format is backward compatible with RIP 1.

Improvements in RIP 2 over RIP 1 include the following:

- Subnet masks are passed as well as IP addresses.
- The first entry in a message can be an "authentication segment," implementing a simple 16-byte clear text password.
- Multicasting is used rather than broadcasting. RIP 1 broadcasts routing tables to all hosts and routers on the local network, whereas RIP 2 uses the multicast address 224.0.0.9 instead, meaning all RIP-2 routers but no other devices. This reduces the processing load on hosts, which do not need routing messages.

RIP 2 is still not a vast improvement on RIP 1—it still has hop count limitations, and also suffers from slow convergence.

RIP Summary

RIP is simple, with only two message types (request and response) and with only one table per router. It gives acceptable performance in small Autonomous Systems (ASs) where failure is rare. RIP is inadequate for large, complex ASs because DVs converge too slowly, causing congestion and packet loss, and larger networks fail more often because there are more links and routers that are capable of failing. RIP does not properly address issues identified by the Type of Service field in the IP header. These and other problems led to the use of routing protocols based on link-state algorithms.

Link-state Interior Routing

Link-state routing is completely different from the distance-vector approach. In link-state interior routing, each router maintains a Link-State Database (LSD), which contains the state of every link in the network—a picture of just what the network looks like. The LSD holds the state of each directed link in the network, consisting of the cost of the link and other information, such as whether it is broken and the number of the message that reported its state.

If a link's source router finds that a link has either failed or come up again, it floods the network with a message about the link status. Messages about a given link are numbered, so a router can tell an old message from a new one. When a link breaks, the database is updated via flooding of information; hence, a new set of routing tables is compiled comparatively quickly. This provides much faster convergence than is possible with distance-vector protocols. Based on information in the LSD, each router uses Dijkstra's algorithm to compute the least cost route to every other router, hence deciding the link on which to send a datagram on its least cost route. This approach overcomes the problems of RIP.

OSPF Design

Open Shortest Path First (OSPF) was developed openly by the Internet Engineering Task Force (IETF). Within OSPF, the HELLO protocol determines whether each neighboring link is broken, establishing and testing neighbor reachability. Every router sends HELLO messages to all its neighboring routers; a neighboring router may be attached either via a point-to-point link or via a LAN. The HELLO protocol also elects one designated router per LAN, which avoids routers on a LAN talking to each other needlessly, and ensures that databases are synchronized between routers on a LAN.

In OSPF, a large AS is divided into independent "areas." It is possible to have a separate set of routes for each IP type of service, where the link weight can be throughput, round-trip time, reliability, and so on. Routes can be chosen to balance network load, subnet addressing can be used, and password authentication is implemented, much as in RIP 2. As with RIP 2, OSPF uses multicasting, not broadcasting. OSPF data is encapsulated directly within IP datagrams; hence, OSPF does not employ either TCP or UDP. IP indicates that it is carrying OSPF data by setting the protocol field to 89.

LSD synchronization is essential in order for OSPF to work correctly; otherwise datagrams may be sent along incorrect routes. The objective of synchronization is to flood link status around the network as quickly as possible. All records are synchronously aged and dropped on expiry, and they are also protected by a checksum (the same format as IP).

Fig. 6.28 Seven-node network used to demonstrate Dijkstra's algorithm. The weights on links could indicate, for example, delay in milliseconds, or the reciprocal of link capacity in Gb/s.

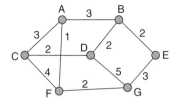

Furthermore, all router messages are password authenticated. Synchronization is achieved in OSPF by several packet types:

- *Encoded database description packets* describe the link-state database contents to a peer OSPF router. They list records to be requested later, and are exchanged when adjacency between routers is initialized. They contain message sequence numbers to identify which packets contain the latest information.
- *Link-state acknowledgment packets* send acknowledgments on a hop-to-hop basis during flooding.
- If necessary, a *link-state request packet* is used to solicit link information.
- State is exchanged using *link-state update messages*, which are sent in response to link-state request packets, and a link-state acknowledge packet is sent on receipt of this.

Dijkstra's Algorithm

Dijkstra's algorithm operates on a graph made up of nodes (or vertices) interconnected by links (or edges)—see Figure 6.28. Here, the nodes represent routers, and the links (or edges) represent transmission links between routers. The objective is to determine the shortest paths from one named node to all other nodes.

It is a classic shortest path algorithm with execution time that is approximately proportional to the square of the number of nodes. It is recommended for use in OSPF, where the LSD defines the graph. It is assumed that although the length or "distance" between all adjacent pairs of nodes is specified (Figure 6.28), this rarely represents physical distance—for example, it could be delay in milliseconds, or cost in arbitrary units. Links are often weighted with 1 for OSPF, so that the total distance is measured in hops.

The minimum distance from designated node i to all other nodes is sought. Each node j is labeled with:

- Whether it is "scanned," indicating distance definitely minimized
- $d(i, j)$, which is the minimum distance recorded so far from node i (the designated node) to the present node j
- A pointer to the next node on the route back to i

$l(j, k)$ is the length of the link from j to k. The algorithm now follows:

1. Mark all nodes as being "unscanned" with a null pointer.
2. Set $d(i, i) \leftarrow 0$ (distance from node i to itself is zero).
3. Set $d(i, j)$ to infinity for all j not equal to i (any distance computed later in the algorithm will be an improvement over infinity).
4. Set $d(i, j) \leftarrow l(i, j)$, and set the pointer at j to i, for all nodes j directly adjacent to i.
5. Mark node i as having been "scanned."
6. If all nodes are "scanned," terminate the algorithm.
7. Find the unscanned node j with the smallest $d(i, j)$.
8. **Scanning:** for all nodes k immediately adjacent to j, if $d(i, j) + l(j, k) < d(i, k)$, set $d(i, k) \leftarrow d(i, j) + l(j, k)$, and set the pointer at k to j (Figure 6.29).
9. Mark node j as having been "scanned" and go back to step 6.

Fig. 6.29   In Dijkstra's algorithm, $d(i, k)$ is the existing value of the distance from i to k, which may be over more than one hop. j and k are directly connected, with distance $l(j,k)$. If the route from i to k via j is shorter than $d(i, k)$, then the route via j is recorded as being preferable.

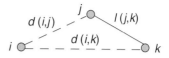

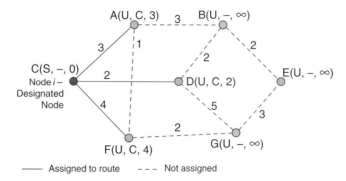

Fig. 6.30   Initialization of Dijkstra's algorithm, steps 1, 2, and 3. Node C is scanned. The notation is ("S" or "U," pointer, distance), where "S" denotes a scanned node, and "U" denotes unscanned. Links in dotted lines have not been assigned to a route, whereas those in bold lines have.

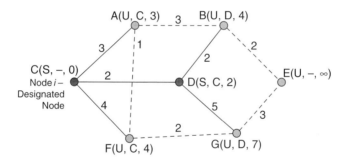

Fig. 6.31   Node D, with distance 2, the lowest unscanned node, is scanned. Distances for B and G are updated.

Example of Dijkstra's Algorithm   Figures 6.30 through 6.36 illustrate a simple example of Dijkstra's algorithm. The algorithm terminates when there are no unscanned nodes, and the distances and routes computed are shown in Table 6.10. In some cases, alternative courses of action must be considered, based on two equal distances. The choices made do not affect the distances computed, although they may affect the routes. Four links—AB, AF, DG, and EG—are not used at all, since this algorithm simply computes the shortest path and does not attempt to use network resources evenly.

Hierarchy in an Autonomous System

With OSPF, an Autonomous System is divided into areas (e.g., Figure 6.37), where one area is the backbone area and all other areas connect to it. The route between any two areas must pass through the backbone area, so that no two areas can be connected unless one is the backbone area. These areas must be defined by the network administrator, which may be troublesome for large ASs with many areas.

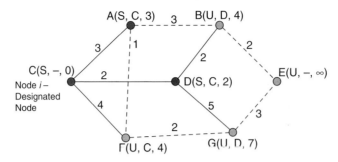

Fig. 6.32 Node A, with distance 3, the lowest unscanned node, is scanned. No distances are updated. F had distance of 4 both via the direct route and via A; it is unimportant whether its distance is updated.

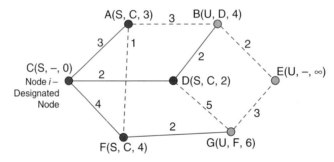

Fig. 6.33 Either F or B may be scanned, since they both share the lowest distance of 4. It does not matter which is chosen—the distances ultimately computed are identical but the routes are different. G's distance is reduced from 7 to 6.

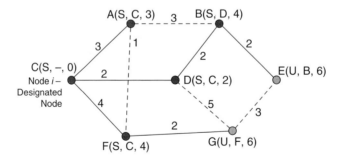

Fig. 6.34 Node B, with distance 4, the lowest unscanned node, is scanned. E's distance is reduced to 6.

Area-border routers (connected to the backbone area) summarize the calculations made within areas, for other routers to use; manual configuration of these is time-consuming and error-prone. Dijkstra's algorithm carries out routing calculations only within a single area, thus tackling its scalability issue.

In Figure 6.38, the backbone area cannot see the details of area A's topology; AB2 and AB4 are equally acceptable routers if A3 is the destination. If link a1 fails (Figure 6.38), area A is split. A1 and A3 update link states A1-A3 and A3-A1 and send them to AB2 and AB4 respectively, either by flooding or by contact with an adjacent router. AB4 must always be

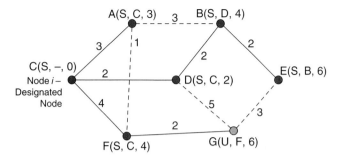

Fig. 6.35 Either E or G may be scanned, since they both share the lowest distance of 6. It does not matter which is chosen—the distances ultimately computed are identical but the routes are different. No distances are updated.

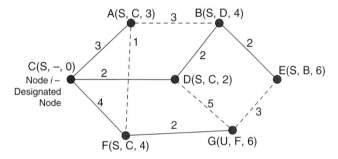

Fig. 6.36 Node G, with distance 6, the lowest unscanned node, is scanned. No distances are updated. Arguably, scanning G is not necessary; however, the algorithm specifies that this step must be carried out.

Table 6.10 Dijkstra's Algorithm Results

Node	Distance	Route
A	3	CA
B	4	CDB
C	0	C
D	2	CD
E	6	CDBE
F	4	CF
G	6	CFG

Fig. 6.37 Division of an autonomous system into three areas, when using OSPF. Information from one area is summarized by area-border routers before being passed into another area. All areas must be adjacent to the backbone area.

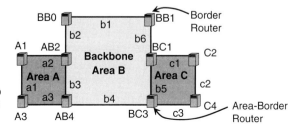

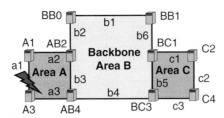

Fig. 6.38   Area A is split, although the backbone routers are not aware of this.

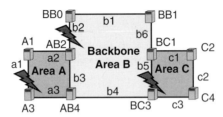

Fig. 6.39   The backbone area is split—links b2 and b5 have failed.

used by the backbone to route a packet to A3, and OSPF ensures that this happens by updating AB2's routing table to route to A3 via AB4.

If links b2 and b5 now fail (Figure 6.39), the backbone area is split. BB0 knows immediately of the failure of b2, and BC1 tells BB0 (via BB1) that b5 has failed. BC1 also advertises to BB0 a route with destination BC3 via area C. OSPF hence creates a virtual backbone link through area C, which enables router BB0 to communicate with router BC3. The only way of reaching area C from BB0 is now via BB1, which is computed by OSPF in the backbone. Once it reaches an area C router (BC1), Dijkstra's algorithm routes via links c1, c2 and c3.

Summary of Interior Gateway Protocols

OSPF is preferable to RIP because it has faster convergence that is loopless. It supports multiple metrics, and hence the ToS field in IP can support multiple paths to the destination by permitting routers to implement load balancing. It also requires less router signaling in large networks and supports another level of hierarchy ("areas") in an AS, permitting much larger ASs.

There are other interior gateway protocols, for example:

- *Interior Gateway Routing Protocol* (IGRP) is based on distance-vector technology. It uses IP, with a protocol field of 88, and is proprietary, owned by Cisco. An improved version exists—Enhanced IGRP or EIGRP.

- *Intermediate System to Intermediate System* (IS-IS) is an ISO-inspired standard for routing ConnectionLess Network Protocol (CLNP—ISO protocol similar to IP). It uses link-state technology, and is similar to OSPF.

Exterior Gateway Protocols

It is not feasible to use OSPF to route over the interconnections between ASs, because there are too many of them (up to 65,535) and because the metrics of ASs (e.g., transit ASs) are often "politically" based. The Exterior Gateway Protocol (EGP) was used in the past, but its backbone architecture has become outdated. Confusingly, EGP refers both to this particular protocol and to the generic class of protocols that set up routes between ASs.

Border Gateway Protocol (BGP 4) is now the standard means of connecting ASs. It is a variant on distance-vector protocols but nevertheless very different from RIP. Border routers send complete source-destination paths, avoiding the count-to-infinity problem by avoiding paths going through the node itself. The receiving router adds its 16-bit AS number and its own metric. To send the path to connected ASs, two conditions must apply. First, the politics must be acceptable; for example, one company may not wish to have its datagrams carried on a network owned by a certain other company, for reasons of confidentiality, or one ISP may not wish to carry traffic from another ISP, since for obvious reasons, it prefers to carry traffic from its own customers. Secondly, the node must not be already in the path (loop avoidance). BGP uses TCP port 179.

Internet Exchanges are a good example of the use of BGP. An Internet Exchange is a point where the networks owned by content providers and ISPs are interconnected to exchange traffic, improving connectivity and providing users with better-quality access to servers and other sites. Routing between these networks is carried out via BGP, so any organization wishing to be connected to an Internet Exchange must have a router running BGP 4, and an AS number.

BGP Routing Table Memory Estimate
A BGP routing table memory estimate is given by:

$$(\text{AS address size}) \times A \times \log_2 A + (\text{network address size}) \times N \text{ bits}$$

A is the number of ASs, and N is number of networks. The mean path length is assumed to be approximately equal to $\log_2 A$, which is exactly true for a tree network. The first term represents the memory required to store paths, while the second term represents the memory required to store next-hop information.

A rough estimate of the router memory size can be obtained. Assume all networks are class B. Assume all the class B address space is used, so $N = 2^{14} = 16,384$. Also assume there is no address aggregation.

$$(\text{AS address size}) \times A \times \log_2 A + (\text{network address size}) \times N$$
$$= 16 \times 65,536 \times 16 + 14 \times 16,384$$
$$= 16,777,216 + 229,376 = 17,006,592 \text{ bits}$$
$$= 2,125,824 \text{ bytes} \approx 2 \text{ Mbyte}$$

From Table 6.2, class B IP addresses begin with "10" binary, and they have a 14-bit net ID. Hence, "10" + 14 bits net ID + 16 bits host ID = 32 bits.

Starting up BGP 4 on a Cisco Router
Figure 6.40 could be the partial contents of a configuration file that starts up BGP 4 on a Cisco router in AS 5555. Static routes are user-configured—they direct datagrams to their destination within this AS, along a specified path. From a border gateway router of AS 5555 with CIDR prefix 209.123.10.0/23, the script declares connectivity to another AS 2222 via its BGP router 1.2.3.4. It then declares reachability to several networks in 5555 via the "network" command. 5555 has only one BGP neighbor specified in the script (i.e., 2222—neighbors are not discovered automatically, and must be specified via the "neighbor" command).

The last statement ensures that BGP advertises 290.123.10.0/23 to other ASs. This is because BGP only advertises routes in the routing table, due to synchronization issues with

```
router bgp 5555  Starts BGP with an AS number of 5555
network 209.123.10.0 mask 255.255.254.0  Prefix announced to peers
. . . many such statements in practice
neighbor 1.2.3.4 remote-as 2222  Peer has IP address 1.2.3.4 and AS no. 2222
. . .         base address      mask         next hop
ip route 209.123.10.0  255.255.255.0  Ethernet0  Static route in routing table
ip route 209.123.11.0  255.255.255.0  Ethernet1  Static route in routing table
ip route 209.123.10.0  255.255.254.0  Null0 254 ——Large distance ensures it is never used
Black hole route – discards packets
```

Fig. 6.40 Starting up a Cisco router running BGP, the text would be placed in a configuration file.

OSPF. The prefixes in the previous two statements do not match the prefix declared in the "network" command—they are more specific, and must be in order to be used for forwarding. This is why the extra dummy routing table is defined in the last line—it matches the "neighbor" command.

Multicast Routing

An IP datagram sent via multicast transmission is received by a group of devices, where each group has a unique class D address. Members of a group may be anywhere on the Internet, but only one copy of an IP multicast datagram passes across a link of any network used in the transmission. The multicast routing protocol ensures that during multicast transmission, routers do not receive multicast datagrams, unless they are involved in the multicast—only group members process multicast datagrams at the IP level. Multicast variants exist of RIP, OSPF and BGP.

Ethernet supports multicasting in layer 2. There is a direct correspondence between layer 2 and layer 3 multicast addresses, so ARP is not needed to translate from layer 3 to layer 2 multicast addresses. Hence multicasting on a single LAN is in principle straightforward.

Multicasting is facilitated by Internet Group Management Protocol (IGMP), which implements signaling between routers, such as messages for joining and leaving multicast groups. IGMP messages are encapsulated within IP datagrams. Multicasting was built into IPv6 when it was designed, but was merely added on to IPv4 as an afterthought.

The Application Layer

This section considers two common applications—e-mail and Web browsing, which make use of the protocols and concepts that have been introduced thus far.

E-mail

In an e-mail system, there are two types of programs, or applications. First, the user agent allows users to compose, read, file, and send e-mails, and this is the only application that many e-mail users are aware of. Second, the message transfer agent moves messages from source to destination, and runs as a background process. Its operation is hidden from the user.

The e-mail message structure consists of a header and the message itself. The envelope is used by the message transfer agent, containing the destination e-mail address, priority, security level, and so on. The message consists of two further subdivisions. The header contains information for the user agent, and the body contains information for the human recip-

Table 6.11 Header Fields for the RFC 822 Message Structure

Header	Meaning
To:	E-mail address(es) of primary recipients
Cc:	E-mail address(es) of secondary recipients
Bcc:	E-mail address(es) for blind carbon copies
From:	Person or people who created the message
Sender:	E-mail address of the actual sender
Received:	Agent ID, date, and time added by each MTA along the route
Return-Path:	Can be used to identify a path back to the sender
Date:	The date and time the message was sent
Reply-To:	E-mail address to which replies should be sent
Message-ID:	Unique number for referencing this message later
In-Reply-To:	Message ID of the message to which this is a reply
References:	Other relevant message IDs
Keywords:	User-chosen keywords
Subject:	Short summary of the message for the one-line display

ient, the most obvious example being a text message. The user agent displays a summary of fields from the envelope and header when displaying the contents of the user's mailbox.

The message structure is formally defined in RFC 822, taking the following form:

- The envelope, defined in RFC 821
- The header fields (Table 6.11)
- A blank line
- The message body

Multipurpose Internet Mail Extension (MIME) is an extension to RFC 822, which is defined in RFCs 2045–2049. One of its most common uses is to send file attachments with e-mails. MIME messages can be sent using existing mail software designed for RFC 822.

First it allows non-ASCII (American Standard Code for Information Interchange) alphabets and characters to be sent, such as French, German, and Chinese. There are two correct ways to send binary files such as executables and images—base64 encoding and quoted-printable encoding. In base64 encoding, the file is divided up into groups of 3 bytes, or 24 bits. Each of these groups is then divided into three smaller groups of 6 bytes, with each of these being encoded as an ASCII character. On the other hand, in quoted-printable encoding, bytes above 127 are encoded as an equal sign (=) plus their value in hex. This is more efficient than base64 if few characters have ASCII values above 127.

MIME also specifies the nature of the message body—for example, text, image, audio, or video. It permits attachments, also known as multipart messages.

Two protocols are used to transfer mail messages—Simple Mail Transfer Protocol (SMTP) and Post Office Protocol Version 3 (POP3). SMTP is defined in RFC 821, and provides transport both between mail servers and from each user machine to its mail server. To deliver e-mail, the source machine establishes a TCP connection to port 25 of the destination machine, while at the destination, SMTP copies each message into the correct mailbox. ASCII commands are defined within SMTP to allow the destination to accept or decline mail, and to acknowledge receipt of a message. Use of ASCII in this way makes the protocol easier to test and debug.

POP3 is defined in RFC 1939. SMTP on its own works well if all machines are always online. However, a TCP connection cannot be set up if the machine is turned off or is off-line, as with a home computer using a dial-up modem. Using POP3, the user machine can read e-mails from the mailbox at the ISP, or delete them, via a TCP connection that is open for the duration of the session. TCP port 110 is used for this purpose.

Web Browsing

This example shows what happens when Hypertext Transfer Protocol (HTTP) is used to retrieve a Web page stored at *www.adotcom.com/path/fermat.html*. Usually this takes place via a Web browser. However, here the interface will be via the `telnet` command, in order to show the commands, and Hyper Text Markup Language (HTML) text that is actually transferred. HTTP is the standard Web transfer protocol, and HTML is the language that Web pages are written in.

When the user requests a new Web page, an HTTP `GET` command is sent to the server *www.adotcom.com*. Port number `80` is used, identifying HTTP as the application. For purposes of illustration, the `telnet` command is used to connect to the remote server:

```
telnet www.adotcom.com 80
```

Then the `GET` command is sent to instruct the server to return the relevant page of HTML:

```
GET/path/fermat.html
```

HTTP is an ASCII protocol that includes other request types besides `GET`, including:

- `HEAD`—read a Web page header
- `PUT`—store a Web page
- `DELETE`—remove a Web page

The server sends the HTML Web page back to the client:

```
HTTP/1.0 200 OK
Date: Mon, 8 Nov 2004, 17:04:39 GMT
Context-Type: text/html
<html>
<body>
<h1>Proof of Fermat's last theorem</h1>
.........
</body>
</html>
```

The code `200` on the first line represents "OK." Other codes are `400` for "bad request," `403` for "forbidden," and `404` for "not found." Usually, a browser formats the HTML text file using the tags enclosed in angle brackets, and then displays the finished document. There is a blank line before the first tag, and more HTML code exists before `</body>`, indicated by the series of dots. However, HTML or browsers are not necessary for HTTP to work, since an application may connect to the server directly. Also, an HTTP server can be accessed manually via telnet (as previously).

Conclusions

Over the last twenty years, TCP/IP has been incredibly successful. The most important protocol in the suite is IP, the Internet Protocol; indeed all data exchanged on the Internet is carried by an IP datagram, a uniform and universal packet format. There is a mapping between IP addresses and link-layer addresses that facilitates the operation of IP with any link-layer technology. This has been a major factor in IP's success, which has allowed IP to operate with several generations of LAN technology. Furthermore, IP has handled the extreme increases in scale that have taken place on the Internet over the last twenty years, although, as the discussion above on IP addressing showed, refinements to the addressing scheme have been necessary.

TCP is also an important protocol, providing end-to-end "data circuits" that are used by many well-known applications, such as FTP and HTTP. It overcomes the main limitations of IP, namely its tendency to drop datagrams or deliver datagrams out of sequence. Routing protocols are also important. RIP, an early Internet routing protocol, described earlier, is now rarely used even though it was popular at one time. This is because, despite its simplicity, it often does not converge gracefully in the event of a fault, often taking several minutes to do so. OSPF is a modern routing protocol that overcomes these problems via a "link-state" algorithm where each node builds up a picture of the entire network in its locality, to assist its routing decisions. RIP and OSPF are designed for routing within an autonomous system—BGP is the standard protocol for interdomain routing.

Acknowledgments   The author wishes to thank Tim Gilfedder of BT for his constructive comments on an earlier draft of this chapter, and Matthew Thomas of the University of Essex for his advice on BGP. Thanks are also due to the students, who have been part of the MSc course EE904 at Essex University, whose comments and queries have helped improve this material over the years.

Resources

Comer, D. E. (2004) *Computer Networks and Internets with Internet Applications, Fourth Edition*, Prentice-Hall.

Fall, K., and S. Floyd (1996) "Simulation-based Comparisons of Tahoe, Reno and SACK TCP, *Computer Communications Review* 26 (July), pp 5–21.

Halsall, F. (2005) *Computer Networking and the Internet, Fifth Edition*, Addison-Wesley.

Hassan, M., and R. Jain (2003) *High Performance TCP/IP Networking*, Prentice-Hall.

Huitema, C. (2000) *Routing in the Internet, Second Edition*, Prentice-Hall.

RFCs (Requests for Comments) from the IETF, available from *http://www.ietf.org/*.

Stevens, W. R. (1994) *TCP/IP Illustrated: The Protocols*, Addison-Wesley.

Tanenbaum, A. (2003) *Computer Networks, Fourth Edition*, Prentice-Hall.

7 Voice over Internet Protocol Networks

Martin J. Reed

Introduction

Voice over IP (VoIP), or IP telephony, is changing the nature of fixed network telephony by providing Internet-based telephony applications and forming the new data/voice convergent backbone network infrastructure for traditional telecommunications operators. Technologically speaking, the components for VoIP were in place as far back as 1974 when the first IP telephony demonstration was made (Gray, 2005). However, it has taken some time for actual IP networks to mature to a state where packet networks can challenge and replace the circuit-switched networks that have been in place since the early days of telephony.

One of the problems has been that IP networks have not been capable of supporting real-time data (mainly interactive voice or video), but instead have been mainly used for non-time-critical data networking. There have been a number of attempts at adding features similar to circuit-switched capability to IP networks that would have allowed real-time services with reasonable quality, but these have mainly become much talked about but not implemented.

However, since approximately the year 2000, three interesting developments have taken place: IP has finally been given some working traffic management capability through new technology (e.g., multiprotocol label switching); IP networks have evolved in bandwidth capability such that voice data is a smaller (rather than larger) component compared to data; and home subscriber access bandwidths carrying both audio and data over the same link are now possible. The last two developments (core and access bandwidth rate increases) have allowed a new breed of operator to sell "free" calls over public IP networks such as the Internet. The fact that voice is a small component compared with the network transmission rates has meant that these systems can operate without any specific quality of service (QoS) support in the network, albeit with lower than toll quality and no guarantees of service reliability.

The first two developments (IP traffic management and high-speed IP core networks) are now allowing operators to move their public-switched telephone network (PSTN) infrastructures from circuit-switched systems, with their many bearer services, toward a single IP bearer. The first operator to do this in a major national network was British Telecom using a next generation network (NGN) architecture that uses the IP multimedia subsystem (IMS).

This chapter gives an overview of the architectures to be provided and then will consider the voice data path, followed by a discussion of signaling and architectural issues.

Simple Architecture Example

The most basic form of VoIP is to consider two telephones connected to an IP network as shown in Figure 7.1. There are two essential aspects to consider. One is the "data path," the way in which the voice signals are converted into packets and sent over the network; the other is "signaling," the way in which the terminals and network are configured to transmit the voice data. The way in which voice signals are transmitted in VoIP is quite

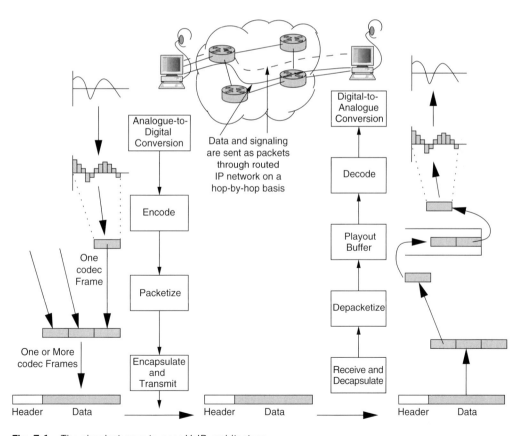

Fig. 7.1   The simplest peer-to-peer VoIP architecture.

different from circuit-switched PSTN communication technology in that IP is connectionless and does not natively support timely delivery of multimedia traffic.

The key element of VoIP is that the digitized and coded voice signal is converted from a continuous bit stream into discrete packets before transmission. The packets are then sent through an IP network where each packet header is examined individually by each router; the router examines its routing table to determine the correct output port (based on destination address) and places the packet in an output queue (or buffer) for the output port. It is this output buffer that gives asynchronous packet-based systems, such as IP, a distinct advantage over synchronous systems, as peak traffic demand loads coming into a router can exceed the output rate as long as the peak load does not cause the finite buffer to be overfilled. This is usually termed statistical multiplexing gain. However, the output buffer is a distinct disadvantage for real-time services such as VoIP, as the interaction of traffic causes the audio packets to be displaced in time or lost entirely if the peak load is too high.

The time displacement is called "jitter," or variable packet delay, and is one of the main disadvantages of using packet networks, such as IP, for voice transmission. There have been a number of attempts to improve IP networks for use with real-time traffic, but none have provided a complete solution using the network layer alone and they do not have mature management solutions.

The simple scenario shown in Figure 7.1 is the example of an Internet peer-to-peer architecture where the caller knows the IP address of the called party. Hence, the signaling can be made simply between the two terminals. In this scenario the terminals are PCs that use standard PC-compatible headsets and sound cards. This scenario has one problem in that the two systems have to address each other using well-known IP addresses or valid DNS host names. In practice, users do not keep host names or IP addresses for long periods of time compared to phone numbers. Also, in this simple scenario, there are no central directory or redirection services that enable users to find each other or move location. One other disadvantage with this architecture is that it does not include an interface to other networks so that it is not possible to communicate with the PSTN.

Integrated VoIP Architecture

Following on from this simple example, a more complete picture of an integrated VoIP system is shown in Figure 7.2. There are, in fact, a number of standardized systems that provide this architecture, including ITU Recommendation H.323, a number of IETF standards (including session initiation protocol—SIP), and the IP multimedia subsystem (IMS). Unfortunately, each of these standards uses slightly different terminology and logically separates the service function into different subsystems. Here the terminology will remain generic; later when the signaling is described, specific terms will be used in their place.

For the IP phones B and C in Figure 7.2, the scenario is similar to that shown in Figure 7.1. However, there is now an added service function that can provide directory services and phone number to IP address or DNS name mapping. This means that IP phones can be used in place of PCs and to users there may not be any apparent differences compared to traditional telephone systems. The service function can participate in the signaling of sessions to a variable degree: It can provide a complete signaling proxy service (as described later) or it can simply provide address mapping so that the terminal can perform peer-to-peer call signaling. The service function can provide many other services, such as multi-party calls and other common telephony applications (and with expansion for new ones).

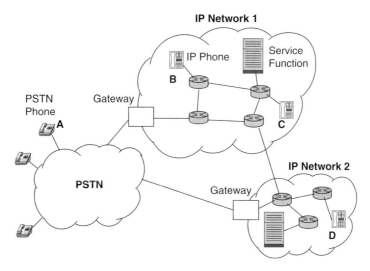

Fig. 7.2 Generic view of VoIP architecture that provides integration with other networks and provides centralized and distributed service functions.

One of the drivers behind IMS, considered later, is that new services can be introduced more flexibly and cheaply in such an architecture compared to traditional telephony systems where the service functions were closely bound to the switching hardware.

The key component that allows integration with existing PSTN networks (and other networks) is the gateway that provides two main functions: mapping between the two signaling standards of each network and conversion between the different media codec standards used in each network. Both of these functions are fairly complex, and providing feature-rich gateway systems has been one of the main efforts made by VoIP vendors (and is one of the competing factors among vendors). However, with the increasing complexity of diverse services possible within a next generation network (NGN), more recent work has been done on standardizing and adding functionality to the service layer.

To allow IP phones to integrate with the PSTN, the VoIP network has to be allocated telephone number ranges within the scope of the PSTN. This fact allows a simple example of the complexity required in the service layer of VoIP systems: The two IP phones B and D in Figure 7.2 wish to start a session. If they both have addresses allocated within the PSTN numbering then, without cooperation at the service layer, it is possible that they will use the PSTN to communicate. Clearly this will be a problem as the media has to pass through two separate gateways and will incur two conversions between codec standards and increased delay due to buffering required for VoIP systems. If there is a compatible direct IP connection, as shown here, then it is clearly better to use it; this requires the service layers of each IP network to cooperate and provide location and path discovery.

There is a related problem in that, if we consider IP networks 1 and 2 to be part of the same VoIP network provider, then if caller D makes a call to caller A, which path is used? Consider if A and IP network 1 are in the same continent but IP network 2 is in another continent. If the VoIP system "hands over" the session to the PSTN in IP network 2, then the VoIP provider has to pay international call charges for the call. However, if the VoIP provider routes the call via IP network 1 then the call rate may be only for a local call in the PSTN. To perform this intelligent routing of calls requires media gateway control protocols.

It is this intelligent routing of calls that allows Internet VoIP providers to offer low call rates for long-distance or international calls to PSTN terminals.

There are a number of business models evolving. One involves offering a free PC-based Internet phone for IP-to-IP calls but charging for calls to the PSTN on a per-minute basis; another model includes IP telephony with broadband access and charges a flat monthly rate for data and telephony calls to the PSTN and other VoIP terminals (extra for certain international and mobile calls).

IP Phones and the Voice Data Path

In many ways the voice circuit path used in VoIP shares the features of most other telephony systems: The voice signal from a microphone is sampled, one or more of the samples are coded in a codec to reduce the bit rate, the output of the codec is taken as a block of samples that are encapsulated in a protocol (or transmission channel), the block is transmitted, the block is decoded in the codec of the receiver, and the analogue waveform is reconstructed at a loudspeaker. This process is shown in Figure 7.1. However, there is one important difference with VoIP compared to most preexisting telephony systems and that is in the nature of the transmission channel, which can have both high loss and variable packet delay (and even packets arriving out of order). Here the issues regarding the codec, packetization, and transmission elements are introduced.

Types of IP Phones

Before considering how a VoIP transmission path works in detail, it is useful to understand the phones that a user may use to access VoIP. VoIP terminals are broadly of three types:

- Hardware VoIP phones
- VoIP softphones
- 2/3 G mobile integrated with VoIP

Hardware VoIP Phones

Hardware VoIP phones look broadly similar to traditional high-functionality PSTN phones. The only obvious difference is that the network connection is either via wired or wireless Ethernet. However, internally there is significantly more complexity in a hardware VoIP phone than in a PSTN phone due to the packet nature of the transport and the greater intelligence required for call signaling.

VoIP Softphones

VoIP softphones (usually simply called softphones) make use of an existing PC soundcard and network connection to provide all the hardware for the "phone" so that all that is provided by the vendor is software. These systems have an advantage in that additional services can be provided with the phone where the PC can be used to provide the user interface. So, for example, softphones often provide address book facilities, "buddy lists" (which notify the user of other commonly contacted people and their "online" status), and instant messaging systems. However, they have a distinct disadvantage in that the audio path is highly variable due to the large variation in PC hardware and the different ways in which users configure sound card gain controls. Later this problem will be considered when echo control is discussed.

To help with the variable audio path, some softphone vendors supply (or highly recommend) headsets with known specifications. Another example of the problem that these phones may have is that if the headset is plugged into the audio sound card, the user will hear a ringing tone only if the headset is worn, or will see the visible warning only if they are looking at the screen. All of these problems can be solved, but users will need to understand that multipurpose platforms such as PCs do not make ideal telephony systems without some careful configuration that is not normally required for hardware phones.

2/3 G Mobile Integrated with VoIP

An interesting development in VoIP phones is in combining the technology with 2/3 generation mobile phones. Such phones typically use Bluetooth or wireless LAN technology when they are close to a home Internet broadband gateway and then use 2/3 G mobile phone technology when away from the home gateway. The move between the two different technologies is termed "vertical handover" and this requires that the VoIP system is integrated into the mobility model of the 2/3 G network control. The vertical handover is not straightforward, as the two networks use very different technologies; one of the problems is equalizing the delay to give inaudible transfer as the user moves between the two systems during a conversation.

The advantages of the system are that a user making outgoing calls gets a low call charging rate when using VoIP and only one phone is needed for all calls. However, because early systems used the 2/3 G network roaming facilities, incoming calls were all routed through the mobile network and billed to callers as such even though both terminals could be using VoIP to access the mobile transport network. Although initial systems were designed to be used with home access gateways, there is increasing interest in using wireless LAN technology in metropolitan areas to replace the 2/3 G infrastructure.

Speech Encoding for VoIP

Summary of Codecs Used for VoIP

Traditional wired, circuit-based telephony uses channels based on 64 kb/s circuits and consequently, while there has been interest in other bit rates for multiplexing and wideband audio, most of the systems have used μ or A-law PCM codecs (ITU-T Recommendation G.711). However, there has been great advance in speech codecs used for mobile telephony and many of these have been put to use in VoIP. Common codecs are shown in Table 7.1. See Cox and Kroon (1996) for an introduction.

Of particular note is that the default standard (G.711) is the same as used in traditional telephony; this is usually mandated as the codec that all VoIP phones must support. However, there are improvements offered by other codecs and so IP phone vendors often offer a number of other choices. In particular, the low bandwidth options have been developed with packet and mobile systems in mind and these give acceptable performance on bit rates of the order of 10 kb/s. One codec of note is the G.722 standard because, although the codec itself is not sophisticated by modern standards, it offers a voice bandwidth of 50–7000 Hz. This is approximately double that offered by traditional telephony that has been locked to the G.711 standard; indeed, this shows one of the advantages with VoIP to be increased flexibility with codecs (and in some cases signaling as we shall see later). One problem with VoIP is that there is a diverse set of terminals and systems that have evolved without a common set of standardization systems. Consequently, a terminal may support only some of the supported codecs.

Table 7.1 Brief Comparison of Some Common Codecs Used in VoIP

Codec (ITU-T standard unless otherwise stated)	Bit rate (kb/s)	Typical delay due to frame size and coding delay (ms)	Comment
G.711	64	Not restricted	The default pulse code modulated (PCM) codec, compatible with all phones, low processing overhead, poor bandwidth utilization.
G.723.1	5.6/6.3	30	A linear predictive analysis by synthesis codec. Supported by most phones (but licensc restricts free use in "free" softphones).
G.722	34–64	3	Wideband speech (50–7000 Hz). Voice quality is probably best in 56–64 kb/s range. Uses adaptive differential PCM and two frequency bands to give approximately twice the bandwidth of the other codecs in this list.
G.729	8	10	Algebraic code excited linear predictive codec. Designed to have low delay.
GSM-FR (ETSI standard 06.10)	13	20	Regular pulse excited long-term predictive codec. Standard is open for free use. (Although the more modern GSM-EFR has better quality it is not available for "free.")

The choice of codec is fairly complex as all codecs have trade-offs between bandwidth, processing delays (packetization and algorithmic), and licensing costs (only G.711 and GSM-FR in Table 7.1 are truly "free"). Indeed, as VoIP has grown to support and use a wide number of different systems the range of codecs in a terminal (and gateway) can be correspondingly wide; consequently, one requirement of the signaling within VoIP is the support of end-to-end negotiation of codec type.

Silence Suppression

An additional feature that the statistical multiplexing of IP benefits from is that, if someone is not talking, in theory it is not necessary to transmit any packets. This is termed silence suppression. For typical speech communication this can give a saving of approximately 50 percent. To operate properly, silence suppression requires a good voice activity detector (VAD) and the injection of comfort noise so that a listener does not notice the change in background noise. VAD requires a complex algorithm to operate properly without clipping the end or beginning of speech. Good VAD comes with codecs such as G.729 and G.723.1; however, G.711 has no such facility and requires additional algorithmic support with reported variable success (James et al., 2004). Indeed, as a good VAD algorithm requires almost as much effort as one of the more complex speech codecs, the use of VAD with G.711 is questionable.

One of the problems with silence suppression is that during periods where someone is not talking packets are not sent; if silence was just played to a receiver then the user would find the change in background noise annoying. Consequently, there is a requirement that the silence suppression insert some representative comfort noise. In practice this does not work particularly well (depending on the background noise level), so many implementers do not support the use of silence suppression. However, in lower-quality, low-bandwidth applications it still has its uses.

Packetization and Protocol Encapsulation

Network and Transmission Layer Protocol

Once a voice signal has been sampled and coded, it is collected into a block that can be transmitted in an IP packet. This is one of the unique differences from traditional telephony in that the transmission channel is not a circuit that is available for voice data for the duration of the call; rather, it is a medium that is shared with other IP communications. Consequently, there is an overhead associated with each packet such that it can be transported to the destination. The protocol generally used for real-time data is the user datagram protocol (UDP). This and the transmission control protocol (TCP) are the most common transmission layer protocols used in IP. Unlike TCP, UDP does not provide any mechanism to guarantee delivery of a packet (or notify of failure); furthermore, UDP does not provide for sequential delivery.

At first sight it might seem that these missing features make TCP a better candidate for voice data; however, TCP uses a mechanism of acknowledgment packets from receiver to sender and retransmission in the absence of acknowledgment. The retransmission process is not appropriate for real-time interactive applications as it is often better to ignore late or missing packets. Waiting for retransmission may incur too much delay; we will see later that the delay budget on VoIP is already quite tight.

Because UDP does not provide sequence indication (and other features needed by VoIP) the audio packets need to be wrapped in an additional protocol: the real-time transport protocol (RTP) was developed for this.

Real-time Transport Protocol

The real-time transport protocol forms an additional header followed by the real-time data, in this case a block of coded voice data as shown in Figure 7.3. Only a summary of the protocol fields are included here; the full standard is in RFC 3550.

Packets sent by a VoIP sender are sent as a regular packet train; however, they may be delayed by differing amounts (jitter) and, as they may pass over different paths, may even be in a different order when arriving at the receiver. At the receiver, the variable delay, out-of-order packets, and missing packets need to be detected and compensated for. RTP allows the receiver to do this. The main features of RTP that are used for detecting packet order/timing problems are the sequence number and time stamp fields. The sequence number is usually incremented by one for each packet; the time stamp is defined as the time that the first octet in the transmitted data block was sampled.

Because time synchronization between the sender and receiver can be difficult, it is only the relative time between the two that is important, so the receiver measures only the time between consecutive packets and from this it can estimate the level of jitter between packets (but not the actual transmission delay). As we shall see, for real-time systems it is this jitter that can be more important than the actual transmission delay for a receiver.

One other field of note in the RTP packet is the payload type. This defines the codec (and thus the type of media) that is transmitted. As the payload type is sent in every packet, it is possible that the sender can change the codec without having to signal this change to the receiver. This is useful as it is possible for the sender to reduce or increase the bandwidth of

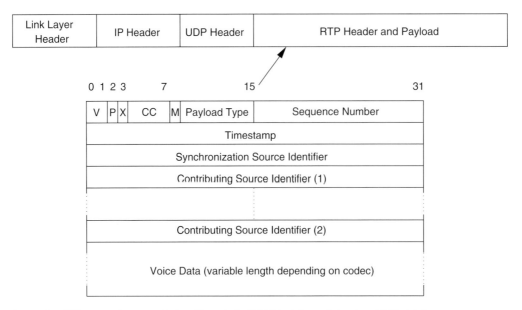

Fig. 7.3 RTP protocol encapsulation (the whole RTP block is contained as UDP data).

the data sent to suit the network; for example, if there are many packets lost, it may be worth using a lower-rate codec in the hope that the path is no longer overloaded. The ability to change codec depending on network condition requires the receiver to notify the sender of problems.

The problems are detected at the receiver by the RTP header information. However, as RTP is only one-way (sender to receiver), there is an additional protocol called real-time transport control protocol (RTCP) that allows (but does not necessarily require) the receiver to send reports to the sender. The payload type field is standardized for an RTP profile for audio and video so that for example 0 specifies G.711 and 18 specifies G.729. Defining these in a fixed manner is clearly not useful for future expansion, as new codecs may follow after the standard; consequently, RTP payload type fields 98 to 127 are reserved for dynamically defined codecs.

RTP (and its partner RTCP) is actually a very feature-rich protocol that allows many other features to be implemented that are not discussed here. It can be noted that there are other fields in the RTP header called source identifiers that allow many communicating sources to be mixed and identified. One synchronization source identifier may be given so that synchronization can be determined relative to one particular RTP stream that may make up a whole application; for example, there may be both audio and video for a video tele-phony application.

This also highlights the fact that, usually, each coded stream is sent individually. For example, in a video telephony application the video is sent in one RTP stream and the audio is sent in another; the reason for this is that each medium often has quite different charac-teristics of regularity of packet transmission; audio is usually continuous bit rate while video is often variable bit rate.

RTP provides the VoIP application with the ability to cope with varying network con-ditions and the problems associated with the inherently nonconnection-oriented nature of

Table 7.2 Header and Data Lengths for a Typical VoIP Packet Using RTP

Header/data	Header/data length (bytes)
IP header (no option fields)	20
UDP header	8
RTP (no contributing source ID—only one source so not needed)	12
Total of IP datagram header	**40**
Ethernet header + CRC	18
Total of IP datagram and Ethernet overhead	**58**
Three G.729 speech frames (30 ms of audio)	30
Total packet length	**88**

IP. However, it does not itself address network QoS issues or even communicate directly with network QoS mechanisms. Network QoS has to be solved using other techniques.

Packet Length

The theoretical range of IP packet lengths that an application can use is relatively arbitrary between the limits of zero and the maximum transmission unit (MTU) which, for IP networks such as the Internet, is typically 1500 bytes. However, for real-time systems such as VoIP, the packet length can significantly impact the performance of the system in a number of ways. The most significant impact of the packet length is in the packetization delay. As we will see later, the delay budget for VoIP can be quite tight so it is important to keep this as low as possible. However, many of the low-rate codecs work on blocks of speech and so limit the smallest data unit that can be usefully sent. For example, the G.729 codec has a 10-byte frame size so the minimum packetization delay is 10 ms (at 8 kb/s).

We could take this as the minimum data frame size as most other codecs have the same or larger frame size. However, in practice most software-based systems have interrupt handling in the order of 10 ms timeframe (this has not changed significantly compared to the comparative increase in network and processor speeds).

Consequently, it is desirable for softphones running on PCs to handle data frames that are less frequent than every 10 ms so as not to overload the processor and bus. This gives rise to a commonly used frame size of 30 ms of speech data (i.e., packing three G.729 frames in each RTP/UDP/IP payload). It is interesting to see how the datagram size is made up using this as an example, and this is shown in Table 7.2. It can be seen that the voice data comprises only about one-third of the total packet size (it would only be about one-seventh if only one G.729 frame was used). Clearly this causes a significant transport overhead. There are header compression techniques that can compress the RTP/UDP/IP header to around 2–4 bytes; however, the compression is only practical for specialist serial systems that compress/decompress the header at each end of the link, or for systems that bulk multiplex voice data streams into a transport tunnel.

IP Transmission Properties

Traditional (wired) telephony has relied on circuits that maintain the delivery of voice data in a reliable manner, with samples being delivered without differing delay and with very low bit error or loss rates. However, IP networks can have very different characteristics of both delay and loss performance. One problem with discussing IP networks is that the

Table 7.3   Approximate Guide to Delay and Loss Characteristics in IP Networks

Network type	Typical one-way delay	Typical delay variation	Typical packet loss
An intranet (nonpublic network on a single site) (James et al., 2004)	10 ms	Less than 10 ms	Less than 1%
Internet (a public network) with sites in remote continents and good access networks	100 ms	Average 10 ms but as much as 100 ms	Of the order of 1% but some bursts of loss in the region of 10–100 ms

protocol is used in very different scenarios. In the two extremes, we have the public Internet, which can have variable (and somewhat uncontrolled) traffic loads, and we have a high-speed local area network that can have carefully controlled traffic loads. Table 7.3 shows some typical delay and loss characteristics found in IP networks; this information is not precise as can be seen from the table, which shows that the values are highly variable depending on the network type.

An additional problem in considering the Internet is that the performance is a combination of many different networks. One important aspect of the performance is the access network used to connect to the Internet. With modem-based systems, the access speed is not high enough to support real-time services such as voice. It is the access network and not the Internet that is the limiting factor here. However, many users are now connecting to the Internet through broadband connections; in this scenario, the Internet is often the limiting factor.

Broadly speaking, the transmission problems can be broken down into, first, a combination of jitter and packet loss and, second, one-way delay. The next discussion takes these in turn and describes why they cause conflicting design problems in VoIP applications.

Coping with Jitter and Lost Packets

As IP networks can have variable packet delays (jitter) and a significant packet loss, there needs to be some mechanism in VoIP applications to deal with these problems. One solution to the problem is to manage the network such that the delays and packet loss are within tightly controlled limits. While this can be done to a certain extent on highly managed intranets, it is proving difficult to see a solution on public IP networks such as the Internet, and even in corporate intranets it is advantageous to relax the level of QoS requirement on the network to simplify the management. Consequently, much of the problem has been moved to the VoIP application itself, in particular the receiving terminal; this is the problem considered here.

IP networks make use of statistical multiplexing to share the bandwidth of the transmission channels between all the flows, and each application may send packets at some indeterminate time. Consequently, it is difficult to predict the future packet arrivals at a router. This could result in more traffic arriving at the output of a router interface than it can cope with. Routers use buffering to cope with the variable load demand and each application flow is multiplexed with packets from other applications.

In the case of VoIP this is a problem as the sender audio generates audio samples at a continuous rate and the receiver needs to play the samples at the same rate—any delay in a packet could lead to an audible gap in the audio. Thus, VoIP receiving terminals

Packets Transmitted at Regular Intervals

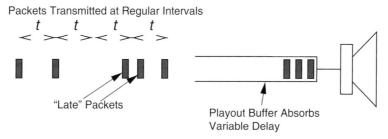

Fig. 7.4 IP networks cause variable packet delay, known as jitter.

have to implement playout buffering to smooth out the effects of network jitter, as shown in Figure 7.4. As the network jitter is of the order of 10 ms (or less) in a well-managed intranet, or in the region of tens of milliseconds in the public Internet, the buffer needs to be as long as this to cope with these variable delays.

Typical values for playout buffers reported in commercial VoIP terminals are in the region of 50 to 100 ms. In practice it is often desirable to implement an adaptive buffer algorithm that adjusts the playout buffer according to jitter measured at the receiver as part of the RTP functions; then if the jitter seems to be large (maybe a session over the Internet), a large buffer is used. Alternatively, it can be considerably shorter if the terminal detects that the session is operating over a well-behaved network.

If the jitter in a packet is greater than the playout buffer can cope with, then there will be a gap in the audio that may be audible. Once the packet has missed its playout slot it does not make sense to play it; else a large gap followed by a sudden influx of "backed-up" packets may cause the playout buffer to be unduly long. Consequently, a late packet can be considered the same as a lost packet; indeed in some forms of analysis a lost packet is considered as a packet with infinite delay.

If there is a gap in the audio, where the gap is greater than a few tens of milliseconds, then it may be audible. The effect is dependent on the speech at the time of the loss, the relationship between how often the loss occurs, and the pattern (if any) of the loss. The simplest approach is just to play silence when there is packet loss; however, this gives quite poor performance. A much better approach is to use some form of packet loss concealment (PLC). This has been proposed for some time as part of the G.711 standard (and is improved in more modern codecs that have more complex PLC algorithms).

Typical PLC techniques involve playing samples from the last received packet to pack out the gap, followed by a graceful move to silence after some period of time (often around 30 ms). PLC can improve the performance considerably with perceived performance for packet loss of approximately 5 percent with PLC compared to packet loss of 1 percent without PLC (Markopoulou et al., 2003).

The Affect of One-way Delay on VoIP Quality

Excessive delay is clearly a problem in an interactive application like speech communication. Recommendation G.114 specifies one-way delay classes as summarized in Table 7.4. It can be seen that the goal of a speech communication system is to achieve the target of less than 300 ms. If the delay becomes too high, users tend to find that they start talking at the same time without realizing it and find that they tend to interrupt or wait long times for replies, which feels unnatural.

Table 7.4   Summary of Recommendation G.114 Linking the Ease of Interactivity with One-way Delay

Class	Delay (ms)	User perception
1	0–150	Acceptable
2	150–300	Acceptable but noticeable to users
3	300–700	Users find it difficult to interact
4	Greater than 700	Unacceptable unless users are trained in "half-duplex" communication

Table 7.5   Approximate Delay Budget for a VoIP System (not including transmission delay)

Source of delay	Approximate delay budget (ms)
Capturing sample from microphone and encoding with G.729	15
Packetization delay with 30-byte frames sent every 30 ms (3 × 10 ms G.729 blocks)	30
Transmission delay	X (where X can be of the order 10–150 ms)
Incoming system queuing/interrupt delay	10
Jitter buffer	50
Total	**105 + X**

At first glance, it might seem that with transmission delays as shown in Table 7.3 there should be no problem with VoIP. However, the overall delay budget tends to be higher than just the transmission delay due to features inherent in VoIP (and other packet-based systems)—in particular, features like the playout buffer. An example delay budget is shown in Table 7.5, giving a delay of 105 ms without including the transmission delay. Following the transmission delays given in Table 7.3, this would give 125 ms delays in an intranet and in the region of 200 to 250 for longer-distance Internet sessions. However, if additional systems are incorporated, such as a VoIP session communicating with a PSTN or mobile user through a gateway, then delays can quite easily reach above the 300 ms target.

It is interesting that while studies show that users are able to notice the reduction from toll quality that this increased delay brings, many users find that they can cope with it. In particular, this is true where users are willing to accept "free" Internet VoIP in return for some loss in quality and may accept delays well above 300 ms.

The affect of delay on the perceived quality of VoIP is further complicated by the problem of echo in the transmission path where the talker hears her voice repeated back to her with some delay. This echo can be very annoying, although users often learn to cope with limited amounts of it. The causes of echo in VoIP systems are generally from two mechanisms: the acoustic/mechanical path between loudspeaker and microphone (at the receiver) and/or the four-wire to two-wire conversion caused by a receiver that is using a two-wire system as found on the PSTN. The latter is always the case if a VoIP session includes a gateway to the PSTN as nearly all current PSTN systems use a four-wire transmission system but convert to two-wire in the local loop. It might be thought that a pure VoIP session is inherently immune to echo problems as it is a four-wire system; however, there may be some echo path through the acoustical/mechanical path at the far receiver.

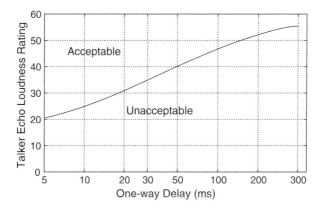

Fig. 7.5 Graph showing the acceptable level of talker echo loudness rating against the one-way delay (as defined in G.131).

The ITU has standardized the annoyance of echo/delay in G.131, which is summarized in Figure 7.5. A simplistic view of this metric is that the talker echo loudness rating (TELR) can be compared to the amount of echo attenuation. In other words, high TELR is equivalent to very low echo, whereas low TELR is equivalent to a high level of echo. The actual calculation of TELR involves the sending level of both ends of the communication together with calculations of coupling losses/gains (see G.131 for more detail).

The line in Figure 7.5 represents the acceptable limit for most users; anything above the line is better. In practice it is not possible to draw an exact line and the result is based on a mean opinion score. Once it is known where a design falls on this graph it is possible to make a decision on whether specialist echo cancelers are required: If the TELR for the one-way delay budget is above the line, then an echo canceler should not be required.

G.131 Standard
Some examples are useful to demonstrate the use of the G.131 standard. For the PSTN in a single country we have typical TELR of 30 dB and a delay of less than 20 ms. This means that echo cancellation is not required (and it is not normally provided in a single country/region PSTN). For a purely VoIP session where we have four-wire communication, the value of TELR might be in the region of 50 dB so that for most pure VoIP sessions the quality might be acceptable if the one-way delay is below 150 ms. From Table 7.3 it would seem that this is true for intranet VoIP and might be true for Internet phone usage if just transmission delay is taken into account. Now consider the case where a VoIP session communicates with a PSTN terminal through a gateway. The TELR now might be in the region of 30 dB and the delay may be 150 ms; clearly now echo cancelation is required.

The examples discussed previously are given using vague values of TELR. TELR is not actually the same as the commonly used "echo return loss" (ERL), which describes how much a terminal naturally suppresses echo. TELR is actually calculated from a summation of terminal/transmission gains (or losses) and the ERL. This can be particularly complicated in VoIP systems as the relative audio gains in elements such as gateways and terminals are not as fully standardized as for PSTN systems. This is particularly problematic for non-trained users who are using a PC-based softphone and nonstandard headsets. For example, it is possible for a user to move the microphone away from his face and compensate for the reduction in audio level by adjusting the PC sound card mixer settings.

This kind of nonstandard system makes accurate calculations of TELR impossible. This issue highlights one of the problems that are still plaguing some VoIP systems (in particular softphone users) in that the PSTN has had carefully matched audio gain, and callers are used to using a handset without having to adjust gain levels. As we move to systems including softphones, callers need to be aware of the need to adjust systems to give acceptable performance. The conclusion is that in most cases IP phones and/or gateways do need some form of echo suppression.

It is interesting that the one-way delay budget in IP telephony is often much higher than would be acceptable for operators providing toll-quality PSTN systems. It seems that this is one area where users may have to reduce expectation in return for lower costs. In the case of Internet telephony users sometimes have to accept quite long delays and audible echo in return for almost "free" VoIP. However, this is one area where regulation poses some problems for operators that are required to meet very tight echo/delay parameters and yet compete with lower-quality Internet VoIP providers that seem to have evaded some of the regulations.

Perceived Quality of Service Depending on the Transmission Path

Measuring the quality of VoIP systems, as with any other user interface system, is complex, as modeling of user perception is an inexact science. One of the standard methods of measuring quality in an audio system is to use subjective testing where each user rates an impairment using a five-point scale. The results from a number of subjects are then averaged to form a mean opinion score (MOS). Running subjective tests for every possible permutation of a design is clearly impractical as it is expensive and slow. Consequently, results of running tests on a number of known impairments are taken and converted into a model that tries to predict an MOS equivalent score given the objective measurements of the impairments. This is termed "objective" testing. In some senses the earlier discussion on the relationship between TELR and delay for acceptable performance is an objective measure (although it is directly derived from subjective experiments). To form an accurate objective model many such impairments need to be combined.

There have been many attempts to model the parameters of the network with telephony applications; one of the most widely used models is the E-model (G.107). The E-model combines a number of different impairments into a set of additive components by mapping each impairment into a range of 0 to 100 and adding and scaling to give an impairment score that is comparable to a mean opinion score. While the E-model has proved highly useful, there is some debate as to its usefulness with the highly nonlinear types of impairments found in VoIP systems. Also, some find that the E-model is a little too pessimistic with the longer delays as found in VoIP.

One of the problems with the E-model derives from the fact that VoIP terminals often perform highly complex processing in the playout buffer to perform packet loss concealment and adaptive buffer management; the E-model derives from using simpler impairment models. Consequently, improved models have been suggested that use perceptual models to predict equivalent MOS scores. Such systems include PESQ as defined in P.862 but even these have problems with some of the new impairments found in VoIP systems that were mostly absent from PSTN systems. For more information, see Takahashi et al. (2004).

Signaling and Control Using H.323 or SIP

The data path discussed earlier is common to most of the VoIP architectures. However, signaling and control have been where there have been many evolving standards, recommendations, or industry de facto standards. Broadly speaking, the earlier standards (H.323 and earlier IETF standards) were mainly concerned with signaling and control within moderate-size VoIP networks that interface to a larger PSTN. However, the later next generation network (NGN) standards incorporating the IP multimedia subsystem are designed to compete with or replace the national/international PSTNs.

This section will discuss three architectures in turn following their time of introduction: H.323, recommended by the ITU, was the first and had the early dominant share of use; session initiation protocol (SIP) (and associated protocols) was a later system standardized by the IETF that is now seen as the main call signaling protocol; the IMS architecture, first standardized by 3GPP and ETSI for mobile systems, is now incorporated into IETF and ITU standard tracks for NGNs. Although it seems that there is a confusing array of standards, the architectures are broadly compatible and some components are standardized with only small differences in multiple standards bodies. For example, the IMS uses a modified form of SIP for its call signaling.

Naming and Numbering Systems Used in VoIP

Users are very familiar with numbering systems used in PSTN telephony (conforming to E.164) and World Wide Web URLs. However, a VoIP terminal may be available from the PSTN and/or an IP network. Consequently, VoIP terminals are identified in both formats and can sometimes be obtained using either. This causes some confusion when it is necessary to convert between them and could also gives rise to uncertainty about the routing of calls. For example, if a VoIP user calls another user using an E.164 address, then, if the end party is also on VoIP, the call should be routed via the IP network, not via the PSTN. Strictly speaking, VoIP terminals are given a uniform resource identifier (URI) not a uniform resource locator (URL).

URLs are location specific and do not perform well if the resource to be identified is moved. To resolve this problem a more generic naming system known as URI has evolved. For example, a URL is usually server specific and may even specify where an object (e.g., Web page) is placed within the server. The problem is that when things change on the server, the URL has to change. A VoIP URI tends to be more like an e-mail address in that it defines a user and the domain (rather than the server) that the user can be located through. In practice this means that the domain name service (DNS) can be used to map between E.164 and URIs; the process for doing this has been named ENUM. A typical VoIP URI might look like

```
myusername@somedomain.com
```

The advantage with this, like e-mail, is that a user can specify a domain name and let the DNS system find the relevant service layer server to direct the call setup.

H.323

The ITU recommendation H.323 is actually an "umbrella" standard that defines a general architecture and specifies a number of other recommendations that provide the individual

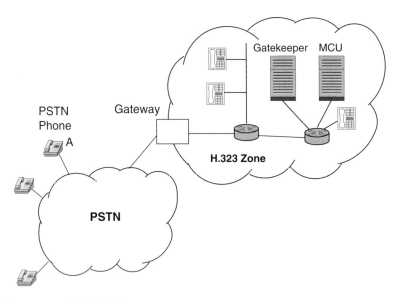

Fig. 7.6 Key components of H.323.

functions for a packet-based multimedia system that can interface with circuit-switched networks such as the PSTN. H.323 is designed with a much wider remit than just VoIP and incorporates audio/video streaming, shared data systems, and audio/video multiparty conferencing.

The architecture of H.323 is summarized in Figure 7.6, which is broadly similar to the generic architecture shown in Figure 7.2. The main new components shown are the gatekeeper and the multipoint control unit (MCU), which provide (or rather aid) the call processing. The protocol stack of H.323 is shown in Figure 7.7, which demonstrates that H.323 is actually a collection of many other standards. The only mandatory parts of H.323 are a G.711 audio system transmitted over IP/UDP/RTP, the call signaling (recommendation H.225), and the call control (recommendation H.245). If only the mandatory parts are used, then neither a gatekeeper, a gateway, nor an MCU is used and the system is effectively equivalent to the basic scenario shown in Figure 7.1. In this case the two endpoints first use H.225 to connect and then use H.245 to negotiate with each other on the capability of each terminal—for example, which codecs to use.

In most H.323 systems it is expected that a gatekeeper is also used. There is at most one of these in each H.323 zone; a zone can be considered a set of terminals managed as one entity that may span a single LAN or up to a whole IP routed domain. As can be seen, the minimalist H.323 system does not need a gatekeeper; however, the gatekeeper performs some important features with regard to signaling and resource management. One of the features that the gatekeeper provides is address resolution from H.323 IDs (e.g., `username@domain`) or E.164 addresses (e.g., telephone numbers) to IP addresses. This allows terminals to be addressed in a flexible format and it may have more than one alias. Hence, a user may be contacted by her E.164-based telephone number or by a more friendly text-based H.323 ID (a URI in IETF language). The gatekeeper can simply provide the address mapping and let terminals perform the signaling directly; alternatively, it can provide admission control based on network availability.

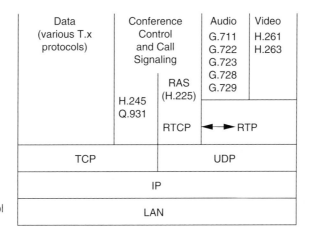

Fig. 7.7 The protocol stack of H.323.

A typical registration, call setup/teardown process, for H.323 can be summarized as follows:

1. *Discovery:* The terminal uses UDP packets specified as part of H.255 Registration Admission and Status (RAS) to find an available gatekeeper using a well-known IP multicast address.

2. *Registration:* The terminal registers its address aliases with the gatekeeper so that the gatekeeper can record the location of an alias address belonging to the network address. The terminal is now ready to initiate/receive call setup requests.

3. *Connection:* The terminal requests an alias to IP address lookup from the gatekeeper. The terminal sends an admission request to the gatekeeper, which checks that there are resources available and the terminal is authorized for the requested connection. The two terminals now have an open signaling channel between them (it may optionally go via the gatekeeper or directly between them).

4. *Capability negotiation:* If the connection is successful, the two terminals negotiate their capabilities; for example, they will need to agree on a common codec.

5. *Transmission:* The media streams can now communicate directly with each-other; for VoIP this uses RTP/UDP/IP.

6. *Termination:* Once one of the calling parties terminates the call, logical resources are freed by the gatekeeper.

The signaling mechanism is necessarily kept quite flexible. For example, in an intranet, the gatekeeper can be responsible for name/address resolution and allocation of network resources, but for scalability the terminals themselves perform the actual call signaling. However, in an environment were there is a need to record calls (e.g., for billing purposes), the gatekeeper acts as a proxy for all signaling so that call start and end times can be centrally recorded.

H.323, with its associated protocols, was one of the first standards that could deliver VoIP and was embedded in systems such as Microsoft NetMeeting. Consequently, there are a large number of systems that support its architecture. The critics, in particular supporters of IETF-based protocols, claim that it is too complex and have suggested alternatives.

Systems Using SIP and IETF Standardized Signaling

Unlike the H.323 recommendation, which clearly defines the VoIP architecture and interrelation between all of the other protocols, the systems based on a signaling protocol called session initiation protocol (SIP) tend to be a looser collection of associated protocols (IMS being the exception). This is in keeping with other areas standardized by the IETF and can lead to some confusion because a "SIP-based system" may use a combination of SIP, session description protocol (SDP), and media gateway control protocol (MGCP). However, the protocols can often be used interchangeably with others so, for example, H.248 (RFC 3525, Megaco) can be used instead of MGCP. Here SIP, SDP, and H.248 will be explained; some of the discussion is also relevant to H.323 as the developments in gateways and gateway/device control described here also apply to H.323.

Session Initiation Protocol

SIP was developed by the IETF and is described in RFC 3261. It is a protocol to create, modify, and terminate sessions between two endpoints; the sessions could be VoIP, multimedia streaming, or interactive applications. Here, only VoIP will be described. SIP can be compared to HTTP in that the message format is ASCII-based, and this is one of the key differences between the equivalent call/media management protocols in H.323 (H.252 and H.245), which use binary formats for their messages. Proponents of SIP claim this is a key advantage as it makes it simpler to debug SIP development and deployment. SIP is a highly flexible protocol that is independent of transport/network protocol and may use TCP/IP or UDP/IP among others. As it is a session layer protocol, and not a media layer-carrying protocol, it is independent of the media transport layer and compatible with many forms of transport; however, the common usage is to control media sessions.

In its simplest form, SIP may be used to perform peer-to-peer call signaling for the architecture shown in the first simple example in Figure 7.1; the call signaling using SIP for this example is shown in Figure 7.8. Note that in SIP-based systems, the end terminal is called the user agent.

Examples of the first two messages are shown next. The first is the invitation that A sends to B:

```
INVITE sip:b@bsdomain.com SIP/2.0
VIA:SIP/2.0/TCPclient.asdomain.com:5060;branch=z9hg4bk74bf9
Max-Forwards: 70
From: A <sip:a@asdomain.com>;tag=8ghyed76sl
To: B <sip:b@bsdomain.com>
Call-ID: 38482762982201885511@ASDOMAIN.COM
CSeq: 1 INVITE
Contact: <sip:a@clientpc.asdomain.com;transport=tcp>
Content-Type: application/sdp
Content-Length: 151
v=0
o=A 2890834526 2890834526 IN IP4 clientpc.asdomain.com
s=-
c=IN IP4 192.0.2.101
t=0 0
m=audio 42242 RTP/AVP 0
a=rtpmap:0 PCMU/8000
```

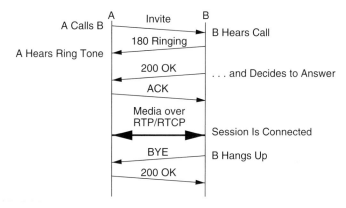

Fig. 7.8 Call signaling using a direct endpoint-to-endpoint path.

B sends notification to A that it has received the invitation from A and is ringing the user at B:

```
SIP/2.0 180 Ringing
Via: SIP/2.0/TCP clientpc.asdomain.com:5060;branch=z9hg4bk74bf9
;received=192.0.3.211
From: A <sip:a@asdomain.com>;tag=8ghyed76sl
To: B <sip:b@bsdomain.com>;tag=6321634344
Call-ID: 3848276298220188511@asdomain.com
CSeq: 1 INVITE
Contact: <sip:b@bsdomain.com;transport=tcp>
Content-Length: 0
```

The invitation message consists of two significant portions: the header, which is the first 10 lines up to the content length field; and the session description protocol (SDP), which comprises the fields that contain single letter codes and associated values. The "o" field describes the originator of the session giving session identifiers and the network address (a DNS hostname in this case). The other field of note in this example is the "m" field, which specifies that the session will be audio communicating on originator port 42242 and that RTP (in Audio/Video profile) will be used to carry it. The last subfield is the RTP payload type, in this case "0," which is the RTP payload type for G.711 audio.

As stated before regarding codecs used for VoIP, due to the many types of codecs that may be supported (or not supported) by a user agent, negotiation to find a suitable common codec may be required. If the caller wishes to accept the call but cannot process the indicated media type, then the caller may respond with a "606 Not acceptable" message, which may list the media types that the caller's user agent can support. It is then up to the calling party to choose an acceptable media type if it can (or return an error).

This simple example is adequate if the end-system address is known. However, in many cases it may not be known. For this reason SIP proxy servers are defined; these can be made well-known through the DNS system so that, in the same manner as e-mail, users may be contacted through a naming format such as

```
username@fullyqualified.domain.com
```

without knowledge of their network address. If SIP proxy servers are used then the first step a user agent needs to perform is to register with the SIP proxy. For security reasons the

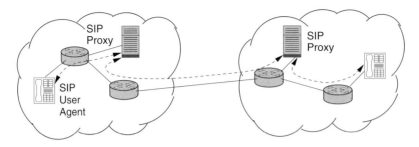

Fig. 7.9 SIP proxy server.

normal way to do this is through a secure version of the SIP protocol (SIPS) where a user-name/password challenge response is used. As part of the registration, the user agent informs the proxy of the aliases (phone numbers or URIs) that it wishes to use. When users wish to make a call to a user at a domain, their user agent contacts the local proxy, which then communicates through the remote proxy (if the called part is on a different domain) to send SIP messages to the remote user agent, as shown in Figure 7.9.

Although SIP is billed as a simple form of protocol, RFC 3261 is 269 pages long and here only an overview is given. Additionally, new additions or modifications are constantly given to improve compatibility or performance. This section has described only the SIP proxy server; however, there are in fact other servers that handle registration and redirection, although in practice these logical servers may be placed on the same physical server. One of the problems with SIP, for an integrated VoIP system, is that it does not sufficiently address the access to, or control of, gateways to other networks (e.g., the PSTN). Hence, additional protocols are required to achieve this.

Gateway Control Protocol (H.248 or RFC 3525, Megaco)

SIP is deemed to be a distributed architecture protocol, as user agents are given enough intelligence to find other user agents and do not necessarily need any other components to make/modify/terminate calls. This is very different from the traditional PSTN model of telephony where the end terminals are "dumb" and all of the intelligence is placed in centralized systems. Proponents of the distributed architecture are keen to show that it is more scalable and "lightweight" compared to the centralized systems. However, proponents of the centralized model point out that (among other points) it is difficult to provide billing systems using the distributed model (and almost impossible for truly peer-to-peer).

This argument can get quite involved; however, it is interesting to note that so-called distributed systems often require the use of some form of central server for systems such as address resolution and signaling proxy as shown for SIP. On the other hand, so-called centralized systems often need replication of central features and caching agents so that the systems are scalable and resilient to failures. Consequently, it is actually difficult to class practical complex systems, such as large telephony networks, as either distributed or centralized because in any system some components will be highly distributed, distributed, or centralized to varying degrees. SIP can certainly be viewed as a highly distributed component and consequently has some difficulty meeting (and in fact it ignores) requirements such as billing, route determination, and resource allocation.

Consequently, the IETF through the Megaco working group, together with ITU Study Group 16, defined an architecture and protocol that have greater centralization so that

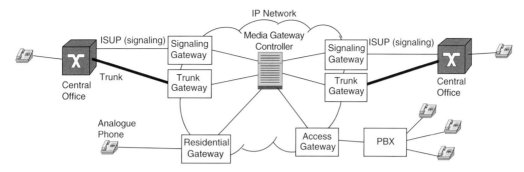

Fig. 7.10 Signaling architecture for H.248 (Megaco); note the media gateways and media paths are omitted (except for the trunk from the PSTN).

systems more compatible with the requirements of traditional PSTN are possible. The result of this collaboration was Gateway Control Protocol in IETF RFC 3525 (often called Megaco from "Media Gateway Control") and ITU recommendation H.248. In particular, the problem they were trying to address was the need for more complex gateway architectures, in particular where many gateways exist and where the signaling and media functions are split apart into separate logical systems.

The signaling architecture for H.248 is shown in Figure 7.10, and, as can be seen there, a wide range of signaling and media interfaces are supported. The main task of H.248 is to define the media gateway controller (MGC) functions and the mapping of signaling between the signaling gateways and the MGC. The other essential components required in the architecture are the media gateways themselves and the transport infrastructure within the IP network; these are omitted in Figure 7.10. The transport is expected to use RTP/UDP/IP and the functions of the media gateways are as described previously, except that here the signaling function is removed to the signaling gateway, which is logically separate.

Another term used for the MGC in some documents is a "softswitch." This is actually quite a good descriptive term because the infrastructure described by H.248 is a system that performs switching between external telephony circuits. The term *softswitch* implies that the acts of actually switching the media and controlling the *switching* are separated. This is seen as a highly useful feature since telecommunication operators had identified problems with tightly coupling the control and switching features in existing PSTN switches. Because this new architecture separates them, the two can evolve separately allowing new signaling/control functionality to be added without needing to upgrade the transport infrastructure.

The range of circuit granularity supported by H.248 is quite wide: from an individual phone connected to a residential gateway (on the customer premises), through to an SS7 signaled PSTN trunk. This allows the H.248 infrastructure to be a drop-in replacement for a traditional PSTN trunking and switching function. Part of the architecture is the ability to transport SS7 ISDN user part (ISUP) signaling over IP so that SS7 trunks can be connected through H.248.

Summary of IETF SIP and Gateway Protocol Signaling
Only been a few of the IETF-based protocols have been explained here (SIP, SDP, H.248/Megaco), but there are many more that support various functionalities such as SIP

integration with SS7 IP/PSTN gateways or alternative media gateway control protocols. Proponents of the IETF standard/recommendation-based systems cite this as a modular system that allows flexible deployment depending on need. However, another view is that there is a confusing array of standards and it is not always clear how they interoperate.

A good example is the fact that it is expected that SIP-based systems for VoIP will use a system such as H.248/Megaco for media gateway control and access, and H.248/Megaco media gateway controllers will use SIP for session signaling between themselves. However, close inspection of the early IETF standards shows that SIP terminal interworking with H.248/Megaco is conspicuously absent. In practice, vendors have provided the packaged solutions using the standards that best suit their customer base. This is certainly an advantage for giving vendors the flexibility they need to differentiate their products. But it is quite a different model than that used by the ITU and needed traditionally by large PSTN operators who use clear architectural standards so that a range of vendors can be used to spread risk and maintain competition.

Summarizing H.323- and SIP-based Systems

It is difficult to draw a simple comparison map for H.323- and SIP-based systems. There have been a number of attempts to do this; however, they quickly date as H.323 and SIP (with associated protocols) are continuously evolving. Broadly speaking, H.323 is often quoted as the older inflexible technology, using a more centralized approach, and giving good compatibility with PSTN. SIP is seen as a highly flexible session control system, a more distributed approach, but with poorer compatibility with PSTN systems. However, this comparison is a little unfair as H.323 has been constantly updated to give an approach that scales well, and SIP is often combined with other protocols as needed by vendors to provide elements such as SS7 PSTN compatibility.

One thing is clear: SIP is often termed as much simpler than H.323 (and this is undeniably true), but, once all the additional functionality required for a full VoIP-based telephony system is added to SIP through additional protocols, the difference in complexity becomes arguable. The debate has become somewhat political in nature between the academics, operators, vendors, and standards bodies; in the end H.323- and SIP-based systems can provide comparable functionality.

Both technologies can be said to evolve from single domain IP multimedia systems that can connect to the PSTN or other networks. They were never originally designed to be a replacement for the PSTN systems themselves, and attempts to make them scale to such sizes have not been well received by operators. Consequently, operators wanting to replace traditional PSTN infrastructures with an IP bearer technology have looked to other solutions that build on H.323, SIP, related protocols, and experience with the systems required for national/global PSTN networks. The next section deals with one such solution.

Next Generation Networks and the IP Multimedia Subsystem

Next Generation Networks

The term Next Generation Network (NGN) was partly generated as a specific term by the ITU-T in 2003 to investigate developments of a new telecommunications architecture and was the focus of ITU-T Study Group 13. Out of this investigation, and discussion with operators, vendors, and regulatory bodies, came the conclusion that there was a need to have a clearly standardized model for a new generation of network and service architectures, resulting in recommendations Y.2001 and Y.2011.

Although the term NGN can sound vague, through standardization it has become a clearly defined architecture. The main aim is stated in Y.2001 and can be summarized as follows:

- It is a packet-based network that can use multiple transport network technologies.
- The transport network has QoS capabilities.
- Service-related functions are separated from the transport technologies.
- The access and core networks are clearly separated so that users can have a choice about who delivers the services.
- Generalized mobility is supported so that users can have ubiquitous access to services.

These aims are designed to deliver carrier-grade services scalable to large-scale networks. This has not been the aim of the earlier discussed VoIP technologies; however, much of the work that has gone into the earlier work on VoIP has been incorporated in NGN. The functional blocks of NGN are shown in Figure 7.11.

One of the main high-level features is that the transport functions and service functions are both logically and physically separated; this answers one of the problems with older

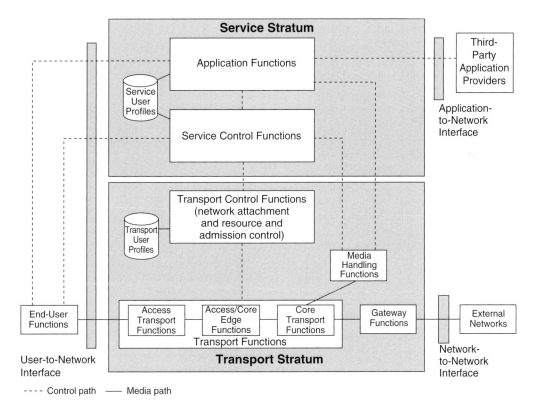

Fig. 7.11 Functional blocks in the NGN architecture.

telecommunications architectures where adding new services often meant upgrades to part of the transport infrastructure. However, it also introduces a new capability for external application providers to build applications and interface them to an operator's network using an application to network interface (ANI). Additionally, the access and core transport functions are clearly separated so that it is possible to split these functions between different providers in a clearly defined manner.

Overview of the NGN Infrastructure

The NGN infrastructure is complex. Rather than explain every functional block in detail, the following is a summary of the remaining functions shown in Figure 7.11. The three external interfaces (UNI, NNI, ANI) are very clearly identified so that existing systems can integrate with them. In the case of the UNI and NNI, there are already existing and working standards in telecommunications to support the wide range of existing terminals and network interfaces. The ANI concept is newer and so there are more developments expected here, with the concept allowing new services to be developed that were not even conceived when the architecture was defined.

An example of a new service might be a system to allow instant messaging to be delivered to a customer through the NGN when an order is complete. Previously to NGN, interfacing with an operator's network would require knowledge of telecommunications protocols and possibly some proprietary interface between the service provider and the operator. The ANI has been developed using open standards defined using open services access (OSA) and the Parlay application programming interface in collaboration with 3GPP and ETSI to allow straightforward and portable applications to be developed.

The NGN provides user profiles for both the service and transport layers. These are possibly multiple databases that link user information to the services that they use and the capability of the network to provide the services. Some of the services may require the network to process the media; consequently, a media handling function is defined that is logically attached to the core network (although they may be provided by servers connected to the core network rather than by the transport hardware). An example of a media handling function might be mixing audio streams for multiparty conferencing or providing audio/video streams on demand. Following the earlier discussions on VoIP architectures, gateway functions are defined to process media and signaling between the NGN and external networks that may be, for example, another NGN or a "legacy" PSTN.

The concept of the NGN is necessarily vague when summarized, as done here, because it is designed to support a wide range of terminals, external networks, and services. The concept becomes clearer when an application is considered, and one of the main applications that the NGN was designed to incorporate was carrier-grade telephony. The NGN architecture does not specify the details of how telephony is controlled and carried; rather it delegates this to another infrastructure called the IP multimedia subsystem (although in theory other systems could be used).

The IP Multimedia Subsystem

The IP Multimedia Subsystem (IMS) was originally proposed for mobile systems and is specified by 3GPP in TS 23.228. However, the IMS has been incorporated into the model of an NGN for handling multimedia sessions such as telephony. IMS uses mainly IETF-based standards; for example, SIP is the main session layer protocol. Investigation of the IMS infrastructure shows that it addresses many of the problems that have been described before in that it provides a scalable architecture by using distributed systems where needed, but there are many centralized features that support billing and many of the traditional telephony applications such as call forwarding or centralized answering services.

IMS Protocols and Features

The IMS uses two of the protocols described earlier: SIP and H.248/Megaco. It is interesting to note that when reading the SIP and Megaco RFCs, both mention that they are in some way complementary but fail to describe how they interwork. Now with the IMS there is a mechanism for them to be used together: SIP is used by end terminals and internally by the IMS for session layer control; the media gateway control function terminates the SIP session for sessions requiring end points in external networks and uses H.248/Megaco. In the following discussion an overview will be given with simplifications to give a broad understanding rather than protocol/architectural detail. The reader is pointed to the 3GPP standards for a full overview (in particular 3GPP TS 23.228).

Some of the features of the IMS are:

- *User equipment (UE):* The terminals can be of a wide number of types including 3G mobile telephones (using packet interfaces), VoIP phones attached to xDSL, and such. The UE uses SIP to communicate to the IMS.

- *Call session control function (CSCF):* This provides the centralized point in the IMS where SIP sessions are routed to application servers, gateways, media servers, or other end points. It takes on the role of a SIP proxy as described earlier as well as providing many other functions. (In actual fact, the CSCF is composed of serving-CSCF, proxy-CSCF, and interrogating-CSCF, but it is described only generically here for simplicity.)

- *Media resource function controller (MRFC):* This controls the media servers that can provide simple media services such as announcements or more complex processing such as mixing of streams for multiparty conferencing.

- *Home subscriber server (HSS):* This is a database (or collection of databases) that store information for each user considered to belong to this IMS. The HSS contains information such as the end user's registration details (including current network attachment address), personal preferences (e.g., voice mail greetings), and telephony services used (e.g., redirection). Note that the mobility model used means that the details are still held in the home network even if the user roams to other IMS-controlled networks.

- *Media gateway control function (MGCF):* This is essentially an H.248 media gateway controller as it uses H.248 to control the signaling and media streams between external networks and the IMS through the signaling/media gateways. SIP is used by the IMS to communicate to the MGCF via a breakout gateway control function (BGCF) (not shown in the diagram in Figure 7.11).

- *Application servers:* There may be many application servers and a user session may use them as necessary (according to subscription). Examples of application servers for telephony are redirection services, caller ID, answering services, and so on.

- *External third-party applications:* As described for the general NGN, there is a standard interface (the ANI) that allows external applications to be provided. So a user could subscribe to most of the services given as standard with the main IMS operator, but may subscribe to a more sophisticated voice mail system provided by a third party.

Call Setup Example

To see how some of these components are used, a simplified example is given for the call setup between a VoIP terminal attached to its home network and an external PSTN terminal:

1. A VoIP user dials a telephone number and a SIP INVITE message is sent to the CSCF (which acts as a SIP proxy).

2. The CSCF looks up the user's profile in the HSS and obtains filter criteria that describe the application services that this user is subscribed to.

3. The CSCF applies each filter in the order specified by the HSS. Each filter may specify an application server that is required to implement the service features of the application. For example, if call redirection is enabled, this would be detected and appropriate action taken.

4. The CSCF sends the SIP INVITE message to its intended destination. To route the signaling, a DNS lookup of the number is made to find the destination network. If this was a user in an external IMS-operated network then this lookup would give the CSCF of the external network to which the SIP message would be sent. However, in this case the DNS lookup returns a number that is in the PSTN.

5. The CSCF forwards the SIP INVITE to a breakout gateway control function. The BGCF determines the optimum gateway to use for this call to the PSTN. For example, if the IMS were to span two continents, and the call is to the remote continent, then it may be better to route the call via IP to the gateway in the remote continent and use the NGN to transport the long-haul portion rather than incur international PSTN charges.

6. The BGCF forwards the SIP INVITE to the chosen MGCF.

7. The MGCF terminates the SIP INVITE and, through the control gateway, converts the SIP session request into the appropriate ISDN user part (ISUP) of the signaling system 7 (SS7) signaled PSTN (or other protocol as required).

8. The MGCF examines the media requirements in the SIP INVITE and selects appropriate media transcoding. For example, the VoIP phone may use G.729 codec but the PSTN will use G.711 (in most cases). The media gateway performs this transcoding as well as any buffering required to smooth out the jitter in the RTP stream from the VoIP phone.

9. The signaling is now in the external network domain, and the appropriate responses are read by the control gateway and converted to appropriate SIP messages.

10. As all the SIP messages are passed through the CSCF, any decisions and timers associated with billing may be triggered in the appropriate telephony application server.

The call example is for when a user is connected to the home network. The home network is where the user profile is stored in an HSS. As the IMS comes from a standards body that is concerned mainly with mobile networks, it is not surprising that it offers good mobility support. Mobility can either mean full roaming (analogous to a cell phone moving between cells but still maintaining a call) or a simpler change of point of network attachment. An example of change of point of attachment is a user with a laptop who may move the laptop and expect to make calls from the new location; however, the user is not expecting to roam as she will not be making calls from the laptop while traveling.

This model of mobility expects that a user may attach to a foreign network and register with it using SIP REGISTER as he would when he attaches to the home network. However, the registration contains the home domain information and the IP network that the UE is connected to in the foreign network. The foreign network will identify this terminal as an unknown device and (assuming there are arrangements between the home/foreign network operators) will simply send SIP messages to the home network for processing. As part of the registration it will send the point of attachment (IP address) to the home network where it will be stored in the HSS. This allows the location of a mobile user to be determined.

This model of mobility (using the session layer) has significant improvements over IPv4 network layer mobility as, although the signaling has to pass through the home network, the media does not and can make use of the routing features in the foreign network to find a direct path to the called party. In network layer mobility models using IPv4, there can be the problem that traffic for the roaming terminal has to pass through the home network; the IMS session control obviates this problem.

As can be seen from the example, the IMS is a complex architecture; however, it has been recognized that this level of complexity will be required to provide a replacement that matches the complexity within existing PSTN networks. There have been a number of attempts to evolve the PSTN architecture that have not been adopted, but now the NGN/IMS combination seems to provide a new architecture that is being adopted because it is arriving when there is a strong need for it.

Network Layer Support for VoIP

IP evolved as a data networking protocol for nonreal-time applications and most users work as a client within a client-server relationship. Consequently, there are many problems to be solved to give real-time transport capability to an application, such as VoIP, that works in a peer-to-peer relationship (or sender-receiver as opposed to client-server). Two of the problems considered here are the provision of QoS-enabled traffic engineering and traversal of network address translators.

QoS Support in IP Networks for Multimedia

Users of the switched circuit based PSTN are used to reliable QoS support. Once a call has been placed it is quite rare (by design) that it is interrupted or that quality is degraded (or changes) during the call. However, with Internet telephony this is not the case. It is possible that the network conditions may change during the call so that quality may become poor or the call may be lost completely. The problem might be stated (as a gross oversimplification) that Internet telephony has no notion of the "network busy tone"; the call will be placed and transport may be attempted although the network (the Internet) may not support the call.

It is highly likely that the Internet, as a mostly "free" use network, will remain a best-effort network as the business model is related to connectivity and not the number of bits sent or the transport quality (beyond basic traffic flow). Consequently, operators that want to implement NGN networks as replacements or alternatives to a circuit-switched PSTN have to implement QoS in their networks, and in fact the NGN model requires QoS support.

There are three main systems proposed to work with IP to provide some degree of QoS and in order of development these are Integrated Services (IntServ), Differentiated Services (DiffServ), and multiprotocol label switching (MPLS). Strictly speaking MPLS is not an IP QoS mechanism but rather an underlying network architecture that can transport IP such that traffic engineering is supported.

Integrated Services

Integrated Services (IntServ) was proposed by the IETF in RFC 1633 and requires routers to identify flows needing QoS and give these flows an appropriate level of QoS by managing the scheduling in the forwarding buffers. The management of QoS is carried out by the sender of the flow, identifying the nature of the flow in a FlowSpec, which is sent along the path of the flow. The FlowSpec specifies the requirements of the flow; it does not actually reserve any resources in the routers. The role of actually reserving resources is carried out by the receivers that send a message upstream to the sender with the resources that are to be allocated. The receiver allocated method was specifically designed to support multicast (point-to-multipoint) such that different receivers could request different resources from the network.

The resources allocated in the routers are the manner in which a particular flow is scheduled in the outgoing buffer of a router. This scheduling process is actually fairly complex; it is not straightforward to just express traffic parameters as a bandwidth (bits per second). Rather, it is a complex mixture of delay requirements and relationships between peak and average data rate. The usual approach (as used in the IntServ model) is to express this as a token bucket that provides a convenient model for describing traffic flows in packet networks.

IntServ requires a signaling system to send the traffic specification from the sender to the receiver and for the resource allocation from the receiver to the sender. The signaling system used for this is the Resource ReSerVation Protocol (RSVP); the operation is shown in Figure 7.12. RSVP operates by sending a traffic specification in the PATH message; this is noted by each router along the path. If a receiver wishes to receive the stream with specific QoS, then it sends a resource specification in the RESV message, which passes along the same route that the PATH message is on. Then as each router receives the RESV message it allocates the resources and passes the RESV message upstream.

The IntServ model received a lot of attention as it seemed to provide a solution to QoS in IP networks. However, the system requires each router to classify each packet as it is forwarded to see if it matches the state of a resource allocated flow; furthermore, the signaling messages are required to be continuously sent (approximately) every 15 seconds since it uses a soft-state mechanism. This gives rise to a scalability problem as maintaining such state is a significant load on a major VoIP trunk, which may be required to monitor and control the forwarding of a large number of flows. Consequently, the consensus is that IntServ is not suitable for providing QoS support for large-scale VoIP deployment.

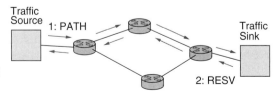

Fig. 7.12 Signaling paths used in RSVP to advertise and allocate resources for IntServ.

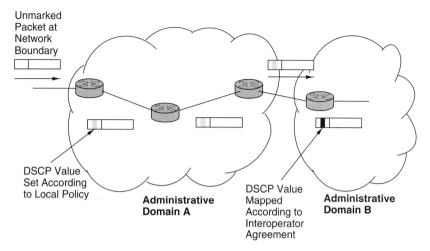

Fig. 7.13 DiffServ uses packet marking rather than signaling.

Differentiated Services

One of the scalability problems of the IntServ model is due to the fine granularity required by the model such that each application flow has to be considered individually. Additionally, users make low-level transport related resource allocations in network elements in an unmanaged manner; this is considered unacceptable by most network operators that want to maintain tight control over traffic trunks. Consequently, the Differentiated Services (DiffServ) architecture was proposed by the IETF in RFC 2475 whereby traffic is allocated to a small number of classes that are given different levels of relative priority.

The model is illustrated in Figure 7.13 showing how packets are "marked" at the network boundary. The marking in the packets is called a differentiated services code point (DSCP) and is at most six bits, giving up to 64 classes of service although, in practice, far fewer are generally used. As DiffServ was developed after IPv4 was defined, there is no DSCP field in an IPv4 header. Instead the IP type of service field and IP precedence bits are reused for DiffServ giving some backward compatibility with the meaning of the IP precedence bits.

Each DSCP is mapped to a particular per-hop behavior (PHB) in a router and typically these are given different priority in router queues so that, for example, best-effort data services are dropped in favor of VoIP packets that need to be delivered with higher QoS. Unlike IntServ, the QoS settings in the router (the PHB) are configured by a management action, not by a signaling protocol. Consequently, operators have greater control over the process.

DiffServ certainly has some advantages over IntServ and has been widely deployed in managed corporate networks to give real-time services improved QoS. But DiffServ is unlikely to be deployed in the public Internet as there is not a clear model for managing it in such a manner that the service can be billed by service providers.

One problem with DiffServ derives from the fact that it is a priority controlled mechanism; applications can be given greater priority than other applications through appropriate PHBs. However, if users (or external service providers) mark packets there is nothing stopping them allocating all of their packets the highest priority and thus bringing the whole network back to the level of a best-effort network again. A partial solution to this is through careful policing allocation of packets at the edge of the network. However, this requires

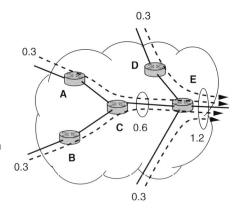

Fig. 7.14 Simple example of problem with DiffServ with inadequate ingress policing; each link is assumed to have a capacity of 1.0 and traffic levels are indicating offered ingress traffic.

management and classification of packets at the network ingress on a per-flow basis to check that they meet (and do not exceed) the service level agreements. A contrived example to demonstrate the problem with DiffServ is shown in Figure 7.14, which shows a case where ingress traffic can cause congestion at a downstream node.

In this case, the egress from router E requires a bandwidth of 1.2 units although it only has 1 unit capability. The problem is caused by inadequate policing of incoming traffic or a network that is underprovisioned. As there is no mechanism for signaling capabilities of routers or traffic allocated to routers (or current traffic levels), it is difficult for the DiffServ model to police incoming traffic to make sure that it does not exceed capabilities at other points in the network.

Although DiffServ has its critics, it has been widely deployed. However, analysis shows that its main successful uses are to limit certain traffic types such as users trying to use excessive peer-to-peer traffic. By giving this "bad" traffic lower priority than "good" traffic, the higher priority traffic gets reasonable quality of service if the network is suitably underprovisioned. In particular, this mechanism works well for TCP traffic where there is no notion of "offered" traffic levels; instead the applications use the TCP flow control mechanism to match the traffic level to the available capacity so an application given lower quality will naturally back off its sending rate. However, VoIP uses RTP/UDP/IP, which does not have any flow control and is usually a constant bit rate service. If the VoIP traffic is a small portion of the network load, then giving this high priority will just cause a graceful reduction in the lower priority of TCP flows to make room for it. But if an operator wants to use the IP network to carry mainly VoIP traffic, then it is important that the ingress traffic does not exceed the capacity at any point in the network. Without predefined traffic trunks, however, it is difficult to engineer the traffic to meet this constraint. Consequently, DiffServ alone is not expected to be enough to deliver a large carrier-grade VoIP transport network to match the quality of service of traditional circuit-switched PSTN transport networks.

Multiprotocol Label Switching

Having seen that IntServ can deliver individual flow guaranteed QoS but is not scalable, and that DiffServ is highly scalable but not necessarily capable of delivering the guaranteed QoS required by services such as large-scale telephony networks, clearly another solution is required. The solution of choice for NGN was identified as multiprotocol label switching (MPLS) as defined in RFC 3031 (and many other associated standards). MPLS evolved as a solution to problems identified with carrying IP over asynchronous transfer

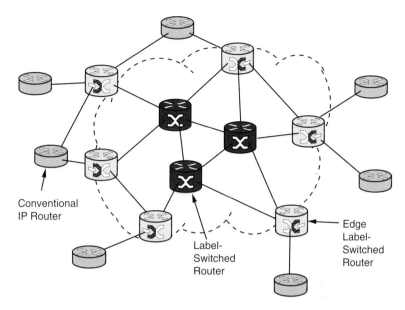

Conventional
IP Router

Label-
Switched
Router

Edge
Label-
Switched
Router

Fig. 7.15 MPLS architecture.

mode (ATM) networks in that ATM control is not directly compatible with IP and, with the need for IP at the edge (not ATM), the advanced QoS mechanisms that ATM offered were not widely used except in carrier networks. Consequently, solutions were presented to use the efficient switching capability of ATM switches but to remove the ATM control functionality and replace it with an alternative control mechanism that became known as MPLS.

MPLS has become much more than an efficient mechanism to manage ATM switches in an IP-centric manner. Rather, it can be seen as a general methodology for controlling circuit-switched networks in a manner that is compatible with IP (or other network layer protocols). Indeed, it is called "multiprotocol" because it can use multiple link layer protocols to carry labeled traffic of a number of higher network layer protocols (although IP is the main network layer supported). One link layer commonly used is Ethernet through the addition of label headers between the link layer and network layer headers. The infrastructure of an MPLS network is shown in Figure 7.15.

The main components are label-switched routers (LSR) that forward packets based on the incoming label number and not the network layer header (as performed by IP routers). As packets are forwarded in an LSR, they have the original label removed and a new one added according to the label expected by the next LSR for the particular flow. Packets in the external networks are mapped into a label-switched path (LSP) at an edge label-switched router (ELSR), which is notionally a combination of a standard IP router and an LSR.

One of the key elements of MPLS is that the control function is separated from the transport switching function. This is quite different from IP where the control function allocates routes that are closely bound with the forwarding function. Consequently, MPLS maps far better to the NGN traffic engineering requirements. MPLS allows the routing of traffic according to pre-allocated network capability and expected traffic demands, rather than a shortest distance path based on destination address. In practice an NGN operator will allocate long-term MPLS label-switched paths through the core network such that the label-switched paths form traffic trunks of a known capacity that VoIP (or other traffic) can be mapped into. Now it is possible to use DiffServ at the edge of the MPLS network to prioritize traffic that is entering a traffic trunk. As the label-switched paths have reserved capacity along planned routes, it is possible to check that the paths do not exceed capabilities at intermediary switches.

The allocation model used in the routers (actually MPLS switches) is usually organized such that the capacity for real-time traffic is reserved, but if it is not being used by real-time traffic, best-effort (Internet type) traffic can make use of the available free space. It is interesting to note that the protocol of choice for controlling the label-switched paths is RSVP (with traffic engineering extensions), but it should be noted that the use of RSVP in this context is not between end systems but only between edges of the MPLS network. Consequently, the RSVP signaling is at the granularity of a whole traffic trunk, not a single VoIP call, and thus does not suffer the scalability problems that occur with the IntServ architecture.

Traversal of Network Address Translation

Here we consider a problem that is particularly applicable to Internet VoIP systems. These systems typically allocate well-known user names (or telephone numbers) to clients using a central location database to map these user names to end system IP addresses. Typically, when a user turns on his IP phone (which may be a softphone on a laptop), he registers with the central location database and this database stores the current IP address and port. This allows a number of features to be performed; the most important is that users can query the location database to find out where to send signaling messages to call another user. However, it is also used for features such as notifying users when other users in their "buddy list" are available (i.e., have registered). This process is complicated by the nature of address space allocation in the modern Internet and in particular the widespread use of private address spaces behind network address translators.

The Internet is generally thought of as a public IP network where the nodes attached to it are given public IP addresses. However, many nodes are not attached directly using public networks for two main reasons:

- As the number of nodes has grown, there are not enough IPv4 addresses (or suitable network address ranges) for users or organizations to have the number they need.
- Small companies or home users often connect using an access model that allocates only one IP address, but they may want to connect multiple machines to this access network.

As a result of the described developments, organizations and small users often allocate the internal machines to a private address space and use only one (or a few) public address at the border with the Internet. Private addresses (e.g., in the range 192.168.0.0 to 192.168.255.255) are reserved for private use and are not deemed to be routable on the Internet. Indeed it is possible that many machines (in different networks) may use the same private address space IP address. Consequently, gateways are placed at the border between the Internet and the private address space networks; these gateways perform network address translation (NAT) between the address space of the Internet and the private network.

The operation of NAT is shown in Figure 7.16 for a client in a private network communicating with a server on the public Internet. The main point to note is that client X in the private network sends and receives packets from the public addressed network using

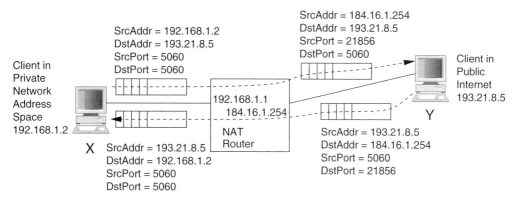

Fig. 7.16 Packet flow through NAT.

the public address of the Internet connected server Y; however, Y sends and receives packets from the public address of the NAT router. The NAT router maintains a table of ports for incoming/outgoing traffic and binds these ports to a particular client in the private network. As the NAT router forwards packets outward it changes the source address from the private address to the public address of the router; on incoming packets it is the destination address that is changed as packets are forwarded to the internal client. It is common to find such NAT functions in company border gateways and in home user access routers.

In the predominantly client-server model used by data applications, NAT performs very well for providing Internet server connectivity to multiple private network clients. This is because the server does not need to know the address of clients in advance of the request for data and is quite happy to forward packets to the NAT router oblivious to the fact that the packet is actually destined for one of many machines beyond the NAT router.

The NAT router makes the system transparent as it controls the mapping of incoming/outgoing ports to particular internal clients. However, NAT is a problem for systems that try to place servers inside the private addresses network or for peer-to-peer applications such as VoIP, as an Internet connected client cannot know in advance which port to use to connect to the Internal device and it cannot use the internal client IP address directly as it is not routable in the Internet. In particular, returning to the discussion at the start of this section, if a client is using NAT, how can it register a public address with a central location database? As a result of this problem VoIP systems require NAT traversal mechanisms if clients are placed in private address spaces (as many are).

NAT traversal has proved to be a difficult problem for systems such as VoIP as there are many NAT systems available and solutions that work well for one type of NAT system do not work with others. Consequently, VoIP systems are often billed as having "intelligent firewall traversal" systems that can probe the network to test it for the type of NAT (if any) and use the appropriate technique to traverse the NAT router. The details of NAT (and firewall) traversal are too complex to fully describe here, but one technique is summarized and one other is briefly mentioned. The discussion will be based around two types of NAT:

- *Full-cone NAT*: For this type, the NAT router permanently (or rather semi-permanently) allocates a port for a particular internal client address and port. This means that any external client can send data to a port on an internal client using the allocated port on the NAT gateway. Following the example in Figure 7.16, Y can send traffic to X by sending packets to port 21856 at the NAT router. For a full-cone NAT, another external client can send data using the same port.

- *Symmetric NAT:* For this type, ports are allocated at the NAT on a per external destination basis so that if client Y sends traffic to port 21856 to access X, another Internet client will not be able to use this port and is likely to need to use a different port and/or IP address.

The problem that NAT causes for Internet telephony is that, if an Internet client needs to call another client behind a NAT router, then it cannot directly send packets to the private client as it does not know its address and does not know the port to send packets to on the NAT router. The Internet client needs to learn this port mapping information so that it can inform the central location database as part of the registration process to allow other clients to call it. For the case of a full-cone NAT (and many other types of NAT except symmetric), simple traversal of the user datagram (STUN) protocol through network address translators is a useful NAT traversal solution. STUN requires that an external STUN server be on the Internet and that clients be adapted to use STUN; in practice many VoIP phones are compatible with this (or similar strategies).

A client uses STUN messages sent and received to and from a STUN server to determine the type of NAT that it is using. The STUN messages contain the IP addresses that the message was sent from or to in the message body so that the client can tell if a NAT system has changed the header. Taking the two NAT examples considered here of a full-cone NAT and symmetric NAT and an Internet connected STUN server, the tests can be described in a simplified form as:

1. Client sends a STUN message to a STUN server that replies to the client with the address and port it was given in the MAPPED-ADDRESS field. If the MAPPED-ADDRESS contains the same address and port as the client uses, then it is not behind a NAT router and may stop testing. Else, it is behind a NAT router and continues in test 2 (also recording the MAPPED-ADDRESS for use later in test 3).
2. Client sends a STUN message to a STUN server with a CHANGE-REQUEST field that makes the STUN server respond to the client using a different address/port on the STUN server. If the client receives a reply, it knows that it is behind a full-cone NAT and stops testing. Else, it continues to test 3 recording the alternative STUN server address/port it received in the reply.
3. Client sends a STUN message to the STUN server on the alternative address/port it received from test 2. If the client receives a reply with the MAPPED-ADDRESS different from that received in test 1, then it knows that it is behind a symmetric NAT.

The tests are actually longer than shown here, but the simplified example allows the issues to be discussed.

If the STUN tests determine that the client is behind a full-cone NAT, then the client can register the MAPPED-ADDRESS it found in test 1 with the central location database. Test 2 shows that a second client can contact the first client using the discovered NAT address/port. However, if the tests determine that the client is behind a symmetric NAT, then it is not possible for the client to register an address with the central location database. The reason for this is that the MAPPED-ADDRESSs in test 1 and test 3 are different showing the NAT does not permanently register a port for use of the internal client but rather dynamically allocates external addresses and ports with each new outgoing client data flow.

The previous discussion shows that STUN does not provide a solution for every type of NAT. One study has shown that about one-third of residential access routers use symmetric NAT and that most corporate firewalls use this type of NAT, so additional techniques are required. One solution is to use a proxy server for all signaling and voice data transport as specified in a system such as traversal using relay NAT (TURN). This is one of the techniques that can cope with all types of NAT as long as the TURN server resides on the public Internet. By placing the TURN server on the public Internet, clients do not have to discover the addresses/ports at the border NAT but, rather, rely on all clients registering with a TURN server that has an address/port that is well known to all clients. The only problem with the TURN solution is that it requires a high-performance server that can cope with forwarding all the signaling and data for the VoIP system.

The NAT solutions presented here are quite different. STUN requires relatively little support from the provider as the STUN server has a relatively low throughput because it deals only with clients discovering the type of NAT they are using and determining the attached address; but STUN does not always work. However, TURN can provide a solution for most NAT types at the cost of a high-bandwidth server. Consequently, a practical solution has to combine these (and other) solutions so that low-effort solutions (e.g., STUN) are used first before the high-effort solutions such as TURN. Automatic detection and configuration systems are proposed in systems such as Interactive Connectivity Establishment (ICE), which standardize the NAT traversal systems and mechanisms to automatically pick the best for the provider and user.

The discussion here has concentrated on NAT traversal. However, most NAT systems are also part of a firewall that may limit traffic in and out of a network border. In fact many corporate firewalls may restrict all UDP traffic, which would seem to restrict the use of VoIP through these firewalls. However, Internet VoIP providers are constantly working to keep ahead of such restrictions. One technique for traversing firewalls is to tunnel VoIP traffic through a protocol that is likely to be allowed through the firewall; commonly this is HTTP using TCP as the transport to a specialist VoIP proxy that converts the HTTP tunneled traffic into conventional RTP/UDP traffic. Users who are making use of firewall traversal systems should be asking themselves if they are breaking any site or service provider rules by using such systems as the firewalls may be in place to stop users from using disruptive (or potentially insecure) real-time services.

Conclusions

VoIP is viewed by some as a disruptive technology that challenges the traditional circuit-switched PSTN. This is in part due to the widespread use of VoIP over the Internet to provide "free" telephony. However, while Internet telephony has been taking off, the traditional PSTN operators have been working on an alternative architecture to circuit-switched bearers using a mostly IP-based bearer in the form of the NGN architecture. The advantage of the NGN architecture is that it can provide services in a highly flexible manner and, through providing a network that has QoS mechanisms, NGN-based networks will be able to provide multimedia services that will be difficult to provide reliably on a best-effort network such as the Internet. In particular, the notion that the Internet is "free" is clearly false; users have to pay for access to the network (either through access costs or for peering arrangements). Consequently, VoIP telephony is just one of the services provided by an NGN architecture network and many users see this provided as part of the access cost.

In the mid-1980s, convergence between voice, multimedia, and data was a much-discussed issue and ISDN seemed to provide the architecture by which this could be delivered. However, ISDN and its circuit-switched architecture never reached the market dominance it needed to become a true success (although there were some exceptions in certain countries). It is interesting that, twenty years later, by providing the telephony in the form of VoIP in the NGN architecture, we have finally seen the convergence between voice, multimedia, and data that has been promised for so long.

References

Cox, R. V., and Kroon, P. (1996) Low bit-rate coders for multimedia communication, *IEEE Communications* 34(12), pp. 34–41.

Gray, R. M. (2005) The 1974 origins of VoIP, *IEEE Signal Processing* 22(4), pp. 87–90.

James, J. H., Chen, B., and Garrison, L. (2004) Implementing VoIP: a voice transmission performance progress report, *IEEE Communications* 42(7), pp. 36–41.

Markopoulou, A. P., Tobagi, F. A., and Karam, M. J. (2003) Assessing the quality of voice communications over Internet backbones, *IEEE/ACM Transactions on Networking* 11(5), pp. 747, 760.

Narbutt, M., Kelly, A., Murphy L., and Perry, P. (2005) Adaptive VoIP playout scheduling: Assessing user satisfaction, *IEEE Internet Computing* 9(4), pp. 28–34.

Takahashi, A., Yoshino, H., and Kitawaki, N. (2004) Perceptual QoS assessment technologies for VoIP, *IEEE Communications* 42(7), pp. 28–34.

Zeadally, S., Siddiqui F., and Kubher, P. (2004) Voice over IP in intranet and Internet environments, *IEE Proceeding Communications* 15(3), pp. 263–269.

8 Next Generation Network Management

Tony Richardson

Introduction

Next Generation Network (NGN) is the title that has to been given to a range of network capabilities aimed at updating and converging many present-day networking capabilities. Such existing network capabilities include:

- Mobile networks
- Fixed networks
- The PSTN
- ISDN
- Internet linkage

Because of the emphasis on convergence, in some circles the development is referred to as the "Converged Network." However, this term is often being used within individual industry initiatives. For NGN to be a complete success it will need to adopt pan-European and preferably full global deployment. For this reason common standards will be needed. Already, there are a number of standards bodies involved in such developments under the banner of "NGN"; further, in this chapter a summary of some of the present activities (at the time of writing) is provided. A major feature of the convergence to be provided by NGN is based on the deployment of a common transport infrastructure. This transport will be built on IP technology. However, in a number of respects this emphasis on the common IP network infrastructure, and the term "NGN" itself, has led to a number of present industry confusions.

In fact, NGN as being defined at present will consist of a number of capabilities and areas of convergence, not all of which will be restricted to the transport network infrastructure (i.e., levels 1 to 4 in OSI communication layers). For example, another major feature of NGN is separation of the network infrastructure from the control and applications. This will provide a much more productive (and hopefully cost-effective) environment for the supply of new applications. Many see this flexible applications environment as providing the real power and business benefit of NGN. To give a better feel for this, a summary of some of the major features of NGN (as specified at present) is provided a little later in this chapter.

Having introduced the subject of NGN, the main feature of this chapter will not be NGN itself (although some outline and references to NGN material is covered), but rather the extremely important concern of the *management* of NGN. As we will see in the material that follows, NGN will be required to provide a highly flexible and adaptive environment that will need to support many of today's telecommunications services and future services in an increasingly integrated and converged fashion. Coupled with this is the need for these services to be increasingly personalized to individual customer needs. A third strand is the need to provide these services in a timescale that is much in advance of present day service offerings. The only way this will be achieved is by supplying a management environment that is fully aware of all these (and many other) factors and able to respond and manage these service needs in a fully integrated and equally flexible fashion. So, it is the management of NGN that forms the main content of this chapter.

NGN Features

The "converged" Next Generation Network will have the following capabilities:

- It will provide a multiservice, multiprotocol, multiaccess, IP-based network that is secure, reliable, and trusted.
- Multiservices will be delivered by a common Quality of Service (QoS) enabled core network.
- Multiaccess will be provided by several access networks to both fixed and mobile terminals.
- It will consist of not just one converged IP network, but different networks that interoperate seamlessly.
- This will be a major enabler for service providers to offer real-time and non-real-time communication services between peers or in a client-server configuration.
- Nomadicity and mobility of both users and devices will eventually be supported, as will intra- and internetwork domains, and eventually be fully supported between fixed and mobile networks.

In summary, essentially, from a customer's perspective, NGN will provide "my communications services—always reachable, everywhere, using any terminal."

NGN Characteristics

These characteristics are derived from ITU Y.2001 and Y.2011 (and previously from Global Information Infrastructure, GII).

- Personalization
  - Services tailored to individual (or group) needs
- Unified service characteristics
  - Decoupling of service provisioning from network configuration

- Provisioning based on open interfaces
- Converged mobility services between fixed and mobile networks
- Unified user service experience, with mass personalization capabilities
- A variety of customer identification schemes
- Independence of service-related functions from underlying transport
  - Separation of "Service Stratum" from "Transport Stratum" (see ITU-T NGN Architecture later in this chapter)
- Support for a wide range of services, applications, and mechanisms
  - Voice, media, data
- Networking
  - Packet-based transfer
  - Separation of control functions among bearer capabilities, call/session, and application/service
  - Broadband capabilities with end-to-end QoS
  - Interworking with legacy networks via open interfaces
  - Access to different SPs, independent of any access or transport technology
  - Support for multiple last mile technologies
- Compliance with all regulatory requirements (e.g., emergency communications, security, privacy, and so on)

NGN Architecture

An overview of the NGN architecture, as defined at present by ITU-T, is shown in Figure 8.1. This shows the high-level view of the NGN architecture, the major parts of which are listed on the next page.

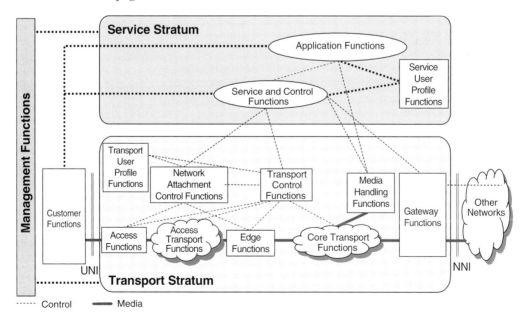

Fig. 8.1   NGN architecture.
Source: FGNGN-OD-00097—NGN Release 1 scope document.

- Transport stratum: various functional areas to provide access, control, and interworking of the common IP transport infrastructure
- Service stratum: supporting the various NGN applications
- Gateways to other networks
- Customer profile functions (supporting service personalization and such)
- Comprehensive management functions to manage the complete NGN environment

Details of the NGN management environment and how this may be provided are described in the remainder of this chapter.

Requirements for Management of NGN

As can be seen from the outline of the NGN architecture, the management of NGN will require a very complex and flexible management environment. Items that will need to be managed in an integrated fashion include the following network, service, and control characteristics:

- Packet-based data transfer
- The separation of control functions among bearer capabilities, call/session, and application/service
- Support for complex value chains, multiple trading partners, and business models (e.g., business-to-business interactions)
- Support for a wide range of services, applications, and mechanisms (including real-time, streaming, nonreal-time, and multimedia services)
- Broadband capabilities with end-to-end QoS
- Interworking with legacy networks via open interfaces
- Generalized mobility, with converged services between fixed/mobile
- Unrestricted access by users to different service providers
- Variety of customer identification schemes (e.g., role-based access)
- Personalization of services to meet customer needs
- Independence of service-related functions from underlying transport
- Support of mass customization in a component-oriented environment
- Compliant with all legal and regulatory requirements (emergency communications, security, privacy, lawful interception, and so on)

NGN Management Architecture

As previously outlined, the management environment for NGN will need to be comprehensive, integrated, and extremely flexible. The approach taken by ETSI in developing an architecture for NGN management is shown in Figure 8.2.

This management architecture is composed of a number of "function sets" of management:

- *Resource Management:* These management functions provide the direct management of NGN resources. These resources may be parts of the transport infrastructure or parts of the service environment, as is indicated by the representation of the actual transport and services resources in Figure 8.2.

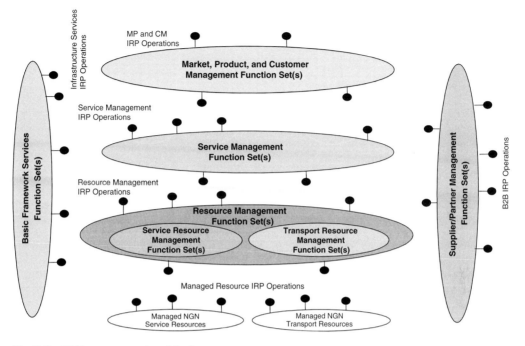

Fig. 8.2 NGN management architecture.
Source: ETSI TS 188 001—(TISPAN) OSS Architecture Release 1.

- *Service Management:* These provide management functions to support various forms of service management (e.g., control of QoS).

- *Market, Product, and Customer Management:* These management functions support direct customer facing management operations (e.g., Customer Relationship Management—CRM).

Together with these "layers" of management, Figure 8.2 shows two additional orthogonal areas:

- *Supplier/Partner Management:* This is the linkage within the architecture to support value chain interactions (at all layers).

- *Basic Framework Services:* This is concerned with the way in which items within the various function sets interoperate. These services allow an Service Oriented Architecture (SOA) form of interaction to take place, as outlined later in this chapter.

IP Multimedia Subsystem

No discussion of NGN would be complete without mention of the IP Multimedia Subsystem (IMS). These specifications are being developed by the Third Generation Partnership Program (3GPP), a technical standards group addressing third generation mobile standards. The specifications represent early forms of NGN development and are now being incorporated into service provider and vendor delivery plans, as outlined in the following text.

Background

Service providers are developing a path from circuit-switched technology to NGN packet-switched, IP-based technology. Reasons for this include the need to reduce operational costs, enhance revenue, and to cope with various forms of disruptive competition (e.g., supply of VoIP), and market saturation of existing services.

Standards Development

IP Multimedia Subsystem specifications are being developed by 3GPP and will form a major part of a future NGN environment, supporting fixed/mobile convergence, common service platform environment, and so on. IMS as defined at present is aimed at the cellular environment. The incorporation of IMS into the wider NGN environment will require extension to support other access technologies such as xDSL.

IMS Outline

IMS is a collection of core network functional entities for the support of services based on the Session Initiation Protocol (SIP). IMS supports the registration of the user and terminal device at a particular location in the network. As part of registration, IMS supports authentication and other security arrangements. IMS utilizes SIP-based control. The services supported by IMS may include multimedia session services and some nonsession services such as presence services or message exchange services. (Presence services are ones that are dependent on location information, for example: Where is the library nearest to my present location?)

In addition to services for the user, IMS defines a number of network reference points to support operator-provided services, and supports various application services via the services support architecture. IMS supports operation and interworking with a variety of external networks via defined reference points, as well as defined reference points for the collection of accounting data in support of charging and billing operations.

IMS also supports defined reference points to the underlying transport infrastructure for the enforcement of QoS negotiated by session signaling and for flow gating. These reference points also support the exchange of information in support of correlation of charging between IMS and the underlying transport.

Structure of IMS

IMS consists of a number of function blocks and associated reference points, main items of which are identified in Figure 8.3.

Some of the main items are:

AS: Application Server.
UE: User Equipment.
CSCF: Call Session Control Function—establishes, monitors, supports, and releases multimedia sessions and manages the user's service interactions.
HSS: Home Subscriber Server—provides AAA functions, maintains subscriber profile data, and is a repository for user-related information.
S-CSCF: Serving Call Session Control Function—handles session states in the network.

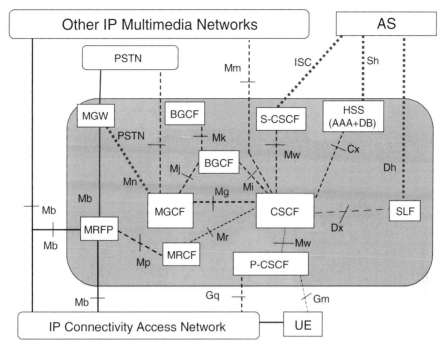

Fig. 8.3 IMS structure.
Source: 3GPP Technical Specification Group Services and System Aspects; IP Multimedia Subsystem.

P-CSCF: Proxy-Call/Session Control Function—the access point for entering the IMS domain, user authentication, outbound/inbound SIP proxy server.
MRFC and MRFP: Multimedia Resource Function Controller and Multimedia Resource Function Processor—delivers media streams such as announcements.
MGCF and MGW: Media Gateway Control Function and Media Gateway—provides control and gateway functions for all media.

As such, the preceding functions may be thought of as supporting 3 layers:

1. Interface to Applications Services (on the AS)—top
2. Control layer functions—middle
3. Interface to networks infrastructure—bottom

Other Standards Bodies

IMS incorporates the specifications from number of other standards bodies—for example, IETF—SIP, SDP (Session Description Protocol), Diameter (user authorization and subscriber service profile access), and AAA (Access, Authorization and Accounting. Correspondingly, IMS is being accepted and incorporated into specifications being produced by other standards bodies—for example, 3GPP2, ETSI/TISPAN (European Telecommunications Standards Institute/Telecommunications and Internet Converged Services and Protocols for Advanced Networking), and OMA (Open Mobile Alliance).

Service Creation/Migration

Because of investment made in present GSM and 3G networks, it is unlikely that full IMS will be deployed to the terminal for some time yet. This requires operators to integrate new and legacy technologies and span both fixed and mobile devices. It will be important that the same applications can be supported on different networks using both SS7 and SIP signaling.

Mass-market telephony services may be migrated to IMS by providing them via the SIP AS. However, the migration of advanced data and voice services will require more complex solutions.

A number of alternative Service Logic Execution Environments (SLEEs) exist, such as JAIN SLEE, J2EE, and Parlay/OSA. Portability of service logic across the various platforms is an important consideration if continuous development effort is to be avoided.

Service Enablers

A number of basic service facilities will be provided that may then be combined and built on to provide more complex services. Examples of these service enablers include: Third Party Call, Call Notification, Short/Multimedia Messaging, Account Management, and so on.

Charging

The complex mix of service capabilities will require an appropriately comprehensive means to set rates and charge for services. In this respect, there will be increasing need to determine rate in real time as the nature of the service changes and also to rate and charge for various different forms of content. Both 3GPP and OMA are developing specifications and standards in this area.

Future Directions

Full IMS will take many years to implement. IMS should be viewed as a means to create more personalized services that integrate new and legacy features that can be accessed over multiple access networks. The real value of IMS will depend on the manner by which underlying capabilities of the core network are brought together to deliver customer-centric services. These objectives are very much in line with the overall strategy for NGN, and it is expected that IMS will become an initial and cornerstone development in the delivery of NGN. IMS will be at the heart of convergence between mobile, fixed, and broadband technologies.

IMS may also be a catalyst for providing a more open environment where service providers (SPs) can operate with other SPs to provide a richer set of customer services. In addition, many third-party suppliers may also provide capability into this environment. Such an approach will perhaps be needed for future cost-effective telecommunications supply if disruptive Internet technologies (such as VoIP) are not otherwise to prevail. Such an approach will also require the support of considerably more complex value chains that exist at present within the telecom industry.

A number of early service provider deployments of IMS functionality are planned for through 2007, with vendor selections taking place at the present time. The initial focus of deployments will probably be on reproducing traditional telecom services and fixed mobile convergence (FMC). But emphasis is likely to move over time to the IMS service layer where the true impact of IMS will be measured in the rapid supply of flexible multimedia services that are tailored to individual customer needs.

Service Oriented Architecture

In a Service Oriented Architecture (SOA), the various component parts of a system are represented through well-defined interfaces and are called "services." These service interfaces may be defined using any particular implementation technology, but for maximum flexibility, they are best defined in a technology-neutral form, thus allowing multiple forms of the same interface to be implemented in different technologies.

The particular feature of an SOA is that interoperation between the services happens only at run time of the system. This means that the same system component (and associated service) can be invoked by many other parts of the running system as and when needed. This feature is referred to as "loose coupling," since there is not a rigid linkage between services at design time of the system (which would be referred to as "tight coupling").

The SOA design approach complements very well with the move toward the design of systems based on the use of "components." The concepts here are that (like Lego blocks) the system can be made to operate by quickly assembling a collection of components (which in theory can be obtained from many different sources). The component-based approach leads to more system flexibility ("unplugging" and "plugging in" new components as changes in system operation are required). Other benefits include more reuse of software across system developments, faster time to delivery, and (in theory) reduced cost of development. Many technologies now support the component-based approach to system development, including Java/J2EE and JINI.

In addition to well-defined interfaces associated with the services, a number of other system infrastructure capabilities are needed to support SOA operation; for example, some form of "repository" is needed that services can use when exchanging information about their interfaces. Because it provides this very necessary indirection capability, there are a number of practical concerns associated with the repository, including interface naming and addressing, throughput/performance/availability/reliability, and so on.

The SOA model supports well the concept of distributed processing where various parts of the system are distributed over several machines (and perhaps geographical locations). The "intelligence" of the overall system is thus achieved by the interaction of these various distributed parts. This differs from other system models such as manager/agent where one part of the system (manager) has more control over the other part of the system (agent). Hence, the SOA approach follows well the trend toward more distributed intelligence in the management environment (e.g., in Network Element technology).

The approach also supports well situations where parts of the system may be provided by multiple suppliers. Hence, overall system operation may be achieved at run time by linking together these variously provided services. Such an approach is followed in the move to Web Services operation, where the Organization for the Advancement of Structured Information Standards (OASIS) has defined an SOA framework for linking together Web-based services to provide overall system capabilities (including the use of the WSDL language for interface definition, XML for transfer syntax, SOAP for communication services, and UDDI for repository/directory services).

NGN Management and TMN

The Telecommunication Management (TMN) framework and its associated collection of recommendations were developed a while back by the ITU-T (in Study Group 4). They represent an open-systems method of providing management of telecommunication systems. TMN is expressed in a comprehensive set of specifications within the ITU-T recommendations. A full description of TMN is out of the scope of this chapter but a few recommendations that provide higher-level information are provided in the References section at the end and ITU-T of this chapter (e.g., ITU-T M.3010, M.3100, ITU-T M.3400). A summary diagram demonstrating some of the TMN concepts is shown in Figure 8.4.

In many ways, NGN Management represents a migration away from the principles expressed in TMN. The main changes include the move from specific TMN implementation technologies to describe managed objects and the services/protocols of interaction between the "manager" and the "agent" (such as GDMO and CMIS/CMIP) to a more technology-neutral and subsequent series of technology-specific implementations. In addition, the well-defined and static TMN reference points (such as "q" and "x" as shown in Figure 8.4) are replaced by the more dynamic and flexible SOA form of operation. (See the overview of SOA discussed earlier.) However, some concepts in TMN remain useful in the move to NGN management such as that of "layering" of management (from business management through service management to element/resource management), which still exists as a useful concept. (See ETSI NGN Management Architecture elsewhere in this chapter.) The detail of specific mapping and migration from TMN to NGN management remains a current and active work item with the ITU-T.

Example of NGN Management Framework: TeleManagement Forum "NGOSS" Applied to NGN Management

The TeleManagement Forum (TMF) has developed a framework formed of a number of technologies that are suitable for the implementation of much of an NGN management environment; this framework is collectively referred to as New Generation Operations Systems and Software or NGOSS. An outline of some of the major features of NGN Operational Support Systems (OSSs) (or Management Systems) is presented below along with the associated linkages to NGOSS.

To fulfill the NGN business vision and to maximize system development speed and operation efficiency, design of the OSS architecture should have the following characteristics:

- All systems must embrace a similar component-based, Service-Oriented technology-neutral Architecture or SOA.
- They should utilize a common information architecture across all management applications so that:
  - Information can be shared across multiple areas of management.
  - A framework can be provided for policy-based management for existing or yet to be defined services.
  - Capabilities can be developed for collecting end-to-end service measurement data.
- OSS solutions must be developed according to a common business process framework.

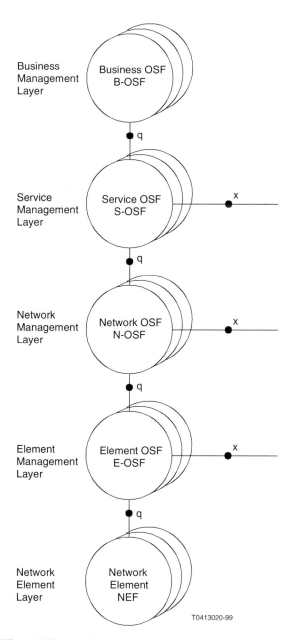

NOTE 1 – Additional or alternative layers are permitted.
NOTE 2 – Other interactions may also occur between nonadjacent layers.

Fig. 8.4 TMN layered architecture.
Source: M.3010—Principles for a telecommunications management network.

- From a network technology support point of view, the system must separate management of service components from management of network technology specific components.

Applying TMF Technologies

NGN management solutions require the integration of business and networking concerns. NGOSS provides many enabling capabilities:

- eTOM (enhanced Telecoms Operations Map): This is how the organization does business.
- SID (Shared Information and Data) framework: This is how the solution is modeled and its behavior orchestrated.
- TNA (Technology Neutral Architecture): This is how the OSS solutions are architected.
- NGOSS Lifecycle and Methodology (with view-centric and model-driven approach): This is how the solution is developed, managed, and used.

A brief overview of the application of each of these items is presented in the following text.

Enhanced Telecom Operating Map

The enhanced Telecom Operations Map (eTOM) provides a common service provider a business process framework. This enables the processes an SP operates to be well defined and the interactions between these processes to be fully expressed. The eTOM can then be used as the basis of other aspects of OSS design (e.g., shared information and software components, as indicated later).

The eTOM is defined at a number of levels of detail and a high-level view is shown in Figure 8.5. The view of eTOM in Figure 8.5 shows it to consist of three main parts:

- *Operations Processes:* These are the processes that drive the operating SP organization.
- *Strategy, Infrastructure, and Product:* These processes provide the "back office" support functions (e.g., defining products offered to customers, putting the network infrastructure in place, and so on).
- *Enterprise Management:* These provide many of the SPs support processes (e.g., financial and human resource control and so on).

In addition to these major areas of business processes, Figure 8.5 also shows layering of the processes from those that are more customer facing (at the top) to those that are more concerned with management of resources and managing supplier/partner relationships (toward the bottom).

It should be noted that there is a strong resemblance between this layering of processes and the layers in the ETSI OSS architecture shown earlier. This is not a coincidence as the eTOM was one of the frameworks used to design the ETSI OSS Architecture.

From this high-level view of eTOM, more detail is added as the business processes are further decomposed and additional detail is defined. For example, the "Level 1" view of eTOM is provided in Figure 8.6. This includes a number of vertical "flow through" processes such as:

- *Fulfillment:* All the process elements needed to fulfill delivery of a customer service.
- *Assurance:* All the process elements needed to keep the service operating.

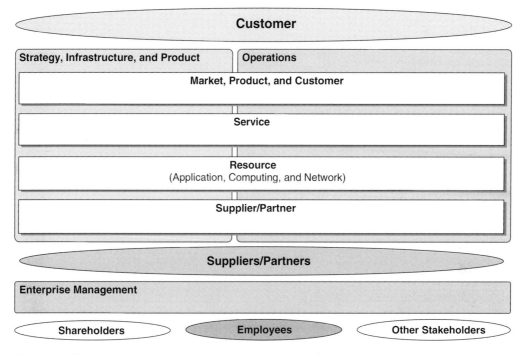

Fig. 8.5 eTOM overview.
Source: TMF enhanced Telecom Operations Map (eTOM)—GB921.

- *Billing:* This involves collecting usage information, applying rating, generating a customer bill, and so on.

Over the past few years, eTOM has become a de facto standard business process framework for the telecommunications industry. Recently it has also being accepted by ITU-T as an International Recommendation (M.3050).

Apart from its application in NGN management specification, eTOM is also used for many other telecommunications business purposes (for example, in OSS procurements, in bringing together business process/OSSs following mergers and acquisitions, and so on).

Shared Information and Data Framework

Within the NGOSS framework, a close partner of the eTOM business process framework is another framework for Shared Information and Data (known as SID). The SID provides a common representation of information and data that is shared across multiple areas of management. This means that, for instance, rather than having several form of representation for "customer details" in marketing systems, inventory systems, configuration systems, and so on, one form of information representation is defined by SID. This makes it much easier for different areas of management within a telecommunications management environment to cooperate as they all share the same view of how this information is represented.

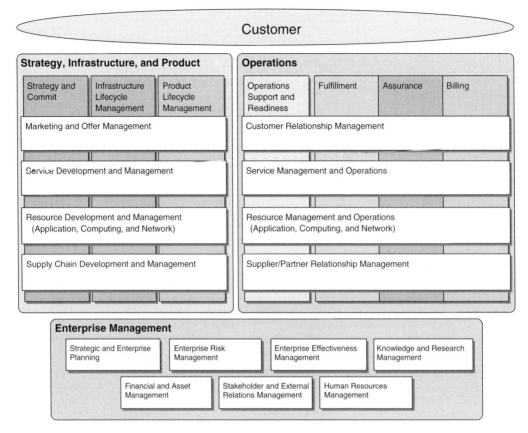

Fig. 8.6   eTOM—Level 1 view.
Source: TMF enhanced Telecom Operations Map (eTOM)—GB921.

In terms of linkage to the eTOM, the SID defines much of the shared information that is manipulated and operated on as the business processes defined in the eTOM carry out their tasks. SID makes a distinction between "information" and "data." In SID terms, *information* is represented in a technology-neutral way using UML (Unified Modeling Language). Specific data implementations of this information may then be produced from this common neutral form depending on the chosen implementation technology (e.g., XML, CORBA/IDL, and so on). The important point here is that they all have the same information form even though they may be finally realized in differing implementation technologies—which obviously has many benefits from an interoperability point of view. A high-level view of the SID is shown in Figure 8.7.

This shows the SID to consist of a number of information domains (e.g., market/sales, product, service, and so on). Each of these domains is further subdivided into what the SID calls Aggregated Business Entities (ABEs). So, for example, the ABEs for resource include topology, performance, configuration, and so on, and each of these is in turn represented by detailed UML diagrams. The high-level mapping between eTOM and SID is shown in Figure 8.8.

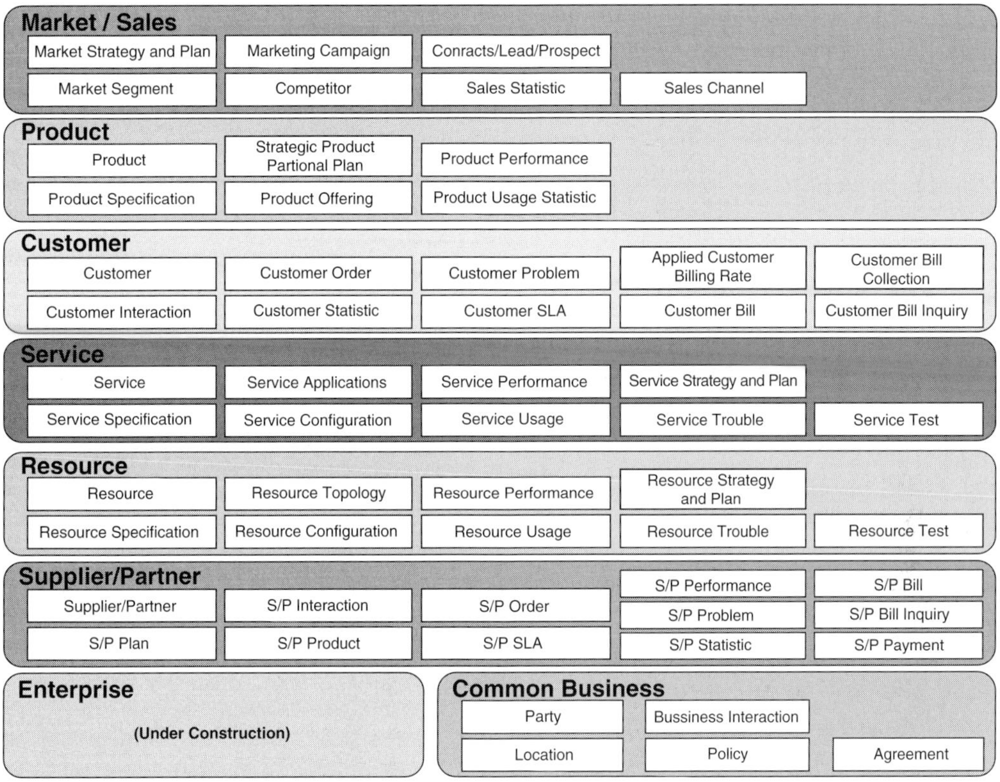

Market / Sales
| Market Strategy and Plan | Marketing Campaign | Conracts/Lead/Prospect | |
| Market Segment | Competitor | Sales Statistic | Sales Channel |

Product
| Product | Strategic Product Partional Plan | Product Performance |
| Product Specification | Product Offering | Product Usage Statistic |

Customer
| Customer | Customer Order | Customer Problem | Applied Customer Billing Rate | Customer Bill Collection |
| Customer Interaction | Customer Statistic | Customer SLA | Customer Bill | Customer Bill Inquiry |

Service
| Service | Service Applications | Service Performance | Service Strategy and Plan | |
| Service Specification | Service Configuration | Service Usage | Service Trouble | Service Test |

Resource
| Resource | Resource Topology | Resource Performance | Resource Strategy and Plan | |
| Resource Specification | Resource Configuration | Resource Usage | Resource Trouble | Resource Test |

Supplier/Partner
Supplier/Partner	S/P Interaction	S/P Order	S/P Performance	S/P Bill
			S/P Problem	S/P Bill Inquiry
S/P Plan	S/P Product	S/P SLA	S/P Statistic	S/P Payment

Enterprise

(Under Construction)

Common Business
| Party | Bussiness Interaction | |
| Location | Policy | Agreement |

Fig. 8.7 SID.
Source: TMF SID Model—Phase III: Concepts, Principles, and Business Entities—GB922.

Technology Neutral Architecture

The Technology Neutral Architecture (TNA) provides the OSS services to allow software components to interoperate to provide the overall OSS functionality. As such, the TNA provides the services needed to support an SOA form of OSS operation, as indicated in Figure 8.9.

Each software component has its interface represented by an NGOSS contract. A software process (or management function) may then make its interface available to another software process by "exporting" the contract into a registry (e.g., as Process B has done in Figure 8.9). Another process that wishes to interoperate with this process may do so by importing the contract from the registry and then using this to invoke operations in the cooperating process (e.g., as Process A has done to invoke operations in Process B in Figure 8.9). Various forms of system services are needed to support this interaction model. In the TNA these are broadly classified into basic and framework services.

The preceding model appears very simple, but in practice it may be quite complex to implement. For example, a more complex trading mechanism may be used to exchange contract details; issues of throughput and reliability need to be taken into account when implementing the repository or trading service; repositories may need to be "federated" together, and so on. More details on these and other issues are contained in the appropriate TMF documentation.

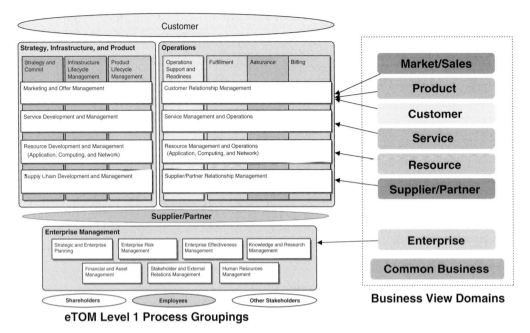

eTOM Level 1 Process Groupings

Fig. 8.8   SID/eTOM mapping.
Source: TMF SID Model—Phase III: Concepts, Principles, and Business Entities—GB922.

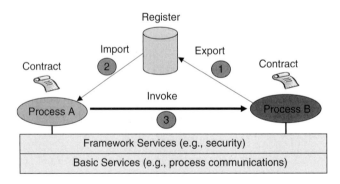

Fig. 8.9   TNA operation.
Source: TMF Technology Neutral Architecture—TMF053.

A point to note is that the NGOSS Contract supports various aspects of the NGOSS Lifecycle. This makes contracts a very comprehensive definition of the software component's interface and therefore extends it beyond a simple Application Programming Interface (API) that is often provided in other distributed processing solutions.

An outline of the NGOSS Lifecycle showing the four different viewpoints is shown in Figure 8.10. These are expressed by the Lifecycle and are represented in NGOSS contracts.

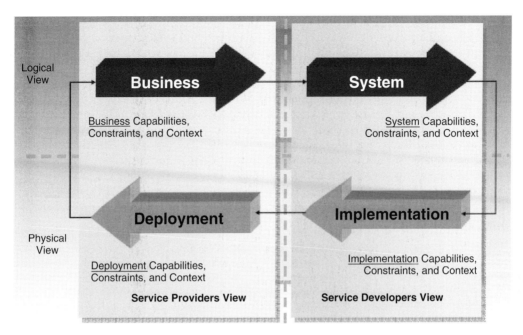

Fig. 8.10 NGOSS lifecycle.
Source: TMF NGOSS Lifecycle Methodology—GB927.

Another point to note with the NGOSS contract is that, like the SID, the contract is defined in a technology-neutral form. This means that the same NGOSS contract may also be deployed in multiple implementation technologies (e.g., J2EE, XML, and so on). As with the SID, this provides many benefits in terms of system interoperability.

As mentioned at the commencement of this section, and demonstrated by the various descriptions above, many of the TMF's NGOSS frameworks (e.g., eTOM, SID, TNA, and Lifecycle) may be combined to provide the comprehensive and flexible management environment needed to manage NGNs.

Additional Remarks

The information contained in this chapter is an attempt to provide a high-level overview of NGN and NGN management. It is based on the author's understanding of the status of the various items of technical development at the time of writing.

To provide an appropriately comprehensive overview, many diverse areas have been covered that are not yet fully integrated within the industry. In addition, this has been a snapshot of many ongoing developments at the time of producing this chapter. In order to provide the breadth of coverage, many sources of industry material have been used to support the information provided and, where possible, associated references have been made to this source material. While every effort has been made to accurately present this information, the author takes no responsibility for any inaccurate items, incorrect references, or subsequent material updates. In addition, no responsibility is taken for any possible consequential losses resulting from the use of this text.

Acknowledgments   This chapter has built on information provided in a number of industry reports and also specifications from industry groups and standards bodies.

The author would like thank in particular: ITU-T, ETSI, 3GPP, and TMF for permission to use diagrams from some of their specifications.

The author would also like to thank a variety of industry report providers who have provided useful information contributing to the creation of this chapter.

3GPP™ TSs, and TRs are the property of ARIB, ATIS, ETSI, CCSA, TTA, and TTC, who jointly own copyright to them. They are subject to further modifications and are therefore provided to you "as is" for information purposes only. Further use is strictly prohibited (© European Telecommunications Standards Institute 2005). Further use, modification, or redistribution is strictly prohibited. ETSI standards are available from *http://pda.etsi.org/pda* and *http://www.etsi.org/servicesproducts/feestandard/home.htm*.

References

ETSI TISPAN DTS 08004: OSS Requirements and Priorities September 2005.
ETSI TISPAN TS 188 001: OSS Architecture Release 1, June 2005.
ETSI TISPAN TR 118 004: OSS Vision (06/2005).
ITU-T M.3010: Principles for a Telecommunications Management Network, February 2000.
ITU-T M.3060: Principles for the Management of NGNs.
ITU-T M.3100: Generic Network Information Model, July 1995.
ITU-T M.3400: TMN Management Functions, February 2000.
ITU-T Y.110: Global Information Infrastructure Principles and Framework Architecture, June 1998.
ITU-T Y.2001: General Overview of NGN, December 2004.
ITU-T Y.2011: General Principles and General Reference Model for NGNs, October 2004.
3GPP TS 23.228 V6.10.0: Technical Specification Group Services and System Aspects; IP Multimedia Subsystem (IMS), December 2006.
TMF enhanced Telecom Operations Map (eTOM): GB921.
TMF NGOSS Lifecycle Methodology: GB927, December 2004.
TMF SID Model—Phase III: Concepts, Principles, and Business Entities: GB922, July 2006.
TMF Technology Neutral Architecture: TMF053.

9 Customer Premises Networks

David K. Hunter

Introduction to Local Area Networks

A Local Area Network (LAN) typically interconnects personal computers, servers, and peripherals, and often connects to an outside Wide Area Network (WAN), usually the Internet. LANs allow information, hardware, and software to be easily shared between different users. Generally they are high-speed networks that cover a small geographical area. For example, a LAN is often installed within a department of a company, university, hospital, research center, or manufacturing site.

LANs often allow expensive hardware such as disk arrays or laser printers to be shared by many users. Many home users have small LANs, employing Ethernet and Wi-Fi technologies, which will be discussed here. Besides the users' computers themselves, a LAN's key components are:

- *Network Interface Cards (NICs):* These provide the electrical interface between each network device and the LAN. They convert between the parallel data format inside the computer and the serial data format used by the LAN. The 48-bit Medium Access Control (MAC) address (or Ethernet address) of the interface is programmed into it by the manufacturer. Very often, a network interface is provided with new computers, and a separate NIC is not necessary.

- *File servers:* These permit LAN users to share the use of a high-capacity disk, allowing them to use it as if it were their own. Users may also share files, although attempts by more than one user to update the same file simultaneously must be prevented.

- *Web servers:* These are computers that store pages formatted in Hypertext Markup Language (HTML) and use Hypertext Transfer Protocol (HTTP) to send those pages to users who request them. Users accessing these pages do so via a Web browser, either on the LAN or outside.

- *Print servers:* These permit LAN users to share one or more printers, which would be uneconomical to dedicate to just one user often because they are very expensive and offer special facilities. Spooling software must be implemented to queue print jobs when necessary. Many printers include this functionality and can be connected directly to a LAN, without a separate print server.

- *Routers and modems:* These provide an interface with other networks in this context, especially the Internet. This permits LAN users to communicate with other computers that are not on the LAN. Modems for communication with the Internet via a telephone line are discussed in the following text.

- *Hubs:* These interconnect multiple network devices to form a star network, transferring information as necessary between the devices. Their operation is explained in more detail later.

- *Transmission media:* These carry information within the LAN between devices. Twisted pair cables are the cheapest option, although they cannot carry high-speed data over long distances. Coaxial cable is more expensive but offers better performance, while optical fiber gives the highest throughput but is the most expensive. Radio has many advantages, permitting users to be mobile while not requiring expensive wiring installation. However, the transmission quality can be variable. Wi-Fi, a wireless LAN technology, is discussed in the following text.

Many of these components are illustrated in Figure 9.1.

LANs and WANs differ in structure, technology, and protocols, although they are often interfaced to one another. The principal characteristics of both are summarized as follows.

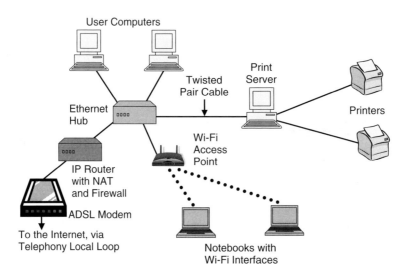

Fig. 9.1 A small customer LAN installation employing Ethernet and Wi-Fi. The ADSL modem, firewall, NAT, router, Ethernet hub, and Wi-Fi access point are often all combined into one box.

- LANs typically cover only a single building or cluster of buildings. WANs cover a large geographical area (up to the whole globe), and require the crossing of public rights of way.
- LANs are usually owned and installed by the same organization that owns the attached devices. WANs rely at least in part on circuits provided by a common carrier.
- Particularly if lightly loaded, LANs can use a simple random access protocol. WANs often have complex access protocols.

The different characteristics of both types of network suggest different technological solutions for each. LANs usually operate over a short distance (less than, say, 2.5 km), with a low limit on the number of stations (typically up to 1000) and with transmission rates between 10 Mb/s and 1 Gb/s. LANs are packet based, and often switching is implemented using a distributed switching protocol. Bandwidth allocation is flexible so that capacity is shared effectively and fairly among users. However, it is also often inefficient, particularly if the access protocol is simple.

WANs, on the other hand, can cover the entire globe, interconnecting many LANs and with no real limit on the number of end stations. WANs may be interconnected by private circuits using SONET/SDH switching and multiplexing technology with transmission rates up to 622 Mb/s, or they may employ a Virtual Private Network (VPN) with peak data rates up to 1 Gbit/s, also carried via a SONET/SDH physical layer. In this case, complex routing and traffic engineering may be employed in order to ensure fair and efficient distribution of resources, while still providing an appropriate degree of flexibility.

The Ethernet Hierarchy

Ethernet packets are known as frames. All Ethernet implementations have similar frame formats (Figure 9.2). Each Ethernet network interface has a globally unique 48-bit MAC address. The most significant 24 bits are unique to a particular manufacturer, while the remainder are assigned by the manufacturer so that each interface has a unique MAC address.

The preamble and Start of Frame Delimiter (SFD) ensure that the receiver can detect exactly when the Ethernet frame starts, and synchronize its clock. The Ethertype field

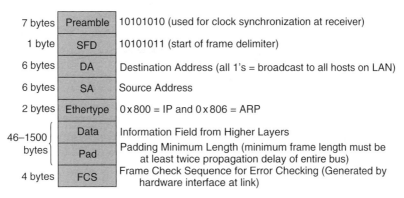

Fig. 9.2 Basic Ethernet frame format. The data field can carry up to 1500 bytes, which explains why IP datagrams of this size are common on the Internet.

indicates which protocol in layer 3 is being carried in the data field—two common values are 0x800 for IP and 0x806 for ARP, although others exist. Ethernet cannot correct bit errors in a frame; however, it may be able to detect them using the Frame Check Sequence (FCS), which is computed by means of a Cyclic Redundancy Check (CRC). It drops frames containing one or more errors found in this way, and such lost frames are resolved elsewhere, often by Transmission Control Protocol (TCP) in layer 4.

A Medium Access Control (MAC) protocol controls access to the transmission medium and decides when each network device may transmit. Because of the MAC protocol used in Ethernet, the frame length must be at least 64 bytes, plus the preamble and SFD. This corresponds to a data field length of at least 46 bytes, since the remaining fields are in total 18 bytes long (see Figure 9.2). Therefore, if the data field is shorter than 46 bytes, it is padded out to that length to prevent the frame becoming too small. The length of the data field cannot exceed 1500 bytes, a value that was fixed in the 1970s when Ethernet was designed, and was dictated by the amount of buffer RAM available.

The original Ethernet was implemented on a broadcast bus made from coaxial cable (Figure 9.3a), operating at 10 Mb/s (IEEE 802.3 standard). Each device on the LAN had equal status, with no master device, and a random access protocol was used where anything transmitted was received by all other devices on the network. In this version of Ethernet, the MAC protocol is Carrier Sense Multiple Access with Collision Detection (CSMA/CD), where a network device sends a frame only if the medium appears to be silent, in order to reduce collisions with frames from other stations. Nevertheless, two devices may transmit virtually simultaneously, with each being unaware that the other is transmitting.

In this case, one of them senses that a collision has taken place and ceases transmission. It then issues a "jam" signal to ensure that all other devices are aware of the collision, and the frame is retransmitted after a random interval. The collision probability rises rapidly with increasing propagation delay (i.e., bus length), which clearly prevents scaling to physically large networks.

CSMA/CD works reasonably well for light traffic, although it cannot carry heavy loads because frequent collisions occur. The original 10 Mb/s CSMA/CD Ethernet was based on a bus constructed from coaxial cable, with taps for plugging in devices. Its maximum length was 500 m, or up to 2500 m if several buses were interconnected by repeaters. This limit was imposed by the CSMA/CD protocol—due to the propagation delay along the bus, there are certain timing restrictions that must be enforced for the protocol to operate satisfactorily. A bus was often not the most convenient way of wiring up a network within a building, and often the whole LAN could be brought down by a single fault.

A star topology was later adopted, permitting use of existing twisted pair cable intended for telephones. Twisted pair cable is less expensive; however, coaxial cable can carry higher bit rates over longer distances. When telephone wiring is installed, it is always in a

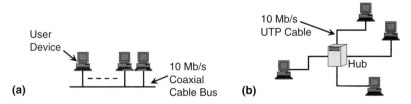

Fig. 9.3 Simple Ethernet LAN topologies: (a) a coaxial cable bus network, as used in the original 10 Mb/s CSMA/CD Ethernet; (b) a star network, interconnecting network devices via a hub in the center.

star configuration to permit the use of a central private exchange, with a connection to each extension. A hub is at the center of the Ethernet star (Figure 9.3b)—this is a multiport repeater: When it receives a frame on one port, it is repeated to all others. The hub makes no routing decisions and has no knowledge of each frame's intended destination, since each frame is sent out on every port. If two network devices transmit at once, there is collision, just as with the original bus network.

Unlike an ordinary hub, a switching hub or Ethernet switch stores frames and forwards them to the correct destination based on the destination address in the header, since there is no need to forward every frame to all ports. The switching hub learns the Ethernet address served by each port, by inspecting the source addresses of incoming frames. If it does not know which port to forward on, it forwards onto all ports. Collisions no longer take place, and, unlike CSMA/CD, there is no need to enforce a maximum distance. Ports need not all be at the same bit rate, and multiple frames can be in transit simultaneously through the switching hub, which increases throughput. In this case, although the CSMA/CD protocol from IEEE 802.3 is not used, the framing format (Figure 9.2) is retained.

Ethernet has adapted over the years to higher and higher bit rates, overtaking other, newer, technologies that have been introduced. Indeed, this is a significant reason for its success. With each new standard, Ethernet bit rates increase by a factor of ten, with 100 Mb/s Ethernet being the first new standard to be developed. It is also known as Fast Ethernet. The 100 Mb/s Ethernet frame format, procedures, and interfaces are the same as for the 10 Mb/s version. It is entirely based on a star topology, involving multiple parallel twisted pair cables or optical fiber transmission media—coaxial cable is not included in the standard. Optical fiber offers the greatest speed and throughput, but it is the most expensive transmission technology. The twisted pair cable option can only extend to 100 m. A hub can be used with CSMA/CD, operating in exactly the same way as at 10 Mb/s. Alternatively, a switching hub may be used, offering switching but without CSMA/CD.

A 1 Gb/s Ethernet (Gigabit Ethernet) retains the frame structure and procedures of the 10 Mb/s version. CSMA/CD is preserved in the standard, although it reaches the limits of its operation because it does not scale well to high bit rates or long link lengths. The minimum frame size is increased to 512 bytes, representing a possible overhead. For example, if a 64-byte packet is transmitted, it is followed by a carrier extension of 448 bytes. Gigabit Ethernet operates primarily in switched mode, always using a star topology. It can use single-mode fiber, multimode fiber, or twisted pair cable. It is often deployed, for example, in an academic environment for campus backbone networking, or for high-performance computing.

The 10 Gb/s Ethernet provides point-to-point connectivity, with CSMA/CD disabled. There are two versions:

1. LAN PHY supports existing Ethernet LAN applications, and uses pairs of optical fibers—either multimode (300 m reach) or single mode (10 km reach).
2. WAN PHY allows 10 Gb/s Ethernet terminals to be connected through 10 Gb/s SDH/SONET; it uses pairs of single-mode optical fibers, reaching up to 40 km.

Token Ring and Other Technologies

Because of Ethernet's success, other LAN technologies are no longer widely used. Nevertheless, it is instructive to consider alternative approaches briefly. In a token ring network, network devices are interconnected via a ring topology. Information is transmitted sequentially around the ring, one device after another, unlike Ethernet, where information is always transmitted in an unscheduled, random fashion. A unique token is passed via the ring between devices, where each may only transmit when it has the token. After receiving a frame, the destination device copies information from the ring and the source device removes the frame from the ring when it replies. After that, the source device sends the token to the next node. Other ring technologies based on similar concepts exist, such as the Fiber Distributed Data Interface (FDDI) and the Cambridge Ring. In an alternative configuration, the token bus protocol involves passing tokens sequentially in a set pattern around devices connected to a bus network.

Wireless LANs—Wi-Fi

Unlike Ethernet, in wireless LANs network devices are not physically fixed, so it is possible for a user to physically move around, or "roam," within the reception area while using a device. The reception area is usually within a single building, and a wireless LAN can be viewed either as a replacement for an Ethernet LAN or as an extension to it. The objective is often to replace office wiring, to ensure access to the Internet without physical restrictions, or to permit communication without prior setting-up or planning, for example, in business meetings. The latter is known as "ad hoc" mode, where an access point is not required for communication to take place and the devices communicate directly with one another.

Wireless LANs are especially appropriate where installing wires would be difficult, or where cables would restrict users. In fact, wiring is often a significant portion of the cost of wired LAN installation. Wireless LANs avoid the need for unsightly cables and can be used where the infrastructure does not exist or is at risk of failure. Furthermore, once the wireless access point has been installed, additional users may be added without increasing the cost of the network. Adapters are necessary for user computers to access the wireless LAN, although many computers now have such adapters already built into them. The software on each user's computer is unaware that a wireless connection, rather than a wired LAN such as Ethernet, is being used.

Wireless devices are often battery powered, so it is essential that they conserve power, and consequently their protocols are specially optimized to do this. Wireless protocols are designed to be compatible with existing applications, higher-layer protocols, and networks. Wireless transmission does have disadvantages, such as not offering the same transmission quality as wired networks and often interfering with other electronic equipment, for example, in hospitals.

The IEEE 802.11 wireless LAN standard is commonly known as "Wi-Fi," and is now widely implemented. It uses microwave radio in the 2.4 GHz band, which is license free, with various versions of the standard supporting transmission up to 54 Mbit/s and beyond. Ad hoc networking is supported in the standard, allowing devices to communicate without an access point, although this mode is not always supported in practice. Infrastructure networks are more commonly used, where an access point communicates with the mobile devices and is also connected to the Internet, often via Ethernet. Indeed, IEEE 802.11 is fully integrated with the other IEEE 802.x LAN standards, including Ethernet. There is also another common wireless LAN standard, Bluetooth, but this is not designed for the applications discussed here.

Wi-Fi's basic transmission protocol is called Carrier Sense Multiple Access with Collision Avoidance (CSMA/CA). While it has been designed to cope with the particular challenges of wireless transmission, it does have some features in common with CSMA/CD, which was considered earlier. When a network device is ready to send, it starts sensing the medium. If the medium is free for the duration of an interval known as an Inter-Frame Space (IFS), then it may start sending. If the medium is busy, the device has to wait for a free IFS; then the station must additionally wait for a random back-off time. By doing this, it implements the "collision avoidance" part of the protocol. If another device occupies the medium during this back-off time, the back-off timer stops in the interests of fairness. The timer starts again at an interval of length IFS after the channel becomes free.

Transporting IP

Ethernet interworks very straightforwardly and successfully with TCP/IP. Encapsulation is a critical feature of such an interface, and an important aspect of protocol layering. At the source, a packet from a higher layer is inserted into the data field of the layer below, and at the destination the data field is removed and presented to the layer above. Thus a hierarchy of packet headers is in fact transmitted by the physical layer.

Each packet header contains a field that states from which protocol in the layer above the payload came and, consequently, to which protocol it should be sent at the destination. In an Ethernet packet (known as a frame—layer 2), this field is called the Ethertype field.

The data field in the data link layer is limited in size by the protocol concerned, limiting the size of the IP datagram that may be placed within it. This size limit is called the Maximum Transmission Unit (MTU), and some common values are shown in Table 9.1. The MTU refers to one link; however, the path MTU refers to an entire path, which may consist of many links and is the minimum value of all the MTUs along the path. Hence the path MTU is the maximum size of IP datagram that can be sent over a given path.

As an example, consider sending a large file over an Ethernet LAN. When the file is sent, the following processes take place:

- *Layer 5:* File Transfer Protocol (FTP) ensures that the next portion of the file is placed in the TCP send buffer.
- *Layer 4:* TCP removes the information from the buffer in blocks of 1460 bytes. This size is dictated by the path MTU of 1500 bytes that was determined by TCP at the start of the connection. TCP then adds a 20-byte segment header, making each segment 1480 bytes in length. The port field in the TCP header is set to 20, indicating FTP in the layer above.

Table 9.1   Default MTU Sizes for Common Link Layer Technologies

Link layer protocol	Default MTU
Point to Point Protocol (PPP)	296
X25	576
Serial Line Internet Protocol (SLIP)	1066
Ethernet	1500
Fiber Distributed Data Interface (FDDI)	4352
4 Mb/s token ring	4464
Token bus	8166
16 Mb/s token ring	17,914
Hyperchannel	65,535

- *Layer 3:* IP adds its header, which is also 20 bytes in length, resulting in a 1500-byte datagram, the maximum size that can be encapsulated in the Ethernet data field. The protocol field in the IP header is set to 6, indicating TCP in the layer above.

- *Layer 1/2:* Ethernet encapsulates the datagram within a frame, which is of total length 1526 bytes, including the preamble and SFD. The Ethertype field is set to 0x800, indicating IP in the layer above. The frame is transmitted across the network to the other user.

When the frame is received, the reverse process takes place. The Ethernet frame is received, and the IP datagram is extracted (decapsulated). The TCP segment is then extracted from the data field of the IP segment, and the data from it is placed in the TCP receive buffer, for FTP to take when it is ready. Each protocol knows, from the appropriate field in its header, which protocol to hand its data field to in the layer above.

Network Address Translation and Firewalls

This section describes two techniques that are deployed at the interface between a customer's network and the outside Internet; in fact, they are often combined into a single device or integrated into a router. A firewall blocks undesired communication between hosts on the customer's network and the outside Internet. Network Address Translation (NAT) allows many IP addresses to be used within the customer's network, while the outside Internet believes that the customer has only one IP address. Hence this is a solution (albeit a short-term one) to the shortage of IPv4 addresses. This is especially appropriate for business users, who require multiple "always-on" connections that always need an IP address, and also for those home users who have their own networks with several network devices.

Hence, with NAT, each customer has a unique IP address that is used by the outside Internet, and each computer on the site (i.e., the customer's network) is assigned an IP address that is not available outside the site. All computers on a site have different IP addresses. However, without affecting their operation, the same IP address may be used many times on other sites. The address ranges 10.0.0.0/8, 172.16.0.0/16, and 192.168.0.0/16 are reserved for such computers, and exclusively allocated for use with NAT.

Suppose a computer on a customer's site has IP address 10.0.0.1, and it sends a datagram to the outside Internet. Suppose also that the customer's true IP address is 202.43.61.212. Then NAT will translate the source address of 10.0.0.1 in the outgoing datagram into 202.43.61.212. That is, the IP source address header field is rewritten. Suppose that a datagram is returned, destined for 10.0.0.1—it has been sent to 202.43.61.212. So, how does NAT know which computer on the site to send it to?

Whenever a datagram leaves the site, NAT looks up a table that defines a new source port in terms of the original IP address and original source port (for example, Table 9.2). The table illustrates how two computers on the same site may use the same port number. If a datagram from 10.0.0.1 is to leave the customer's network, and has a source port number of 4005, then the source address is replaced with 202.43.61.212 and the source port number is replaced with 10001, before the datagram is sent into the outside Internet. Also, the IP header checksum must be recomputed. If no entry exists for particular values of the original IP address and original source port, a new entry is created and a new port number is assigned.

Table 9.2 Sample NAT Table

Original IP address	Original source port	New port number
10.0.0.3	4005	10000
10.0.0.1	4005	10001
10.0.0.1	4009	10002

When a datagram arrives from outside, the incoming port number is used to index the right-hand column of the table, to find the corresponding original IP address and port. These are inserted into the datagram's header, the header checksum is recomputed, and the datagram is passed to the customer's network.

Although NAT is a quick fix to the problem of IP address depletion, it assumes that UDP and TCP are the only layer-4 protocols in use, so that if another protocol is introduced, NAT will not work without being changed. Furthermore, for each protocol that sends IP addresses in the body of the text, NAT must be specially modified to function correctly. This was necessary for FTP, and is inconvenient if new protocols are devised. For these and other reasons, many in the IP community regard NAT as being an unsatisfactory and inelegant quick fix to the IP addressing problem, which has delayed the deployment of IPv6.

Thus far, it has been assumed that all sessions are initiated by a computer on the site, but if a server is on the site, the first incoming datagram in a session may not have a corresponding table entry since it is from an external host, and it will not be possible to forward it to the correct computer. In this case, when an application on the Internet looks up the domain name of a computer on the customer's network via DNS, an entry is placed in the NAT translation table.

NAT can be implemented via either software or hardware, although software implementation is only appropriate for relatively low-speed links such as 10 Mb/s Ethernet. It is also used by ISPs offering ADSL to customers—each customer has its own private address (such as 10.a.b.c), and when traffic leaves the ISP and enters the main Internet, network address translation takes place. Thus the ISP and all its ADSL customers are treated as one company using NAT.

A firewall is often deployed with NAT in one networking device—both are widely used in customer premises equipment. It is placed between an organization and the outside Internet, and is an important way of screening unwanted communication between computers inside the organization and those outside. A firewall must be placed on each external network connection, preventing outsiders from overwhelming the organization's network with unwanted traffic, attacking its computers, or probing them. Firewalls are based on packet filters; in the simplest case, they are configured by the network administrator to block packets based on their IP source address, IP protocol field, and TCP/UDP port field. Hence, it is specified exactly which applications, from exactly which hosts, are blocked. Instead of specifying individual addresses, IP prefix notation may specify groups of IP addresses to be blocked.

This approach of specifying the datagrams *not* allowed in is flawed, since it is impossible to specify every possible type of datagram that should be filtered. Hence, very often, instead of starting with the assumption that most datagrams are acceptable, and only those that are *not* allowed in should be explicitly specified, the packet filter assumes that *no* datagram should be allowed in unless it has been specified that it should. However, there is now a further complication. A client inside the organization's network may wish to access a server outside. The server will reply using the arbitrary port number that was specified by the client. To avoid such datagrams being blocked, the firewall must examine outgoing packets so that it can deduce what port numbers to accept on incoming datagrams. Such a firewall is called a stateful firewall.

Very often, routers contain a special board that implements packet filtering in hardware, based on a table defined by the network administrator. Also, some firewalls operate at the application layer and inspect, for example, files transported via FTP or e-mail in order to detect viruses. To permit the network administrator to observe trends and examine the events leading up to a security breach, firewalls keep a log on disk of any attempts to bypass security. Unlike the TCP/IP or Ethernet protocols, the implementation of packet filters in firewalls is not standardized, and both their capabilities and the interface provided for the network administrator vary widely.

Modems

56K Modems

Within a computer or other network device, information is sent without modulation—one voltage level represents a "zero," and another represents a "one." While this format works well for transmission over small distances, it is inappropriate for sending information over a telephone line, where modulation must be used. The word "modem" is short for "modulator/demodulator," and this section will discuss the principles of modem operation, with particular reference to 56 Kbit/s modems, which are usually referred to as "56 K modems." These permit computer data to be carried over 64 kbit/s telephony channels, which has enabled the public telephone network to be used for Internet access as it has developed.

Three basic types of modulation are used in modems:

1. *Amplitude shift keying (ASK):* The amplitude of an audio tone, known as the carrier signal, is set to one value for a "zero" and another value for a "one."
2. *Frequency shift keying (FSK):* The frequency of the carrier signal is set to one frequency for a "zero" and another frequency for a "one." It was used in early 300-baud modems, but to achieve higher speeds, more sophisticated techniques are necessary.
3. *Phase shift keying (PSK):* The scheme generally used is called Differential PSK (DPSK). For a "zero," the phase of the signal is advanced by 90 degrees, whereas for a "one," it is advanced by 270 degrees. Because the phase is relative to its previous value, there is no need for the receiver to maintain a reference carrier signal, simplifying the demodulation circuitry.

It was assumed in these descriptions that the symbol rate (also known as the baud rate) was equal to the bit rate (i.e., each symbol transmitted—change of amplitude, frequency, or phase represented either a single "zero" or a single "one"). When the bit rate is higher than the baud rate, multiple bits are transmitted by each symbol and higher bit rates are possible. This is crucial to the operation of modern high-speed modems. For example, phase modulation can be used with phase changes of 45, 135, 225, or 315 degrees, so that two bits are sent for each phase change. This is known as Quadrature PSK (QPSK).

To realize a further increase in throughput, ASK and PSK can be combined to implement Quadrature Amplitude Modulation (QAM). For example, 16-QAM has four bits per symbol and can transmit at 9600 bits/s over a link that can handle 2400 baud. More bits per symbol are possible (for example, 6 bits per symbol, 64-QAM), but transmission errors become increasingly frequent, and appropriate coding must be used to detect or correct these. This has resulted in Trellis Coded Modulation (TCM).

Many modem standards have been defined by the ITU-T over the years, operating at bit rates from 300 bits/s to 56,000 bits/s. As a result, it is common for modems to support more than one bit rate. When setting up a connection, there is a training phase to determine the highest bit rate that the connection can support. With analogue transmission of the type discussed, the limit on transmission speed that can be reached due to the physics of the local loop is approximately 35,000 bits/s, and in that respect, the most advanced standard in the ITU-T series is V.34bis, which achieves 33,600 bits/s by having 14 bits per symbol at 2400 baud. So how do 56 K modems work?

The 56 K modems only permit transmission at 56,000 bits/s from the ISP to the customer—because the connection at the ISP end is purely digital, higher-speed transmission is possible. On the other hand, the customer can only transmit at 33,600 bits/s (using the V.34bis standard), since the customer connection is analogue. However, this is not a difficulty since most applications (such as Web browsing) require a higher downstream bandwidth than upstream. V.90 is the ITU-T standard that implements this, and there is a further standard (V.92) that can transmit upstream at 48,000 bits/s if the line is of sufficient quality, and also has a shorter training phase.

Furthermore, the throughput of modems can be greatly improved by using data compression. The improvement depends on the type of data, but throughput can be at least doubled in many circumstances. V.42bis is a data compression standard for modems, and is used in conjunction with V.42, which is a standard for error correction. V.44 is a later data compression protocol that is implemented within V.92.

ADSL Modems

Although 56 K modems have allowed the public telephone network to be used as the Internet has developed, the rapid development of applications and services has led to the need for faster modems. The twisted pair copper wires used to connect telephones to the central office (or exchange), in fact, can carry much higher data rates providing carefully chosen modulation methods are used. Several versions of Digital Subscriber Line (DSL) technologies are used, although Asynchronous Digital Subscriber Line (ADSL) is the most common. Just as for 56 K modems, asymmetric transmission speeds are used, with faster transmission downstream than upstream.

There are several reasons why ADSL has been offered for high-speed Internet access by telecommunications operators:

- Without it, for example, with 56 K modems or Integrated Services Digital Network (ISDN), the access network is a bottleneck.
- The local loop is mostly based on old category-3 twisted pair cables, and it is desirable to re-use these existing cables if possible. ADSL is designed to operate over the existing twisted pair local loop, avoiding the need to replace it.
- The Internet is growing and has been doing so for many years.
- Cable companies are competing by offering broadband Internet access.

ADSL signals bypass any 3400 Hz low-pass filters at the exchange as the data is fed directly into a high-speed connection to the ISP. Hence, the entire frequency spectrum of the

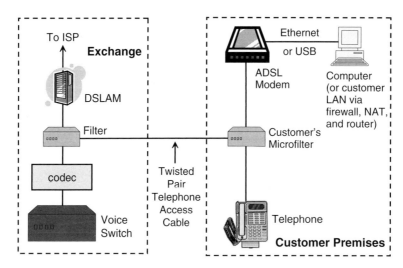

Fig. 9.4 Configuration of ADSL and related equipment, between the local exchange and the customer. At either end, a filter separates the telephone signal (<3400 Hz) from the ADSL data signal (>26,000 Hz).

local loop is available, and transmission is limited only by the physics of transmission. The possible bit rate decreases with increasing distance between the subscriber and the exchange, so that ADSL offers either low-speed service to customers in a wide radius or high-speed service to customers in a small radius. It works over category-3 twisted pair cables between the customer and the exchange, even those that were installed decades ago, and permits simultaneous use of the telephone on the same line. It offers a much higher data rate than a 56 Kb/s modem or ISDN.

Internet data is transmitted over 250 parallel FDM channels, each modulated on a different carrier using QAM. The data rate of each channel is adjusted according to perceived channel line quality, adapting automatically to noise on the line, such as radio interference. With ADSL, there are usually 32 upstream channels, with the remainder being used for downstream communication. Hence, the first letter in ADSL stands for *asymmetric*.

On short runs over high-quality cable, 8 Mb/s downstream and 1 Mb/s upstream is possible. A typical ISP offers 512 Kb/s downstream and 64 Kb/s upstream, with 2 Mb/s downstream and 512 Kb/s upstream premium service. Most domestic Internet users require a much greater downstream data rate than upstream—for example, when browsing the Web or downloading music.

The subscriber's telephone and the voice switch at the exchange use frequencies below 3400 Hz (Figure 9.4). The ADSL modem and the corresponding equipment in the exchange, called the Digital Subscriber Line Access Multiplexer (DSLAM), use only frequencies above 26,000 Hz. Hence, both can coexist on the same line without conflict, providing appropriate microfilters are installed to separate the frequency bands.

Conclusions

This chapter has examined the networking technologies that would be deployed on customers' premises. Ethernet has been assumed as the Local Area Network (LAN) technology, simply because it has been so successful, and is now used almost universally. Ethernet has been in existence in one form or another for more than twenty years, and although other LAN technologies have been proposed, Ethernet has eclipsed them all. This is because, first, Ethernet is simple, flexible, cheap, and easy to maintain. It works well with TCP/IP, which is the dominant protocol suite above the data link layer (because, for example, Ethernet and IP are both connectionless protocols). Perhaps most important, Ethernet has evolved and adapted to leapfrog new technologies that were introduced. FDDI, Fiber Channel, and ATM were all superior to Ethernet when they were introduced, but new, faster versions of the Ethernet standard were devised that overtook them. Also, wireless LANs are now popular, and are widely deployed in offices, homes, airports, and cafés—wherever portable access is required. Clearly these scenarios cannot be conveniently fulfilled by Ethernet.

Early modems operated at 300 bits/s, and a succession of ITU-T standards has steadily improved this up to 56,000 bits/s. ADSL is now widely deployed, with downstream bit rates of 2 Mbit/s being common and higher bit rates being possible close to the exchange. Related technologies such as VDSL off bit rates the customer of 24 Mbit/s and above, where an optical fiber link exists between the exchange and a roadside cabinet, with the remaining link to the customer using existing twisted pair cable. Finally, future potential deployment of Fiber to the Home (FTTH) via Passive Optical Networks (PONs) could offer even higher capacities.

Firewalls are one of the most important defenses that a company or individual can have against malicious hackers and other attackers from outside. These prevent unwanted (and potentially unsafe) communication from taking place across the firewall. Other security techniques, such as encryption, also exist. Network Address Translation (NAT) provides a user with many IP addresses while only using up one IP address on the outside Internet. It is a quick and inelegant fix to the shortage of IP addresses, and this problem will only be resolved fully by the introduction of IPv6.

Resources

Cheswick, W. R. (2003) *Firewalls and Internet Security: Repelling the Wily Hacker*, Addison-Wesley.

Comer, D. E. (2004) *Computer Networks and Internets with Internet Applications, Fourth Edition*, Prentice-Hall.

France, P. (2004) *Local Access Network Technologies*, IEE.

Halsall, F. (2005) *Computer Networking and the Internet, Fifth Edition*, Addison-Wesley.

Perlman, R. (2000) *Interconnections: Bridges and Routers, Second Edition*, Addison-Wesley.

Tanenbaum, A. (2003) *Computer Networks, Fourth Edition*, Prentice-Hall.

Part Four

Cellular Telephone Networks

As digital circuit-switched networks were developed during the 1970s, mobile networks soon followed to meet a variety of purposes, delivering both public and private services. The public land mobile networks (PLMN), which now reach up to 99 percent of the population in some countries, use a landline component that employs many of the basic principles of landline telephony, either based on circuit or packet techniques. However, they require additional functionality to enable radio connections to be made to mobile phones and to keep track of where the network users are located. The chapters in Part 4 that cover cellular telephone networks concentrate on the methods that have been developed to achieve this additional functionality.

Just as major change is reshaping fixed, or landline, telephone networks, so, too, is massive change being seen in mobile networks with third-generation networks now being widely introduced. The following chapters track the evolution of cellular systems and describe the developments in the European and North American markets. Most emphasis is placed on the first- and second-generation systems (1G and 2G), with a briefer summary of the later systems. Further coverage to 2.5-generation and third-generation "broadband" systems will be given in Volume 3 of *The Cable and Telecommunications Professionals' Reference*, which will deal with broadband data.

In Part 4, we describe how cellular networks have evolved (Chapter 10), the basic principles of the 1G and 2G systems (Chapter 11), and the main system types (Chapter 12, with emphasis on GSM). A brief summary of the other systems beyond 2G is given in Chapter 13 and Chapter 14 describes Bluetooth systems.

10 Evolution of Cellular Systems

Yael Baruch and Daniel Mallah

Introduction

This chapter provides a brief introduction to the history and to each of the generations of the cellular system, explaining their main features, differences, advantages, and disadvantages. It also discusses how the cellular market has developed in both Europe and the United States, taking account of services, operators, transmission rates, and user equipment. Also, as an introduction to the second and later generations of cellular systems, the basic principles of multiple-access techniques are described.

The first radiotelephone service and the basic principles of cellular systems were introduced in the United States in 1949 by Bell Laboratories. It was a service that connected between a user in a car and the fixed public network. Later, in the 1960s, Bell launched a new system called Improved Mobile Telephone Service (IMTS). This new system was a great improvement over the first system as it offered features such as direct dialing and greater bandwidth. But the technology for building true commercial networks offering mobile services to the public was not practical until the early 1980s. These analogue systems are considered the first generation (1G) of cellular systems.

By the early 1980s, different first-generation cellular networks had been developed in countries across the world. Generally, each country had a distinctive cellular standard appropriate for the telecommunications authority, or it adopted a proprietary standard. Due to the fact that frequency, allocations, and other basic parameters (such as channel spacing) had often been set at the national level and not coordinated between countries, these local factors resulted in different standards being adopted by different countries.

In addition, in those countries where the standard was late to develop and advances in technology were introduced, new features established by these technologies were inserted into the standards. This further increased the diversity among national standards.

The second-generation (2G) cellular networks developed during the late 1980s and were introduced commercially at the beginning of the 1990s. They were based on a digital air interface and circuit switching core network. The move to digital technology brought new access methods: first "time division multiple access" and later "code division multiple access." These are both described later in this chapter. The two main technologies of the second generation are Global System for Mobile (GSM) communication and the Code Division Multiple Access system One (CDMAOne). GSM was introduced commercially in the beginning of the 1990s and CDMAOne was introduced in the middle of the 1990s.

Just a few years after the deployment of the second-generation systems, market demand surpassed the capacity limit of this technology (more market demand than the systems were able to support), and mobile voice and basic data services had become an integral part of user requirements. Fundamental market demand and the urgent need to increase capacity (for data and voice) led to the deployment of the next generation. The earlier intention had been to move directly from second generation to third generation. Although the third generation was a solution to these demands, it was not ready for deployment. To provide a quick response to market demands and to provide a solution to problems associated with the second generation, emphasis was placed on improving the second generation, and as a result the "Low" 2.5-generation system was developed.

Later on, the profitability of the third generation was still not clear. Massive investments were already made in second-generation and in Low 2.5-generation systems. Many network operators preferred an evolutionary approach rather than reinvesting in a new infrastructure for the third-generation, considering the following:

- Huge investments required for third-generation deployment
- Problems with frequency band availability
- Unproved potential of new services

This led to the development of the "High" 2.5-generation systems that in some places replaced the move to third-generation systems.

The third generation was designed to provide a much fuller variety of broadband data services. Operator licenses required for the new spectrum essential to third generation were approved in 1998/1999. This spectrum was first used for additional capacity for second-generation and 2.5-generation systems. In 2000, the first version of third-generation specifications was ready and in 2000 and 2001, the first systems were deployed in Korea and Japan.

From 2000, increasingly more third-generation systems were deployed. The assumption was that the third-generation systems would eventually comprise the majority of the market. In 2006, all cellular technologies (first-, second-, 2.5-, and third-generation) coexisted, with the most common cellular technology being GSM (of the second generation). Fourth- and fifth-generation technologies are now in the design stage.

Generations of Cellular Systems

First-generation Systems

The first-generation (1G) wireless networks primarily targeted voice and carried a very low speed data signal. These analogue systems saw two key improvements: the invention of the microprocessor and the digitization of the control link between the mobile phone and the cell site. Despite these improvements, deployment and services remained very expensive

Table 10.1   Main First-generation Technologies

MPS	U.S. and the Middle East
N-AMPS	U.S. and the Middle East
TACS	Europe
NMT	Nordic countries
C-450	Germany
MCS	Japan
CNET	Germany, Portugal, and South Africa
MATS-E	France

and mainly targeted business needs. A variety of 1G standards were adopted around the world. These main technologies and their region of use are summarized in Table 10.1. The systems are described more fully in the next chapter.

Cellular systems have two major differences compared to fixed landline networks:

- *Capability for mobility:* This is an advantage over the fixed networks and represents a new concept.
- *Limited capacity:* This is a disadvantage compared to fixed networks.

Mobility means the ability to move while communicating. This feature greatly increases the complexity of the network and requires a wireless connection to the user. Wireless connections have a limited range and a limited amount of bandwidth in comparison with a fixed connection.

Second-generation Systems

The second-generation (2G) cellular systems are digitally transmitted over the air and are based on a circuit-switching core network. Additional spectrum was made available to meet the growing demand. From the start of their deployment, the 2G systems rapidly superseded their analogue counterparts and were deployed worldwide. With their deployment, mobile service became common and a huge number of users became cellular subscribers.

The 2G system GSM has been described as the fastest-growing technology of all time. For example, by the end of 2003, there were more than 1 billion cellular subscribers compared to 306 million in 1998.

The systems' emergence coincides with, and was driven by, advances in digital processing techniques, faster microprocessor speed, and reduced power consumption. The 2G systems positioned the technology far ahead of the 1G systems, as they provided cheaper deployment, additional capacity, a variety of services (especially data services), and worldwide commonality. Roaming gives the 2G a great advantage over the 1G. This feature enables users to receive services in other operators' areas (or/and countries) due to the seamless transfer between 2G networks of different operators (providing that the proper agreements have been signed between the operators).

With the development of the 2G systems, the opportunity was taken to develop common standards and systems across several countries. One such system is the Global System for Mobile communication and another is the Code Division Multiple Access system One. The generations of these two technologies compete to take the lead as the world standard. Today GSM has overcome this competition (i.e., there are more GSM networks than CDMAOne networks). The main technologies of this generation are GSM,

CDMAOne,[1] and D-AMPS (Digital Advanced Mobile Phone Service,[2] which is also known as TDMA).

Second-and-a-half generation Systems

This generation is divided into two parts: Low two-and-a-half generation (L 2.5G) and High two-and-a-half generation (H 2.5G). The "Low" and the "High" relate to the rate of the air interface.

The 2.5G systems use packet technology in their core network and in their air interface to improve the network ability to support a wider range of data. User equipment began using packet technology in 3G systems.

These systems are based on the same infrastructure and the same frequencies as the 2G systems and usually coexist using the same core network. In these common systems, the 2G usually carries the voice services and the 2.5G carries the data as an overlay system. Examples of these kinds of overlay data systems are:

- GPRS for GSM, where GPRS stands for General Packet Radio Service
- EDGE for GSM, where EDGE stands for Enhanced Data GSM Environment
- CDPD for D-AMPS, where CDPD stands for Cellular Digital Packet Data

The move to 2.5G was used as a test case for 3G misgivings by testing the required type of services, the service directions, the market demand required to cover the necessary investment, the target audience for the services, and so on. The 2.5G improved the quality of services, increased capacity, and provided the foundation for new data service capabilities at a lower cost.

The main technologies of this generation are:

Low 2.5G: GPRS, HCSCD, IS136+, IS95+ (where HSCSD stands for High-Speed, Circuit-Switched Data and IS stands for Interface Specification)
High 2.5G: EDGE, 1xRTT (where RTT stands for 1x (single-carrier) Radio Transmission)

A short description of these systems is given in Chapter 12.

Third-generation Systems

The entire telecommunication industry, including the manufacturers, operators, and national and regional standard-setting bodies, made a concerted effort to create a common standard for the 3G cellular system. As a result, the ITU began work on a common world cellular standard.

In 2000, after the ITU had invested more than ten years of work in collaboration with many other entities, the technical specifications for the 3G systems were unanimously approved. A general name for all the standardized 3G systems was defined by the ITU under the brand of IMT-2000 (which stands for International Mobile Telecommunications 2000), and the spectrum between 400 MHz and 3 GHz was declared as suitable for 3G systems

---

[1] CDMA is the technology of Code Division Multiple Access, while CDMAOne, CDMA2000, and so on, are names given to particular standard versions of CDMA systems from different generations. The CDMA system family has also become known as CDMA, which can, on occasion, lead to some confusion.
[2] Both GSM and D-Amps use the Time Division Multiple Access method (described later in this chapter), whose common abbreviation AMPS is TDMA. The D-AMPS system has also become known as TDMA, which can, on occasion, lead to some confusion.

(IMT-2000). This was the outcome of the mutual effort made by entities inside the ITU (ITU-R and ITU-T) and outside the ITU as 3GPP and 3GPP2 (where 3GPP stands for Third-Generation Partnership Project). See references (3GPP).

The goals of 3G mobile systems were faced with several challenging technical issues, such as:

- A global target of complete interoperability and interworking between mobile systems (wide mobility).
- Cheaper development and cheaper equipment.
- The possibility to create complex and progressive systems and services in a short time.
- Defining a system capable of supporting broadband data and multimedia services that are not possible with 2G systems.
- Bringing new types of services and abilities to the consumers while providing an enhanced user experience.
- Providing significant capacity, especially in urban centers, supporting a greater number of voice and data customers.
- Obtaining higher data rates at a lower incremental cost than 2G.

The new features and capabilities of the 3G provide operators with the opportunity to enhance the relationship they already enjoy with their customers and drive new revenue opportunities by encouraging additional traffic, stimulating new usage patterns, and strengthening customer loyalty.

The main 3G technologies are the Universal Mobile Telecommunications System (UMTS) and CDMA 2000.

Fourth-generation Systems and Beyond

Usually the next-generation plan is designed in parallel with the deployment of the last-generation systems. Therefore, before 3G networks were fully launched and utilized, various study groups considered the character of the next generations of cellular technology: 4G and beyond. Unlike the 3G, a single global vision for the 4G has not yet been developed. Several standard technologies have been already developed to improve 3G characteristics, and some technologies are still in the study stage for other possible systems.

An example of standard technologies used today for 3G improvements are High Speed Downlink Packet Access (HSDPA) for UMTS and Evolution Data Only/Evolution Data Optimized (EVDO) for CDMA2000 (3GPP2). Other 4G systems are all IP (ALL-IP) and 3GSM. The ALL-IP network will probably be one of the main 4G systems. The main goal of this generation is to provide seamless services across both wired and wireless networks worldwide, anywhere, any time (universal mobility), with new services that will change customer lifestyles.

The number of users per geographical area (network capacity) and the available spectrum per user (user capacity) in cellular systems were and still are lower than in the fixed networks. As the cellular generations progressed, coding methods improved and the cellular networks increased their capacity, and the difference in capacity between fixed and cellular became less significant. In the future, each user will have his or her own IP number and the fixed networks will have the ability to identify the user independently with its connection point (portability). As the portability feature of fixed networks develops and the additional capacity of the cellular networks grows, the differences between the networks will decrease.

Evolution of Cellular Markets

Evolution of Generations (1G to 3G)

Three generations of cellular systems have been introduced since their first creation, and currently all three coexist. The majority of today's market consists of 2G systems, with the 1G almost gone and 3G rapidly becoming established.

Third-generation systems, CDMA2000 and UMTS, are expected to become the dominant wireless platform worldwide within a few years and take market leadership from 2G GSM. In 2006, UMTS is the most widely deployed 3G technology in the world. The 3G system will not necessarily replace previous generations, but may coexist with them and share a single common core network.

The migration phases of the main cellular system technologies are presented in Table 10.2. The operator can choose to go directly from the second generation to any of the other possibilities. For example, a GSM operator can migrate its network to GPRS and then to UMTS; others can go directly to EDGE without continuing to UMTS, and so on.

Table 10.2 Cellular Technologies by Generation

2G	2.5 Low	2.5 High	3G	Higher
GSM	GPRS	EDGE	UMTS	HSDPA
D-AMPS	GPRS, CDPD	EDGE, IS136HS	UMTS	HSDPA
CDMAOne	IS-95B	1xRTT	CDMA2000	EVDO, EVDV

Sometimes 1xRTT is described as high 2.5G and sometimes as 3G. It is also known as "IS-2000," "MC-1X," and "IMT-CDMA MultiCarrier 1x." CDMA2000 is the 3G of CDMAOne. Sometimes the term is related to the first 3G implementation of CDMAOne (in the year 2000) and sometimes to all 3G systems of CDMAOne, which can include: 1xRTT, EVDO (or 1xEV-DO), and EVDV (or 1xEV-DV).

With the advance of the generations (1G, 2G, 3G, and beyond), call quality improved, coverage and capacity increased, privacy was enhanced, and user bandwidth increased. New technologies and new concepts became integrated, networks and the number of subscribers grew phenomenally, access devices developed, and traffic increased. The services and devices include intelligent content and the addition of new business models. Emphasis and revenues changed from access (airtime) to services. Table 10.3 summarizes the evolution of features among the cellular generations.

In mid-2006, two main standards were competing for 3G:

- UMTS (almost 75 million customers)
- CDMA2000 (more than 25 million customers)

The 3G deployments could also be endangered by the fast adoption of Wi-Fi and the emerging Wi-Max. For slow mobility or quasi-static usage, Wi-Fi provides semi-mobile Internet (a service similar to 3G service).

Evolution of Transmission Rate

As the generations have advanced, user capacity has increased, as illustrated by Table 10.4. For example, CDMAOne (2G) has 8 to 10 times the capacity of AMPS (1G), and GSM (2G)

Table 10.3   Summary of Migration Features

1G	2G and 2.5LG	3G
Voice	Voice and data	Voice, data, and video
Analogue	Digital	Digital broadband
Circuit switch	Circuit switch and packet switch	Circuit switch, packet switch, and IP multimedia, or packet switch and IP multimedia
Low transmission rate	Medium transmission rate	High transmission rate
Voice-oriented approach	Data and voice approach	Lifestyle change
Heavy and large equipment	Small, variety, full of features	Lighter, PDA, PC, sophisticated
Voice services	Voice, data, and basic multimedia services	Voice, data, and advanced real-time multimedia applications
Business oriented	User oriented	User oriented
Not secured	Secured, good privacy	Better secured, better privacy
Slow growth	Exponential growth	Fast growth
Private interfaces	Open and standardized interfaces	Open and standardized interfaces
Managed by operators only	Operators and content providers	Operators and content providers

Table 10.4   Different User Rates Supported by the Four Generations

Generation	Bandwidth	User data rate
1G	10–30 KHz	Up to 9600 bps
2G	Up to 1.25 MHz	Up to 384 kbps
3G	2.5 MHz	Up to 2.4 M
4G	More than 10 MHz	8–10 M

has 4 to 5 times the capacity of AMPS. Table 10.4 presents rate values for their different generations.

Operators and Networks

GSM Operators

GSM (2G European system), D-AMPS (2G U.S. system), and Japan's PDC system all have clear evolution paths to UMTS (3G). Some CDMAOne operators have also chosen to deploy UMTS (e.g., in Korea). The evolution of the GSM family networks has a clear and standard path. The advance from one core network system to another is built in steps, with the operators mainly required to add/remove entities and upgrade software. Therefore, GPRS is an elegant method of migrating from GSM to UMTS (usually it is used as an intermediate step). For UMTS, operators needed to purchase new spectrum. At the beginning when the complete spectrum was not utilized, some operators used the 3G spectrum to increase 2G capacity. Table 10.5 explains the operator upgrade options, depending on specific conditions.

GSM Evolution

Table 10.6 shows the evolution of GSM standards. The first GSM release was introduced in 1991 with further releases created up to 1998. The first standards release of GPRS and EDGE

Table 10.5   Conditions for Upgrading the Existing GSM Family Network to a Higher Generation

Operator system	Conditions	Upgraded system
GSM	No UMTS (3G) spectrum and/or requires medium capacity and/or does not have urgent need for IMS	GPRS or EDGE
GPRS/EDGE	Has new UMTS spectrum, immediate need for additional capacity to deliver new services especially IMS	UMTS
GPRS/EDGE	Has unpaired spectrum, immediate need for additional capacity to deliver new services, especially IMS	UMTS
UMTS	Requires enhancement of the user data rate, especially for IMS	HSDPA

Table 10.6   GSM Standards Evolution

Years	Release	Additional Services
1991–1993	GSM Phase 1–2	Speech, SMS
1994–1998	GSM Phase 2+	Data, full and half rates
1994–2000	GPRS/EDGE	Enhanced data and packet services
2000–today	UMTS/EDGE	Packet, real-time and multimedia services

Table 10.7   CDMA Standards Evolution

First deployment	Release	Notes
1993	CDMAOne (IS95)	Also known as IS-95A
1997	IS-95B	
2000	1xRTT, CDMA2000	
2002	EVDO Release 0	
2006	EVDO Release A, EVDV	
2007	CDMA2000 Release B	Ready to ALL-IP

was in 1994 and further versions of EDGE are still being created. The first UMTS standard was introduced and deployed in the year 2000. In the 3GPP standard, the GPRS and EDGE versions are called GSM release 4 and above. In this book, we distinguish between the legacy GSM, GPRS, and EDGE.

D-AMPS Operators

In the beginning of the 1990s, D-AMPS (called IS-136 as well) was the most common mobile system in the United States. Most of the operators decided to upgrade the D-AMPS system to EDGE over the existing spectrum. Others decide to migrate to CDPD or to GPRS systems. The migration paths from these systems are the same as described for GSM operators.

CDMA Evolution

Table 10.7 shows the growth and changes of CDMA.

Evolution of Services

From the basic voice service available in 1G and 2G, the Short Message Service (SMS) mainly marked the path to additional services. However, the increase in the use of services started mainly with the implementation of the GPRS (L2.5G) and the EDGE (H2.5G) networks. The migration to 3G enabled real multimedia services. 3G brought a huge variety of services to users.

These sophisticated services created a new paradigm: the use of content providers. Usually the operators preferred to have a content provider (i.e., service providers who could handle the types of services that required large-scale and variable information). This included companies that deal with video and audio content such as Yellow Pages.

In certain cases, it has been suggested that services and applications be differentiated. Services are the operator portfolio choices to users, such as service pricing model, roaming, and interoperability between terminals. Applications are built on the service and are related to user content and operator content (e.g., user bill) and so on.

Among 3G-enabled services, one can find:

- Video conversation
- Video clips download
- Ring tone download
- Mobile Internet
- Mobile TV

All these 3G services require completely different terminals usually including larger screens having a higher resolution, cameras, and so on. The question as to whether there will be a convergence or competition between "smart-phones" and "personal digital assistants" (PDA) was still open in 2006.

Although the take-up of 3G is driven by the introduction of 3G terminals and services, its in-depth adoption and the lifestyle changes will be driven by the richness and user-friendliness of the 3G services. The mobile terminal is a vital tool for business contacts, and in many respects business customers use their handsets in a manner very similar to consumers. However, the growth of 3G services (e.g., video calling and video streaming) has initially been slow.

Evolution in European and U.S. Markets

Although the cellular markets in the United States and in Europe started more or less at the same time (in the second half of the 1980s), they evolved rather differently. Initially, the take-up of cellular services was slow, governed mainly by their high costs. The real takeoff was triggered at the beginning of the 1990s by the move to digital technologies. At this point, Europe and the United States evolved along different paths. Europe evolved from analogue to GSM while the United States evolved mainly to D-AMPS and then to CDMAOne. Deployment of GSM in Europe was about two years ahead of CDMAOne deployment in the United States.

The European Telecommunication Standard Institute (ETSI) developed GSM as the common standard for the European market, and in Europe cellular networks were based mainly on this standard. Due to its success, GSM has become a world standard and has been deployed in the United States as well. The adoption of this standard worldwide has made GSM the most widespread 2G system.

The U.S. market had several 2G digital systems. One such system was D-AMPS upgraded from the 1G system, AMPS. During the deployment of GSM in the European market, Qualcomm developed CDMAOne (Code Division Multiplex Access) for the U.S. market.

Evolution of the Mobile Market

Following are some of today's enhanced data services that are driving new revenues:

- MMS (Multimedia Messaging Service)
- Downloadable ring tones, music, voice mail open messages, images and games, and news and information sources
- Mobile chat
- Internet-style portals

With continued growth and evolution in mobile systems, handsets are becoming more diverse and sophisticated and include additional features, such as:

- Color screens
- Built-in cameras
- PDA-like functions
- Bluetooth wireless connectivity
- MMS support
- High-speed data access
- Downloadable applications (e.g., games)
- E-mails and file download
- Video

The Subscriber Identity Module (SIM) and its evolving cards allow the operators to manage user-specific information. SIM gives the user more security and privacy. An important feature is the possibility to store applications on the card for network services, mobile banking, ticketing, and so on. Following are examples of these types of applications:

- Broadcast of specific information (news, sales, messages from the network)
- Mobile banking
- Credit (buying products, prepaid services)
- SMS alarms (e.g., from a car as a thief enters it, when it is being driven too fast, when mechanized problems arise, and so on)
- Information services (explanations in tourist sites, museums)
- Automatic communication to specific service centers depending on user location (i.e., taxis, hospital, guard center)

To guard user privacy, all of these applications are given only by request and after user permission.

Cellular Systems Multiple Access Techniques

Each individual wireless cell has sufficient bandwidth capacity to provide many channels and therefore to handle numerous simultaneous users. As different users start and end their calls, the channels are reallocated according to user demand. Protocols are therefore required to allow the system to detect a user wanting to initiate a call and to allocate a free channel to that user for the duration of the call.

Three basic multiple-access methods are currently in use in cellular systems. They are the following:

- Frequency division multiple access (FDMA)
- Time division multiple access (TDMA)
- Code division multiple access (CDMA)

A mobile system can use a single method or a combination of two methods.

Frequency Division Multiple Access

Frequency division multiple access (FDMA) is the common access technology for analogue 1G systems. This method uses a carrier frequency band that is divided into separate sub-channels, each with its own frequency band within the overall bandwidth. Each channel is assigned to a single cellular user and can carry a voice conversation. Until the initial call is completed or until it is handed off to a different channel, this channel is not available to other users. A "full-duplex" FDMA transmission requires two channels: one for transmitting and the other for receiving. AMPS, the most widely installed analogue technology in the United States, uses FDMA access as does TACS, the European analogue system. D-AMPS and GSM also use FDMA but combine it with TDMA to increase the number of calls that can be handled on a single channel.

Figure 10.1 illustrates a cellular system with frequency division mode. Each cell has its own frequencies and the frequencies can be repeated according to the cell repeat patterns rule (see Chapter 11). Figure 10.2 describes the basic principle of FDMA:

- Each frequency is assigned to a single user.
- The assigned channel is not available to other users until the initial call is completed.

Time Division Multiple Access

Time division multiple access (TDMA) improves channel capacity by dividing each frequency channel into time slots. As in FDMA, the available spectrum is divided into

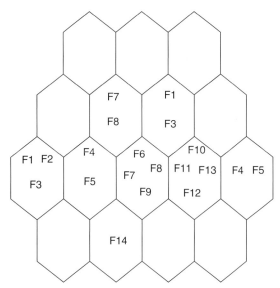

Fig 10.1 Frequency division in a cellular system.

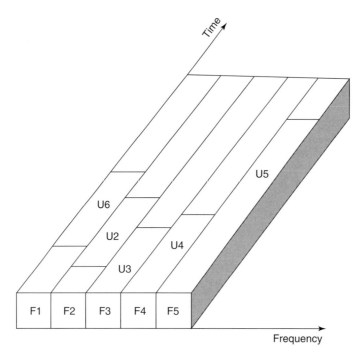

Fig 10.2 Basic principles of FDMA.

frequency channels. But unlike FDMA these frequency channels are split into a number of time slots and each user is assigned a specific time slot. This allows other users to share the same frequency channel and results in improved channel utilization.

TDMA is the common access technology for the 2G cellular systems. These are the digital cellular systems. For example, the following cellular systems use TDMA access mode:

- Digital Advanced Mobile Phone Service (D-AMPS)
- Global System for Mobile communications (European standard GSM)
- Digital Enhanced Cordless Telecommunications (DECT)
- Personal Communication System (PCS)

Each of these systems performs TDMA in somewhat different ways. For example, D-AMPS divides the frequency channel into three time slots and GSM divides each frequency into eight time slots.

Figure 10.3 describes the basic principles of TDMA. Each frequency is divided into time slots. In this example, there are three time slots (as in the D-AMPS system). Each user has a dedicated single frequency and is allocated a specific time slot (TS) for the duration of the conversation. This figure shows two frequencies, F1 and F2. Each frequency is divided into three time slots (TS1, TS2, and TS3). The time units go from T0 to T4. Each unit of time (T) is equal to the duration of three time slots (TS). For example, T1 is equal to the total elapsed time of TS1, TS2, and TS3.

Six user calls occupy frequency F2. At the beginning, the three users, U1, U2, and U3, occupy TS1, TS2, and TS3, respectively. The first user (U1) starts a call at T0 and terminates it at T2. Therefore, the total time for the call occupies three time slots of the channel (the first: TS1 at T0, the second: TS1 at T1, and the third: TS1 at T2). During this time, two

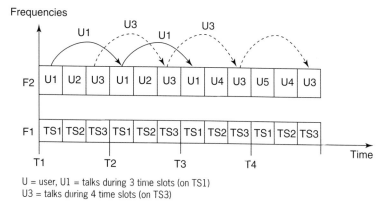

Fig 10.3 Basic principles of TDMA.

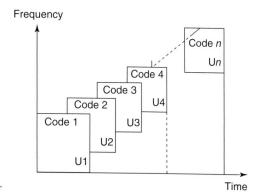

Fig 10.4 Basic principles of CDMA.

other users are also conversing. When U1 terminate its call, the fifth user receives TS1 and starts a call at Time T3, and so on.

Code Division Multiple Access

CDMA mode is based on spread spectrum technology and uses a spreading code, called a PN (pseudo-noise) codes. It uses encrypted transmission and its original application was for military use. The CDMA was not approved immediately by the industry standards committee because at that time (1989) they had just resolved a two-year debate between the TDMA and FDMA modes and they were not eager to immediately consider another access technology. In 1993, TIA/EIA (Telecommunications Industry Association/Electronic Industries Association) issued a temporary standard called IS-95. In 1995, the first CDMA standard called IS-95A or CDMAOne was presented. The IS-95B is the Low 2.5G generation.

The following is a simple explanation on how information is sent via CDMA technology (see Figure 10.4). A unique PN code is assigned to each user and this code is used to modulate the user's digital voice or data signal. The coded information from all the different users is then spread over all the channels and is transmitted over the whole CDMA spectrum. At the receiving end, the desired signal is recovered according to the user's PN code. This code differentiates between the desired user call and the other calls. CDMAOne has better performance and better capacity than the other second-generation systems, including GSM. Additional details regarding coding, PN codes, and CDMA technology are described in the next chapter.

Since all the transmitted information on the air is coded by a strong code (approximately trillions of possible frequency-sequencing codes), CDMA has a high level of security, safety, privacy, and good resistance to cloning. This was the main reason that ITU selected a wideband version of CDMA as the radio interface for 3G systems. CDMAOne operates on 800 MHz and 1.9 GHz, ultra-high-frequency (UHF) bands with a channel bandwidth of 1.25 MHz.

References

CDMAOne: Further information about CDMAOne can be found in the following locations: *http://www.3gpp2.org/* and *http://www.itu.int* (for imt2000).
D-AMPS: Digital AMPS, ANSI Interim Standards IS-54 and IS-136; *http://www.eia.org/*, *http://www.tia.org/index.html*, *http://www.ansi.org/*.
GPRS, EDGE: Detailed information about GPRS and EDGE can be found in the 3GPP Technical Specifications Series 41–55 under the name "GSM only (Rel-4 and later)." See *http://www.3gpp.org/*.
GSM: Detailed information about GSM can be found in the 3GPP Technical Specifications Series 00–12 under the name "GSM only (before Rel-4)." See *http://www.3gpp.org/*.
IMT-2000: *http://www.itu.int/home/imt.html*.
3GPP: *http://www.3gpp.org/*.[3]
3GPP2: *http://www.3gpp2.org/*.[3]
UMTS: Detailed information about UMTS with GPRS and EDGE can be found in the 3GPP Technical Specifications Series 21–35 under the name "3G/GSM R99 and later."

---

[3] GPP™ TSs and TRs are the property of ARBIS, ATIS, ETSI, CCA, TTA, and TTC, who jointly own the copyright on them. They are subject to further modifications and are therefore provided "as is" for information purposes only. Further use is strictly prohibited.

11 Principles and Services of First- and Second-generation Cellular Systems

Yael Baruch

Introduction

Although first-generation (1G) systems are analogue and second-generation (2G) systems are digital, the underlying principles of the two have much in common in terms of architecture and functionality. In this chapter, the system principles of both generations are explained first and then the main services are described. In fact, the 1G systems have largely been replaced by the 2G. As there are also more of the latter, only a brief description is given of 1G systems. Nonetheless, the discussion provides a good illustration of the factors that have to be taken into account in cellular system design.

More detailed coverage is given in Chapter 12 in the systems description. The entities' names used in the general description (principles and services) are taken from the 2G Global System for Mobile (GSM) communication system. It should be noted that systems based on Time Division Multiple Access (TDMA) technologies (such as D-AMPS and GSM) are very close to the following description, while systems based on CDMA technology (such as CDMAOne) have some differences.

Principles of First- and Second-generation Cellular Systems

Principles of Operation

Network Configuration

A public cellular network is known as a Public Land Mobile Network (PLMN). The area of the cellular network (PSTN area) is the geographical area in which a PLMN provides communication services to mobile users. The user can receive or set a call to another network that may be a fixed network (such as a PSTN, or public data network) or another PLMN.

A PLMN consists of:

- Mobile Switching Centers (MSCs)
- Base Station Subsystem (BSS) that includes:
  ○ Base Station (BS)
    ▪ base transceiver stations
    ▪ base station controllers
- Mobile Subscriber (MS)

In a PLMN, the area to be covered is divided up into a number of small areas called cells, with a single radio Base Transceiver Station (BTS) positioned to provide radio coverage to each cell. One or more base transceiver stations are connected to one base station controller. Each Base Station Controller (BSC) is connected by a fixed link to a mobile services switching center that is generally a digital telephone exchange with special software to handle the mobility aspects of its users.

A PLMN can consist of a number of MSCs, each with its own base stations and interconnected by means of fixed links. The MSCs interconnect to other external networks for both outgoing calls to and incoming calls from the users located in their area. Figure 11.1 shows a typical network arrangement.

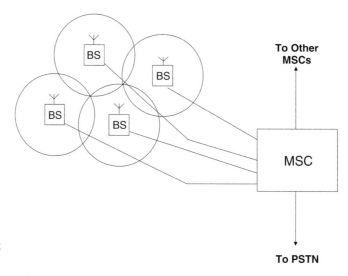

Fig. 11.1 Cellular network configuration.

A PLMN includes a number of radio frequency channels for use across its coverage area. This number is dependent on the channel spacing defined by the technical standard, the amount of spectrum made available by the licensing authority in the area, and the amount of spectrum allocated to the PLMN operator. The radio frequency channels are grouped together into a number of channel sets. These sets are then allocated to the cells, one set per cell, on a regular basis across the whole coverage area. The network will therefore reuse each channel many times. The method of radio planning and allocation of channels to the cells is described later in this chapter.

Signaling

Generally, at least one channel is set aside in each cell to carry signaling information between the network and the mobile stations. In the base station to mobile direction (the downlink), overhead information about the network operating parameters, including an area identifier code, is broadcast to all mobiles located in the cell's coverage area. In addition, specific commands are transmitted to individual mobiles in order to control call setup and to update the mobiles' location. In the mobile to base station direction (uplink), the signaling channel is used by the mobiles to send location-updating information, mobile-originated call setup requests, and responses to land-originated call setup requests.

Location Registration

When a mobile is not engaged in a call, it tunes to the cell-signaling channel that monitors the downlink signaling information. As the mobile moves around the network and the signal from the current cell falls below an acceptable threshold, the mobile will need to retune to the signaling channel of another cell. When the mobile retunes in this way, it reads the overhead information broadcast by the new cell and updates the operating parameters as required. The mobile also checks the location information being broadcast by the new cell and, if this differs from the previous cell, the mobile automatically informs the network of its new location by means of an interchange on the signaling channel (Figure 11.2). This location registration procedure enables the network to recognize all the users allocated in its

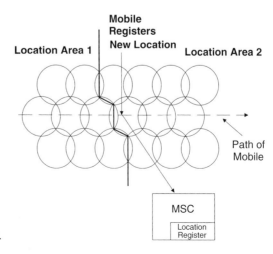

Fig. 11.2   Mobile location registration.

area in a specific time. The network also updates its database of all mobiles in the location area for the downlink call setup procedure.

Call Setup

The signaling procedures for call setup depend on the technical standard of the technologies used in the PLMN. However, the general procedure described here is based mainly on the GSM PLMN names and procedures, but it holds true for many other 1G and 2G PLMNs. When the user wishes to make a voice call, after the "call initiation" key is pressed (i.e., the SEND button), the mobile will transmit an access request to the network on the uplink signaling channel. This may be preceded by the mobile rescanning to verify that it is operating on the signaling channel of the nearest base station.

If the network can process the call, and it finds the user is authorized to act in this network place, the base station will send a voice channel allocation message that commands the mobile to switch to a designated voice channel. This is one of the channels allocated to that cell. The mobile retunes to the channel indicated, and the network proceeds to set up the call to the desired number. As part of the call setup procedure, the network will validate the mobile requesting the call to verify that it is a legitimate customer. To carry out this validation, many networks incorporate specific security features that depend on many kinds of checks (e.g., the phone is not stolen; the user is authorized to make the type of call requested; payments are up to date).

When the network receives a call for a mobile (e.g., from the PSTN), it will first check the location database to determine the location area in which the mobile last registered. Paging calls to the mobile are transmitted on the downlink signaling channels of all the base stations in the identified location area, and the network waits for a response from the mobile. If the mobile is turned on and receives the paging call, it will acknowledge to its nearest base station on the uplink signaling channel. The base station receives the acknowledgment on the uplink signaling channel. Upon receiving the acknowledgment, the base station sends a voice channel allocation to the mobile and informs the network that the two sides of the call can be connected. When the call is terminated, the voice channel is released and becomes available to other users.

In-call Handover

At all times during a call (whether downlink or uplink), the base station currently serving the mobile monitors it signal (strength and/or quality). If the signal falls below a predesignated threshold, the network will command neighboring base stations to measure it (Figure 11.3a). If another base station is receiving the mobile's call with a stronger signal than the current base station, a signaling message is sent to the mobile from the current base station commanding the mobile to a new voice channel, namely a free voice channel from those allocated to the neighboring cell. The mobile changes to the new allocated frequency channel (of the neighboring cell) and simultaneously the network diverts the call to the new base station responsible for the neighboring cell (Figure 11.3b). The measuring process and new cell selection may take several seconds, and the user should be unaware of the brief break in transmission during this process.

Power Control

The goal of power control is to control the transmission power levels in such a way that acceptable Quality of Service (QOS) of the call will be maintained together with the lowest

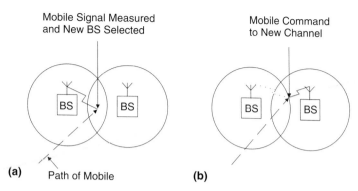

Fig. 11.3   In-call handover.

possible transmission powers. The power control reduces the average power level of the network and of the mobile subscriber. In this way it minimizes interference, preserves signal quality for the user, improves overall system quality, and increases battery life of the mobile. Therefore, power control is an essential radio resource management method used in the cellular networks.

Although the size of a cell may be from under one to tens of kilometers across, it is not necessary for a mobile to transmit on full power at all times to maintain an acceptable call quality and a satisfactory signal level at the base station's receiver. A fixed transmit power level would use more energy than necessary when the mobile subscriber is close to the base station, and/or not use sufficient energy power when it is far from the base station. Achieving suitable power at any moment is possible because most cellular standards incorporate mobile power control. There are some differences between technologies regarding this process. For a more specific illustration, see the power control section in the GSM system description in Chapter 12.

A mobile subscriber close to the base station can transmit using very little power (the minimum necessary for proper signal quality). To ensure the same signal quality when the mobile moves further away, or when more interference occurs in its area, the power should be increased. To cope with the random changes in the radio channel and interference, the power control should be checked and adjusted, even if the mobile subscriber remains in the same place or at the same power level as before. Therefore, the power control is a loop process, between the base station and the MS, and can continually reach the optimal transmission power in the presence of random changes. A simplistic description of power control is as follows:

1. The base station commands the mobile to transmit at a specific power level that will maintain the received signal level within prescribed limits.
2. The mobile subscriber adjusts to this power of transmission.
3. The power level is decreased/increased by the MS depending on specific power control methods and algorithms used in the system (for example, after a fixed time, or because the mobile moves closer to/further from the base station).
4. The MS reaches the required power.
5. The base station informs the MS that it has reached the required power level.
6. The process goes back to step 2.

Power control consumes radio signaling resources. Therefore, extensive power control should be balanced with other procedures that require signaling radio resources as there are cases when the control feedback bandwidth is limited.

It is important to note that power control is an essential radio resource management method in CDMA cellular communication systems, much more so than in TDMA-based systems. The reason is that, in CDMA systems, the cochannel interference is the primary capacity-limiting factor.

Radio Network Design

This subsection is particularly relevant to cellular networks using TDMA technology, rather than CDMA-based systems. This is because the CDMA networks require one frequency per cell only and therefore are almost unaffected by the frequency reuse problem and their network planning is much simpler.

As previously described, cellular radio reuses the same radio channels in different cells. Due to this reuse, two mobiles using the same channel in different cells may interfere with each other. This phenomenon is known as cochannel interference. The key objective of planning a cellular radio system is to design the cell repeat pattern and frequency allocation in order to maximize system capacity while controlling cochannel interference to within acceptable limits.

Cell Repeat Patterns

In a TDMA-based network, the radio channels are grouped together into a number of channel sets. These sets are allocated one set per cell on a regular basis across the whole coverage area.

The network plan has to be selected so that the number of channel sets (N) fits together with minimum gaps and/or overlaps. Only a specific number of channel sets (N values) achieve this objective. Typical arrangements of interest to cellular radio are $N = 4, 7$, and 12, as shown in Figures 11.4(a), 11.4(b), and 11.4(c), respectively. The value of N has a major effect on the capacity of the cellular system. As the number of channels sets decreases, the number of channels per cell increases; hence the system capacity increases. The capacity reached by $N = 4$ is thus higher than the capacity reached by $N = 7$.

For example, if there is a total of 140 available channels, a 4-cell repeat pattern would provide 35 channels per cell, while a 7-cell pattern would provide 20 channels per cell. On this basis, the smallest possible value of N seems desirable. However, as N decreases, so the distance between cells using the same channels decreases, which in turn increases the level of cochannel interference. The repeat distance D and the cell radius R are related by the geometry of the cell pattern, as shown in Figure 11.5 and given by Equation 11.1.

$$Reuse\ ratio\ \frac{D}{R} = \sqrt{3N} \tag{11.1}$$

In practice, in a real network, it is not possible to achieve a regular cell pattern. This is because radio propagation at the frequencies used by cellular radio systems is affected by the terrain and by buildings, trees, and other landscape features. Computer-based radio planning tools are commonly used, enabling the automatic assignment of the best frequency based on predicted interference. It is not, therefore, necessary to stick to a rigid and regular repeat pattern.

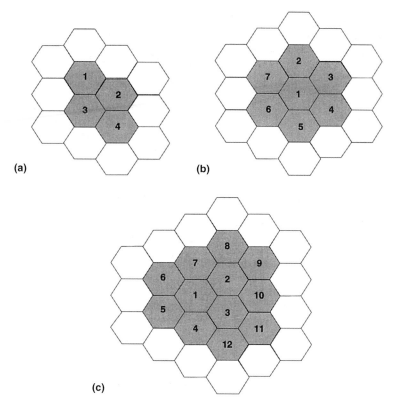

Fig. 11.4 (a) Cell repeat patterns equal to four ($N = 4$). (b) Cell repeat patterns equal to seven ($N = 7$). (c) Cell repeat patterns equal to twelve ($N = 12$).

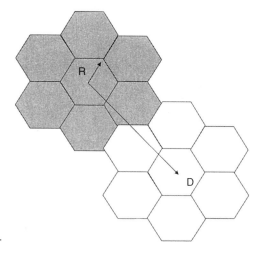

Fig. 11.5   Frequency reuse: D/R ratio.

It is not enough to be satisfied with the theoretical calculations, however. A series of on-site tests are performed to check the interferences, the gaps, and the overlays between the frequency channels, and corrections made accordingly.

Cochannel Interference

Generally, a mobile will receive a desired carrier signal (C) from the base station serving the cell in which it is located, in addition to cochannel interfering signals (I) from other cells. The Carrier-to-Interference ratio, C/I, is related to the reuse ratio, D/R. Cellular radio systems are designed to tolerate a certain amount of interference, but beyond this, speech quality will be severely degraded. The lower limit on C/I effectively sets the minimum D/R ratio that can be used.

The two key factors ensuring that good-quality transmission can occur between a mobile and base station are these:

- The desired signal strength is sufficiently large, that is, above the receiver threshold sensitivity.
- The interference level is low enough to give an adequate C/I ratio.

Both of these factors depend on the radio propagation between the mobile and the base stations.

Radio Propagation

Radio propagation refers to the radio signal's behavior during transmission. Obstacles can influence the propagation, whether they are static (such as buildings, trees, and mountains) or moving (such as moving cars). The propagation waves are assumed to be bidirectional, linear, and varying with time. The same physical principles and rules apply to propagation of radio waves, for all the wireless technologies (cellular, pagers, fixed wireless, and so on). Communication quality between a base station and mobile subscriber depends on (among other things) the degree of downfall of the received signal.

The propagation phenomena can be divided into three types of downfall degrees:

- *Multipath fading:* This is when the transmitted signal suffers a large scale of path loss. It occurs when the radio signal meets obstacles like buildings and suffers from progressive loss, and the received signal is very different from its source.
- *Shadowing:* This is when the transmitted signal suffers from a medium scale of loss. Shadowing is the loss of signal strength. Usually this loss happens if the mobile subscriber is far from the antenna, more than a few tens or hundreds of meters.
- *Multipath propagation:* This is when the transmitted signal suffers a small scale of loss and causes instability of the signal's phase and amplitude.

In urban areas, the effects of propagation phenomena can be much higher than in rural areas. Following is the explanation of the same propagation phenomena, but from a cellular point of view. There are a number of elements that contribute to the signal strength received by a mobile. For a line of sight path, there is a free space path loss that is related to the radial distance between base station and mobile. In addition to this loss, for a nondirect line of sight path, there will be a diffraction loss resulting from obstructions in the path. In general, multiple signals arriving at the mobile due to reflections from buildings and other terrain features will also affect signal strength. This multipath effect will result in either a constructive or destructive addition of signals.

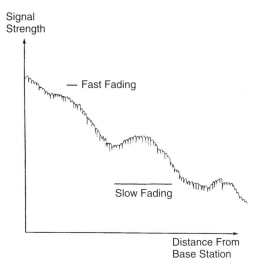

Fig. 11.6   Fading effects.

Due to these factors, as a mobile moves within a "cell," it will experience a varying signal, as shown in Figure 11.6. Fast fading is caused by the multipath effect and occurs with only a small movement of the mobile. This is also known as Rayleigh fading. Slow fading is mainly caused by terrain features and occurs over large distances of hundreds of meters.

Path loss also depends on the type of terrain, for example, urban terrain with dense buildings, rural terrain with trees, or open water. Mobile height and base station height above ground level also affect propagation, although mobile height is generally not a variable.

Predicting path loss is an essential part of radio planning. Empirically based formulae are used due to the large number of contributing factors. The most widely used formula is the Hata model, which is based on the propagation measurement results of Okumura, et al.

Hata's basic formula for the total path loss, L_p, is given by Equation 11.2, where:

- f_c is the carrier frequency in MH.
- h_b is the base station antenna height.
- h_m is the mobile antenna height.
- R is the radial distance in kilometers.
- $a\,(h_m)$ is the mobile antenna height correction factor.

$$L_p(dB) = 69.55 + 26.16 \log(f_c) - 13.82 \log(h_b)$$
$$-a(h_m + (44.9 - 6.55\log(h_b))\log R$$

(11.2)

Correction factors can be used to take into account the type of terrain.

Practical Radio Planning

With a propagation model, both the desired signal strength and the interference level for all locations in a cell can be calculated. Generally, this is done with a computer-based tool that can draw on a database of cell site information and terrain data. Some advanced tools can also take into account diffraction losses.

Table 11.1   Typical Power Budget (TACS)

Signal strength budget	Downlink	Uplink
Transmitter power (EIRP)	50 dBm	39 dBm
Receiver sensitivity (including antenna gain)	106 dBm	113 dBm
Fade margin (90 percent of area)	4 dB	4 dB
Required receiver level[1]	102 dBm	109 dBm
Maximum path loss (allowing for fading)	152 dBm	148 dBm
Interference budget		
C/I threshold	17 dB	17 dB
Target C/I for planning (90 percent of area)[1]	25 dB	25 dB

[1]Key planning parameters.

For practical purposes, a planner will aim to achieve the required signal strength and C/I ratio of more than 90 percent of the cell coverage area by varying:

- Antenna heights
- Transmitter power
- Frequency allocations
- Additional factors, as appropriate

To simplify calculations, an allowance for Rayleigh fading and shadow fading is usually made within the system power budget. A typical power budget is shown in Table 11.1.

Adding Capacity

Once a cellular network has been planned to provide overall coverage, there are a number of ways to increase capacity. A simple and cost-effective option is to allocate additional radio channels to existing cells. However, this can only be done by an extension band. Other alternatives involve rearranging the cellular plan, either by cell splitting or by sectorization.

Cell splitting is achieved by adding additional base stations to an existing cell and dividing the cell into a number of smaller cells, as shown in Figure 11.7. It is then necessary to reallocate the radio channels.

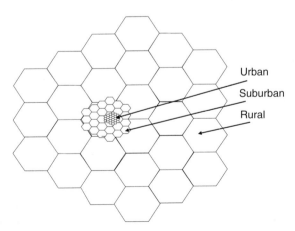

Fig. 11.7   Cell splitting.

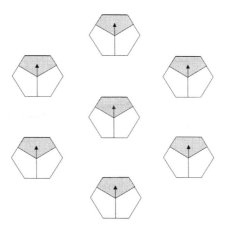

Fig. 11.8 Sectorization.

By repeatedly splitting cells, cell size, and, hence, system capacity, base stations can be tailored to meet the traffic capacity requirements demanded by customer behavior in all areas. In practice, as cell sizes decrease, propagation effects, particularly in city areas, cause an increase in cochannel interference even if the repeat pattern is maintained. In addition, it becomes increasingly difficult to find suitable base station sites. Base stations must be accurately positioned in order to keep to a regular pattern. The cost of providing and maintaining a large number of individual base stations is also a factor. Therefore, cell sectorization, in addition to cell splitting, is commonly used in urban areas.

In a regular cellular layout, cochannel interference will be received from six surrounding cells that use the same channel set. One way of significantly cutting the level of interference is to use several directional antennas at the base stations. Each antenna illuminates a sector of the cell and a separate channel set is allocated to each sector. The most common method of sectorization uses three 120-degree sectors, as shown in Figure 11.8. Sectorization reduces the number of prime interference sources because most of the surrounding cochannel cells are not directed at the desired cell. A disadvantage of sectorization is that the channel sets are divided between the sectors, resulting in fewer channels per sector, which reduces trunking efficiency, which means that the total traffic that can be carried for a given level of blocking is reduced. However, this effect is offset by the ability to use smaller cells, resulting in a significant increase in total capacity.

Network Planning

When planning a cellular radio network, an operator has to consider a wide range of factors, including:

- What geographical coverage is required?
- What traffic density and profile are expected?
- What level of service and quality will be offered?
- What services will be provided?
- What level of investment will be needed and how will it be phased?

Answering these questions will determine the numbers and locations of base stations together with the number of radio transceivers in each base station needed to cope with the expected demand. It is particularly important at this stage to have defined coverage objectives and to have identified the key target areas for service, for example, cities, towns, motorways, roads, railways.

Next, the BSCs, which are needed to control the base stations, must be planned. A major consideration in determining the location of BSCs is the transmission costs in connecting to each BTS, and a major factor in these costs is whether the operator is permitted to self-provide microwave links to the BTSs. Self-provision of links gives a much lower running cost, at the expense of additional capital expenditure and planning overhead.

BSC and Transmission Network Planning

The number of BTSs to be supported by a BSC depends on several considerations: expected traffic levels, the type of BTS and BSC, and transmission costs. The answer can range between a few BTSs (in very high traffic areas) up to 128 BTSs or higher. However, it is important to decide, when planning the network, which BTSs will be parented on to which BSC.

There are three main alternatives for connecting BTSs to BSCs: star, ring, or a combination of the two.

Star Configuration   Figure 11.9 shows a star configuration. This is simple to implement and, depending on traffic load, consists of a minimum of one 2 Mbit/s path between each BTS and the BSC. The main advantage to this type of configuration is that it is easy to augment routes. Growth at one particular site has no particular impact on growth at other sites. However, the main disadvantages are:

- *Cost:* For self-provided links, it may not be possible to achieve line of sight between every BTS and the BSC, so the more expensive option of leased circuits would be needed.
- *Reliability:* Only one route exists between each BTS and the BSC, which means that if a link fails so will the BTS.

Ring Configuration   Figure 11.10 shows a ring configuration. This has the main advantage that each BTS has two routes to the BSC, thus giving an increase in reliability. However, as demand increases, unsuspected hot spots may develop, so careful management of traffic volumes is very important.

Ring-star Combination   To produce a general method of linking BTS to BSC a mixture of the star and ring methods is preferable, as shown in Figure 11.11. To allow control of capacity, digital cross-connect (DCX) equipment is used to reroute traffic when necessary.

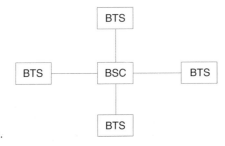

Fig. 11.9 BTS star configuration.

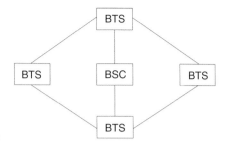

Fig. 11.10   BTSs ring configuration.

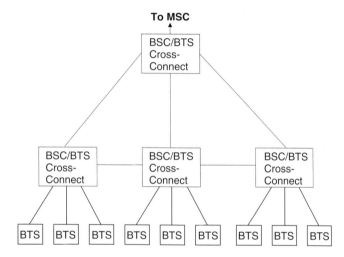

Fig. 11.11   Star BTSs within a ring of cross-connects and BSCs.

MSC and the Trunk Network

The number and locations of MSCs are determined by the geographical distribution of traffic and the balance between the costs of transmission within the cellular network and the costs of interconnect to the other networks. Because one of the functions of the BSC is to provide traffic concentration, high-capacity connections are needed from the BSCs to the MSC. In some cases, MSC and BSC can be collocated, or alternatively they are connected by a high-capacity, leased link with route diversity.

The trunk network is used for carrying inter-MSC traffic and for interconnecting to other networks, such as the national PSTN/ISDN. The interconnection of MSCs, given the distances involved, is achieved by using high-capacity, leased bearers. Routing schemes are used to ensure that costs are optimized while maintaining the required quality of service. In large networks, a transit-switching layer may also be used to separate purely transit traffic from the cellular switches. Standard telecommunications traffic planning practices are used to develop routing rules for inter-MSC routes and to ensure an acceptable grade of service.

Numbering

One of the important tasks when developing the network-routing strategy for both customer traffic and signaling information is to develop a detailed numbering plan. Numbering plays an important part in cellular networks and it is essential to ensure that sufficient numbers

Table 11.2   Comparison of 1G Cellular System Parameters

Technology	AMPS	TACS	NMT 450	NMT 900	C-450
Frequency band	800 MHz	900 MHz	450 MHz	900 MHz	450 MHz
Channel spacing	30 kHz	25 kHz	25 kHz	12.5 kHz	20 kHz
Speech modulation	FM	FM	FM	FM	FM
Deviation	12 kHz	9.5 kHz	5.0 kHz	5.0 kHz	4.0 kHz
Signaling	Direct FSK	Direct FSK	Audio FFSK	Audio FFSK	Direct FSK
Signaling bit rate	10 kb/s	8 kb/s	1200 bit/s	1200 bit/s	5280 bit/s
"Overlay" signaling	No	No	No	No	Yes

are available for both the current and future telecommunications services. There are various types of numbers used within a cellular network:

- *International Mobile Subscriber Identity (IMSI):* The IMSI is the number that the network uses to identify the mobile, and is normally invisible to the customer. The IMSI is held in the mobile station in the Subscriber Identity Module (SIM) card.

- *Mobile station number:* This number is the telephone number of the customer (e.g., +972 3 1234567 for international and 03 1234567 for national). The HLR holds the mapping between the mobile station number and the IMSI.

- *Roaming number:* This is used to route incoming calls to a customer when he or she is in a different service area. For instance, where the customer has moved to another network, the HLR will allocate a temporary roaming number so that the network can route calls to her.

- *Global title:* This is a number conforming to ITU-T Recommendation E.214 used to route MAP SCCP information to appropriate nodes.

- *Point codes:* They are used to route signaling information through the No 7 network.

Comparison of First-generation System Parameters

Table 11.2 presents the four 1G cellular systems: AMPS, TACS, NMT, and C-450 and their basic system parameters.

First- and Second-generation Services

A PLMN provides the user with a wide range of services and facilities that, aside from some that are specific to mobile situations, make use of access to other existing networks (like fixed telephone networks and data networks). The primary purpose of all cellular radio networks is to offer speech telephony service to their customers. In addition, most networks offer a range of supplementary and value-added services to enhance the basic product.

In analogue systems, basic telephony is provided directly by the audio path between mobile and network. Other than some linear speech processing to increase the channel's signal to noise performance, the audio path is transparent across the speech band, allowing other sounds (tones, nonvoice signals, and such) to pass through undistorted. By contrast, digital systems such as GSM use a speech coder tailored to voice characteristics. They therefore provide a fully acceptable telephony service, but nonvoice signals can suffer distortion across the nonaudio, transparent path.

As analogue systems are being phased out, improvements and additional services are no longer being introduced. On the other hand, 2G systems are common and usually share the same core network with more progressive systems or overlays from 2.5G. For example, many GSM and GPRS systems share the same core network. Therefore, it is easy to give some additional services that were developed or expanded in GPRS to the GSM subscribers without large investment. This is true as long as GSM can support them. A broadcast service is an example of this.

The following examples give a description of basic services from each type of system. In practice, there are far more services than those described in the following sections.

Supplementary Services

Supplementary services are provided by means of enhancements to the basic call processing software in the MSCs. Many of these services have specific relevance to the cellular radio user and, in the main, they parallel services that are becoming increasingly available on the fixed telephone networks. Typical services are:

- *Call divert:* All calls are diverted to the specified number, which may be another mobile from the same network or for termination on another network. This is of use if the user wishes to make calls but not receive them on his mobile.

- *Divert on no answer:* Calls are diverted to the specified number when the user does not answer within (for instance) 20 seconds. This is of use, for example, if a mobile is left switched on in an unattended vehicle.

- *Divert on mobile unreachable:* Calls are diverted to the specified number if the network cannot contact the mobile owing to its being turned off or out of range. This is of particular use in a cellular system where, for example, users are not available at all times, and where coverage is not continuous. This service is often combined with the divert-on-no-answer service.

- *Divert on busy:* Calls are diverted to another place (i.e., voice mail or another number) when the mobile is already engaged on a call. As an alternative, networks also provide call waiting.

- *Divert on no answer:* Calls are diverted to another place (e.g., to a voice mail service or to another number) after a predefined number of rings without an answer from the mobile.

- *Call waiting:* If a call is received when the mobile is already engaged on a call, the user is informed that a second call is waiting and can choose to place the first call on hold while dealing with the second.

- *Multiparty calling:* The mobile user may set up calls to several other parties and connect them in a conference. This service can also be used to make inquiry calls while holding the original call.

- *Call barring:* The mobile user may bar calls to certain destinations from being made, such as international calls, premium-rate calls, and so on.

Value-added Services

Value-added services are normally provided by means of peripheral units attached to the cellular network, or to the fixed network with which the cellular network interconnects. In some countries the prevailing regulatory regime will influence what services may be offered and in what manner. However, the following are typical.

Messaging Services

Voice messaging is commonly available in association with cellular networks. Used in conjunction with the call-divert supplementary services, the messaging service can pick up calls when the user cannot, and the caller can leave a message for later retrieval by the user (voice mail). Some services allow the user to be alerted to the receipt of messages by means of a radio paging service (i.e., a beep tone) or by a ringback on the cellular network itself. In addition to voice messaging, GSM networks incorporate the Short Message Service (SMS), which effectively turns a GSM mobile into a two-way alphanumeric pager with forced message delivery and message delivery confirmation.

Information Services

In the past, voice information services were usually available on fixed networks only, normally carrying a premium call charge. Today, the 2G systems commonly carry information services. These services (such as travel and weather information) are of particular value to a mobile user and the networks operators make these services more readily accessible. Some of the information services are dependent on the mobile's current location.

Data Services

Analogue Systems (1G)

In analogue cellular systems, the transparent audio path between the network and mobile can be used not only for voice communication but also for nonvoice communication such as data using in-band modems and facsimile. In order to be used with a mobile, data modems and fax machines, which are designed for PSTN use, have to be adapted for connection to the mobile by means of a special interface. Such interfaces are available for a range of mobiles, and often permit automatic call establishment and clear down under the control of the modem or fax machine.

The data rate achievable over a 1G cellular radio channel is less than that over a direct PSTN path, mainly due to the limited bandwidth available on the analogue channel and to the spread of delay that is affected by the audio processing in both mobile and base station. However, data transmission up to 4800 bit/s (using V32) or 9600 bit/s (V32bis) can be achieved quite commonly on cellular networks, as well as fax up to 9600 bit/s.

The radio link between a cellular network's base stations and a mobile station is a notoriously hostile environment for data transmission. Disturbance and interruptions come from a variety of sources, such as variability of the radio signal strength, noise and interference, and intentional breaks due to signaling interchanges between base station and mobile for handover and power control. In order to transmit data reliably over such a path, error control of some form is essential. The simplest form of error control is a Layer 2 protocol, such as ITU-T V42, which is built into most modems as standard. Facsimile transmission over cellular has benefited by the increasingly widespread adoption of Group 3 error-correcting (ECM) fax machines and the availability of portable machines suitable for vehicle use.

Digital Systems (2G)

In 2G systems that share their network with other, more progressive overlay systems (such as GSM sharing with GPRS, or D-AMPS sharing with CDPD), the data services are carried normally by the more advanced system and the older system is used for voice only. This is the most efficient division of functionality because of the big difference between the data-handling capabilities of the two network types. In the following section, the use of GSM illustrates this principle. Most other 2G systems adopt a similar strategy.

GSM Services and Features

The telecommunications services provided by the GSM PLMN are divided into two main groups, called basic services:

- *Bearer services:* Telecommunications services that give the user the capacity needed to transmit appropriate signals between certain access points (user-network interfaces).

- *Teleservices:* Telecommunications services that provide the user with necessary capacities, including terminal equipment functions, to communicate with any other users.

Supplementing and/or modifying the basic services (bearer services and teleservices), the supplementary services complete the offer to the user. Bearer services are described in 3GPP Specification GSM TS 02.02; teleservices are described in 3GPP Specification GSM TS 02.03; supplementary services are described in 3GPP Specification GSM TS 02.04 and the GSM 02.80 series. An overview of the basic services is given in Tables 11.3 and 11.4.

In addition to speech, GSM offers a wide range of data bearer services up to 9.6 kb/s suitable for connection to circuit-switched or packet-switched data networks. GSM also supports Group 3 facsimile as a data service by use of an appropriate converter. A comprehensive range of supplementary services are offered by GSM, including call forwarding, call barring, multiparty service, advice of charge, and others. A full description is provided in the GSM Recommendations.

An important feature of GSM is the SMS. This allows transmission of alphanumeric messages of up to 160 characters to or from a mobile via a service center. If the message

Table 11.3   Basic Bearer Services

Data circuit duplex asynchronous 300–9600 bit/s (T/NT)
Data circuit duplex synchronous 1200–9600 bit/s (T/NT)
PAD access circuit asynchronous 300–9600 bit/s (T/NT)
Data packet duplex synchronous 2400–9600 bit/s (T/NT)
Alternate speech/data (T/NT)
Speech Followed by data (T/NT)

T/NT = transparent/nontransparent

Table 11.4   Basic Teleservices

Telephony
Emergency calls
Short message mobile terminated point-to-point
Short message mobile originated point-to-point
Short message cell broadcast
Alternate speech and facsimile group 3
Automatic facsimile group 3
Voice group call service
Voice broadcast service

cannot be delivered, for example, due to the mobile being switched off or being outside of the coverage area, the message is stored at the service center and retransmitted when the mobile registers again. Received messages can be displayed on the mobile and stored in the SIM for future reference. A related service is cell broadcast, which allows messages of up to 93 characters to be sent to all mobiles within a specific geographical area, for example, to deliver traffic or weather reports.

Useful 3GPP Documents

Tables 11.5 and 11.6 list useful documents describing GSM and GPRS systems. It is important to mention that 3GPP™ TSs and TRs are the property of ARBIS, ATIS, ETSI, CCA, TTA, and TTC who jointly own the copyright in them. They are subject to further modifications and are therefore provided "as is" for information purposes only. Further use is strictly prohibited. Table 11.6 summarizes the 3GPP specification for GSM technology.

Table 11.5   Useful 3GPP Documents

TS 22.001-600	Principles of circuit telecommunication services supported by Public Land Mobile Network (PLMN)
TS 22.002-600	Circuit Bearer Services (BS) supported by Public Land Mobile Network
TS 22.003-600	Circuit Teleservices supported by a Public Land Mobile Network
TS 22.004-600	General on supplementary services
TS 23.002-710	Network architecture
TR 21.905-720	Vocabulary for 3GPP Specifications
TS 01.01-700	Technical Specifications and Technical Reports for GERAN-based 3GPP system
TS 01.02-601	General Description of a GSM Public Land Mobile Network
TS 22.101-800	Service aspects; Service principles
TS 01.02-601	General Description of a GSM Public Land Mobile Network
TS 02.07-800	Mobile Station (MS) Features
TS 23.060-710	General Packet Radio Service (GPRS); Service description; Stage 2
TS 42.068-800	Voice Group Call Service (VGCS); Stage 1
TS 42.069-700	Voice Broadcast Service (VBS); Stage 1

Table 11.6   3GPP Specification for GSM Technology

Subject of specification series	*GSM only (before Rel-4)*
General information *(long defunct)*	00 series
Requirements	01 series
Service aspects ("stage 1")	02 series
Technical realization ("stage 2")	03 series
Signaling protocols ("stage 3")—user equipment to network	04 series
Radio aspects	05 series
Codecs	06 series
Data	07 series
Signaling protocols ("stage 3")—RSS-CN	08 series
Signaling protocols ("stage 3")—intra-fixed-network	09 series
Program management	10 series
Subscriber Identity Module (SIM), Cards. Test specs.	11 series
OAM&P and Charging	12 series

Acknowledgments   Parts of this chapter are based on earlier work by Malcolm Appleby and Fred Harrison in Edition 2 of the *Telecommunications Engineer's Reference Book*. Sections related to GSM and its subsequent generations are based on 3GPP documentation. 3GPP™ Technical Specifications and Technical Requirements are the property of ARBIS, ATIS, ETSI, CCA, TTA, and TTC, who jointly own the copyright in them. They are subject to further modifications and are therefore provided "as is" for information purposes only. Further use is strictly prohibited. Also, the author wishes to acknowledge the valuable help and advice given by Goff Hill in preparing this chapter.

References

CDMAOne: *http://www.3gpp2.org/*.

GSM: Detailed information about GSM can be found in the 3GPP Technical Specifications Series 00–12 under the name, "GSM only (before Rel-4)." See *http://www.3gpp.org/*.

Hata, M. "Empirical Formula for Propagation Loss in Land Mobile Radio Services." *IEEE Trans. on Vehicular Tech.* (August), pp. 317–325, 1980.

Okumura, Y. "Field Strength and Its Variability in VHF and UHF Land–Mobile Radio Service." *Review of the Electrical Communications Laboratory*, 16(9–10), 1968.

12 First- and Second-generation Cellular Systems

Yael Baruch

Introduction

This chapter provides information about specific systems that have been implemented, starting with a summary of the various first-generation (1G) systems; a description of the most widely deployed second-generation (2G) systems follows.

First-generation Systems

Advanced Mobile Phone System

The Advanced Mobile Phone System (AMPS) was developed by Bell Laboratories as a successor to the heavily congested Improved Mobile Telephone System (IMTS) primarily for the North American market. AMPS underwent a long development period and an extended trial (technical and commercial) that not only fixed the system parameters but also contributed to the basic planning rules that hold true for all cellular systems. The system design was first presented in 1979, but it was not until 1983 that operating licenses were issued and true commercial exploitation of the system commenced. The AMPS system produces a high-quality voice circuit with the capability to maintain performance in a high-capacity (poor-interference ratio) configuration.

The first system was made available in Chicago. The Federal Communications Commission (FCC) allocated for AMPS in ITU Region 2 (the Americas) a total of 40 MHz of spectrum from the 800 MHz band, with 30 kHz channel spacing, in common with established PMR practice.

During the 1980s AMPS was operating extensively across the North American continent (United States and Canada). Due to the regulatory conditions in force in the United States, deployment has been in the form of a patchwork of largely independent, stand-alone systems, with two competing systems operating in each license area. Although commercial roaming agreements existed (allowing customers of one operating company to obtain service from another company when they were in a different part of the country), a seamless, nationwide service was not available. The roaming agreement was used for charging the user for airtime. AMPS was also used in a number of Central and South American countries, in Australia, and in some East Asian and Middle East countries (like Israel).

AMPS uses analogue FM for speech transmission, using a total of 40 MHz of spectrum from the 800 MHz band, with 30 kHz channel spacing, but with a wider frequency deviation (12 kHz) than is the norm for a 30 kHz channeling system. Moreover, it offers 832 channels with a data rate of 10 kbps. At the beginning, AMPS implementation included only omni-directional antennas. Later on, when it became obvious that directional antennas would generate better cell reuse, a 120-degree directional antenna was added to the network architecture and a 7-cell reuse pattern was adopted.

Signaling between mobile and base station is at 10 kb/s, with Manchester encoding taking the bit rate to 20 kb/s. The data is modulated on the radio carrier by direct Frequency Shift Keying (FSK). Error control is achieved by multiple repetition (5 or 11 times) of each signaling word, with majority voting applied at the receiver to correct errors. A Broadcast Channel (BCH) block code is also applied to detect any uncorrected errors.

When a call is in progress, the base station transmits a low-level Supervisor Audio Tone (SAT) in the region of 6 kHz. The network uses three different SAT frequencies that are allocated to the base station so that the nearest cochannel base stations (i.e., those most likely to cause interference) have a different SAT from the desired base station. The mobile continuously monitors the received SAT and automatically transmits the signal back to the base station. If the mobile (or the base station) detects a difference between the received SAT and what is expected, the audio path is muted to prevent the interfering signal from being overheard. If the condition persists, the call is aborted.

Total Access Communications System

The Total Access Communications System (TACS) provides 1000 channels, at a 8 kb/s data rate. It uses Frequency Modulation (FM) for transmission, and the traffic is multiplexed using a Frequency Division Multiple Access (FDMA) system.

The United Kingdom adapted the AMPS standard from 1985, when cellular radio was first licensed for operation. The adaptation was necessary to comply with European frequency allocations around 900 MHz, with 25 kHz channel spacing. This required a reduction in frequency deviation and signaling speed.

The AMPS signaling scheme was retained largely unchanged, but several enhancements were introduced, particularly regarding the procedures for location registration, making the standard more suitable for deployment in systems offering contiguous, nationwide coverage. Originally, TACS used all of the 1000 channels (2 × 25 MHz) allocated to mobile services in Europe. However, in the United Kingdom, the licensing authority released only 600 channels (2 × 15 MHz) to AMPS because the remaining channels were reserved for GSM. Subsequently, an additional allocation of channels below the existing TACS channels was made, namely the Extended TACS (ETACS) channels, and the standard was modified accordingly.

TACS equipment availability and cost have both benefited from the standard's similarity to AMPS. TACS systems have been adopted by several European countries (United Kingdom, Ireland, Spain, Italy, Austria, Malta), several Middle East countries (Kuwait, UAE, Bahrain), and the Far East (Hong Kong, Singapore, Malaysia, China). A variant of TACS named J-TACS has also been adopted in Japan.

Nordic Mobile Telephone System

NMT stands for "Nordic Mobile Telephone" (system) and was developed jointly by the PTTs of Sweden, Norway, Denmark, and Finland during the late 1970s and early 1980s. The system was designed to operate in the 450 MHz band and was later adapted for the 900 MHz band. Although NMT was developed after AMPS, it saw commercial service before AMPS, starting in late 1981.

NMT450 uses a channel spacing of 25 kHz, speech modulation being analogue FM with a peak frequency deviation of 5 kHz—the same as standard PMR practice. NMT900 also uses a frequency deviation of 5 kHz, but with a 12.5 kHz channel spacing to double the number of available channels, albeit with a degraded adjacent channel rejection performance that must be taken into account during frequency planning. Signaling is at 1200 bit/s using audio Fast Frequency Shift Keying (FFSK). Error protection of the signaling information is by means of a Hagelbarger convolution forward error correcting code.

NMT was designed from the outset to support international roaming and was first implemented with full four-nation roaming in the four participating countries. Since then, NMT450 has been deployed in many European countries (Austria, Spain, The Netherlands, Belgium, Luxembourg, Switzerland, France, Iceland, the Faroe Islands, Turkey, and Hungary). But, due to differences in frequency allocations in the 450 MHz band between countries, not all networks are fully compatible to allow roaming.

NMT900 was developed as a necessity when maximum capacity was reached in the NMT450 networks. Since 1987 NMT900 has been deployed as an overlay network in several countries.

C450 System

C450 (also known as Netz-C) was developed by Siemens during the early 1980s under the direction of the (West) German PTT, Deutsche Bundespost. Following a trial period, commercial service opened in 1985. C450 has mainly served only the German market, although systems are also operating in Portugal and South Africa.

C450 has a channel spacing of 20 kHz, in common with other mobile services in Germany at 450 MHz, and speech modulation is analogue FM with a frequency deviation of 4.0 kHz. Signaling for call control is transmitted at 5.28 kb/s by direct FSK. Error protection of the signaling is by bit interleaving with a BCH block code backed by an acknowledgment protocol.

In addition, during a call C450 uses continuous signaling between base station and mobile, achieved by time-compressing speech in bursts of 12.5 ms, with each burst being compressed into 11.4 ms. This process opens up slots of 1.1 ms duration every 12.5 ms and the signaling data is inserted into these slots and extracted by the receiver. The receiver also time-expands the speech back to its original form.

This continuous signaling serves several purposes. First, it allows the base station to send power control and handover messages to the mobile without disturbing the voice channel. The data is then checked for jitter, enabling the determination of channel quality that will indicate whether a handover is required. The time delay between a base station transmitting a data burst and receipt of the response from the mobile is measured at the base station. This is used to calculate the distance between the base station and the mobile. This distance is also taken into account in handover determination. Lastly, the data is used as a timing reference by the mobile to lock its internal clocks.

C450 contains a number of advanced features that were made possible by the application of developments in technology. Although speech transmission is analogue, it can be regarded as a hybrid technology system. Several system characteristics, such as time-slotted signaling channels and continuous signaling during calls, have been carried through into the GSM system design.

Narrow-AMPS System

In 1991, the Narrowband Advanced Mobile Phone System (Narrow-AMPS or N-AMPS) evolved from the U.S. AMPS specification (EIA-553). The first N-AMPS standard (sometimes known as IS-88) identified parameters required to begin the design of NAMPS radios, such as radio channel bandwidth, modulation type, and message format.

N-AMPS technology is based on the AMPS system and increases its capacity while adding extra functionality. It allows a cellular provider to support three cellular voice calls in the same space required by AMPS for a single cellular voice call. Moreover it has more advanced features than AMPS, such as ESN authentication, caller ID, and short messaging.

Mobile Cellular System

The first commercial mobile cellular system (MCS) was deployed in 1979 in Japan. Several different cellular systems have migrated from the original MCS system, such as MCS-L1 and MCS-L2. In addition to this family, additional analogue cellular systems were deployed in Japan (systems such as JTACS and NTACS).

MCS-L1 was the first cellular system in Japan. It was operated and developed by NTT. The system performed in the 800 MHz band, in the channel bandwidth of 25 kHz. The MCS-L2 system, developed due to the limited capacity of the MCS-L1 system, uses the same frequency bands as the MCS-L1 system but the radio channel bandwidth was reduced from 25 kHz to 12.5 kHz with 6.25 kHz interleaving. MCS-L2 had better performance than the previous system.

Cellular NET

Cellular NET (CNET) was first deployed in Germany in 1985. This analogue cellular system was used in Germany, Portugal, and South Africa. The principal objective of the CNET system was to bridge the gap of cellular systems in Germany until the digital European system, GSM, could be introduced. The CNET system performed at 450 MHz with 4.44 MHz transmit-and-receive bands with frequencies of 461.3 to 465.74 MHz and 451.3 to 455.74 MHz. The primary channel bandwidth is 20 kHz with 10 kHz channel interleaving.

MATS-E

The MATS-E system, used in France, includes many of the features used in different cellular systems. It used frequency bands of 890 to 915 MHz and 935 to 960 MHz and the channel bandwidth.

Global System for Mobile Communications

The Global System for Mobile (GSM) communications is a digital public cellular network using Time Division Multiple Access (TDMA) techniques for multiplexing and transmission within the frequency bands of 900 MHz, 1800 MHz, and 1900 MHz. It is used widely in Europe and today it is the most common technology of cellular networks in the world. The following description provides a summary of the network entities, their functions and objectives, and the services and facilities that the network offers.

The GSM Standard

The Special Mobile Group (SMG) Technical Committee of the European Telecommunication Standards Institute (ETSI) specified the GSM Standard. The GSM Standard started from the first version called GSM "Phase 1" and was progressively upgraded as follows:

- Phase 2
- Phase 2+ Release 1996
- Phase 2+ Release 1997
- Phase 2+ Release 1998
- Phase 2+ Release 1999; GERAN-based system

Between Phase 1 and Phase 2+ 1998, additional features were added, rates become higher, the air interface was approved, and Phase 2+ Release 1999 had the characteristics of a new system, purely GERAN-based (GERAN stands for GSM EDGE Radio Access Network).

   Therefore, the explanation of 2G GSM in this chapter is related to the first 4 releases (GSM Phase 1 until GSM Phase 2+ Release 1997). The next versions (Phase 2+ Release 1999 and up) are described in Chapter 13 (2.5G systems).

Basic GSM Architecture

The basic GSM architecture consists of:

- Mobile switching centers (MSC)
- A Base Station Subsystem (BSS) that includes
  - Base Transceiver Stations (BTSs)
  - Base Station Controllers (BSCs)
- Mobile Subscriber (MS)
- A variety of registers and a network management system (see Figure 12.1)

The mobile subscriber, sometimes called a mobile station, consists of mobile equipment and a Subscriber Identity Module (SIM). In addition to these functional entities, GSM also defines the following interfaces:

- Radio interface (Um).
- The interface between the MSC and BSC (A interface).

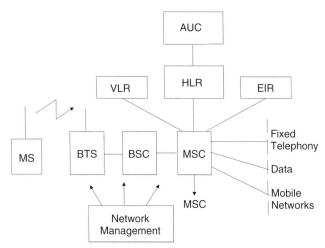

AUC = Authentication Center, BSC = Base Station Controller, BTS = Base Transceiver Station, EIR = Equipment Identity Register, HLR = Home Location Register, MS = Mobile Station, MSC = Mobile Switching Center, VLR = Visited Location Register

Fig. 12.1   GSM architecture.

- A signaling interface that allows roaming between networks: GSM signaling is based on the ITU-T No. 7 signaling standard and is defined in the Mobile Application Part (MAP).

Base Station Subsystem

The base station subsystem (BSS) is the centrally located physical equipment used for radio coverage over a determined geographical zone called a cell. It contains the equipment required for communicating with the user equipment called mobile station. The BTS and BSC together form the BSS and carry out all the functions related to radio channel management. This includes management of radio channel configurations, allocating radio channels for speech, data and signaling purposes, controlling frequency hopping, and power control. The BSS also manages speech encoding/decoding, and channel encoding/decoding. The specification for the equipment forming the BSS is found in 3GPP specification: series TS 11.20 through TS 11.30. The series of specifications, definitions, and specifications for the interfaces between the various components of the BSS is found in specification: series TS 08.

Mobile Services Switching Center

The Mobile Switching Center (MSC) is a center containing all the switching functions required for mobiles located in an associated geographical area. This is called an MSC area. The MSC is mainly concerned with the mobility management functions of its subscribers. These include authentication and registration of the mobile customer, location updating, call setup and release, and all the necessary radio resources (especially location and handover). In addition, the MSC handles the interworking with other networks (such as the PSTN or ISDN). This requires specific functions associated with the MSC, known as interworking functions (IWF). More about IWF can be found in 3GPP specification GSM TS 09.04 through GSM TS 09.07.

Home Location Register

The Home Location Register (HLR) is the master subscriber database used for managing mobile subscribers. The number of HLRs in a PLMN varies according to the characteristics of the PLMN itself. The HLR is responsible for:

- Carrying subscriber information such as the International Mobile Subscriber Identity (IMSI), the mobile station ISDN number, subscription levels, supplementary services, and so on.
- Handling part of the mobile location information allowing incoming calls to be routed to the MSC for the said mobile, such as current location (or most recent location) of subscribers.
- Any administrative action by the network operator on subscriber data.
- Visitor Location Register (VLR) address.

The organization of subscriber data appears in 3GPP Specification GSM 03.08.

Visitor Location Register

The visitor location register acts as a temporary subscriber database for all roaming subscribers within its coverage area and contains information similar to that in the HLR. The VLR also contains other information needed to handle incoming/outgoing calls, such as the location area in which the mobile has been registered, the temporary mobile subscriber identity, and so on. This information is received via a dialogue with the HLR associated with the mobile subscriber. When a roaming mobile enters the MSC area, the associate MSC warns the VLR of this situation. The mobile receives a Mobile Subscriber Roaming Number (MSRN) or visited network address that serves to route incoming calls to that mobile.

The procedure for exchanging information between VLR and HLR is described in 3GPP GSM TS 03.12.

Authentication Center

To guard against fraud, the Authentication Center (AUC) works closely with the HLR and provides information to authenticate all calls.

Equipment Identity Register

The Equipment Identity Register (EIR) is used for equipment security and validation of different types of mobile equipment. This information can be used to screen mobile types before they access the system. For example, it can be used to identify stolen mobile equipment, unauthorized users, or a fault that could disrupt the network.

Mobile Station

Hundreds of GSM handset models are available at a vast array of prices. They range from entry-level products to high-end business solutions with an enormous variety of features. The mobile station comprises:

- Varieties of mobile equipment depending on the services the MS can support
- The Subscriber Identity Module

The MS is used by the subscriber to gain access to the telecommunications services offered by the GSM PLMN. There are various types of MS, such as vehicle-mounted stations, portable stations, and handheld stations. The specific aspects of the man–machine interface are covered in 3GPP specifications GSM TS series 11.

The main features of the MS are:

- Automatic calling, repeat-call attempt restrictions
- Display of called number
- Indication of call progress signals
- Country/PLMN indication
- Country/PLMN selection
- Keypad
- International Mobile Equipment Identity
- Short message indication and acknowledgment
- Short message overflow indication
- ISDN "S" terminal interface
- International access function
- Service Indicator (SI)
- Auto-calling restriction capabilities
- Emergency call capabilities
- Dual Tone Multi Frequency (DTMF)
- Subscription identity management
- On/off switch
- Support of encryption A5/1 and A5/2
- Short Message Service Cell Broadcast (SMSCB)
- Service provider indication
- Support of the extended short message cell broadcast channel
- Support of additional call-setup MMI procedures
- Network identity and time zone
- Ciphering indicator
- Network's indication of alerting in the MS
- Network-initiated Mobile Originated (MO) connection
- Support of Localized Service Area (SoLSA)
- Support of supplementary services
- Abbreviated dialing
- Fixed-number dialing
- Barring of dialed numbers
- DTMF control digits separator
- Selection of directory number in messages
- Last numbers dialed
- Mobile equipment/SIM lock
- Service dialing number

Subscriber Identity Module

The subscriber identity module (SIM) is the entity that contains the identity of the subscriber. When placed in mobile equipment, together the module and the equipment become a mobile station (MS). The primary function of the SIM is to authenticate the validity of an MS when accessing the network. In addition, it provides a means to authenticate the user and may also store other subscriber-related information or applications. Subscription entitlements are not stored in the SIM but in the network. The SIM is a removable module containing the International Mobile Subscriber Identity that unambiguously identifies a subscriber. Without a valid IMSI, GSM service is not accessible (except for emergency calls). The SIM provides storage for subscriber-related information and protection against unauthorized use, and may also contain applications. The basic security aspects supported by the GSM SIM are:

- Authentication algorithm (A3)
- Subscriber authentication key (Ki)
- Cipher key generation algorithm (A8)
- Cipher key (Kc)
- Control of access to SIM stored data and functions performed in the SIM.

An algorithm A38 may perform the combined functions of A3 and A8. There are two types of SIM parameters:

- *Frequently updated parameters:* These include location information, cipher key, and cipher key sequence number, which are updated after each call termination.
- *Parameters permanently burned on the SIM:* These include IC card identification.

The following are several examples of the SIM storage capability:

- *Administrative information:* This indicates the SIM mode of operation (e.g., normal, type approval).
- *IC card identification:* This is a number that uniquely identifies the SIM and the card issuer.
- *SIM service table:* This indicates the optional services that are provided by the SIM.
- *International Mobile Subscriber Identity:* This unambiguously identifies a subscriber.
- *Location information:* This comprises the Temporary Mobile Subscriber Identity (TMSI) and Location Area Information (LAI).
- *Cipher key and cipher key sequence number:* The cipher key is a sequence of symbols needed to encrypt or decrypt carried information.
- *Broadcast Control Channel information:* This is the list of carrier frequencies to be used for cell selection.
- *Forbidden PLMNs:* These are held by the SIM to avoid unnecessary registration attempts.
- *Language preference:* This indicates the Man–Machine Interface (MMI) language(s) preferred by the subscriber.

GSM Air Interface

The GSM air interface (Um) provides the physical link between the mobile station and the network. Some of the main characteristics of the air interface are given in Table 12.1.

As already described, GSM is a digital system employing time division multiple access (TDMA) techniques and operating in the 900 MHz band. Two frequency bands have been made available for use throughout Europe by the GSM system, namely:

- 890 MHz to 915 MHz for the mobile to base station (uplink) direction
- 935 MHz to 960 MHz for the base station to mobile (downlink) direction

Table 12.1   Main Characteristics of the Air Interface

Frequency Band Mobile—Base	*890–915 MHz*
Frequency Band Base—Mobile	*935–960 MHz*

124 radio carriers spaced by 200 kHz
TDMA structure with 8 time slots per radio carrier
Gaussian Minimum Shift Keying (GMSK) modulation with BT = 0.3
Slow frequency hopping at 217 hops per second
Block and convolutional channel coding with interleaving
Downlink and uplink power control
Discontinuous transmission and reception

These 25 MHz bands are divided into 124 pairs of carriers spaced by 200 kHz. In addition, carriers have been specified in a pair of extension bands: 872 MHz to 888 MHz and 917 MHz to 933 MHz. In the early 1990s, extra spectrum was made available in the U.K. and other countries in the 1710 MHz to 1785 MHz band paired with 1805 MHz to 1880 MHz. The GSM specification was adapted to cover this range in a modified standard known as DCS 1800. The GSM standard is now also being used at 1900 MHz in North America in a variant known as PCS 1900.

Each of the radio carriers is divided up into eight TDMA time slots with a length of 0.577 ms such that the frame length is 4.615 ms. The recurrence of each time slot makes up one physical channel, such that each carrier can support eight physical channels, in both the uplink and downlink directions. The time slot allocation in either direction is staggered so that the mobile station does not need to transmit and receive at the same time. Data is transmitted in bursts within the time slots and a number of different types of burst can be carried, as shown in Figure 12.2. The normal burst has a data structure as shown. It consists of 148 bits of which 114 bits are available for data transmission; 26 bits are used for a training sequence (allowing the receiver to estimate the radio propagation characteristics and to set up a dispersion equalizer), 6 bits as tail bits, and two stealing flags. These physical channels therefore provide a data throughput of 114 bits every 4.615 ms or 24.7 kb/s.

The bursts modulate one of the RF carriers using Gaussian Minimum Shift Keying (GMSK) modulation with a Base Station Transmitter index of 0.3. The allocation of the carrier can be such that frequency hopping is achieved (i.e., consecutive bursts of a physical channel will be carried by differing RF carriers). This hopping is performed every TDMA frame, or every 4.615 ms, and provides extra protection against channel fading and cochannel interference.

A number of logical channels can be carried by the physical channels described before. These are summarized in Table 12.2.

There are two categories of traffic channels:

- Speech, whether full rate using 22.8 kb/s or half rate using 11.4 kb/s
- Data, providing a variety of data rates

There are four basic categories of control channels:

- BCCH, the Broadcast Control Channel
- CCCH, the Common Control Channel
- SDCCH, the Stand-alone Dedicated Control Channel
- ACCH, the Associated Control Channel

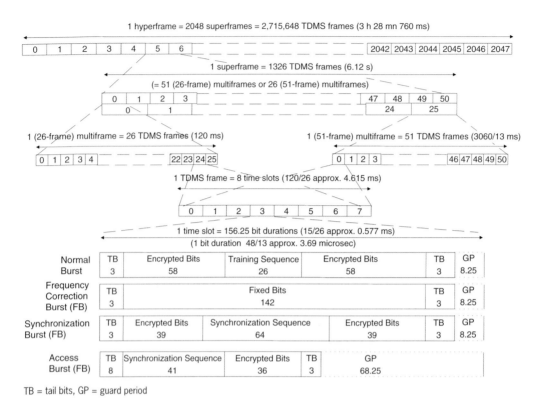

TB = tail bits, GP = guard period

Fig. 12.2 GSM timeframes, time slots, and bursts.
Source: 3GPP Specifications GSM Recommendation 05.01

Table 12.2 GSM Logical Channels

Traffic Channels (TCH)		Control Channels (CCH)			
Speech	Data	Broadcast CCH (BCCH)	Common CCH (CCCH)	Stand-alone dedicated CCH (SDCCH)	Associated CCH (ACCH)
Full-rate TCH/F	TCH/F9.6	Frequency Correction	Paging Channel		Fast (FACCH)
	TCH/F4.8	(FCCH)	(PCH) Random		
	TCH/F2.4	Synchronization (SCH)	Access (RACH)		Slow (SACCH)
Half-rate TCH/H	TCH/H4.8		Access Grant		
	TCH/H2.4		(AGCH)		

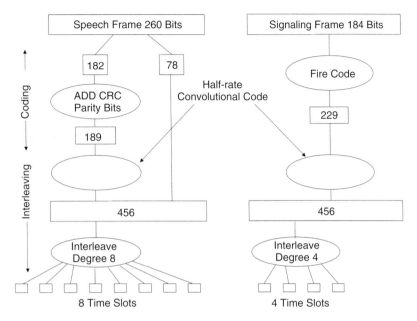

Fig. 12.3   GSM channel coding and interleaving.

These are further divided into channels with specific purposes. For a detailed description of these channels, refer to the 3GPP Specifications GSM series TS 05.

Each of these logical channels is mapped onto the physical channels via an appropriate burst type, as shown in Figure 12.3. TDMA frames are first built up into 26-frame or 51-frame multiframes, such that individual time slots can use either of the multiframe types, and then into superframes and hyperframes. The TCH and the associated ACCH use the 26-frame structure, while the BCCH and CCCH use the 51-frame structure. The SDCCH may occupy one physical channel, providing eight SDCCH, or may share a physical channel with the BCCH/CCCH. Typical arrangements for allocating the eight physical channels could be:

- 7 channels TCH and SACCH + 1 channel BCCH/CCCH/SDCCH.
- 6 channels TCH and SACCH + 1 channel BCCH/CCCH + 1 channel SDCCH

Each cell must have at least one physical channel assigned to the BCCH/CCCH. Where there are two or more carriers per cell, the non-BCCH carriers may have all eight channels allocated to TCH.

Frequency Hopping

Frequency-hopping capability is optionally used by the network operator on all or part of the network. The principle of slow frequency hopping is that every mobile transmits its time slots according to a sequence of frequencies that it derives from an algorithm. Frequency hopping occurs between time slots; therefore, a mobile station transmits (or receives) on a fixed frequency during one time slot ($\approx 577\,\mu s$) and then must hop before the time slot on the next TDMA frame. Due to the time needed for monitoring other base stations, the time allowed for hopping is approximately 1 ms, according to the receiver implementation. The receive-and-transmit frequencies are always duplex.

The main advantage of this feature is to provide diversity on one transmission link (especially to increase coding and interleaving efficiency for slowly moving mobile stations) and also to average the quality on all the communications due to interference from diverse sources. It is implemented on all mobile stations. In addition, frequency hopping gives higher privacy and security.

GSM Speech Coding and Channel Coding

The speech coder is a Regular Pulse Excited Linear Predictive Coder (RPE-LPC) with long-term prediction. This provides a net bit rate of 13 kb/s. It is a block-based coder where the input samples are analyzed in blocks with a 20 ms duration. A half-rate speech coder that will effectively double the system capacity of GSM and an enhanced full-rate coder to provide better-quality speech have also been specified.

Before being assembled into the time slots and frames, the digital speech and signaling data is encoded and interleaved. The speech coder output is divided into three classes of bits and the most sensitive bits are encoded by adding parity check bits followed by a convolution coder. Signaling data is encoded using a FIRE code. A process of interleaving is then used to spread the data blocks over a number of bursts.

For speech, an interleaving degree of 8 is used (i.e., the speech block is spread over 8 bursts), while an interleaving degree of 4 is used for signaling. This overall process is shown in Figure 12.4 (GSM signaling model) and the combined use of coding and interleaving provides good protection of channel data from the fading, dispersion, and interference effects on the radio path. With the addition of frequency hopping and diversity techniques, the GSM air interface is particularly robust.

One of the penalties for this is an increase in overall transmission delay. The speech coder contributes about 25 ms and the channel coding and interleaving a further 37 ms. The rest of the transmission delay budget allows for analogue to digital conversions, 16 kb/s transmission, and switching in various parts of the network. The overall one-way transmission delay thus amounts to around 90 ms. Such a delay means that echo control is necessary even on national calls.

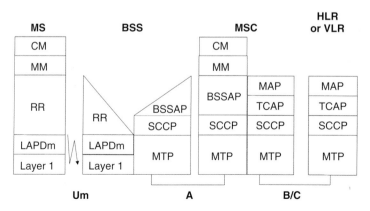

Fig. 12.4 GSM signaling model.

GSM Signaling

Signaling in the GSM system consists of two very different parts. The first part is a specific system signal at levels 1, 2, and 3. This is applied to the radio path MS-BTS. The second part is based on the pertinent CCITT Recommendations (Signaling System No 7) and will be described separately in the next section. Figure 12.4 shows the overall signaling model. The air interface uses LAPDm layer-signaling protocol, which is also used for the A-bis, BTS-to-BSC interface.

The Layer 3 protocol consists of three sublayers: radio resource management (RR), mobility management (MM), and connection management (CM). Radio resource management deals with managing the logical channels, including paging, channel assignments, handover, measurement reporting, and other functions. The mobility management layer contains functions necessary to support the mobility of the user, including authentication, location updating, attach and detach of IMSI, and registration. The connection management layer deals with call control, establishing and clearing circuits, management of supplementary services, and the Short Message Service.

An example of the signaling messaging for establishing a mobile originating call is shown in Figure 12.5. The following are the key events.

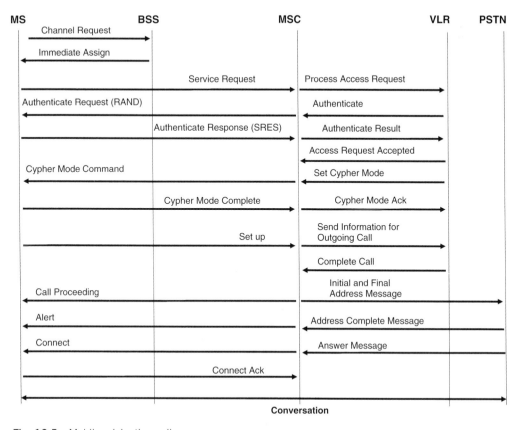

Fig. 12.5 Mobile originating call.

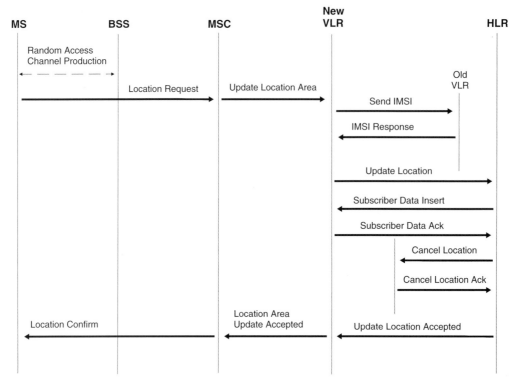

Fig. 12.6   Location updating.

- Request and assignment of a channel between the MS and the BSS
- A service request procedure that accesses the VLR
- An authentication and ciphering exchange that validates the mobile user and sets the encryption cipher
- Call setup that includes the sending of dialed digits and establishing the connection

Location updating is shown in Figure 12.6. An update request is indicated by the mobile and passed to the VLR in the new location area. The new VLR requests the IMSI from the old VLR and then signals the new location to the HLR. The HLR provides the subscriber data to the new VLR and cancels the subscriber entry in the old VLR. Finally, a confirmation message is sent back to the mobile.

There are, of course, many other signaling exchanges dealing with mobile terminating calls, supplementary services, and short message service. The scope of this chapter limits a detailed discussion of these cases.

Signaling System No 7

The International Standard Institute ITU-T designed the specification for this common channel signaling system. Signaling System No 7 (SS7) was originally known as CCITT No 7, often abbreviated to C7 or SS7, and it has become the world standard for interexchange telephony signaling. SS7 is used between digital exchanges for call setup and supervision;

in the SS7 signaling network, these exchanges are described as Service Switching Points (SSP), which is used as a high-level communication link on a dedicated 64 kb/s data channel. It is also used to enable SSPs to interrogate remote databases known as Service Control Points (SCP).

Signals may be routed between SSPs and SCPs over direct connections or transmitted via Signal Transfer Points (STP). In addition to call setup, it supports other functions such as: ISDN calls (including call classification), user-to-user data, Calling Line Identification (CLI), network address extension, distinctive ringing, call diversion, database interrogation, and more. A single SS7 64 kb/s link can support call setup for about 2000 telephone channels (about 60×2 Mbit/s PCM links). The signaling messages do not need to be carried along the same route as that carrying the related telephony traffic.

SS7 employs a functional concept embracing processor function, user function, and message transfer function. This allows changing to any one function without significant impact on the others. It involves, for example, flexible arrangements in the message transfer function, enabling an error control option to be exercised without involving the exchange call processing subsystem. The BSC-to-MSC air interface and the MSC-to-register interfaces employ ITU-T No. 7 signaling using the Message Transfer Part (MTP), Signaling Connection Control Part (SCCP), Transaction Capabilities Part (TCAP), and Mobile Application Part (MAP).

GSM Message Transfer Protocol

This is the core of SS7 and comprises the three lower levels of the protocol:

- *MTP level 1:* This is the 64 kb/s data link.
- *MTP level 2:* This inserts signaling units for transmission, each with a forward and backward sequence number and acknowledgment bits. The forward serial number ensures message transmission in the correct order. The backward serial number is used to acknowledge messages received from the far end.
- *MTP level 3:* This provides message routing between signaling points in the SS7 network.

MTP level 3 also reroutes traffic away from failed links and signaling points and controls traffic when congestion occurs. In addition, it allows user parts to communicate with each other by means of addressed messages.

GSM Security Features

The information on the air interface needs to be protected to provide user data confidentiality (including speech) and to prevent fraudulent use of subscriber and mobile identities. The basic mechanisms employed for this are user authentication and user data encryption. Each mobile user is provided with a subscriber identity module that contains the IMSI, the individual subscriber authentication key (Ki), and the authentication algorithm (A3).

After the mobile user has made an access and service request, the network checks the identity of the user by sending a random number (RAND) to the mobile. The mobile uses the RAND, Ki, and A3 algorithm to produce a signed response (SRES). This response is compared with a similar response calculated by the network, and access continues only if the two responses match. The SIM also contains a cipher key generating algorithm (A8). The MS uses the RAND and A8 to calculate a ciphering key (Kc) that is used to encrypt and decrypt signaling data and user data.

The authentication center (AUC) is responsible for all security aspects, and its function is closely linked with the HLR. The AUC generates the Ki's and associates them with IMSIs, and for each IMSI, it provides the HLR with sets of RAND, SRES, and Kc. The HLR then provides the appropriate VLR with these sets and it is the VLR that carries out the authentication check. Authentication of mobile users can be carried out on both mobile-originated and mobile-terminated call setup, on location updating, and on activation of supplementary services. As the authentication sets are used up in the VLR, further sets are requested from the HLR.

An additional GSM security feature is the equipment identity register. This enables the monitoring of mobile equipment IMEI (International Mobile Equipment Identity) used to validate mobile equipment, preventing nonapproved, faulty, and stolen equipment from using the system. It can also be used to monitor and test mobile performance. This range of security features provides a high degree of protection for the user and the network operator.

Code Division Multiple Access System

Code Division Multiple Access System (CDMAOne) is the basis for many of the commercial 2G cellular systems around the world and is also used in many other types of communications systems, including Personal Communications System (PCSs), fixed wireless (wireless local loop), Global Positioning System (GPS), and the OmniTRACS satellite system for transportation logistics.

Principles of Code Division Multiple Access

CDMA systems allow many data signals to be multiplexed and transmitted over a wireless channel at the same time and in the same frequency band without interfering with each other. In a CDMAOne system, data signals to be transmitted are modified through the use of a pseudo-noise code (called a PN code).

Whereas, in time-division systems, channels are distinguished according to which time slot they occupy, in code-division systems, channels are distinguished according to which PN code they use. PN codes used in a system are orthogonal, or nearly so. Orthogonal means that the codes should not correlate among themselves nor should they be time-shifted versions of each other. Therefore, each signal with a unique PN code can be detected from other signals. When a particular PN code is auto-correlated, the result is "high" (or "unity"), but when cross-correlated with other PN codes in the same set, the result is zero, or very low. PN-coded sequences are typically generated using one or more shift registers that have particular feedback connections.

CDMAOne systems use three types of PN code:

1. *Walsh code:* CDMAOne use of the Walsh code is of 64 different orthogonal codes used on the:
   a. Downlink, for differentiate user (spreading). It is a code channel.
   b. Uplink, not to differentiate users but for modulation.

2. *Long PN code:*
   a. On the uplink, it is used for identifying the mobile station (spreading). A long PN code is a 42-bit code.
   b. On the downlink, it is used for data scrambling.
3. *Short PN code:*
   a. On the uplink, the mobile uses this code for extra signal robustness without offset.
   b. On the downlink, it is used to identify the cell (the base station)—different base stations will have different offset in the code. It is a 16-bit short PN code.

The way to distinguish between two different calls (users) in the received signal is shown in Appendix 12.1.

Whereas analogue and TDMA systems have a definite limit to the number of simultaneous users in any cell (defined by the frequency allocation, the reuse pattern, and the time slot structure), CDMA technology does not have a fixed limit as such. However, as the number of users in a cell increases, the effective levels of noise interference increase, and this reduces the quality of the calls. Thus, the factors of coverage, quality, and capacity are tightly interrelated and must be balanced against one another to reach the required level of system performance. The transmit rate of one CDMAOne channel is about 192 kbps. The number of time slots in one CDMAOne channel is 64, and the average data rate for each user is up to 3 kbps. (See Appendix 12-2.)

CDMAOne Technology

CDMAOne is a particular form of spread spectrum radio access technology based on CDMA. It was originally developed for military applications where considerations of high resilience, tolerance to jamming, and resistance to detection and interception were paramount. Now it is the basis for many of the commercial 2G cellular systems around the world. CDMAOne is also known as IS-95, which refers to the original ITU IS-95 wireless interface protocol.

CDMAOne defines the structure of the CDMA channels, power control, call processing, hand-offs, and registration techniques for system operation. It allows voice and data communications on either a 30 kHz AMPS radio channel (when used on the 800 MHz cellular band) or a 1.25 MHz CDMAOne radio carrier with frequency band of 800 MHz or 1.9 GHz. Many operators also provide circuit-switched data connections at 14.4 kbps.

The radio channel allows multiple users to communicate on the same frequency at the same time and carry pilot, synchronization, paging, access control, voice, and data signals simultaneously. The technology can have several control channels on the same frequency simultaneously with other voice channels because of the many codes that enable the CDMAOne technology to discriminate between signals.

CDMAOne technology uses two voice coding schemes, one with a bit rate of 8 kb/s and the other with a bit rate of 13 kb/s. The lower bit rate coder offers higher system capacity but at lower speech quality. In both cases, the bit stream from the vocoder is spread up to 1.25 Mbit/s before being transmitted over the radio carrier. CDMAOne is also able to change the data rate for each user dependent on voice activity (using variable rate speech coding), and by this means it can reach about 3.8 kbps (instead of 3 kbps) average data rate per user. This allows more users to share the radio channel, but it leads to higher interference and slightly lower voice quality.

Since all mobiles share the same radio channel, a call can be smoothly handed off between two cells (so-called *soft handoff*) by arranging for both cells to listen to the same spreading code and selecting whichever one gives the best-quality channel at any one time. By spreading the signal across a wide bandwidth, the impairments of multipath propagation and delay spread are reduced. Highly accurate power control of mobile stations is essential to avoid a mobile near the base station swamping the receiver with a strong signal. CDMAOne base stations send power updates to mobiles every 1.25 ms to achieve this. While this is a complex solution the advantage is that mobiles only transmit the minimum power at any one time, helping to conserve battery and to reduce interference.

It should be noted that the term CDMA is sometimes used as a general name for systems related to this access technology, including cellular systems and/or other types of systems. For example, the term CDMA is sometimes used to describe the Wideband Code Division Multiple Access (W-CDMA) technology that is the air interface of the 3G of GSM (UMTS), and of the third generation of the Japanese system Freedom of Mobile Multimedia Access (FOMA). However, although the technologies are similar, they are incompatible. Moreover, other applications of CDMA, like GPS (which conforms to ANSI-J-STD-008) and Omni-TRACS, are entirely different from CDMA cellular applications.

Advantages of CDMAOne

The CDMAOne generations have several advantages over other 2G systems.

Capacity
Capacity is one of the main advantages of CDMAOne technology. It can accommodate more users per MHz of bandwidth than any other technology of the same generation. For example, 2G CDMAOne has 4 to 5 times the capacity of 2G GSM (but not more than 3G UMTS), and has 8 to 10 times the capacity of 1G AMPS.

Call Quality
CDMAOne has better call quality, with better and more consistent sound, as compared to AMPS and GSM systems. A reasonable call can be produced with lower signal levels, less interference, and therefore better quality. Moreover, the soft handoff feature (where the mobile subscriber is hooked into both cells during transfer from one cell to the next) reduces the probability of dropped calls.

Frequency
CDMAOne technology requires one frequency only per cell, compared with maximum possible frequencies in TDMA technology. Therefore, the frequency reuse plan is much easier to handle.

Coverage
CDMAOne gives better coverage and requires fewer antenna sites. Cell size and system coverage depend mainly on:

- Power used by the handset and the base station
- Terrain
- Frequency (the higher the frequency, the smaller the coverage)
- Efficiency of the algorithms used to reduce the noise

CDMAOne covers larger areas and consumes less power than the other technologies and so requires fewer cell sites in a given area. The fewer cells in the systems, together with smaller number of frequencies, means fewer antenna sites are needed. This advantage is very important nowadays when permission to install antennas is a sensitive subject and difficult to obtain. In addition, fewer sites mean a significant economic advantage over TDMA-based standards, or the older cellular standards that used Frequency Division Multiple Access (FDMA).

Multipath Performance

CDMAOne has better multipath performance than AMPS and GSM systems. When a radio signal is transmitted to a receiver, it can take a direct route or it can take a reflected path. The signal arriving via the reflected path arrives later because of the greater distance traveled, and this can interfere with the correct operation of the receiver. Sometimes there can be several reflection paths, giving rise to a multipath effect. In FDMA and TDMA this phenomenon is a disadvantage, and causes interference. However, in a CDMAOne receiver the reflected signals can be faithfully received, adding to the strength of the received signal.

A significant influence on the deployment of CDMAOne compared to GSM is that most of the CDMAOne technologies are patented and require licenses whereas GSM is an open standard and can be almost freely used.

The standard specification of CDMAOne and its family relates only to the radio part (the air interface); the core network depends on the choice of each operator. The GSM standard, on the other hand, is a specification of an entire network infrastructure (core network, radio access network, and mobile subscriber, and the signaling and data interfaces between them).

Versions of CDMAOne

The family of CDMAOne systems includes the following cellular versions:

- Temporary version: IS-95 (announced in 1993).
- CDMAOne or IS-95A: (announced in 1995).
- IS-95B is a 2.5G system; other 2.5G systems are 1xRTT, IS-95+, and CDMA2000 1x.
- 1xRTT is a High 2.5G system and sometimes considered a 3G system.
- The CDMA2000 family:
  - CDMA2000 is a 3G system, sometimes called IS-2000.
  - CDMA2000 1xEV-DO is the standard for 1x Evolution-Data Optimized, originally 1x Evolution-Data Only, also referred to as 1xEV-DO, EV-DO, EVDO, or just DO.
  - CDMA2000 1xEV-DV (1x Evolution-Data/Voice).

Information regarding the preceding systems is given in Chapter 13.

CDMAOne Architecture

The system architecture of CDMAOne is similar to that of the GSM architecture, which was described at the beginning of this chapter as part of the 1G and 2G cellular systems. Figure 12.7 describes the CDMAOne architecture, with its corresponding interfaces.

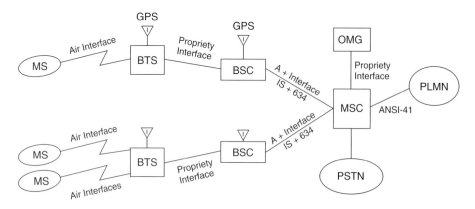

Fig. 12.7 CDMAOne architecture.

The interfaces used are as follows:

- *IS-634:* This is an open interface between MSC and BSC.
- *A-bis or A+:* This is an interface between BSC and BTS (proprietary interface, depends on the vendor).
- *ANSI-41:* This is an interface with other PLMNs.
- *ISUP: ISDN user part:* This interfaces with the PSTN.
- *IS-95:* This is an air interface between base station subsystems and mobile subscribers.

Uplink and Downlink

The *uplink* refers to transmission from the base station to the mobile station (or frequency reception by the mobile subscriber). The *downlink* refers to transmission from the mobile station to the base station. The uplink frequency is built and arranged differently from the downlink frequency; they essentially carry some symmetrical information (such as traffic channels) and some asymmetrical information (as in a paging system that uses only the uplink).

Downlink

The downlink is sometimes called the forward channel, base station frequency transmission, or mobile station/mobile subscriber frequency reception. On the downlink, the PN codes are used as follows:

- The Walsh codes are used to differentiate users (spreading).
- The short PN code is used to identify the cell (the base station)—different cells will have a different offset in the code.
- The long PN code is used for data scrambling.

The downlink consists of the following channels:

- One pilot channel = one TS
- One sync channel = one TS

- Up to seven paging channels = up to seven TS
- A number of downlink traffic channels—i.e.; the number of the remaining channels, calculated as 64 total downlink channels – 1 paging channel –1 sync channel – (7 – number of paging channels)

The downlink traffic channel is used for information specific to the MS during a call. It carries the control channels and user data (transmission and reception of speech, data, and signaling).

The Digital Control Channel (DCC) carries system and paging information, coordinates access, precision synchronization signal, pilot, extended sleep mode, and others. The DCC channel includes the following:

- A *pilot channel* that is coded by Walsh code number 0: The channel contains PN pilot sequence that is a sequence of all 0's (or 1's). This channel is an unmodulated spread spectrum signal. The pilot channel is transmitted by the base station all the time on one of the active downlink CDMAOne channels. It is used for coherent detection and synchronization (timing and phase reference) for the MS operating within the coverage area of the transmitted BTS. It includes cell identifiers and sector identifiers and allows the MS to detect the strength of received signals and references for phase shift.

- A *synchronization channel* (SYNC) uses Walsh code 32 (a sequence of 32 zeros, following by 32 ones). This channel is used to achieve critical time synchronization data for the MS. With the synchronization messages transmitted on the synchronization system, the MS is able to synchronize immediately to the CDMAOne network. The SYNC channel is an encoded, interleaved, spread, and modulated spread spectrum signal that is used by MS inside the BTS coverage area.

- A *paging channel* that is used for MSs in no-communication stage: A 1.23 MHz CDMAOne signal can carry one to seven paging channels. The paging channel uses Walsh code 1 up to Walsh code 7. The paging channel is also an encoded, interleaved, spread, and modulated spread spectrum signal. The base station transmits overhead information and specific messages to the MS. It provides the MS with access information (such as assigned traffic channel), pages the MS and system information (such as neighbored cells and base stations).

- The *Analogue Control Channel* (ACC) is similar to DCC but contain less information and very basic data messages. The CDMAOne system allows analogue (EIA-553) and dual-mode (CDMAOne) subscribers to use the same analogue control channels.

The downlink structure is shown in Figure 12.8.

Uplink

The uplink is sometimes called reverse channel, MS frequency transmission, or base station frequency reception. The uplink channels are coded with PN long code. The unique MS equipment identifier is part of this PN code.

On the uplink, the PN codes are used as follows:

- The Walsh codes are used for modulation (and not to differentiate users).
- The short PN code is used by the mobile for extra signal robustness without offset.
- The long PN code on the uplink is used to identify the mobile station (spreading).

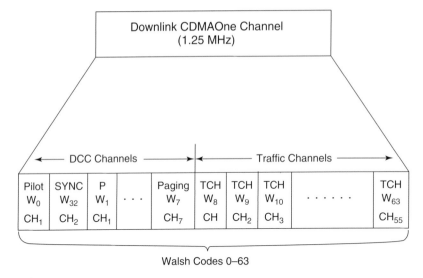

Fig. 12.8   Downlink structure.

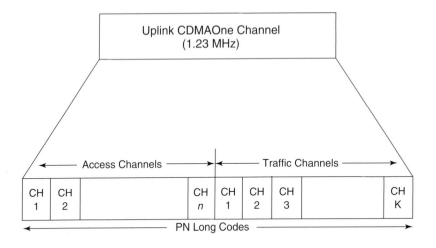

Fig. 12.9   Uplink structure.

The uplink consists of the following channels:

- *The access channel:* This is used by the MS to initiate communication and to respond to paging channel messages.
- *A reverse traffic channel (TCH):* This is the same as for the downlink traffic channel.

The uplink structure is shown in Figure 12.9.

Digital Advanced Mobile Phone System

The Digital Advanced Mobile Phone System (D-AMPS) is a 2G development of the AMPS standard. It is a digital cellular network using time division multiple access techniques for multiplexing and the 1900 MHz frequency band for transmission within the frequency bands of 800 MHz range for the existing AMPS and 1900 MHz for AMPS and D-AMPS. The D-AMPS standard enhances the basic architecture and signaling protocol of AMPS, and adds security and applications features such as short messaging service. It is used mainly in the Americas, particularly in the United States and Canada. A variant of the standard has also been developed for the 1900 MHz personal communications system frequency band.

During the 1990s, D-AMPS replaced the AMPS systems. Today, operators of this system prefer to convert them to GSM 2.5G systems (as GPRS or EDGE), or change them directly to 3G CDMA200. The same D-AMPS system and/or versions of it go by other names such as Digital AMPS (DAMPS), Digital A, IS-54, IS-54B, IS-136, IS-136, TDMA, ETDMA, E-TDMA, Digital AMPS (D-AMPS), North American Digital Cellular (NADC), ANSI/TIA/EIA-627, and United States Digital Cellular (USDC).

The latest version of D-AMPS combined the IS-54 and IS-136 standards. The evolution of D-AMPS is as follows (later on, each version will be explained):

- IS-54 was the first American 2G standard (CDMAOne systems were deployed a few years later).
- ANSI/TIA/EIA-627: After the IS-54 standard was approved by the American National Standards Institute (ANSI), it became an American National Standard called ANSI/TIA/EIA-627.
- IS-136 or TDMA is an improved version of IS-54.
- ETDMA stands for Extended TDMA. This version extends the IS-136 standard.

IS-54 and ANSI/TIA/EIA-627

IS-54 stands for Interim Standard 54, which is a mobile communication standard employing digital technology. It was the first mobile communication system to have a provision for security, and the first to employ TDMA technology. Actually, TDMA mode was initially defined by the IS-54 standard and later on adopted and specified in the IS-13x series of specifications of the EIA/TIA. IS-54 was deployed in the American market in the late 1980s, a few years before the deployment of GSM in Europe. It is a dual-mode, digital cellular standard (AMPS and D-AMPS) specified first by the Telecommunications Industry Association (TIA) and the Electronics Industries Association (EIA).

When an interim standard becomes an American National Standard (after ANSI approval), the IS designator is dropped and a new name that includes the ANSI abbreviation is assigned. The ANSI term for IS-54 is ANSI/TIA/EIA-627. Yet this standard is usually called IS-54. IS-54 maintains compatibility with AMPS (it is a digital extension of AMPS) and, on the other hand, increases capacity by digital means using the TDMA access mode. TDMA in this case tripled call capacity.

The first revision of the IS-54 specification (Rev. 0) identified only the basics needed to begin designing TDMA systems and equipment. The basic specification includes parameters such as time slot structure, type of radio channel modulation, and message formats. IS-

54 security features use the CAVE (Cellular Authentication, Voice Privacy, and Encryption) algorithm for authentication and CMEA (Cellular Message Encryption Algorithm) for encryption.

IS-136 or TDMA

IS-136 is an improved version of IS-54 known also as TDMA. IS-136 added a number of features to the original IS-54 specification, including:

- Text messaging
- Circuit-switched data (CSD)
- Improved compression protocol
- Utilization of time division multiplexing for both voice and control channel transmissions
- An extra digital control channel to go with the Digital Traffic Channel (DTC) that allows a mobile telephone to operate in a single digital-only mode
- Several messaging applications (like SMS and text messaging) implemented in a fashion nearly identical to GSM
- Longer standby time
- Support for small, private, or residential systems that can coexist with the public systems
- Over-the-air activation
- Expanded data applications
- Easier adaptation of existing AMPS systems

There are several advantages in the adaptation of the AMPS standard. One important advantage is the possibility to use the AMPS base stations. Since the radio channels of IS-136 continue to use the same bandwidth as AMPS (30 kHz), the base stations of AMPS can be replaced by TDMA radio units. This is a tremendous economic advantage. Another advantage is the use of new dual-mode mobiles that can operate on either IS-136 digital traffic (voice and data) channels or existing AMPS radio channels. This allows a single mobile telephone to operate on any AMPS system and use the IS-136 system whenever it is available. That increases the available capacity and widens the coverage area.

Extended TDMA

Hughes Network Systems developed extended TDMA (ETDMA) in 1990 as an extension to the existing IS-136 TDMA industry standard. ETDMA has the same channel structure and radio channels as the IS-136 TDMA. Its receivers can operate in triple mode: AMPS, TDMA, or ETDMA. The important differences between ETDMA and IS-136 are:

- *Increased channel efficiency:* This is achieved by a combination of low bit rate speech coder, voice activity detection, and interference averaging.
- *Improved system capacity:* The half-rate speech coder (4 kb/s) reduces the amount of transmitted and received information and therefore allows more users to be added. This is done by adding users during voice silence time. (The gap between words and syllables during conversation time is called voice silence time.)
- *Assisted handoff technique:* This is a technique to achieve more accurate and timely handoffs during voice silence time. During this time the mobile receiver can monitor

transmissions from other base stations and relay back to the network the results of the measurements taken. This supplementary information is used by the network to qualify handoff measurements made by the network, and results in more accurate and timely handoffs.

- *Improved radio communications performance:* Triple mode compared to dual mode in IS-136 means that ETDMA can assign an AMPS channel, a TDMA full-rate channel, a TDMA half-rate channel, or an ETDMA channel. IS-136 can assign AMPS channels and TDMA full-rate channels.

D-AMPS Technology

D-AMPS uses a 30 kHz channel like AMPS and incorporates digital speech encoding with TDMA. To get more capacity in D-AMPS, each of the 30 kHz AMPS channels is divided into 3 full-rate TDMA channels, each supporting a single voice call. When half rate is used, each of the full-rate channels is subdivided into two half-rate channels. Thus, D-AMPS could have 3 full-rate TDMA channels or 6 half-rate TDMA channels. Therefore, D-AMPS provides 3 to 6 times the capacity of AMPS traffic channels. The division of the original AMPS 30 KHz channels is shown in Figure 12.10.

As explained earlier, the D-AMPS digital radio channels are divided into frames with 6 time slots. A time slot is of 6.67-ms duration, and includes 324 information bits used for a 48.6 kb/s carrier bit rate. The information bits in each time slot are used as follows:

- 260 bits are used for the 13 kb/s full-rate coded traffic information (voice and data):
  - 5.05 kbps error detection bits and correction bits
  - 7.95 kbps full-rate digitally coded speech

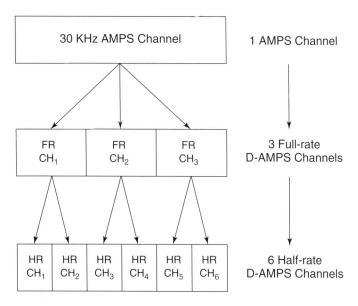

Fig. 12.10   Division of the AMPS 30 KHz channels to D-AMPS time slots.

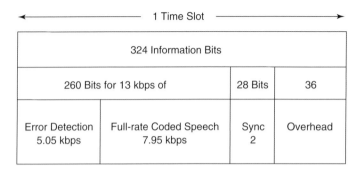

Fig. 12.11 D-AMPS time slot bit separation.

- An allocation of 28 bits is reserved for synchronization. The synchronization sequence is known by all receivers so that each receiver can achieve frame alignment and establish synchronization to its own preassigned time slots.
- The remaining bits are overhead.

This explanation about D-AMPS time slot bits is shown in Figure 12.11.

The voice coder produces a bit stream at 7.95 kbps with a speech frame length of 20 ms. Error correction (channel coding) is applied, giving a user data rate of 13 kb/s, after which the bits are interleaved and mapped onto the time slots. The half-rate coding reduces the overall bit rate for each call to 6.5 kbps, and provides almost the same quality as the 13 kb/s rate. The total carrier bit rate is 48.6 kb/s, which includes some signaling overhead.

The modulation scheme of the traffic channels is $\pi/4$ differential quaternary phase shift keying ($\pi/4$ DQPSK). The DQPSK modulates at a 24.3 k baud channel rate. This modulation technique gives 1.62 bpspHz bandwidth efficiency, which is better than GSM by about 20 percent. Control channels use Binary Phase Shift Keying (BPSK) modulation, the same modulation as in AMPS. The primary downlink and uplink control channels of D-AMPS use the same signaling techniques as AMPS to maintain compatibility and the ability to support either analogue AMPS phones or D-AMPS phones. The MS frequency transmission (uplink) is 824–849 kHz; base station frequency transmission (downlink) is 869 to 894 kHz.

Each base station has its own preassigned color code. This is the digital verification color code (DVCC) that is used to identify the signals directed to a specific base station and to suppress and ignore incoming interfering signals received from other cells (i.e., other base stations). There are 256 8-bit color codes. Neighboring base stations have different color codes.

To avoid simultaneous transmission and reception by the mobile subscriber, the time slots related to the forward and reverse channels (up- and downlinks) are time related.

Comparison of GSM, CDMAOne, and D-AMPS Parameters

Table 12.3 summarizes some technical parameters of the three main 2G cellular system technologies (GSM, CDMAOne, and D-AMPS).

Table 12.3 Comparison of GSM, CDMAOne, and D-AMPS Parameters

	Parameters	GSM	CDMAOne	D-AMPS
1	Access mode	TDMA/FDD	CDMA/FDD	TDMA/FDD
2	Carrier (channel) spacing	200 KHz	1.25 MHz	30 KHz
3	Time slots (TS) in a frame—full-rate	8 TS	64 TS	3 TS
4	Time slots (TS) in a frame—half-rate	16 TS	128 TS	6 TS
5	Downlink frequency (base station transmit to mobile subscriber)	925–960 MHz 1805–1880 MHz	869–894 MHz	869–894 MHz
6	Uplink frequency (mobile subscriber transmit to base station)	880–915 MHz 1710–1785 MHz	824–849 MHz	824–849 MHz
7	Frequency separation	45/95 MHz	45 MHz	45 MHz
8	Carrier (channel) total bit rate	270.833 kbps	1228.8 kbps	48.6 kbps
9	Full-rate coded traffic information (voice and data)—total sum of parameters: 10, 11	22.8 kbps	About 77 kbps	13 kbps
10	Error detection and correction bits	13 kbps		5.05 kbps
11	Full-rate, digitally coded speech	9.8 kbps	About 8/13 kbps	7.95 kbps
12	Modulation technique	GMSK	QPSK/BPSK	π/4 DQPSK

Acknowledgments Parts of this chapter are based on earlier work by Malcolm Appleby and Fred Harrison in the Second Edition of the *Telecommunications Engineer's Reference Book*.

The author also wishes to acknowledge the valuable help and advice given by Goff Hill in preparing this chapter.

References

CDMAOne: *http://www.3gpp2.org/*.
GSM: *http://www.3gpp.org/* Detailed information about GSM can be found in the 3GPP Technical Specifications Series 00–12 under the name "GSM only (before Rel-4)."

Appendix 12.1: Simple Explanation of CDMA

The way to distinguish between different users (calls) on the CDMAOne channel can be compared to distinguishing two groups of people (group A and group B) that are identical to each other. Group A is at one side of the room and group B is at the opposite side of the room. Each group includes people that know only one language and do not understand other languages. This means that each person in group A has only one person in group B that can understand him/her because of their unique language. For simplification, these two people will be called *partners*.

All the people in the room talk together at the same time and the room is very noisy. When the partners want to talk between themselves, they focus on the partner's voice and all the other voices around (in different languages) will be considered as noise. If the noise is not too loud, they will succeed and understand each other.

CDMAOne technology is of course much more efficient in identifying the user's data and making it comprehensible. If the noise is very loud (too many people speaking together), CDMAOne technology will also have a problem in differentiating the information.

In CDMAONE technology:

- Group A is the base station, and group B is the mobile subscriber.
- The PN codes are like the different languages.
- The partners are:
  ○ First person (represented as the base station by group A): This person can be another mobile subscriber or other type of subscriber like a fixed subscriber.
  ○ Second person: This person is a mobile subscriber (represented by group B).
- The partners perform a call between them.

Appendix 12.2: Average Data Rate for Each User

The transmit rate of one CDMAOne channel is about 192 kbps. The number of time slots in one CDMAOne channel is 64. For example, if a CDMAOne channel has only traffic channels, then the transmission rate provides for 64 coded traffic channels. Therefore, the output for all users cannot exceed this rate. To obtain a maximum of 64 communication channels for each CDMAOne radio channel, the average data rate for each user should be up to 3 kbps. If fewer than 64 traffic channels are used, the average data rate can be higher.

13 Beyond Second-generation Systems

Goff Hill and Yael Baruch

Introduction

Following the widespread acceptance of 2G systems, the demands for voice service grew to the point where the existing systems could not meet the demand. This provided an incentive to develop systems that could support more voice channels. Developments began on new 3G systems that would offer higher capacity for voice but could also offer a new range of data services. However, it became clear that the new systems would not be ready in time to meet the exponentially growing market demand. A different solution was therefore needed. The required solution had to be simple to develop and it would be adequate, as an interim measure, to offer a marginal improvement over the existing 2G systems.

Chapter 10 described how the Low 2.5G and High 2.5G systems emerged and how the market and other factors influenced the path from the 2G to the 3G systems. It also explained that three types of development were needed: the radio physical layer had to support higher-speed data services; the core network needed a network based on packet transmission and switching; and a new range of subscriber devices was necessary to make it easy to use in more advanced data services.

The new two-and-a-half- and third-generation networks were therefore characterized by factors such as:

- The maximum data rate available to each subscriber
- The maximum number of subscribers that could be supported
- Whether the solution used the existing frequency bands or new ones
- The extent to which roaming was possible

Differences in technology arise according to:

- Whether packet or circuit switching is used
- The modulation and encoding methods
- The multiple access technique used

The next section discusses the possible migration paths of the three main 2G systems (GSM, CDMAOne, and D-AMPS) to the two planned 3G systems (UMTS and CDMA2000). Following this, each of the main systems is described in terms of their evolutionary stage (i.e., 2.5G systems, then 3G systems, then the systems beyond 3G). More detail about 3G systems will be provided in Volume 3 of *The Cable and Telecommunications Professionals' Reference* and will focus on broadband systems for both landline and mobile services.

Evolution from Second Generation to Third Generation

The intermediate steps in evolving from 2G to 3G for each of the main systems are shown in Figure 13.1. The preferred solution for GSM legacy networks (such as exist in Western Europe) is to move first to GPRS and then to UMTS. However, a lower-cost alternative to GPRS called High Speed Circuit Switched Data (HSCSD) has also been adopted in some countries. Whereas GPRS uses a packet-switched core network, HSCSD continues to use a circuit-switched core. A further system known as Enhanced Data Rate for GSM Evolution (EDGE) provides an alternative to UMTS, which continues to use the same bands as GSM. The generally preferred evolution for D-AMPS systems is also to follow the GPRS route,

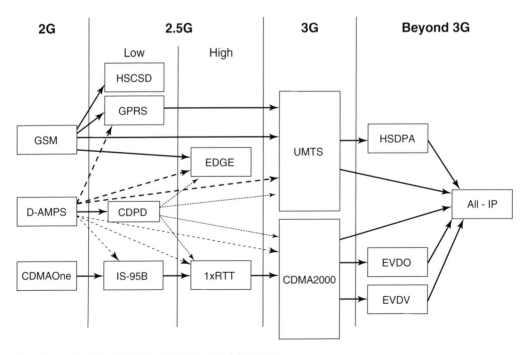

Fig. 13.1 Possible evolution routes beyond 2G systems.

although an alternative overlay system known as Cellular Digital Packet Radio is also possible. However, the overlay approach does not offer an upgrade route to either the EDGE or 3G systems.

The preferred evolution routes for markets that adopt CDMA legacy networks (as in the United States) is to progress to the CDMA2000 system. The main intermediate stage is known as 1xRTT, which offers a radio bandwidth of 1.25 MHz.

Two-and-a-half-generation Systems

General Packet Radio Service

The General Packet Radio Service (GPRS) is a packet-bearer service that uses packet-mode techniques to transfer both high-speed and low-speed data and signaling. It overlays GSM and is considered as an evolutional phase towards the third-generation Universal Mobile Telecommunications System (UMTS). It is the most widespread of the 2.5G systems.

GPRS enables new entities and features to be added to and integrated with an existing GSM network. It does not provide the full capacity and capability of UMTS but, nevertheless, enables advanced data services to be tried out both technically and within the market. GPRS is designed to support higher-quality service levels than GSM to allow efficient transfer of nonreal-time as well as real-time traffic (e.g., voice and video).

The GPRS infrastructure is logically divided into core network, access network, and mobile subscriber domains. The core network domain includes a circuit-switching domain, a packet-switching domain, and an IP multimedia subsystem domain. Whereas in GSM only the circuit-switching domain exists, in GPRS all three domains exist.

GPRS maintains strict separation between the radio subsystem and the network subsystem. This allows the network subsystem to be reused with other radio access technologies and enables independent optimization of network and radio resources. The radio frequency is similar to that used for GSM although the General Packet Radio Service air interface is different.

The air interface is the interface between the access network and the mobile subscriber. It deals with the physical layer associated with the radio-frequency link between the cell phone or other device and the base station, and it allows the frequency band to be used to carry voice and data signals of different bit rates and modulation formats. It also allows the access network to connect with a variety of subscriber devices. The interface defines the frequencies and bandwidth of the radio channels, the encoding and modulation methods used, and the functions associated with the physical layer in the mobile station and in the base station.

GPRS frequency bands are within a range of 380 to 1900 MHz. The widths of the individual bands vary from 5.6 MHz to 75 MHz. The frequency plan is defined in pairs of bands, one being for base station to mobile and the other from mobile to base station.

The modulation scheme may be either Gaussian MSK (GMSK) or 8-PSK, depending on the type of channel. The modulation rate is approximately 270.83 ksymbol/s. Details of the modulation and encoding scheme are specified in 3GPP specifications TS 45.004.

As in GSM, the Time Division Multiple Access (TDMA) method is used. This includes eight basic physical channels per carrier where carriers are separated by 200 kHz. A physical channel is therefore defined as a sequence of TDMA frames, number of time slots, and a frequency-hopping sequence.

Many different types of channel are defined for carriage over the air interface. Traffic channels are designed to carry encoded voice or data, with data rates varying from 2.4 kb/s to 43.2 kb/s. They may be defined as full rate or half rate (to allow more users to be supported), adaptive, cell, or packet.

The Special Mobile Group (SMG) Technical Committee (TC) of the European Telecommunication Standards Institute (ETSI) specified all generations of the GSM standard and the initial definitions for 2.5G. In July 2000, the work was continued as part of the 3G Partnership Project (3GPP). A new technical specification group was formed, called GERAN (GSM/EDGE Radio Access Network), with the responsibility to continue the GSM and EDGE work.

High-speed, Circuit-switched Data

The High-speed, Circuit-switched Data (HSCSD) system is an enhancement to the GSM system that has been defined by ETSI. It has not been widely adopted and today there are only a few networks. This standard is based on circuit-switched data transmission with a rate of 14.4 kilobits per second (kbps). It uses four time slots and allows theoretical speeds of up to 57.6 kbps (similar to one ISDN B-Channel).

HSCSD is cheaper to deploy (from GSM) than GPRS because it is based on circuit switching and is well suited for connection to circuit-switching systems (like the PSTN). As such, it was well adapted to speech. However, HSCSD cannot support advanced services in the same way that GPRS and EDGE can.

The standard allows up to four channels to be combined: on 900 MHz systems the channels are limited to 9.6 kb/s, which gives a system capacity of 38.4 kb/s. On 1800 MHz systems, the channels may be at the higher rate of 14.4 kb/s, yielding a 57.6 kbps data transfer. Because it uses a dedicated circuit-switched communications channel, HSCSD is able to provide a guaranteed quality of service, making it more attractive for real-time applications.

As explained in the introduction, HSCSD took the wrong marketing direction by focusing on voice and circuit switching, and by connecting with "old" systems rather than the new packet direction of the market.

Enhanced Data Rate for GSM Evolution

Enhanced Data Rate for GSM Evolution (EDGE) is a High 2.5G cellular system. It was designed as an enhancement to GSM networks, to provide additional capacity and advanced data services in situations where it was impractical to deploy UMTS. (See Table 10.5 in Chapter 10.) In principal, it would be possible to upgrade EDGE to UMTS later but in practice, once installed, the cost benefit of the upgrade may not be attractive.

The main advantage of EDGE is its high speed, which approaches that of UMTS, yet it uses the same frequencies as GSM, and does not require new radio bandwidth (which UMTS does). EDGE provides up to three times the data capacity of GPRS enabling three times more subscribers than GPRS or three times the data rate per subscriber. Further advantages over other 2.5G systems are that it includes advanced quality of service mechanisms and that a single common core network can support GSM, GPRS, and EDGE radio systems.

EDGE uses the same frequencies as the GSM network. It employs an 8-level Phase Shift Keying (8PSK) modulation scheme to allow speeds of up to 48 kb/s per time slot. When all eight GSM time slots are used, this gives a maximum bit rate of 384 kb/s per user transmission per time slot. This is three times more than the amount of information that is transferred by GPRS. In good radio propagation conditions, in theory, speeds of up to 69.2 kb/s per time slot may be available (or up to 553.6 kb/s) over all channel data transmission (384 kb/s user data transmission).

Cellular Digital Packet Data

Cellular Digital Packet Data (CDPD) uses spare bandwidth that was allocated to AMPS mobile phones developed in the early 1990s. It never quite gained widespread acceptance, perhaps because its consumer offerings were limited.

CDPD is an overlay IP-based packet data network deployed in the United States. It is based on TDMA principles and provides enhanced capability compared to D-AMPS. It supports data transfer rates of 19.2 kb/s and its operating frequency is in the range of 824 to 894 MHz. A principle limitation of CDPD is that it has no easy migration path to UMTS or EDGE systems, or to the CDMA family. CDPD has been used as an overlay on analogue systems such as AMPS as well as over D-AMPS.

TIA/EIA IS-95B: Interim Standard 95B

The first-generation CDMA standards were developed by the Telecommunications Industry Association/Electronic Industries Association. They were given the designation IS-95 and IS-95A and called "CDMAOne" by the CDMA Development Group. These form the basis of many of the CDMAOne systems around the world.

The IS-95B standard, which is regarded as a Low 2.5G system, extended the data capability of CDMAOne systems to a maximum of 115.2 kb/s, with 64 kb/s being commonly used. The system operates in the 800 MHz band and the 1.9 GHz PCS band and was designed to allow dual-band operation.

1xRTT Radio Transmission Technology

The 1xRTT Radio Transmission Technology system has been developed as part of the CDMA 2000 standards family and is regarded as a High 2.5G system. 1xRTT was designed to support advanced data services over an IS-95 network using the same channel bandwidth as CDMAOne (radio bandwidth of 1.25 MHz) but is able to carry more voice channels and support higher data rates. 1xRTT refers to single-carrier Radio Transmission.

The spectrum and modulation techniques are optimized for mixed traffic but voice is given priority. However, when a 1.25 MHz band is optimized for data only, higher data rates can be achieved. Initial releases of the 1xRTT system support up to 144 kb/s in theory, but in practice rates around half that are more typically achieved.

Third-generation Systems

3G systems provide a significant evolutionary step in terms of capacity, data speeds, and new service capabilities compared to 2G mobile networks. They are designed to provide a link between cellular wireless networks that have traditionally been designed to carry voice services and the Internet, which is able to provide a wide range of data and multimedia services. This new capability not only can allow new services to be offered over mobile networks, but also improves customer support or user experience. This allows operators to develop their brand of product and drive new revenue opportunities. A further aim is to allow operators to reduce their capital and operational expenditures.

Efficient operation of data services requires a broadband connection to users that is significantly faster than the data speeds available from 2G and 2.5G networks. This calls for core networks that are based on packet switching and air interfaces that can provide enhanced bandwidth to support the faster services. Because real-time services (e.g., voice and video) must be carried over the same network as data services, quality of service control mechanisms are needed.

Also, because 3G systems aim to allow services such as TV and video on demand, high-speed data, and Internet access to be carried, it has been necessary to develop a new range of mobile devices. This includes mobile phones and personal digital assistants with larger screens and wireless cards for laptops and other devices.

The two main 3G standards that have been developed are the Universal Telephone Mobile Service (UMTS) and CDMA2000. A number of bodies work together to develop the standards and specifications. The ITU is developing a framework wireless access standard known as IMT-2000 designed to enable various terrestrial and satellite systems to be linked together, and it is working toward agreements on spectrum allocation as well as technical specifications for radio and network components. The Third Generation Partnership Project (3GPP), whose key members include standards bodies from China, Europe, Japan, Korea, and the United States, oversees the development of the UMTS standards. A sister project, known as 3GPP2 focuses on the CDMA family of standards. A key input to the considerations of 3GPP is made by various market groups, including the GSM Association, which provides a view on the market requirements from the operator community, the CDMA Development Group, and the IPv6 Forum.

The UMTS standards development is overseen by 3GPP and is an open standard while the CDMA2000 standard is overseen by 3GPP2. The two initiatives are described below. Further information and detailed specifications for 3G systems can be found via the Web sites of both the ITU and 3GPP.

Universal Mobile Telephone Service

Univseral Mobile Telephone Service (UMTS), an evolutionary step beyond GSM, GPRS, and EDGE, employs techniques giving backward compatibility to these systems. The air interface standard UTRAN (UMTS Terrestrial Radio Access Network) conforms to the spectrum agreed to through the ITU IMT-2000 framework standards, adopting a 5 MHz channel carrier bandwidth (which is much higher than the earlier 2G bandwidths). Spectrum allocations may be in either paired or unpaired bands. The wider available bandwidth allows operators to hold their costs down while offering much-increased bit rates. The paired bandwidth arrangements mean that symmetrical data rates are available, which are well suited to new applications such as real-time video, unlike most landline systems.

The end-to-end specification of UMTS inherently provides for international roaming, and many of the back-office systems designed for 2G and 2.5G can continue to be used by operators as they migrate from 2G to 3G.

Code Division Multiple Access 2000

CDMA2000 is a family of standards that have evolved from the earlier CDMAOne systems and conform to the IMT2000 framework standards developed by the ITU. The development of the CDMA2000 standards is overseen by the 3GPP2 partnership. Early CDMA2000 systems are designed to support mobile data communications at speeds ranging from 144 Kbps to 2 Mbps. The 1xRTT standard (generally viewed as a 2.5G standard as it operates in the same frequency band as the earlier IS-95 systems) was an early version within this family.

Three-and-a-half-generation Systems

High Speed Downlink and Uplink Packet Access

High Speed Downlink Packet Access (HSDPA) and High Speed Uplink Packet Access (HSUPA) are evolutions of the UMTS third-generation wireless interface standard. These technologies are leading to wireless access systems providing data speeds that are comparable to wired broadband.

HSDPA improves the spectral efficiency of UMTS networks to allow a doubling of capacity. It also delivers faster data rates, typically of the order of 1 Mb/s (with theoretical peak rates of up to 14.4 Mb/s) and reduces latency in the downlink. Enhancements to the uplink packet access scheme will follow to allow uplink speeds up to 5.8 Mb/s as the standards work develops.

HSDPA and HSUPA enable individuals to send and receive e-mail with large file attachments, play real-time interactive games, receive and send high-resolution pictures and video, download video and music content, or stay wirelessly connected to their office PCs—all from the same mobile device. Initial services have been aimed at laptop users using an HSDPA PC card with HSDPA-compatible handsets following.

Evolution Data Only and Evolution Data and Voice

Evolution Data Only (EVDO) is a development beyond the CDMA 1xRTT system that is optimized to carry data rather than voice or data. It uses the same 1.25 MHz channel frequency bands as the IS-95 and 1xRTT systems that support multiple voice channels and medium-speed data services. Transmission rates up to 2.4 Mb/s are provided, with average throughputs of 300 to 600 kb/s. The Evolution Data and Voice (EVDV) system is a further enhancement of 1xRTT that supports both voice and data services, with data rates up to approximately 2.7 Mb/s.

Fourth-generation Systems

The global vision for 4G systems is still being developed and a range of views are currently being debated to determine what should be included. From a functional point of view, the next generation of network seems likely to provide a new air interface that supports higher data rates, perhaps up to 100 Mb/s, to be all IP-based, and to support worldwide roaming capability. However, at this stage these are targets.

Other topics that are being considered include:

- The network should be a "pervasive" and cater to the convergence of technologies such as 3G cellular and wireless LANs (WLANs).
- The use of cognitive radio: This is a means of monitoring the radio path and adjusting spectrum usage and transmitter power, or modifying the network to get the maximum efficiency from a variable-quality radio medium.
- The next generation should be packet switching only, with low-latency transmission.

The 4G system, in principle, will allow high-quality video transmission and high-quality audio to be carried. It is also expected to deliver more advanced versions of the improvements embodied within 3G, such as enhanced multimedia, smooth streaming video, universal access, and portability across all types of devices.

Transmission methods under discussion include the use of Orthogonal Frequency Division Multiplexing (OFDM) and Orthogonal Frequency Division Multiple Access (OFDMA) to better allocate network resources to multiple users.

References

Detailed information about GPRS and EDGE can be found in the 3GPP Technical Specifications Series 41–55 under the name "GSM only (Rel-4 and later)" (see *http://www.3gpp.org/*). Detailed information about UMTS with GPRS and EDGE can be found in the 3GPP Technical Specifications Series 21–35 under the name "3G/GSM R99 and later."

CDMA Development Group: *http://www.cdg.org/*.
GSM Association: *http://www.gsmworld.com/index.shtml*.
IMT-2000: *http://www.itu.int/home/index.html*.
IPv6 Forum: *http://www.ipv6forum.org/*.
ITU: *http://www.itu.int/home/index.html*.
3GPP: *http://www.3gpp.org/*.
3GPP2: *http://www.3gpp2.org/*.

14 Bluetooth

Ian J. Dilworth

Introduction

In the years near 950 AD, a Viking king named Harold Bluetooth (*Harald Blåtand* in Danish) united Denmark and Norway. The initial study on wireless connectivity of electronic devices was conducted by Ericsson Mobile communications in Sweden, and so it was natural for them to adopt Bluetooth as the name to "unite" wireless connectivity between devices within a "personal operating space," defined, at first, by a 10-meter circle. Bluetooth is aimed at providing wireless connectivity between personal electronic devices from 1m to less than 100 m distances. Since Bluetooth is a radio system, the distance criteria depend on the transmitted power, the receiver sensitivity, propagation path characteristics, and what is in proximity to the antennas used. (Also see this chapter's last section, Power Consumption.)

Bluetooth technology provides such wirelessly connected devices the possibility of creating an ad hoc network, often referred to as a Wireless Personal Area Network (WPAN). In 1998, the Bluetooth Special Interest Group (SIG) was formed. Since then Ericsson, 3Com, IBM, Intel, Lucent, Microsoft, Motorola, Nokia, Toshiba, and thousands of others have worked together for the purpose of developing a global wireless network standard that could be used worldwide. Bluetooth standard 1 (IEEE 802.15.1) and then standard 2 (often referred to as enhanced data rate or EDR) are the resulting compatible technologies (Bing, 2002, and Nathan, 2001).

From a commercial standpoint, Bluetooth has not been adopted as rapidly as predicted by the original business plans and, perhaps most significantly, the price of a Bluetooth transceiver has not yet met the business plan design target of $5 (TX/RX). Nevertheless, at the time of this writing, equity prices of many Bluetooth IC producers have risen many fold in the space of just a year. This has largely been a result of the adoption of Bluetooth in mass-produced cellular telephones. The primary application of Bluetooth 1 (721 kbps data rate) so far, the wireless connection of a duplex earpiece/microphone to a mobile telephone, can transmit up to 721 Kbps in one direction and 57.6 Kbps in the other. If the user calls for the

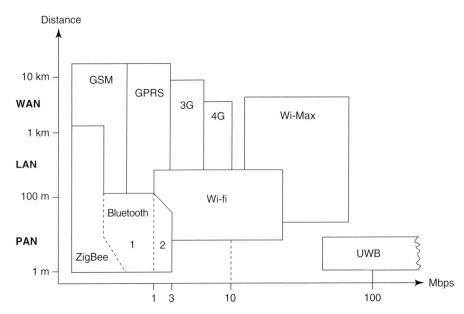

Fig. 14.1   Relative approximate data rates versus distance covered for various modern digital wireless communications systems.

same speed in both directions, Bluetooth 1 can establish a link with 432.6 kbps capacity in each direction. However, it is likely that Bluetooth-connected computer peripherals (mouse, keyboard, stereo audio, wireless printer, mobile phone, PDA, and so on) may all be required to operate simultaneously and this could embarrass a Bluetooth 1 system. Bluetooth 2 (EDR) provides a 3 Mbps data rate, which has adequate capacity for such applications in a standard 7 user Bluetooth piconet.

Future developments of Bluetooth are planned. The next Bluetooth enhancement is code-named "Lisbon." Its primary aims are increased security and usability. Following this is project Seattle, primarily aimed at increasing the available data rate beyond that of Bluetooth 2. For example, one identified application is streamed video, which requires more bandwidth. The SIG has decided that this development depends on the adoption of UWB, which is currently a somewhat contentious technology because of its spectrum requirements. For those interested, Lisbon and Seattle are keywords with which to search and monitor developments. In March 2006, the SIG announced its intent to work with UWB manufacturers to develop a next-generation Bluetooth technology providing UWB speeds. Figure 14.1 illustrates relative data rates and ranges of Bluetooth 1 and 2 compared to existing technologies such as GSM, GPRS, 3G, Wi-Fi, Wi-Max, and the proposed UWB.

Purpose and Role

The main aims of Bluetooth are to provide a universal radio interface for ad hoc wireless connectivity at low cost, delivering modest speed (1–3 Mbps), short range (1–100 m), and low power consumption. Infrared (IR) interconnections between devices operating at modest data rates have been available for many years and indeed these are incorporated into many portable devices. But the properties of IR radiators and receivers are not as unidirectional or as tolerant of obstacles as the worldwide radio frequencies used for Bluetooth. A universal wireless interface tolerant of obstruction and not requiring physical alignment was sought.

In addition, IR devices are not suited to connecting many devices simultaneously. The Bluetooth SIG expanded the original concept to become a standard under IEEE 802.15 (WPAN). The SIG published version 1.0 of the Bluetooth specification in July 1999, version 1.1 in November 2001, and a higher data rate version 2.1 in November 2004. It was only in 2006 that Bluetooth 2.1 products were appearing in volume. Thousands of manufacturing companies have become involved in the official Bluetooth certification program, the main aim of which is to guarantee interoperability between equipment from different vendors. Version 2 Bluetooth (2005), which can operate at 3 MBps, is designed to be backwardly compatible with version 1.

The ubiquity of lightweight portable computers laptops, notebooks, PDAs, mobile telephones, peripherals such as printers, and so on, continues to spur the need and convenience offered by wireless networks. In general such networks can be divided into Wireless Local Area Networks (WLAN) and Wireless Personal Area Networks (WPAN). Important system parameters include the coverage area, the data transmission rate, and the power consumption. Major worldwide standards of WLAN include IEEE 802.11 and the European High Performance Radio LAN (HIPERLAN, introduced in 1991 and standard-approved in 1996). Although the performance and technical features incorporated into HIPERLAN are in many ways superior to some IEEE standard offerings, it is the rapid introduction and adoption of the IEEE standards and their market presence in terms of hardware that appear to have won the worldwide market. Bluetooth has become the WPAN worldwide standard. ZigBee is also a WPAN system and standard that is now emerging from various false starts. It is also a technology that has advantages over Bluetooth for some applications, especially those that require low data rate and long range. ZigBee is briefly described at the end of the chapter.

Although the Medium Access Control (MAC) layers of IEEE 802.11 WLAN and Bluetooth WPAN are implemented in separate ways, both of them provide services for voice (synchronous) and data (asynchronous) transmission together with symmetric and asymmetric data on the uplinks and downlinks.

The IEEE 802.11 WLAN standard has existed since 1990 and is used as a universally accepted broadband LAN. It covers just the physical layer (PHY) and medium-access control layer and fits seamlessly into the other 802.x wired LANs, such as IEEE 802.3 (Ethernet) and IEEE 802.5 (token ring). Unlike IEEE 802.11, the initial purpose of developing Bluetooth technology was to replace a wire cable with RF wireless links. Bluetooth's flexible ad hoc architecture, low cost, and low-power features match many present applications for short-range 1–3 Mbps communications.

Bluetooth Physical Layer

The Spectrum Employed Worldwide

The electromagnetic spectrum employed by Bluetooth uses a worldwide unlicensed radio frequency band 83.5 MHz wide in the United States and most of Europe, starting at 2400 to 2483.5 MHz. Current exceptions are that only 37 MHz is available in France (2446.5–2483.5 MHz) and only 30 MHz from 2445 to 2475 MHz in Spain. In Japan, the band is currently restricted to 26 MHz (2471–2497 MHz). These frequency allocations are classified as Industrial Scientific and Medical (ISM) bands.

Frequency Hopping

The physical connection of Bluetooth uses frequency hopping at 1 MHz-spaced hops and the actual carrier is Spread Spectrum (SS) modulated. This is known as Frequency Hopped Spread Spectrum (FHSS) and is used to improve immunity from interference—mainly from IEEE802. Wi-Fi WLAN shares the same spectrum as other users of the ISM band—for example, transitorily operated microwave ovens.

There is a fundamentally sound reason for this approach. WLANs occupy significantly more bandwidth than 1 MHz and employ SS techniques, in which a pseudo-random code (which is noise-like in its spectrum) is used to modulate the information transmitted. At the receiver the same pseudo-random code has to be available to demodulate it successfully—called correlation. As a result, many WLANs using different pseudo-random codes can transmit simultaneously without mutual interference. A Bluetooth frequency-hopped system appears as an uncorrelated noise-like signal to a WLAN receiver and, what is more, its signal energy is spread over the entire band of the WLAN by the decorrelated demodulation process. So, in practice, mutual interference amounts only to a raised noise floor for both systems. A raised noise floor simply reduces the available Carrier to Noise (C/N) and hence reduces the range possible. This is the main effect of mutual interference. An exception can occur when signals are so strong, because of the close proximity of the antennas, that they overwhelm a receiver.

Bluetooth's frequency hopping rate is chosen to be 1600 hops per second with a 625 μs dwell time per hop. The minimum dwell time of 625 μs corresponds to a single time slot. Part of this hop timing is taken up by the guard time of 220 μs, allowing the hardware-based frequency synthesizer time to settle. The transmission channel is derived from the time slots, spread over 79 or 23 frequencies. The frequency hopping scheme uses time division multiplexing (TDMA), as illustrated in Figure 14.2, where the frequency step $n = 1$ MHz.

Radio-propagation Considerations

The most important parts of any radio link are the receive-and-transmit antennas, together with their location. After all, the most excellent radio transceiver will be absolutely useless without an effective antenna. Bluetooth has a limited Effective Isotropic Radiated Power (EIRP). This means we cannot just increase our transmitter power to overcome interference

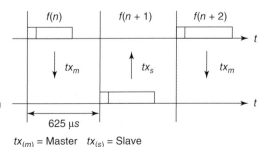

Fig. 14.2 Frequency hopping from packet to packet in a TDM scheme for a master and a single slave ($n = 1$ MHz).

$tx_{(m)}$ = Master $tx_{(s)}$ = Slave

or increase range. However, EIRP implies that we can use any combination of antenna gain and transmit power so long as we do not exceed the allowed EIRP; that is, EIRP = transmit power (dBm) + antenna gain (dB). For the majority of Bluetooth applications, the antenna gain will, for practical reasons, be 0 dBi or probably less because of interactions with its surroundings. The path loss to the receiver will be approximated by the following systems equation (all quantities in dB power):

$$[Prx] = [Ptx] - [\text{path loss}] + [\text{gain TX antenna}] + [\text{gain RX antenna}] - [\text{medium losses}]$$

If we connect high-gain antennas to Bluetooth, it is easily possible to exceed the maximum permitted radiated power. Ranges of more than 1 km have been achieved using such methods. (See the section on Bluetooth Security later in this chapter.)

The path loss aspect is readily calculable. The imponderable from the Bluetooth systems designers' viewpoint is the [medium losses] quantity to employ.

Medium Losses and Radio Propagation

A frequency of 2.4 GHz corresponds to a wavelength $\lambda \sim 12.5$ cm. Such wavelengths interact strongly with object sizes of $\lambda/2$ or greater and it is possible to fabricate efficient antennas just $\lambda_o/4$ long or $\lambda_o/4(\varepsilon_r)^{0.5}$ if a dielectric of $\varepsilon_r > 1$ is employed (λ_o is the free space wavelength) (Fujimoto et al., 1987). The human body and especially the human head absorb electromagnetic radiation throughout the electromagnetic spectrum. The whole human body acts as a lossy, partly conductive, saline-drenched mass. At Bluetooth frequencies the human head, for example, has an average dielectric constant of ~47. That corresponds to a wavelength for Bluetooth in the Brain of $\lambda_o/(\varepsilon_r)^{0.5} \sim 1.8$ cm. So it is not surprising that having traveled several equivalent wavelengths through the head the frequencies used by Bluetooth can introduce up to ~20 dB of absorption through electromagnetic interaction mainly with head fluids. The torso and limbs also interact at Bluetooth frequencies to an extent that depends on the direction of the path between the transceivers. This is, of course, particularly pertinent in the common application of head- (ear-) mounted Bluetooth transceivers or, indeed, where human beings are in close proximity with the application's antenna. Close proximity in this context means in practice within, say, 2 wavelengths. In this range the human body can introduce at least +/−6 dB of influence on signal levels and more as the distance from the person to the antenna decreases.

The wavelength of Bluetooth is 12.5 cm. This wavelength is significantly scattered by objects of a similar electrical length, especially metal objects. Hence propagation is not restricted to "line of sight." Nevertheless, as a design rule of thumb, assuming propagation is from an earpiece-mounted Bluetooth to a belt-held Cellular transceiver, to assume ~15 dB extra loss above free space loss in systems calculations.

Free space loss (FSL) is just the inverse square law $= 20\log_{10}(\lambda/4\pi d)$ dB, where the distance d = meters and $\lambda \sim 0.125$ m, in the case of Bluetooth.

In practice, losses are always more than predicted from the above free space formula. For example, a Bluetooth transceiver mounted in a PCMCIA or USB on a laptop to another laptop will, as a rule of thumb, suffer +10 dB extra loss over free space compared to the above free space loss calculation. Scattering, reflections, and human electromagnetic interactions make predicting coverage an inexact science even when detailed electromagnetic modeling is performed for a particular physical and electrical arrangement. In practice that means +/−6 dB is about the level of confidence one can assume for any modeled scenario however detailed it may be. More significant, such detailed electromagnetic modeling reveals that standing waves can occur, which result in >30 dB variation in signal level within one-half wavelength of movement (that is, just ~6 cm for 2.4 GHz Bluetooth) in all three dimensions. This is when signals vectorially add to produce deep troughs, which can result in complete signal loss and which periodically repeats at half-wavelength intervals in three-dimensional space.

Model and System Design for Typical Bluetooth Propagation

In an indoor environment, a simplified propagation model obtained by empirical results can be used as a starting point (Gibson, 1996). According to many of the available published models, the path loss is proportional to the squared distance from the antenna at very short ranges (<8 m). Due to reflection effects, the path loss increases as a function of $(distance)^n$ at a distance of a few meters from the antenna (>8 m), where n is a factor best found by measurement. In free space $n = 2$. However n values of 2 to 6 can be experienced in practice. The following path loss (L_p in dB) is only a good starting point in system design where d refers to the distance in meters between the TX and RX ("transmitter and receiver") or vice versa:

$$L_p = \begin{cases} -40 - 20\log(d) & \text{when } d \le 8\,\text{meters} \\ -58.5 - 33\log(d/8) & \text{when } d > 8\,\text{meters} \end{cases}$$

Using these empirical models, the loss profile can be plotted versus distance, as shown in Figure 14.3.

This empirical model predicts accurately only when human beings and other objects are not directly in the path. It represents a reliable starting point for system design based on empirical measured data. Unfortunately, the human torso and especially the human head

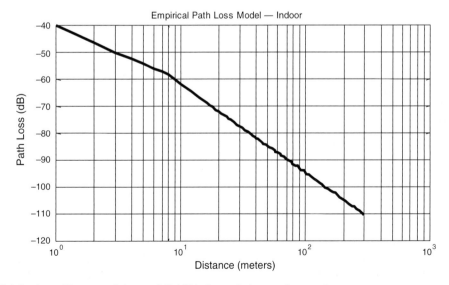

Fig. 14.3   Loss dB versus distance at 2.4 GHz for an indoor environment.

can introduce up to −20 dB extra path loss, depending on the propagation direction (this also applies to mobile telephones at 900 and 1800 MHz). The ~40 dB power control used in GSM mobile radio systems masks such effects to a large extent. Bluetooth's low-radiated power is surely not a health hazard, and the power levels even of class 3, 100 mW, are much less than those of current mobile telephones. Human body absorption is nevertheless a very relevant parameter in most system designs involving Bluetooth applications. Consequently designers and manufacturers seek to maximize their Bluetooth antenna efficiency even in the presence of lossy objects such as human beings, because maximizing efficiency results in increased signal strength and, hence, system robustness. There is no simple solution to this problem; however, antenna diversity can certainly help.

Antennas and Diversity

Antenna diversity proves to be very effective in systems such as Bluetooth, Wi-Fi, and cellular radio, which suffer multipath propagation. Paradoxically cellular radio actually relies on multipath propagation in order to work! Diversity antennas provide major benefits. Systems that use some form of diversity can typically provide a 6 dB to 12 dB or greater diversity gain (the improvement of SNR) for the worst 1 percent of cases (Gibson, 1996). This has the added benefit for many applications that battery life will increase due to the increased efficiency of the radio link. In general, the disadvantage of diversity is that the antenna system occupies significantly more 2-D and 3-D space.

The outputs of diversity antennas can be selected or combined to maximize the received signal-to-noise ratio. Diversity-combining methods include Maximal Ratio Combining (MRC), Equal Gain Combining (EGC), and Interference Rejection Combining (IRC), which are mainly applicable in Wi-Fi systems. There are just two methods that are practicably applicable to Bluetooth. These are space and polarization. Of these, the latter is by far the most practical and effective in the small physical volume required of a Bluetooth antenna (Boley, 2004). The available diversity dimensions are space, time, and polarization.

Spatial Diversity

Spatial diversity uses two or more antennas, which are separated in space. It can be readily implemented in various ways, the simplest of which is at least two antennas spaced by more than a half-wavelength in any (x, y, z) 3-D plane. Spatial diversity can be employed to effectively and significantly combat multipath fading for both indoor and outdoor wireless communication. Reflection, refraction, and scattering all occur, and the received signal arriving at each antenna produces signals at the antenna outputs, which differ in amplitude and phase. The receiver makes use of the diversity branches and generates an output through a selection or combining process, or simply by switching to the strongest signal. Spatial diversity, combined with diversity-combining techniques, can also be used in directional interference rejection. A dual antenna is practically all that is possible with Wi-Fi and is probably not feasible at all with most Bluetooth applications due to its desired small size. A future exception might be in body-worn applications where Bluetooth transceivers and antennas will be sewn into clothing for various applications.

Polarization Diversity

Unlike spatial diversity, polarization diversity is based on the fact that signals at different polarization states encounter statistically independent multipath reflections. Multipath propagation inevitably results in polarization changes to the transmitted signal. Thus the performance of the receiver can be significantly improved by using receive antennas (even a single antenna), which respond to orthogonal polarizations.

A useful dual-polarization antenna design is the so-called "inverted F." The length of the antenna is approximately $\lambda/4 = 3$ cm in air. If a low-loss dielectric is used, the length is reduced by $\lambda_o/(\varepsilon_r^{\wedge 0.5})$. However, the antenna's bandwidth is then reduced. In the future, Bluetooth products may well start appearing sewn into clothing. Fabric-mounted antennas in the form of an inverted F can offer good performance when incorporated in clothing for Bluetooth and also at cellular telephone frequencies (Massey, 2005). Do not be surprised if a body-worn WLAN and Bluetooth appear sewn into your next jacket!

Diversity-combining Techniques

Diversity-combining techniques can play a significant role in improving the performance of a link. In general, two diversity-combining algorithms, so-called MRC and IRC, have been used to obtain improved performance through canceling interference. However, these only find practical application in Wi-Fi networks and in general they are not useful for Bluetooth class 1 and 2 but are possibly for class 3, 100 mW.

Hop Selection Scheme

The Bluetooth specification defines a small cell, a so-called *piconet*, which forms a star connectivity with a master transceiver and up to 7 slave transceivers, all of which may be connected to the master transceiver. It is limited to 7 slaves because only 3 bits are used for the active member addressing (AMA). However, parked member address (PMA) occupies 8 bits, and so up to 255 of these can exist. Figure 14.4 illustrates the basic master–slave star arrangement. (See also Figure 14.7.)

The Bluetooth specification assigns a specific frequency-hopping pattern for each piconet. The pseudo-random hopping pattern is determined by the piconet identity and the master transceiver. The native clock of the master also defines the phase of the hopping sequence. All other Bluetooth devices within the piconet are called slaves, and these use the master identity to select the same hopping sequence and so synchronize with the master-imposed, frequency-hopping regime.

In a system employing all of the 79 (1 MHz) frequency channels that are available, these are arranged in odd and even classes. The overall hopping pattern is divided into 32-hop segments. Each 32-hop sequence starts at a point in the spectrum and hops over the pattern that covers 64 MHz because it hops either on odd or even frequencies. The selection mechanism is shown in Figure 14.5. In the first block, the identity selects a 32-hop pseudo-random

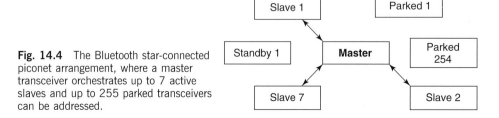

Fig. 14.4 The Bluetooth star-connected piconet arrangement, where a master transceiver orchestrates up to 7 active slaves and up to 255 parked transceivers can be addressed.

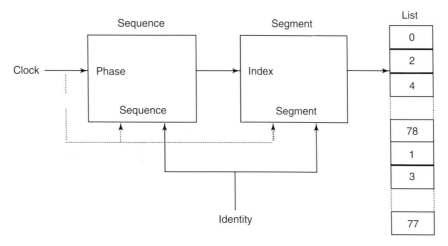

Fig. 14.5 Piconet hop selection arrangement and mechanism where the sequence of frequency hopping and channels is arranged.

subsequence that is indexed by a number. When the clock runs, the second block produces the hopping sequence and uses the corresponding carriers according to the index. After completion of each segment, the sequence is altered and the segment is shifted 16 frequencies in the forward direction. In this way, segments "slide" through the carrier list to maintain the average time each frequency is used at an equal probability (Lansford et al., 2001). Note that a change of identity or clock will change the sequence and segment mapping.

Every Bluetooth device has a unique address that falls into a range of addresses it has established for a particular device type and application. The SIG-approved device applications or profiles are expanding all the time. When a device is first turned on, it polls asking for a response from any units with an address range. This range also depends on the unit's communication range, which is dependent on the proximity of objects and people as well as the power level (i.e., radio propagation). Assuming another Bluetooth presence is detected, the result is a "network" of at least two devices and thus a piconet.

Having established a connection by means of the unique device addresses, then, even if one of these devices should receive a signal from another (star-based) piconet system, it will ignore it since it is not from within this established network. Once a network is established, the devices begin talking among themselves and operate in synchronism with the adopted and established base station transceiver. Each piconet hops randomly through the available frequencies, and individual piconets hop randomly independent of others.

To mitigate against mutual Bluetooth piconet interference (both transmitted and received), each network hops in frequency 1600 times a second. Hence it is unlikely that any two networks will be on the same frequency at the same time. If it turns out that they are, then the resulting error requires a retransmission. Acknowledge Receipt (ARQ) and Forward Error Correction (FEC) are both built into the protocol (see Figure 14.7 later).

Bluetooth uses packet-based transmission. Each packet occupies one basic time slot of $625\,\mu s$ that can be extended to three slots ($1875\,\mu s$) and/or five slots ($3125\,\mu s$). Each packet is sent in one frequency hop irrespective of how many slots it lasts. This means that Bluetooth never changes its frequency when sending multislot packets. Adopting this frame format and a frequency-hopped TDMA (FH/TDMA) access mechanism allows a master to poll multiple slaves at different data rates. This may be required, for example, if voice and data applications exist within a single piconet or when addressing Bluetooth 1 devices with Bluetooth 2.

Modulation Formats

The signal bandwidth of the RF channel is limited to 1 MHz in Bluetooth 1 and to 3 MHz in Bluetooth 2. For robustness, Bluetooth 1 uses two-level Gaussian-shaped Frequency Shift Keying (GFSK) modulation with a bandwidth time product of BT = 0.5 and a nominal modulation index of 0.28 to 0.35. This modulation scheme allows the implementation of very low-cost transceiver hardware. Bluetooth 2 can deliver up to 3 Mbps by transmitting exactly the same as version 1 with just the payload part of the packet being transmitted differently by using more complex modulation schemes, which produce more bits per symbol (BPS). The I–Q maps of the modulation states for GFSK, $\pi/4$ DQPSK, and 8 DPSK are illustrated in Figure 14.6.

Note that the amplitude and phase space between discrete modulation states gets smaller in the Bluetooth 2 modulation schemes, making uncertainties in receiver detection more sensitive to hardware linearity, thermal noise, and, of course, interference. This means that the noise margin is reduced compared to GFSK. The price to be paid for Bluetooth 2 higher data rates is essentially the requirement of a strong signal level. Hence, on average the available range, all else being equal, is somewhat shorter for Bluetooth 2.

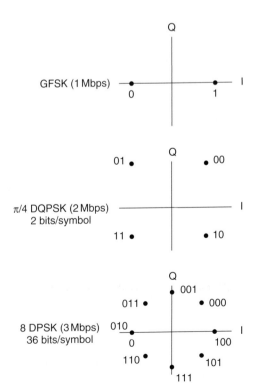

Fig. 14.6 I and Q state diagrams for Bluetooth 1 and 2 modulation schemes, yielding, from top to bottom, 1, 2, and 3 Mbps.

Table 14.1   Bluetooth Transmitter Power Classes

Power Class (~Range)	Maximum Output Power (P_{max})	Nominal Output Power	Minimum Output Power (P_{min})	Power Control
1 (~100 m)	100 mW (20 dBm)	N/A	1 mW (0 dBm)	P_{min} < +4 dBm to P_{max} Optional: P_{min} to P_{max}
2 (~10 m)	2.5 mW (4 dBm)	1 mW (0 dBm)	0.25 mW (−6 dBm)	Optional: P_{min} to P_{max}
3 (~1 m)	1 mW (0 dBm)	N/A	N/A	Optional: P_{min} to P_{max}

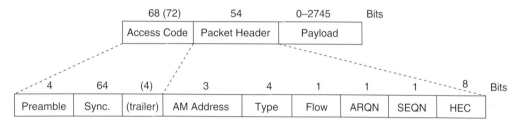

Fig. 14.7   The Bluetooth data transmission packet; the payload is optional.

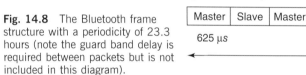

Fig. 14.8   The Bluetooth frame structure with a periodicity of 23.3 hours (note the guard band delay is required between packets but is not included in this diagram).

Transmitted Power Classes

There are three transmitter power classes in all Bluetooth devices, as shown in Table 14.1. The nominal value provides some indication of the expected performance, whereas the maximum value is that allowed in the test specification (Haarsten, 1998). When considering these power levels, it may be useful to remember that a doubling of distance corresponds to 6 dB more loss, assuming free space propagation. The receiver target sensitivity is <−70 dBm. For a 10^{-2} BER. Typical mass-produced hardware currently on the market has a noise floor of −84 dBm (CSR, UK).

The Bluetooth Packet

As we have seen, each packet is sent between master and slaves in a regular 625 μs-frame TDM basis. The packet itself is of fixed format as shown in Figure 14.7 the repetition period of the sequence is $2^{27} - 1 = 23.3$ hours as determined by the master's clock and as shown in Figure 14.8.

Note that the payload part of the packet can vary from nothing to 2745 bits (i.e., a payload does not have to be sent). Also note that the header of 54 bits contains the automatic retransmission ARQ, checksum, and so on.

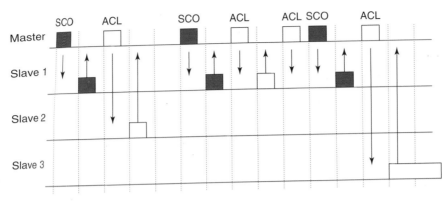

Fig. 14.9 Mixing of SCO and ACL links with one master and three slaves.

Types of Connection

Bluetooth physical links support asynchronous services (e.g., burst type data traffic) and synchronous services (e.g., voice traffic). The two services are defined in Bluetooth specifications:

1. An asynchronous connectionless (ACL) link
2. A synchronous connection-oriented (SCO) link

The SCO link is a point-to-point link between the master and a single slave supported by duplex slots at regular intervals. The ACL link is a point-to-multipoint link between the master and all the slaves in the piconet using the remaining time slots. Three types of SCO packets are defined, primarily for voice communication. Six of the packet formats are defined for the ACL link for asynchronous data communications. Just four types of packets are used for common control of both SCO and ACL links. Figure 14.9 illustrates how mixing the synchronous SCO link and asynchronous ACL link on a single piconet channel operation is arranged.

For SCO links voice is carried at 64 kbps without any error correction. The link is symmetric, circuit switched, and point to point. However, FEC can be invoked if the interval is decreased. An ACL link, on the other hand, supports both symmetrical and asymmetrical links. It is packet switched as opposed to circuit switched and point to multipoint. For ACL, 1-, 3-, and 5-slot data packets are possible. Table 14.2 summarizes the symmetric and asymmetric data rates possible for Bluetooth 1. Note that if no data is transmitted, frame slots do not have to be used.

Table 14.2 Symmetric and Asymmetric Data Rates Possible for Bluetooth 1

Number of slots	FEC Encoded/ unprotected	Symmetric kbps	Asymmetric kbps	Asymmetric kbps
1	FEC	108.8	108.8	108.8
1	Unprotected	172.8	172.8	172.8
3	FEC	256	384	54.4
3	Unprotected	384	576	86.4
5	FEC	286.7	477.8	36.3
5	Unprotected	432.6	721	57.6

Bluetooth 2

A weakness of Bluetooth 1.1/1.2 (which can produce data rates up to ~723 kbps, in one direction, in some applications) is that the data rate is not sufficiently high for sharing among demanding multiple applications. Given the demands on portable data communications and especially Internet-related applications it is likely the data rate will never be high enough (e.g., for streaming video). Nevertheless, in November 2004 a new version of Bluetooth, known as Bluetooth 2, was ratified. The maximum data rate can now reach 3 Mbps—this is the so-called EDR (enhanced data rate). In this scheme Bluetooth 2 transceivers only remain fully active for about a third of the time, which can result in significantly increased battery life despite the three-fold increase in speed, a major bonus. The Bluetooth 2 EDR standard is completely backward compatible and so allows networks to contain a mixture of EDR and standard Bluetooth 1 devices. To achieve the 3 Mbps data rate, the modulation schemes incorporated *just for the payload part of the packet* into Bluetooth 2 are more complex. Bluetooth 2 can run simultaneously with Bluetooth 1 because it is *only* the payload part of the packet that is different.

Bluetooth 1 and Bluetooth 2 Data Packets

Bluetooth data packets are made up of four elements:

1. *Access Code:* This is used by the receiving device to recognize the incoming transmission.

2. *Header:* This describes the packet type and its length.

3. *Payload:* This is the data that is required to be carried. (This is where Bluetooth 2 has a different modulation scheme and specification.)

4. *Inter-Packet Guard Band:* This ensures that transmissions from two sources do not collide and enables the receiver to hop in frequency in 1 MHz steps and settle. It is 220 μs long.

The elements are illustrated in Figure 14.10.

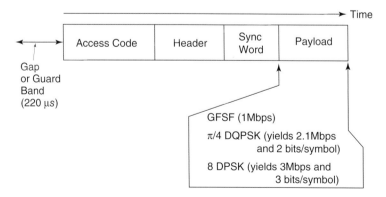

Fig. 14.10 Bluetooth packet structure: Bluetooth 2 packet differs from Bluetooth 1 only in the sync word bits and the payload modulation scheme.

In the original Bluetooth 1 standard, all the elements of the transmission are transmitted using Gaussian Frequency Shift Keying (GFSK) where the carrier is shifted by +/–160 kHz. By so doing, it indicates a one or a zero and in this way one bit is encoded per symbol. Bluetooth 2.0 uses a variety of forms of modulation. GFSK is still used for transmitting the access code and header and so compatibility is maintained. However, other, more complex forms of modulation can be used for the payload. To achieve this, there are two additional forms of modulation that have been introduced. One of these is mandatory, the other is optional. A further change is the addition of a small guard band between the header and the payload. In addition to this, a short synchronization word is inserted at the beginning of the payload.

The Bluetooth 2 specification defines ten new packet formats for use with the higher data rate modulation schemes and five for each of the EDR schemes. Three of these are for the 1-, 3-, and 5-slot asynchronous packets used for transferring data. The remaining two are used for 3- and 5-slot extended Synchronous Connection Oriented (eSCO) packets, which use bandwidth that is normally reserved for voice communications. The format for these packets does not incorporate FEC, so the QOS may be reduced. If the signal is relatively weak and so FEC is required, then the system switches back automatically to the slower standard data rate of Bluetooth 1. It is, of course, necessary for the packet type to be identified so that the receiver can decode data correctly, including the type of modulation employed. An identifier is therefore included in the header, which is sent using the normal (robust) Bluetooth 1 GFSK modulation. The packet header used for the original version of Bluetooth used 4 bits. This is insufficient for the additional information that needs to be sent in Bluetooth 2. Thus it was not possible to change the header format because backward compatibility would not then be possible. Instead, different link modes are defined. When two Bluetooth 2 devices communicate, the messages are used in a slightly different way to indicate that EDR modes are possible.

Overall, the advantages of Bluetooth 2 come at the expense of a requirement for a stronger received signal than that required for Bluetooth 1. Also, the hardware needs to be more linear in phase and amplitude, which has been a manufacturer's headache. In practice, all these requirements result in the range to be expected being slightly less than Bluetooth 1, assuming the antennas and the radio propagation paths are identical.

Mandatory Bluetooth 2 Modulation Format

The first of the modulation formats that must be included on any Bluetooth 2 device gives a twofold improvement in the data rate and thereby allows a maximum speed of 2 Mbps. This is achieved by using π/4 Differential Quaternary Phase Shift Keying (π/4 DQPSK)—a form of modulation significantly different from the GFSK that is used on previous Bluetooth standards. The Bluetooth 2 standard uses a form of phase modulation, whereas previous ones used only frequency modulation. Using quaternary phase shift modulation means that there are four possible phase positions for each symbol. Accordingly, this means that two bits can be encoded per symbol, and this provides the *twofold* data increase in data rate over the frequency shift keying used for the previous versions. See the IQ diagrams shown in Figure 14.6.

To enable a full threefold increase in data rate to be achieved, 8-phase differential phase shift keying (8 DPSK) is incorporated, as shown in Figure 14.6. This enables eight positions to be defined with 45 degrees between each of them. By using this form of modulation, eight positions are possible and three bits can be encoded per symbol. This enables the data rate of 3 Mbps to be achieved. As the separation between the different phase positions is much smaller than it was with the QPSK used to provide the twofold increase in speed, the noise immunity has been reduced in favor of the increased speed. So in practice, it can only be used when we have a sufficiently strong and reliable signal.

Networking

When in range of each other, Bluetooth transceivers can set up ad hoc networks. All the transceivers are identical; however, to establish order, one unit acts as a master and then the others must act as slaves. The master transceiver is the one that initiates the connection and this connection of two or more transceivers is known as a piconet. Masters can change roles and become slaves and vice versa. The master's clock is used as a reference. Offsets (+ or −) are added to the clocks of slaves in order to synchronize frames in which packets are exchanged on a TDM basis.

Piconets

In a piconet, the channel and the information are shared between users, and a single master transceiver controls all the traffic. For example, this transceiver allocates capacity in voice links (SCO circuit-switched mode) by reserving slots. In ACL mode (packet-based switching), a polling scheme is used that consists of an access code and header containing the slave's MAC address. It thus should not be possible to have a collision between multiple slaves in a single piconet. When Bluetooth transceivers are not members of a piconet, they enter a so-called standby mode in order to preserve battery power.

Standby Mode

Waking up a Bluetooth transceiver in standby mode requires knowledge of the unit's identity in order to obtain this information from units within range of the paging transceiver. An enquiry access code (EAC) is common to all Bluetooth devices and this is transmitted on all wake-up frequencies. When a unit responds, it does so with its identity and clock phase. Then the paging transceiver can select this particular slave to wake. In standby mode, a transceiver periodically listens for paging by its access code of 72 bits. This is sent by the paging transceiver for 10 mS on different hop frequencies repeatedly. Each transceiver is preprogrammed by its identity with a subset of 32 or 16 or 79 or 23 hopping frequencies on which to listen. The listening interval is 11.25 mS corresponding to 18 frequency slots. In this period, the transceiver listens on one frequency and if it correlates its access code, it wakes up and starts the connection procedure. If not, the standby transceiver moves its listening frequency 1 MHz and the periodic listening process continues.

Scatternets

A scatternet consists of a collection of two or more piconets. The individual piconets within the scatternet do not have to share information. In fact, it is assumed they do not. They do share the available frequencies within the allocated spectrum. It is possible for Bluetooth transceivers to set up peer-to-peer connections independent of a piconet. However, for those wishing to exchange information more widely, a piconet is required where they all share the same 1 MHz spread spectrum channel. In this case, the throughput per user can drastically reduce to kbps as more units join a piconet.

Because piconets are independently hopped in frequency in a pseudo-random way, they avoid mutual interference. So the up to seven users of a single piconet do not have to share their 1MHz allocation. The throughput of piconets connected in a scatternet only slowly reduces with the number of piconets. For example, a scatternet consisting of 10 piconets reduces the throughput in a member piconet by ~10 percent, assuming 300 users. If each belongs to one network and all share the same 3 MHz Bluetooth 2 channel, then the average throughput per user falls to 10 kb/s. Splitting these 300 users into smaller groups, assuming they do not all wish to communicate with each other, allows the throughput per user to increase. For example, with 8 users per 3 MHz channel (master and 7 slaves), each user gets an average 375 kb/s throughput. Of course, given the "bursty" nature of data, individual transceivers will likely experience much more than the average throughput. In reality, collisions can occur even between statistically independent piconet hops, which will reduce throughput. However, in practice, scatternet throughput exceeds that occurring in a single piconet.

Communications between piconets involves slave-slave or master–slave transceiver over-the-air connectivity and this requires synchronism with the channel of the desired piconet. This can only be achieved by knowing the identity and the clock offsets required for each piconet to be connected. However, if the hopping channel is temporarily shared between piconets, then packets can be identified by the unique piconet access codes alone. In fact, a Bluetooth unit may sequentially switch between adjacent piconets on different frames. What is not allowed is synchronization of two or more piconets because the EIRP generated could result in increased interference levels.

Dynamic Scatternet Reconfigurability

It is assumed that slaves and masters will physically be moveable and/or that the propagation conditions will change (e.g., by a person moving) and so a scatternet may need to reconfigure dynamically.

When a Bluetooth transceiver enters a piconet, its frequency offset (+ or −) is adjusted to take account of the local hardware frequency (cheap hardware has relatively poor frequency stability at 2400 MHz). Packet by packet, a slave transceiver can thus operate in several piconets. On leaving a piconet, the relevant master is informed of the break. Similarly a master transceiver can also move to another piconet and act as a slave in the same manner. In this case though, all traffic on its original piconet is halted until it returns. It may sound complicated and involved in words, but it works well enough and dynamic reconfigurability on the short-term time scale of packets is very robust in practice.

Figure 14.11 illustrates how the frequency offset and hop selection sequence can be adjusted for in 4 piconets by incorporating frequency offset in the clocks and in the identity address.

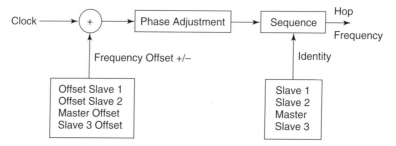

Fig. 14.11 Frequency offset and hop selection sequence can be adjusted for in four piconets by incorporating frequency offset in the clocks and in the identity address.

Interference between Bluetooth and 802.11

The price we pay for unlicensed (zero cost) use of this part of the electromagnetic spectrum is potential interference from others users, who encompass a very wide variety of applications. Consequently robust wireless system design and associated technology is increasingly important to get the best performance out of these unlicensed frequency bands. Short-range wireless systems such as Bluetooth are not as vulnerable in this respect as systems operating over longer paths such as Wi-Fi and WiMax. Nevertheless, in this respect it needs to be appreciated that using a shared band can result in interference and this is a price of not having to pay a license fee. Contrast this with the millions of dollars paid by companies worldwide for exclusive use of licensed frequency allocations for cellular telephones. In fact the interference experienced between Bluetooth and 2.4 GHz Wi-Fi may diminish over the next few years because of a migration of Wi-Fi to the 5 GHz ISM band, although the legacy of 2.4 GHz Wi-Fi will be present for many years.

There are several ways to ameliorate potential interference: system-level approaches that address coexistence through the use of special antenna designs and arrangements with PHY and MAC techniques all offer the potential to dramatically reduce interference between cochannel ISM systems (Henriksson, 2002). The IEEE 802.15.2 Working Group, Bluetooth SIG, and others (Zyren, 1999; Kamerman, 1999) have worked on effective ways to minimize mutual interference, and novel approaches are frequently published. This is an ongoing activity. In the end, the antenna systems employed when matched to the application are often the best solutions to maximize the effectiveness of a system. With Bluetooth, it is invariably the fact that we wish to minimize antenna size and this alone practically limits what can be achieved in terms of directional antennas (Fujimoto et al., 1987). Considering the different characteristics of physical layers in Bluetooth and WLAN operating with IEEE 802.11, the collision of radio signal occurs both in time and in frequency.

To mitigate possible mutual interference, the IEEE 802.15 Working Group has created Task Group 2 (TG2) for developing mechanisms for facilitating coexistence for WLAN and WPAN devices (Batra et al., 2001). Two classes of coexistence mechanisms have been defined: collaborative and noncollaborative (Shellhammer, 2001). With collaborative techniques, it is necessary to establish a connection between two overlay networks. The related information is *exchanged* through the connection so as to reduce the probability of simultaneous transmission. In contrast, the noncollaborative techniques do not need any coordination operation between the two systems. The mobile devices operate independently with one or more coexistence mechanisms to minimize the interference from adjacent interferers. Examples of collaborative coexistence mechanisms are the scheduling scheme, the so-called META (MAC Enhanced Temporal Algorithm) (Treister et al., 2001), and the TDMA (Time Division Multiple Access) scheme (Liang, 2001). One of the two noncollaborative techniques that achieve satisfactory combination performance and true "Coexistence without Compromise" (Ennis, 1998) is Adaptive Frequency Hopping (AFH) (Haartsen, 2000), which works by restricting Bluetooth hopping to a portion of the 2.4 GHz band not occupied by a Wi-Fi network. Typically, a Wi-Fi network may occupy about a quarter of the available band. AFH can allow a Bluetooth-enabled PDA to simultaneously operate on a Wi-Fi network. Antenna diversity is another technique that can help by reducing interference, albeit at the expense of employing more than one antenna (Treister et al., 2001).

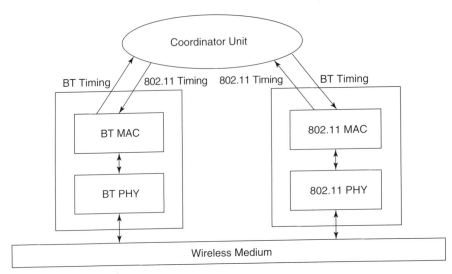

Fig. 14.12   Coordinator Unit implementation for a Bluetooth and Wi-Fi coexistence system called META.

MAC Enhanced Temporal Algorithm

The first collaborative coexistence technique, MAC Enhanced Temporal Algorithm, or META, requires the use of a Coordinator Unit (CU) that monitors the Bluetooth and 802.11 traffic and allows exchange of information between the two wireless networks (Treister, 2001). The CU usually works in the MAC layer and allows precise timing of packet traffic so as to avoid interference between two collocated devices. An implementation diagram is shown in Figure 14.12.

When each 802.11 or Bluetooth device is ready to send packets, it sends a request for permission to access the medium. The CU returns a message to each wireless system when it is allowed to transmit. In order to adapt to such a coordinate mode, the 802.11 channel access mechanism should be modified to reserve BT time slots for collision avoidance. At the same time, the Bluetooth devices should be restricted, transmitting packets only in the reserved time slots. The CU handles the traffic request based on a priority scheme that gives high priority to the time-bound services if two types of devices request simultaneously. The significant advantage of META is that it introduces only limited transmission delay.

Time Multiplex Division Access Scheme

The second collaborative mechanism, Time Multiplex Division Access, or TDMA, divides the 802.11 beacon-to-beacon interval into two subintervals: one for 802.11 and one for Bluetooth (Liang, 2001). Hence, both radio systems can work properly in their own time intervals even if they are located very close to each other. Figure 14.13 shows the two intervals.

The 802.11 interval begins just before a beacon frame that contains a Target Beacon Transmit Time (TBTT). The timing is controlled by the 802.11 MAC layer, which sends a medium-free signal (shown at the bottom of Figure 14.13) to the Bluetooth devices to notify the change-of-medium state. Hence, the Bluetooth devices are able to identity the BT interval and restrict its transmission time. One of the benefits of TDMA is that all the systems

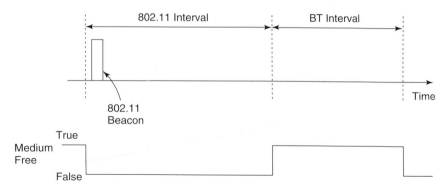

Fig. 14.13 Intervals allow systems to work cooperatively even with close proximity of transceiver antenna.

are synchronized and hence have the exact WLAN and WPAN intervals. Not only is there little interference within the mobile terminal containing both 802.11 and Bluetooth, there is also little interference from other nearby mobile terminals (Batra et al., 2001).

Adaptive Frequency Hopping

Adaptive Frequency Hopping, or AFH, is a noncollaborative approach that has been developed by the 802.15.2 Working Group for Bluetooth devices. AFH is based on the idea of classifying the frequency channels as "good" and "bad" and hopping only among the good channels. The channel selection performs periodically and the channel information is sent to all devices in the network so that they are synchronous with hops according to the updated hopping sequence. Thus, mutual interference can be avoided or minimized. In general, AFH can be divided into four stages: device identification, channel classification, exchange of classification information, and resulting adaptive hopping.

AFH Stages

The device identification stage provides a mechanism for the master of a piconet to identify the AFH capability of the slaves within the same network. The goal of the Channel Classification stage is to determine the quality of each channel. After this stage, each frequency channel of Bluetooth should be classified as good or bad. The channel information composes a classification list sent to each device within the network through the Classification Information Exchange stage. Once every device obtains the updated channel information, a new hopping sequence is generated to replace the original hopping sequence. Therefore, all the Bluetooth devices are able to shift to the new hopping sequence simultaneously without disturbing the upper layer. In order to minimize the modification of Bluetooth, AFH is designed as an additional unit between original hop kernel and frequency synthesizer.

Channel Classification

Obviously, accurate channel classification is essential for operating in AFH mode. It should include a simple (easy to implement), quick, and accurate interference mechanism. A variety of suggested methods are currently being investigated by the 802.15.2 Working Group. These methods can be categorized into the following four groups. According to Ennis (1998), the quality of transmission may be determined as follows:

1. Through the packet loss ratio.
2. Through detecting and assessing the received signal strength.
3. Through detecting the ratio of loss packets of test packets that cover the entire spectrum.
4. Through receiving the 802.11 pass band location for the 802.11 devices.

Once the channels have been classified, a classification list will be used to compile a final list of good and bad channels. A new hopping sequence will be generated according to the classification list.

Exchange of Channel Information

Since the master controls medium access in a piconet, it is easy for it to make decisions on channel classification and the working mode. Once the master decides to enter AFH mode, it sends an AFH mode request message to ask the slaves to prepare to operate in AFH mode. When the master and the slaves have switched to AFH mode, the classification list can be exchanged between them. There are several complementary messages required to support the exchange of channel information, such as AFH mode star/terminate, request of slave's classification list, and return to regular hopping request.

The AFH Mechanism

A block diagram of the adaptive frequency hopping mechanism is shown in Figure 14.14 (Treister, 2001). It consists of three distinct components: the legacy hop kernel, the partition sequence generator, and the frequency remapper. The legacy hop kernel refers to the original hop frequency generator that generates the hopping sequence according to the Bluetooth standard.

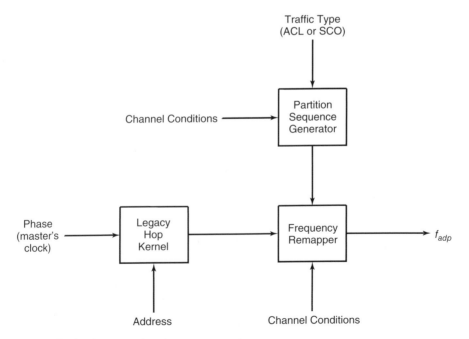

Fig. 14.14   Adaptive frequency hopping arrangement.

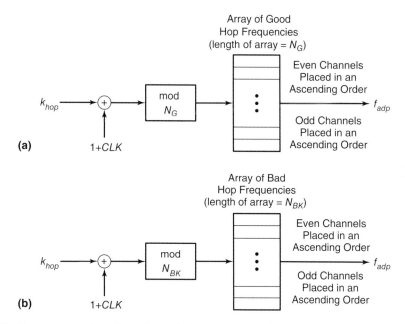

Fig. 14.15 Remapping system allows hopped channel usage to be optimized.

The component called the frequency remapper performs the function of modifying the hopping sequence. The output of the original hopping generator is fed into the frequency remapper with set of good channels and a set of bad channels. The frequency remapper passes the good channels through and remaps a bad channel onto a good channel according to the partition sequence that generates a partition sequence generator. The remap function is illustrated in Figure 14.15 (Treister, 2001).

The reason we still use the bad channels in the adaptive hopping sequence is that the required minimum number of hopping frequency, which is set by local regulation, may be larger than the number of good channels. For example, the minimum number of used channels in the United States is set at 15. Assuming a Bluetooth WPAN is operating with three 802.11b networks that are on a different channel, the number of good channels would be about $79 - (3 \times 23) = 10$. Therefore, some bad channels have to be used to meet the EMC regulations.

In order to enhance performance in the case of using bad channels, the partition sequence generator is applied to regroup the good and bad channels for specific services. As mentioned before, the partition sequence generator generates a partition sequence $(p(k))$ that is fed into the frequency remapper and performs the function of grouping the good and bad channels in clusters. The partition sequence can be modeled as follows:

$$p(k) = \begin{cases} 1 & \text{if } f_{adp} \text{ should be an element of good channels } (S_G) \\ 0 & \text{if } f_{adp} \text{ should be an element of bad channels } (S_{BK}) \end{cases}$$

Note that the partition sequence does not specify the exact frequency in each hop, but only defines the set that the hopping frequency should be within.

In general, the partition sequence depends on the number of good and bad channels available in the band. The adaptive hopping sequence contains only the good channels.

For the ACL link, the major issue is the constraint of the bad channel window so as to avoid timeout for the higher layers of the stack (e.g., TCP/IP). For the SCO link, which offers periodic data transmission, the good hopping frequencies must be distributed among the SCO time slots.

Bluetooth Security

Bluetooth has, from the outset, invoked robust security measures both for authentication and key generation. A powerful E0 cipher is used for encrypting packets, so eavesdropping on a Bluetooth connection is difficult but, alas, not impossible. It was mentioned in the Introduction that if EIRP considerations are ignored, it is possible to employ directional high gain antennas to greatly increase the range of Bluetooth. Nevertheless it still remains necessary to know the address of the Bluetooth in question and which part of the spectrum to use. The main features are these:

- Authentication involves a challenge and response.
- Encryptions involve a streamed cipher.
- Session keys can be changed any time, even during a connection.

Overall, Bluetooth is reasonably secure. But there remain weaknesses, particularly if a Bluetooth-enabled laptop is left working, for example, in a parking lot. Pairs of Bluetooth devices can establish a relationship by user input known as a pass key.

Power Consumption

Bluetooth has, from the beginning, been designed to take into account operation from a battery. That means that all aspects of design and operating protocol are very aware of power consumption and battery longevity. In Bluetooth 2, battery longevity is increased over version 1 by three times in this 3 Mbps enhancement.

Bluetooth 1 and 2 typical power requirements are: transmitter power of 1 milliWatt (0 dBm; −30 dBW) and battery voltage of 1.8 volts. The current consumption is 20 µA sleep, 0.6 mA parked slave, 47 µA standby (no RF activity but connected to host), and 8 to 30 mA transmitting.

In a Bluetooth 2 operating at 100 mW (+20 dBm), the typical consumption is ~130 mA @ 1.8 V ~250 mW. Because there are limitations in reducing the power consumption of Bluetooth, "WiBree" has recently been introduced. The main application of this Bluetooth-compliant variant is very short range communications at reportedly up to 80% reduced power consumption. The main advantage is that very small batteries can be employed so that for example, a headset can be miniaturized.

ZigBee

ZigBee is described in IEEE 802.15.4 as the access mechanism of IEEE 802 wireless networks; it is Carrier Sense Multiple Access (CSMA) with collision avoidance, and uses the 2.4 GHz ISM band. It is designed for lower data rate (e.g., kb/s) than Bluetooth, so its range is greater and it also differs from Bluetooth in that it employs mesh rather star topology. This means, for example, that there can be hundreds of ZigBee nodes, each of which can relay a message that is out of range of the original sender. Thus, ZigBee can readily provide a longer range than Bluetooth simply by means of a chain of transceivers or nodes. In short, a specific application of ZigBee is low data rate security systems in which each node is daisy-chained to another. Home control of low bit rate devices such as central heating room by room is another example. Typical hardware characteristics are a receiver noise floor of −100 dBm (~<1000 m range for 100 mW).

References

Batra, A., Chen, H. K., and Gan, H. (2001) Adaptive Frequency Hopping, *IEEE* 802 (November): *http://www.ieee802.org/15/pub/TG2.html*.

Bing, B. (2002) *Wireless Local Area Network*, Wiley-Interscience.

Bluetooth Core Specification (2003–2006): *http://www.bluetooth.com*.

Boley, D. (2004) Radiowave, Ltd., private communication *www.radiowave.co.uk/*.

Broadband Radio Access Networks (BRAN), High performance radio local area networks (HIPERLAN) Type 2, Systems, overview and references thereof: *http://www.etsi.org/*.

Ennis, G. (1998) Impact of Bluetooth on 802.11 Direct Sequence, *IEEE* 802 (September).

Fujimoto, Henderson, and James, (1987) *Small Antennas*. Research Studies Press.

Gibson, J. D. (1996) *The Mobile Communications Handbook*, CRC Press.

Haarsten, J. (1998) BLUETOOTH—The universal radio interface for ad hoc wireless connectivity, *Ericsson Review*, No 3.

———. (2000) Ericsson Radio Systems B.V., The Bluetooth Radio System, *IEEE Personal Communications* (February).

Henriksson, C. (2002) Reducing Bluetooth interference with diversity techniques, *IEEE 802.11b networks*, Master's thesis, Royal Institute of Technology (May).

Kamerman, A. (1999) Coexistence between Bluetooth and IEEE 802.11 CCK Solutions to Avoid Mutual Interference, *Lucent Technologies Bell Laboratories*, (January).

Lansford, J., et al. (2001) Wi-Fi (802.11b) and Bluetooth: Enabling Coexistence, *IEEE Network* (September/October).

Liang, J. (2001) Proposal for Collaborative BT and 802.11b MAC Mechanisms for Enhanced Coexistence, *IEEE 802* (January): *http://www.ieee802.org/15/pub/TG2.html*.

Massey, P. (2005) Philips Research, Redhill, U.K., Private communication.

Maxstream: *www.maxstream.net/*.

Mobilian Corporation (2001) WiFi (802.11b) and Bluetooth: An Examination of Coexistence Approaches, *Mobilian White Papers* (March).

———. (2000) WiFi (802.11b) and Bluetooth Simultaneous Operation: Characterizing the Problem, *Mobilian White Papers* (November).

Muller, N. J., (2001) *Bluetooth Demystified*, McGraw-Hill.

Shellhammer, S. (2001) Collocated Collaborative Coexistence Mechanism: TDMA of 802.11 and Bluetooth, I (January): *http://www.ieee802.org/15/pub/TG2.html*.

Treister, B., et al. (2001) Adaptive Frequency Hopping: A Noncollaborative Coexistence Mechanism, *IEEE 802* (May): *http://www.ieee802.org/15/pub/TG2.html*.

Treister, B., Gan, H., Skafidas, E., Sapozhnykov, V., et. al. (2001) Clause 14.3 Adaptive Frequency Hopping, *IEEE 802* (August): *http://www.ieee802.org/15/pub/TG2.html*.

ZigBee: *www.802.org/15/pub/TG4.html* and *www.zigbee.org*.

Zyren, J. (1999) Reliability of IEEE 802.11 Hi Rate DSSS WLANs in a High-Density Bluetooth Environment. *Bluetooth '99* (June).

Part Five

Analytical and Simulation Modeling

Cable and telecommunications networks are highly complex, distributed machines that are very expensive to build. In developing new network solutions, many possible pitfalls exist and there are many areas where optimization of cost and performance must be made in order to carry traffic efficiently. However, it is often not possible to test in advance how a new design, or changes to an existing one, will perform under the traffic conditions that will arise in practice. Nevertheless, it is necessary to build confidence into the design by some means.

Analytical and simulation modeling allows computer models to be constructed of the proposed network or system and a range of their characteristic parameters to be defined. The ranges of external conditions expected to be met in real-life situations are also defined, such as traffic load or noise. These are then applied to the model either to determine the required dimensions of the equipment or, given the dimensions, provide an estimate of performance.

By using computer techniques, a wide range of operating conditions can be rapidly assessed in order to build confidence in the proposed solution, before building the complete network or system.

Part 5 discusses methods of modeling and analyzing telecommunications traffic. Chapter 15 covers statistical analysis and provides an introduction to simulation modeling. Chapter 16 describes queuing theory and the characteristics of self-similar traffic. Chapter 17 provides an introduction to teletraffic theory and discusses how this is applied to circuit-switched networks. Chapter 18 introduces particle filtering—a method of applying Bayesian methods to statistics, signal processing, and communications problems, which previously could only be solved using ad hoc methods.

15 Statistical Analysis and Computer Simulation

Fraidoon Mazda and David K. Hunter

Introduction

Vast quantities of data arise in all areas of telecommunications, especially when evaluating the performance of networks and systems. This chapter describes the more commonly used techniques for presenting and manipulating data to obtain meaningful results, and discusses computer simulation, a common source of data and an important way of modeling telecommunications systems.

Data Presentation

Probably the most common method used to present data is by tables and graphs. For impact, or to convey information quickly, pictograms and bar charts may be used. Pie charts are useful in showing the different proportions of a unit.

A strata graph shows how the total is split among its constituents. For example, Figure 15.1 shows that the total revenue obtained by a PTO steadily increases with time, but that only services B and D have growth while service C is reducing and may eventually become unprofitable.

Logarithmic or ratio graphs are used when one is more interested in the change in the ratios of numbers rather than their absolute value. In the logarithmic graph, equal ratios represent equal distances.

Frequency distributions are conveniently represented by a histogram as in Figure 15.2, which shows the number of people using a given service, banded according to age group.

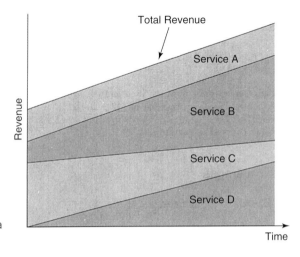

Fig. 15.1 Illustration of a strata graph.

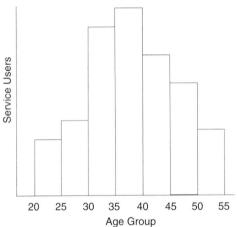

Fig. 15.2 A histogram.

There are very few users below 20 years and above 55 years, the most popular ages being 35 to 40 years. This information will allow the service provider to target its advertising more effectively.

In a histogram, the areas of the rectangles represent the frequencies in the different groups. Ogives, illustrated in Figure 15.3, show the cumulative frequency occurrences above or below a given value. From this curve it is possible to read off the total number of users above or below a specific age.

Approximation and Errors

The values used in statistical analysis are often approximate, even though they may be stated to several decimal places. There are several reasons for this:

1. The information is based on surveys and sampling processes, which are by themselves approximation techniques.

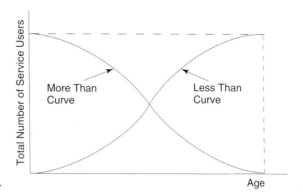

Fig. 15.3   Illustration of ogives.

Table 15.1   Approximating Numbers

Actual number	Approximation	Error spread	Method of representation
36,845.615	36,845.62	±0.005	To two decimal places
36,845.615	36,846	±0.5	To nearest whole number
36,845.615	36,850	±5	To nearest ten
36,845.615	36,800	±50	To nearest hundred
36,845.615	37,000	±500	To nearest thousand

2. Often approximate information will suffice and the cost of greater accuracy is not justifiable.
3. Numbers have been rounded, for example, to group them into classes.
4. Inaccuracies and bias creep in due to human error.
5. The figures often try to predict some future event, for example, the sales of the company in future years, and as such they can only be approximate.

This section considers the methods used to represent numbers that are approximate and the errors that occur when they are manipulated.

Representation of Errors

Table 15.1 shows some of the methods that are used to represent an approximation of the number 36,845.615, and the spread in errors. For example, this number can be represented to the nearest ten by 36,850, resulting in an error of ±5. This method would normally be used if the data being collected and presented was likely to have an error of this magnitude.

Errors in numbers can be represented as absolute or relative, and Table 15.1 shows absolute error. For example, if the actual number is 36,845.615 and it is represented to the nearest hundred by 36,800, then the absolute error is $36,845.615 - 36,800 = 45.615$. If the actual number is not known (i.e., only the approximation of $36,800 \pm 50$ is known), then the absolute error is 50.

For relative error, the absolute error is represented as a percentage of the actual value. So, in the example given earlier, it is $45.615/36,845.615 = 0.12\%$ (or equal to $50/36,800 = 0.14\%$ if the actual value is not known) and is written as $36,800 \pm 0.12\%$ (or $36,800 \pm 0.14\%$).

Accumulation of Errors

When numbers that are approximate are manipulated further, they result in an accumulation of errors. The following shows the methods to be used when carrying out the basic functions of addition, subtraction, multiplication, and division.

The process may seem strange, but can be easily followed if one remembers that errors are always accumulated (added). For addition and subtraction, absolute errors must be used while for multiplication and subtraction relative errors are used.

Addition

When adding two numbers that are approximate, their errors are also added. For example, suppose that the two numbers 265 ± 5 and 512 ± 20 are to be added. The result is $265 + 512 = 777$ and $(\pm) 5 + (\pm 20) = \pm 25$, giving a total of 777 ± 25.

Subtraction

For subtraction, the two main numbers are subtracted from each other, but the errors are added. This means that often the resulting error will be so large, compared to the actual number, that the result becomes meaningless. For example, the result of $512 \pm 20 - 265 \pm 5$ is $512 - 265 = 247$ and $(\pm 20) + (\pm 5) = \pm 25$, giving a total of 247 ± 25.

Multiplication

Relative errors must be used for multiplication. If absolute errors are given they must first be converted into relative errors. The two main numbers are multiplied and the errors are added. For example, to obtain the result of $512 \pm 20 \times 265 \pm 5$, the absolute errors must first be converted into relative errors (i.e., $512 \pm 3.91\%$ and $265 \pm 1.89\%$). The steps in multiplication are now $512 \times 265 = 135{,}680$ and $3.91\% + 1.89\% = 5.8\%$, giving a total of $135{,}680 \pm 5.8\%$. Note that the relative error can be converted back to absolute error by multiplying it by the main number. In the above example $135{,}680 \pm 5.8\%$ equals $135{,}680 \pm 7869.4$.

Division

Relative errors must again be used and the two errors added together. For example, the result of 512 ± 20 (or $512 \pm 3.91\%$) divided by 265 ± 5 (or $265 \pm 1.89\%$) is equal to $1.93 \pm 5.8\%$. Using absolute error this is equal to 1.93 ± 0.11.

Averages

Arithmetic Mean

The arithmetic mean of n numbers $x_1, x_2, x_3, \ldots, x_n$ is given by Equation 15.1, which is written as in Equation 15.2.

$$\bar{x} = \frac{x_1 + x_2 + x_3 + \ldots + x_n}{n} \tag{15.1}$$

$$\bar{x} = \frac{\sum\limits_{r=1}^{n} x_r}{n} \tag{15.2}$$

The arithmetic mean is easy to calculate and it takes into account all the figures. Its disadvantages are that it is influenced unduly by extreme values and the final result may not be a whole number, which can be absurd at times (e.g., a mean of 2.5 men).

Median and Mode

Median or "middle one" is found by placing all the figures in order and choosing the one in the middle or, if there are an even number of items, the mean of the two central numbers. It is a useful technique for finding the average of items that cannot be expressed in figures (e.g., shades of a color). Also, it is not influenced by extreme values. However, the median is not representative of all the figures. The mode is the most "fashionable" item, that is, the one that appears the most frequently.

Geometric Mean

The geometric mean of n numbers $x_1, x_2, x_3, \ldots, x_n$ is given by Equation 15.3.

$$x_g = (x_1 \times x_2 \times x_3 \times \ldots \times x_n)^{1/n} \tag{15.3}$$

This technique is used to find the average of quantities that follow a geometric progression or exponential law, such as rates of changes. Its advantage is that it takes into account all the numbers but is not unduly influenced by extreme values.

Harmonic Mean

The harmonic mean of n numbers $x_1, x_2, x_3, \ldots, x_n$ is given by Equation 15.4.

$$x_h = \frac{n}{\displaystyle\sum_{r=1}^{n} \frac{1}{x_r}} \tag{15.4}$$

This averaging method is used when dealing with rates or speeds or prices. As a rule when considering items such as A per B, if the figures are for equal A's, then the harmonic mean is used, but if they are for equal B's, then the arithmetic mean is used. Thus if a plane flies over three equal distances at speeds of 5 m/s, 10 m/s, and 15 m/s, the mean speed is given by the harmonic mean as in Expression 15.5.

$$\frac{3}{\dfrac{1}{5} + \dfrac{1}{10} + \dfrac{1}{15}} = 8.18 \, \text{m/s} \tag{15.5}$$

If, however, the plane were to fly for three equal times of, say, 20 seconds at speeds of 5 m/s, 10 m/s, and 15 m/s, then the mean speed would be given by the arithmetic mean as in Expression 15.6.

$$\frac{5 + 10 + 15}{3} = 10 \, \text{m/s} \tag{15.6}$$

Dispersion from the Average

Range and Quartiles

The average represents the central figure of a series of numbers or items. It does not give any indication of the spread of the figures in the series from the average. Therefore, measurements of errors made on two circuits A and B may result in the curves shown in Figure 15.4. Both

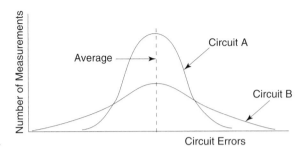

Fig. 15.4 Illustration of deviation from the average.

circuits have the same calculated average errors, but circuit B has a wider deviation from the average than circuit A and, at the top end, its errors may be unacceptably high.

There are several ways of stating by how much the individual numbers in the series differ from the average. The range is the difference between the smallest and largest values. The series can also be divided into four quartiles and dispersion stated as the interquartile range (i.e., the difference between the first and third quartile numbers, or the quartile deviation, which is half this value).

The quartile deviation is easy to use and is not influenced by extreme values. However, it gives no indication of distribution between quartiles and covers only half the values in a series.

Mean Deviation

This is found by taking the mean of the differences between each individual number in the series and the arithmetic mean, or median, of the series. Negative signs are ignored.

For a series of n numbers $x_1, x_2, x_3, \ldots, x_n$ having an arithmetic mean, the mean deviation of the series is given by Equation 15.7.

$$\frac{\sum_{r=1}^{n} |x_r - \bar{x}|}{n} \tag{15.7}$$

The mean deviation takes into account all the items in the series, but it is not very suitable since it ignores signs.

Standard Deviation

This is the most common measure of dispersion. For this, the arithmetic mean must be used and not the median. It is calculated by squaring deviations from the mean, so eliminating their sign, adding the numbers together, taking their mean and then the square root of the mean. Therefore, for the series of n numbers as above, the standard deviation is given by Equation 15.8.

$$\sigma = \left(\frac{\sum_{r+1}^{n} (x_r - \bar{x})^2}{n} \right)^{1/2} \tag{15.8}$$

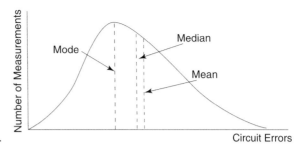

Fig. 15.5 Illustration of skewness.

The unit of the standard deviation is that of the original series. So if the tariff charged is in dollars for a given service by different PTOs, then the mean and the standard deviation are in dollars.

To compare two series that have different units, such as the cost of a service and the quality of that service, the coefficient of variation is used, which is unitless, as in Equation 15.9.

$$\text{Coefficient of variation} = \frac{\sigma}{x} \times 100 \qquad \textbf{(15.9)}$$

Skewness

The distribution shown in Figure 15.4 is symmetrical since the mean, median, and mode all coincide. Figure 15.5 shows a skewed distribution with positive skewness. If the distribution bulges the other way, the skewness is said to be negative.

There are several mathematical ways for expressing skewness. They all give a measure of the deviation between the mean, median and mode, and they are usually stated in relative terms for ease of comparison between series of different units. The Pearson coefficient of skewness is given by Equation 15.10.

$$P_k = \frac{\text{mean} - \text{mode}}{\text{standard deviation}} \qquad \textbf{(15.10)}$$

Since the mode is sometimes difficult to measure, this can also be stated as in Equation 15.11.

$$P_k = \frac{3(\text{mean} - \text{median})}{\text{standard deviation}} \qquad \textbf{(15.11)}$$

Combinations and Permutations

Combinations

Combinations are the number of ways in which a proportion can be chosen from a group. Therefore, the number of ways in which two letters can be chosen from a group of four letters A, B, C, D is equal to 6 (i.e., AB, AC, AD, BC, BD, CD). This is written as in Expression 15.12.

$$^4C_2 = 6 \qquad \textbf{(15.12)}$$

The factorial expansion is frequently used in combination calculations where factorial n is written as in Expression 15.13.

$$n! = n \times (n-1) \times (n-2) \times \ldots \times 3 \times 2 \times 1 \tag{15.13}$$

Using this, the number of combinations of r times from a group of n is given by Equation 15.14.

$$^nC_r = \frac{n!}{r!(n-r)!} \tag{15.14}$$

Permutations

Combinations do not indicate any sequencing. When sequencing within each combination is involved, the result is known as a permutation. Therefore, the number of permutations of two letters out of four letters A, B, C, D is 12 (i.e., AB, BA, AC, CA, AD, DA, BC, CB, BD, DB, CD, DC). The number of permutations of r items from a group of n is given by Equation 15.15.

$$^nP_r = \frac{n!}{(n-r)!} \tag{15.15}$$

Regression and Correlation

Regression

Regression is a method for establishing a mathematical relationship between two variables. Several equations may be used to determine this relationship, the most common being that of a straight line. Figure 15.6 shows the number of defective public telephones that were reported at seven instances in time. This is called a scatter diagram. The points can be seen to lie approximately on the straight line AB.

The equation of a straight line is given by Equation 15.16, where x is the independent variable, y the dependent variable, m the slope of the line, and c its intercept on the y axis.

$$y = mx + c \tag{15.16}$$

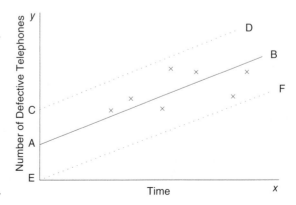

Fig. 15.6 A scatter diagram.

If the line intercepts the y axis on its negative part, c is negative, and m is negative if the line slopes the other way to that shown in Figure 15.6.

The best straight line to fit a set of points is found by the method of least squares as in Equations 15.17 and 15.18, where n is the number of points. The line passes through the mean values of x and y (i.e., $\bar{x}$ and $\bar{y}$).

$$m = \frac{\sum xy - \frac{\sum x \sum y}{n}}{\sum x^2 - \frac{\left(\sum x\right)^2}{n}} \tag{15.17}$$

$$c = \frac{\sum x \sum xy - \sum y \sum x^2}{\left(\sum x\right)^2 - n \sum x^2} \tag{15.18}$$

Correlation

Correlation is a technique for establishing the strength of the relationship between variables. In Figure 15.6 the individual figures are scattered on either side of a straight line, and although one can approximate them by a straight line, it may be required to establish if there is correlation between the reading on the x and y axes.

Several correlation coefficients exist. The product moment correlation coefficient (r) is given by Equation 15.19 or 15.20.

$$r = \frac{\sum (x - \bar{x})(y - \bar{y})}{n \sigma_x \sigma_y} \tag{15.19}$$

$$r = \frac{\sum (x - \bar{x})(y - \bar{y})}{\left(\sum (x - \bar{x})^2 \sum (y - \bar{y})^2\right)^{1/2}} \tag{15.20}$$

The value of r varies from +1, when all the points lie on a straight line and the number of defects increase with time, to −1, when all the points lie on a straight line but defects decrease with time. When $r = 0$ the points are widely scattered and there is said to be no correlation between the x and y values.

The standard error of estimation in r is given by Equation 15.21.

$$S_y = \sigma_y \left(1 - r^2\right)^{1/2} \tag{15.21}$$

In about 95 percent of cases, the actual values will lie within plus or minus twice the standard error of estimated values given by the regression equation. This is shown by lines CD and EF in Figure 15.6. Almost all the values will be within plus or minus three times the standard error of estimated values.

It should be noted that σ_y is the variability of the y values, whereas S_y is a measure of the variability of the y values as they differ from the regression that exists between x and y. If there is no regression then $r = 0$ and $\sigma_y = S_y$.

It is often necessary to draw conclusions from the order in which items are ranked. For example, two customers may rank the styling of a telephone and we need to know if there is any correlation between their rankings. This may be done by using the rank correlation coefficient (R) given by Equation 15.22, where d is the difference between the two ranks for each item and n is the number of items.

$$R = 1 - \frac{6\sum d^2}{n^3 - n}$$

(15.22)

The value of R will vary from +1, when the two ranks are identical, to −1, when they are exactly reversed.

Probability

If an event A occurs n times out of a total of m cases, then the probability of occurrence is stated to be as in Equation 15.23.

$$P(A) = \frac{n}{m}$$

(15.23)

Probability varies between 0 and 1. If $P(A)$ is the probability of occurrence, then $1 - P(A)$ is the probability that event A will not occur and it can be written as $P(A)$.

If A and B are two events, then the probability that either may occur is given by Equation 15.24.

$$P(A \text{ or } B) = P(A) + P(B) - P(A \text{ and } B)$$

(15.24)

A special case of this probability law is when events are mutually exclusive (i.e., the occurrence of one event prevents the other from happening). Then Equation 15.25 is obtained.

$$P(A \text{ or } B) = P(A) + P(B)$$

(15.25)

If A and B are two events, then the probability that they may occur together is given by Equation 15.26 or 15.27.

$$P(A \text{ and } B) = P(A) \times P(B \mid A)$$

(15.26)

$$P(A \text{ and } B) = P(B) \times P(A \mid B)$$

(15.27)

$P(B \mid A)$ is the probability that event B will occur, assuming that event A has already occurred, and $P(A \mid B)$ is the probability that event A will occur, assuming that event B has already occurred. A special case of this probability law is when A and B are independent events (i.e., the occurrence of one event has no influence on the probability of the other event occurring). Then Equation 15.28 is obtained.

$$P(A \text{ and } B) = P(A) \times P(B)$$

(15.28)

Bayes's theorem on probability may be stated as in Equation 15.29.

$$P(A \mid B) = \frac{P(A)P(B \mid A)}{P(A)P(B \mid A) + P(\overline{A})P(B \mid \overline{A})}$$

(15.29)

As an example of the use of Bayes's theorem, suppose that a company discovers that 80 percent of those who bought its multiplexers in a year had taken the company's training course. Thirty percent of those who bought a competitor's multiplexers had also taken the training course. During that year, the company had 20 percent of the multiplexer market. The company wishes to know what percentage of buyers actually went took its training course, in order to discover the effectiveness of this course.

If B denotes that a person bought the company's product and T that he took the training course, then the problem is to find $P(B|T)$. From the data $P(B) = 0.2$, $P(\bar{B}) = 0.8$. Then from Equation 15.29 Expression 15.30 is obtained.

$$P(B|T) = \frac{0.2 \times 0.8}{0.2 \times 0.8 + 0.8 \times 0.3} = 0.4 \qquad \text{(15.30)}$$

Probability Distributions

There are several mathematical formulae with well-defined characteristics that are known as probability distributions. If a problem can be made to fit one of these distributions, then its solution is simplified. Distributions can be discrete, when the characteristic can only take certain specific values, such as 0, 1, 2, and so on, or they can be continuous, when the characteristic can take any value.

Binomial Distribution

The binomial probability distribution is given by Equation 15.31.

$$(p+q)^n = q^n + {}^nC_1 p q^{n-1} + {}^nC_2 p^2 q^{n-2} + \ldots + {}^nC_x p^x q^{n-x} + \ldots + p^n \qquad \text{(15.31)}$$

where p is the probability of an event occurring, q ($= 1 - p$) is the probability of an event not occurring, and n is the number of selections.

The probability of an event occurring m successive times is given by the binomial distribution as in Equation 15.32.

$$p(m) = {}^nC_m p^m q^{n-m} \qquad \text{(15.32)}$$

The binomial distribution is used for discrete events and is applicable if the probability of occurrence p of an event is constant on each trial. The mean of the distribution $B(M)$ and the standard deviation $B(S)$ are given by Equations 15.33 and 15.34.

$$B(M) = np \qquad \text{(15.33)}$$

$$B(S) = (npq)^{1/2} \qquad \text{(15.34)}$$

Poisson Distribution

The Poisson distribution is used for discrete events and, like the binomial distribution, it applies to mutually independent events. It is used in cases where p and q cannot both be defined. For example, one can state the number of times a telephone circuit failed over a given period of time, but not the number of times when it did not fail.

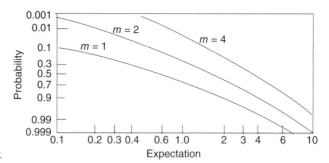

Fig. 15.7   Poisson probability paper.

The Poisson distribution may be considered to be the limiting case of the binomial when n is large and p is small. The probability of an event occurring m successive times is given by the Poisson distribution as in Equation 15.35.

$$p(m) = (np)^m \frac{e^{-np}}{m!} \tag{15.35}$$

The mean $P(M)$ and standard deviation $P(S)$ of the Poisson distribution are given by Equations 15.36 and 15.37.

$$P(M) = np \tag{15.36}$$

$$P(S) = (np)^{1/2} \tag{15.37}$$

Poisson probability calculations can be done with probability charts as shown in Figure 15.7. This shows the probability that an event will occur at least m times when the mean (or expected) value np is known.

Normal Distribution

The normal distribution represents continuous events and is plotted in Figure 15.8. The x axis gives the event (e.g., telephone line failure) and the y axis the probability of the event occurring. The curve shows that most of the events occur close to the mean value, and this is usually the case in nature. The normal curve is given by Equation 15.38, where $\bar{x}$ is the mean of the values making up the curve and σ is their standard deviation.

$$y = \frac{1}{\sigma(2\pi)^{1/2}} \exp\left(\frac{-(x-\bar{x})^2}{2\sigma^2}\right) \tag{15.38}$$

Different distributions will have varying mean and standard deviations, but if they are distributed normally their curves will all follow Equation 15.38. These distributions can be normalized to a standard form by moving the origin of their normal curve to their mean value, shown as B in Figure 15.8. The deviation from the mean is now represented on a new scale of units given by Equation 15.39.

$$\omega = \frac{x - \bar{x}}{\sigma} \tag{15.39}$$

The standardized normal curve now becomes as in Equation 15.40.

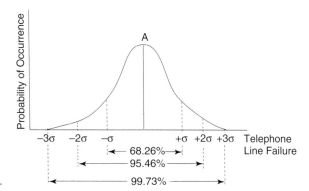

Fig. 15.8   The normal curve.

$$y = \frac{1}{(2\pi)^{1/2}} \exp\left(\frac{-\omega^2}{2}\right) \qquad (15.40)$$

The total area under the standardized normal curve is unity, and the area between any two values of ω is the probability of an item from the distribution falling between these values. The normal curve extends infinitely in either direction, but 68.26 percent of its values (area) fall between $\pm\sigma$, 95.46 percent between $\pm2\sigma$, 99.73 percent between $\pm3\sigma$, and 99.994 percent between $\pm4\sigma$.

Table 15.2 gives the area under the normal curve for different values of ω. Since the normal curve is symmetrical, the area from $+\omega$ to $+\infty$ is the same as from $-\omega$ to $-\infty$. As an example of the use of this table, suppose that 5000 telephones have been installed in a city and that they have a mean life of 1000 weeks with a standard deviation of 100 weeks. How many telephones will fail in the first 800 weeks? From Equation 15.39, Expression 15.41 is obtained.

$$\omega = \frac{(800 - 1000)}{100} = -2 \qquad (15.41)$$

Ignoring the negative sign, Table 15.2 gives the probability of telephones not failing as 0.977 so that the probability of failure is $1 - 0.977$ or 0.023. Therefore, 5000×0.023 or 115 telephones are expected to fail after 800 weeks.

Exponential Distribution

The exponential probability distribution is a continuous distribution and is shown in Figure 15.9. It has the form given in Equation 15.42, where $\bar{x}$ is the mean of the distribution.

$$y = \frac{1}{x}\exp\left(\frac{-x}{\bar{x}}\right) \qquad (15.42)$$

Whereas in the normal distribution the mean value divides the population in half, in the exponential distribution 36.8 percent of the population is above the average and 63.2 percent below the average. Table 15.3 shows the area under the exponential curve for different values of the ratio $K = x\sqrt{x}$, this area being shown shaded in Figure 15.9.

Table 15.2 Area under the Normal Curve

ω	0.00	0.02	0.04	0.06	0.08
0.0	0.500	0.508	0.516	0.524	0.532
0.1	0.540	0.548	0.556	0.564	0.571
0.2	0.579	0.587	0.595	0.603	0.610
0.3	0.618	0.626	0.633	0.640	0.648
0.4	0.655	0.663	0.670	0.677	0.684
0.5	0.692	0.700	0.705	0.712	0.719
0.6	0.726	0.732	0.739	0.745	0.752
0.7	0.758	0.764	0.770	0.776	0.782
0.8	0.788	0.794	0.800	0.805	0.811
0.9	0.816	0.821	0.826	0.832	0.837
1.0	0.841	0.846	0.851	0.855	0.860
1.1	0.864	0.869	0.873	0.877	0.881
1.2	0.885	0.889	0.893	0.896	0.900
1.3	0.903	0.907	0.910	0.913	0.916
1.4	0.919	0.922	0.925	0.928	0.931
1.5	0.933	0.936	0.938	0.941	0.943
1.6	0.945	0.947	0.950	0.952	0.954
1.7	0.955	0.957	0.959	0.961	0.963
1.8	0.964	0.966	0.967	0.969	0.970
1.9	0.971	0.973	0.974	0.975	0.976
2.0	0.977	0.978	0.979	0.980	0.981
2.1	0.982	0.983	0.984	0.985	0.985
2.2	0.986	0.987	0.988	0.988	0.989
2.3	0.989	0.990	0.990	0.991	0.991
2.4	0.992	0.992	0.993	0.993	0.993
2.5	0.994	0.994	0.995	0.995	0.995
2.6	0.995	0.996	0.996	0.996	0.996
2.7	0.997	0.997	0.997	0.997	0.997
2.8	0.997	0.998	0.998	0.998	0.998
2.9	0.998	0.998	0.998	0.998	0.999
3.0	0.999	0.999	0.999	0.999	0.999

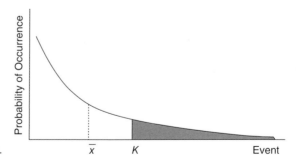

Fig. 15.9 The exponential curve.

For example, suppose that the time between failures of a piece of equipment is found to vary exponentially. If results indicate that the mean time between failures is 1000 weeks, then what is the probability that the equipment will work for 700 weeks or more without a failure? Calculating K as $700/1000 = 0.7$, then from Table 15.3 the area beyond 0.7 is 0.497, which is the probability that the equipment will still be working after 700 weeks.

Table 15.3   Area under the Exponential Curve

K	0.00	0.02	0.04	0.06	0.08
0.0	1.000	0.980	0.961	0.942	0.923
0.1	0.905	0.886	0.869	0.852	0.835
0.2	0.819	0.803	0.787	0.771	0.776
0.3	0.741	0.726	0.712	0.698	0.684
0.4	0.670	0.657	0.644	0.631	0.619
0.5	0.607	0.595	0.583	0.571	0.560
0.6	0.549	0.538	0.527	0.517	0.507
0.7	0.497	0.487	0.477	0.468	0.458
0.8	0.449	0.440	0.432	0.423	0.415
0.9	0.407	0.399	0.391	0.383	0.375

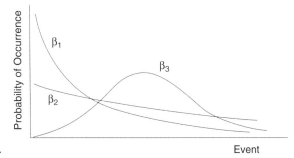

Fig. 15.10   Weibull curves.

Weibull Distribution

This is a continuous probability distribution and is given by Equation 15.43, where α is called the scale factor, β the shape factor, and γ the location factor.

$$y = \alpha\beta(x - \gamma)^{\beta-1}\ \ \exp\left(-\alpha(x - \gamma)^{\beta}\right) \tag{15.43}$$

The shape of the Weibull curve varies depending on the value of its factors. β is the most important, as shown in Figure 15.10, and the Weibull curve varies from an exponential ($\beta = 1.0$) to a normal distribution. In practice β varies from about 1/3 to 5. Because the Weibull distribution can be made to fit a variety of different sets of data, it is popularly used for probability distributions.

Analytical calculations using the Weibull distribution are cumbersome. Usually predictions are made using Weibull probability paper. The data are plotted on this paper and the probability predictions read from the graph.

Sampling

A sample consists of a relatively small number of items drawn from a much larger population. It is analyzed for certain attributes, and it is then assumed that these attributes apply to the total population, within a certain tolerance of error.

Sampling is usually associated with the normal probability distribution and, based on this distribution, the errors that arise due to sampling can be estimated. Suppose a sample of n_s items is taken from a population of s_p items, which are distributed normally. If the sample is found to have a mean of μ_s with a standard deviation of σ_s, then the mean μ_p of the population can be estimated to be within a certain tolerance of μ_s. It is given by Equation 15.44.

$$\mu_p = \mu_s \pm \frac{\gamma \sigma_s}{\sqrt{n_s}} \tag{15.44}$$

The variable γ is found from the normal curve depending on the level of confidence we need in specifying μ_p. For $\gamma = 1$ this level is 68.26 percent, for $\gamma = 2$ it is 95.46 percent, and for $\gamma = 3$ it is 99.73 percent.

The standard error of mean σ_c is often defined as in Equation 15.45.

$$\sigma_c = \frac{\sigma_s}{\sqrt{n_s}} \tag{15.45}$$

Therefore Equation 15.44 can be rewritten as in Equation 15.46.

$$\mu_p = \mu_s \pm \gamma \sigma_c \tag{15.46}$$

As an example, suppose that a sample of 100 items, selected at random from a much larger population, gives their mean weight as 20 kg with a standard deviation of 100 g. The standard error of the mean is therefore $100/(100)^{1/2} = 10$ g, and one can say with 99.73 percent confidence that the mean value of the population lies between $20 \pm 3 \times 0.01$ or 20.03 kg and 19.97 kg.

If in a sample of n_s items the probability of occurrence of a particular attribute is p_s, then the standard error of probability p_e is defined as in Equation 15.47, where $q_s = 1 - p_s$.

$$p_c = \left(\frac{p_s q_s}{n_s} \right)^{1/2} \tag{15.47}$$

The probability of occurrence of the attribute in the population is then given by Equation 15.48, where γ is again chosen to cover a certain confidence level.

$$P_p = P_s \pm \gamma P_c \tag{15.48}$$

As an example, suppose that a sample of 500 items shows that 50 are defective. Then the probability of occurrence of the defect in the sample is $50/500 = 0.1$. The standard error of probability is $(0.1 \times 0.9/500)^{1/2}$ or 0.0134. Therefore, one can state with 95.46 percent confidence that the population from which the sample was drawn has a defect probability of $0.1 \pm 2 \times 0.0134$ (i.e., 0.0732 to 0.1268); or one can state with 99.73 percent confidence that this value will lie between $0.1 \pm 3 \times 0.0134$ (i.e., 0.0598 to 0.1402).

If two samples have been taken from the same population and these give standard deviations of σ_{s1} and σ_{s2} for sample sizes of n_{s1} and n_{s2}, then Equation 15.45 can be modified to give the standard error of the difference between means as in Equation 15.49.

$$\sigma_{dc} = \left(\frac{\sigma_{s1}^2}{n_{s1}} + \frac{\sigma_{s2}^2}{n_{s2}} \right)^{1/2} \tag{15.49}$$

Similarly, Equation 15.47 can be modified to give the standard error of the difference between probabilities of two samples from the same population as in Equation 15.50.

$$P_{dc} = \left(\frac{P_{s1}q_{s1}}{n_{s1}} + \frac{P_{s2}q_{s2}}{n_{s2}} \right)^{1/2} \qquad (15.50)$$

Tests of Significance

In taking samples, one often obtains results that deviate from the expected. Tests of significance are then used to determine if this deviation is real or if it could have arisen due to sampling error.

Hypothesis Testing

In this system, a hypothesis is set up and is then tested at a given confidence level. For example, suppose a coin is tossed 100 times and it comes up heads 60 times. Is the coin biased or is it likely that this falls within a reasonable sampling error? The hypothesis is set up that the coin is not biased. Therefore, one would expect that the probability of heads is 0.5 (i.e., $p_s = 0.5$). The probability of tails q_s is also 0.5. Using Equation 15.47 the standard error of probability is given by Equation 15.51.

$$p_c = \left(\frac{0.5 \times 0.5}{100} \right)^{1/2} = 0.05 \qquad (15.51)$$

Therefore, from Equation 15.48 the population probability at the 95.45 percent confidence level of getting heads is $0.5 + 2 \times 0.05 = 0.6$. It is highly likely, then, that the coin is not biased and the results are due to sampling error.

The results of any significance test are not conclusive. For example, is 95.45 percent too high a confidence level to require? The higher the confidence level the greater the risk of rejecting a true hypothesis, and the lower the level the greater the risk of accepting a false hypothesis.

Suppose now that a sample of 100 items of production shows that five are defective. A second sample of 100 items is taken from the same production a few months later and gives two defectives. Does this show that the production quality is improving? Using Equation 15.50, the standard error of the difference between probabilities is given by Expression 15.52.

$$\left(\frac{0.05 \times 0.95}{100} + \frac{0.02 \times 0.98}{100} \right)^{1/2} = 0.0259 \qquad (15.52)$$

This is less than twice the difference between the two probabilities (i.e., $0.05 - 0.02 = 0.03$), so the difference is very likely to have arisen due to sampling error and does not necessarily indicate an improvement in quality.

Chi-square Test

This is written as χ^2. If O is an observed result and E is the expected result, then Equation 15.53 is obtained.

Table 15.4 The Chi-square Distribution

| Degrees of freedom | Probability level | | | | |
	0.100	0.050	0.025	0.010	0.005
1	2.71	3.84	5.02	6.63	7.88
2	4.61	5.99	7.38	9.21	10.60
3	6.25	7.81	9.35	11.34	12.84
4	7.78	9.49	11.14	13.28	14.86
5	9.24	11.07	12.83	15.09	16.75
6	10.64	12.59	14.45	16.81	18.55
7	12.02	14.07	16.01	18.48	20.28
8	13.36	15.51	17.53	20.09	21.96
9	14.68	16.92	19.02	21.67	23.59
10	15.99	18.31	20.48	23.21	25.19
12	18.55	21.03	23.34	26.22	28.30
14	21.06	23.68	26.12	29.14	31.32
16	23.54	26.30	28.85	32.00	34.27
18	25.99	28.87	31.53	34.81	37.16
20	28.41	31.41	34.17	37.57	40.00
30	40.26	43.77	46.98	50.89	53.67

$$\chi^2 = \sum \frac{(O-E)^2}{E} \qquad (15.53)$$

The χ^2 distribution is given by tables such as Table 15.4, from which the probability can be determined. The number of degrees of freedom is the number of classes whose frequency can be assigned independently. If the data are presented in the form of a table having V vertical columns and H horizontal rows, then the degrees of freedom are usually found as in Equation 15.54.

$$\text{Degrees of freedom} = (V-1)(H-1) \qquad (15.54)$$

Returning to the earlier example, suppose a coin is tossed 100 times and it comes up heads 60 times and tails 40 times. Is the coin biased? The expected values for heads and tails are 50 each so that Expression 15.55 is obtained.

$$\chi^2 = \frac{(60-50)^2}{50} + \frac{(40-50)^2}{50} = 4 \qquad (15.55)$$

The number of degrees of freedom is 1 since once we have fixed the frequency for heads; that for tails is defined. Therefore, entering Table 15.4 with one degree of freedom, the probability level for $\chi^2 = 4$ is seen to be above 2.5 percent (i.e., there is a strong probability that the difference in the two results arose by chance and the coin is not biased).

As a further example, suppose that over a 24-hour period, the average number of accidents that occur in a factory is seen to be as in Table 15.5. Does this indicate that most of the accidents occur during the late night and early morning periods? Applying the χ^2 tests, the expected value, if there was no difference between the time periods, would be the mean of the number of accidents (i.e., 5).

Table 15.5   The Chi-square Distribution

Time (24-hour clock)	Number of accidents
0–6	9
6–12	3
12–18	2
18–24	6

Therefore, from Equation 15.53 Expression 15.56 is obtained.

$$\chi^2 = \frac{(9-5)^2}{5} + \frac{(3-5)^2}{5} + \frac{(2-5)^2}{5} + \frac{(6-5)^2}{5} = 6 \tag{15.56}$$

There are three degrees of freedom; thus, from Table 15.4 the probability of occurrence of the result shown in Table 15.5 is seen to be greater than 10 percent. The conclusion would be that, although there is a trend, as yet there are not enough data to show if this trend is significant or not. For example, if the number of accidents were each three times as large (i.e., 27, 9, 6, 18, respectively), then χ^2 would be calculated as 20.67 and from Table 15.3 it is seen that the results are highly significant since there is a very low probability, less than 0.5 percent, that it can arise by chance.

Significance of Correlation

The significance of the product moment correlation coefficient of Equation 15.19 or 15.20 can be tested at any confidence level by means of the standard error of estimation given by Equation 15.21. An alternative method is to use the Student's t-test of significance. This is given by Equation 15.57, where r is the correlation coefficient and n the number of items.

$$t = \frac{r(n-2)^{1/2}}{(1-r^2)^{1/2}} \tag{15.57}$$

Tables are then used, similar to Table 15.4, that given the probability level for $(n-2)$ degrees of freedom.

The Student's t for the rank correlation coefficient is given by Equation 15.58, and the same Student's t tables are used to check the significance of R.

$$t = R\left(\frac{n-2}{1-R^2}\right)^{1/2} \tag{15.58}$$

Student's *t*-test to Determine Confidence Intervals

Student's t-test can be used to derive *confidence intervals* from a number of samples of a distribution, where, subject to certain assumptions and with a certain probability p, the mean of the distribution lies within it. p is specified before carrying out the analysis. The calculation proceeds as follows. For n samples, calculate the sample mean $\bar{x}$ and sample variance s^2, where:

$$\bar{x} = \frac{1}{n}\sum_i x_i \text{ and } s^2 = \frac{1}{n-1}\sum_i (x_i - \bar{x})^2$$

Assuming that the underlying distribution is normal, the real mean falls within confidence interval $\bar{x} - \varepsilon$ to $\bar{x} + \varepsilon$ with probability p (usually 0.9 or 0.95), where:

$$\varepsilon = t_{n-1,1-p}\sqrt{\frac{s^2}{n}}$$

$t_{n-1,1-p}$ is the Student's t-distribution with $n - 1$ degrees of freedom, which may be obtained from Table 15.6.

Computer Simulation of Communications Networks

Simulation involves using a computer to emulate the operation of a real system, in order to determine its performance. There are several reasons why it is used. First, the system being simulated may not in fact exist because it is still being designed. For example, it may not be advisable to build the system until there is reasonable certainty that it will function correctly, due to the time and expense involved. Alternatively, modeling of the system may be taking place in order to gain an understanding of it—it may not yet be clear exactly how to build it, or indeed it may not yet be technologically feasible to do so. Finally, the real system may be too large or expensive to experiment on, and an alternative is required. For example, a national telecommunications network is difficult to experiment on because it is being used, especially since it would very often be necessary to know how it performs under extreme loads or conditions.

Simulation is a valid alternative or complement to the analytical techniques of Chapters 16 and 17. In all but the simplest systems, approximations and assumptions must be made to make analytical study possible. On the other hand, arbitrary detail can in principle be incorporated into a simulation model; indeed it is possible to simulate systems that are analytically intractable. Very often, both analysis and simulation are used on the same system, facilitating cross-checking between them, where analysis can be used in simpler cases or to generate approximate results.

Designing, writing, debugging, and executing simulation programs can take a long time—often, they take days or weeks in total to execute, since many long simulation runs are often required to characterize the system accurately. The problems of design, writing, and debugging can be ameliorated by simulation packages such as OPNET or NS2, although a substantial investment of time is required when learning how to use them. Once a simulation program has been written, there is no guarantee that it is correct, so effort must be made to validate the simulator and build confidence in it.

It can be difficult to use simulation results to optimize performance of the system being simulated, since it is usually necessary to determine what the system properties should be in order to obtain the desired simulation result such as packet loss or delay. This can be very difficult because a simulator only provides data points, and not a formula indicating trends that can be algebraically manipulated. Furthermore, system characteristics can often be deduced from an analytical formula merely by inspection, whereas this is not so easy to do with a large amount of simulation data.

Pseudo-random Number Generation

Because simulations require stochastic (random) events to be modeled, they are driven by a pseudo-random number generator, generating values of random variables such as call interarrival time or packet length. Hence the simulation results themselves are stochastic in

Table 15.6 Student's *t*-Distribution

Degrees of freedom	0.40	0.25	0.10	0.05	0.025	0.01	0.005	0.0005
1	0.324920	1.000000	3.077684	6.313752	12.70620	31.82052	63.65674	636.6192
2	0.288675	0.816497	1.885618	2.919986	4.30265	6.96456	9.92484	31.5991
3	0.276671	0.764892	1.637744	2.353363	3.18245	4.54070	5.84091	12.9240
4	0.270722	0.740697	1.533206	2.131847	2.77645	3.74695	4.60409	8.6103
5	0.267181	0.726687	1.475884	2.015048	2.57058	3.36493	4.03214	6.8688
6	0.264835	0.717558	1.439756	1.943180	2.44691	3.14267	3.70743	5.9588
7	0.263167	0.711142	1.414924	1.894579	2.36462	2.99795	3.49948	5.4079
8	0.261921	0.706387	1.396815	1.859548	2.30600	2.89646	3.35539	5.0413
9	0.260955	0.702722	1.383029	1.833113	2.26216	2.82144	3.24984	4.7809
10	0.260185	0.699812	1.372184	1.812461	2.22814	2.76377	3.16927	4.5869
11	0.259556	0.697445	1.363430	1.795885	2.20099	2.71808	3.10581	4.4370
12	0.259033	0.695483	1.356217	1.782288	2.17881	2.68100	3.05454	4.3178
13	0.258591	0.693829	1.350171	1.770933	2.16037	2.65031	3.01228	4.2208
14	0.258213	0.692417	1.345030	1.761310	2.14479	2.62449	2.97684	4.1405
15	0.257885	0.691197	1.340606	1.753050	2.13145	2.60248	2.94671	4.0728
16	0.257599	0.690132	1.336757	1.745884	2.11991	2.58349	2.92078	4.0150
17	0.257347	0.689195	1.333379	1.739607	2.10982	2.56693	2.89823	3.9651
18	0.257123	0.688364	1.330391	1.734064	2.10092	2.55238	2.87844	3.9216
19	0.256923	0.687621	1.327728	1.729133	2.09302	2.53948	2.86093	3.8834
20	0.256743	0.686954	1.325341	1.724718	2.08596	2.52798	2.84534	3.8495
21	0.256580	0.686352	1.323188	1.720743	2.07961	2.51765	2.83136	3.8193
22	0.256432	0.685805	1.321237	1.717144	2.07387	2.50832	2.81876	3.7921
23	0.256297	0.685306	1.319460	1.713872	2.06866	2.49987	2.80734	3.7676
24	0.256173	0.684850	1.317836	1.710882	2.06390	2.49216	2.79694	3.7454
25	0.256060	0.684430	1.316345	1.708141	2.05954	2.48511	2.78744	3.7251
26	0.255955	0.684043	1.314972	1.705618	2.05553	2.47863	2.77871	3.7066
27	0.255858	0.683685	1.313703	1.703288	2.05183	2.47266	2.77068	3.6896
28	0.255768	0.683353	1.312527	1.701131	2.04841	2.46714	2.76326	3.6739
29	0.255684	0.683044	1.311434	1.699127	2.04523	2.46202	2.75639	3.6594
30	0.255605	0.682756	1.310415	1.697261	2.04227	2.45726	2.75000	3.6460
infinity	0.253347	0.674490	1.281552	1.644854	1.95996	2.32635	2.57583	3.2905

The top row defines the right tail area, $1 - p$. For example, to determine the 0.05 critical value (95% confidence, or $p = 0.95$) from the *t*-distribution with six degrees of freedom, look in the 0.05 column at the 6 row: $t_{0.05,6} = 1.943180$.

357

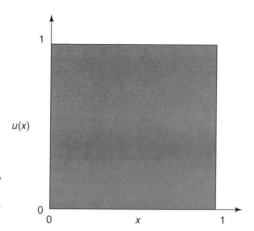

Fig. 15.11 Probability density function of a uniform random variable, which takes on values between 0 and 1.

nature. Computers cannot really generate purely random numbers, unless they tap into a natural source such as radioactive decay. Hence a pseudo-random number generator is required, which generates a deterministic sequence with properties approaching purely random numbers as closely as possible. Take such a sequence of numbers $Z_0, Z_1, Z_2, Z_3, \ldots$ with $0 \leq Z_i \leq M$, which is discussed next, and divide each by M. Each now appears to be independent, identically distributed samples from the uniform probability density distribution $u(x)$ of Figure 15.11, where samples lie between 0 and 1.

Linear congruential generators produce random numbers with a uniform probability density function (or pdf), and are very popular due to their ease of implementation. The basic approach starts with an arbitrary integer Z_0, called the "seed." This is then multiplied by an integer a. The product is divided by another integer M, leaving a remainder Z_1, which is the next term in the series. The following term Z_2 is the remainder when aZ_1 is divided by M, and so on. The generator is known as a multiplicative congruential generator. In some cases a further integer b may be added after the multiplication. This mechanism is then referred to as a mixed congruential generator.

The seed Z_0 completely determines the sequence produced, assuming specific values of constants a, b, and M. In spite of this, it is possible to produce a pseudo-random sequence that is difficult to distinguish from a sequence of purely random numbers, if the constants and the seed are chosen properly.

It is often necessary to generate numbers distributed other than uniformly, and there are three main approaches—inverse transform, composition, and accept–reject.

Inverse Transform Approach

The inverse transform method uses a function that is the inverse of the required distribution function. Starting with a distribution function $F_Y(y)$ for the random variable Y, a new random variable may be formed, $Z = F_Y(Y)$. The distribution function for the new random variable is $G_Z(z) = P[Z \leq z] = P[F_Y(Y) \leq z]$. Assuming that the inverse function F_Y^{-1} exists, the events $F_Y(Y) \leq z$ and $Y \leq F_Y^{-1}(z)$ are the same, since F_Y^{-1} can be applied to both sides. Hence, $G_Z(z) = P[Y \leq F_Y^{-1}(z)] = F_Y(F_Y^{-1}(z)) = z$, where $0 \leq z \leq 1$. This is because $P[Y \leq y] = F_Y(y)$. Thus, the variable Z is uniformly distributed in the range 0 to 1. Hence, to produce random numbers Y with the distribution function $F_Y(y)$, first generate random numbers Z with a uniform distribution; then apply the inverse transformation of $F_Y(y)$—that is $F_Y^{-1}(Z)$—to those numbers.

Unfortunately, the inverse of the distribution function does not always exist, and resort must be made instead to either the method of composition or the accept–reject method. However, the negative exponential distribution has a monotonically increasing distribution function, so the inverse exists and the inverse transform method can be used:

$$f_Y(y) = \lambda e^{-\lambda y}$$

$$\Leftrightarrow F_Y(Y) = \int_0^Y f_Y(y)dy = 1 - e^{-\lambda Y} \quad (F_Y(Y) = Z \text{ is uniformly distributed})$$

$$\Rightarrow Z = 1 - e^{-\lambda Y}$$

$$\Rightarrow e^{-\lambda Y} = 1 - Z$$

$$\Rightarrow -\lambda Y = \ln(1 - Z)$$

$$\Rightarrow Y = -\frac{1}{\lambda}\ln(1 - Z)$$

If a sequence of uniformly distributed random numbers is substituted for Z, the result will be a sequence of exponentially distributed random numbers. A simplification is possible, since if Z is uniformly distributed, $1 - Z$ will also have a uniform distribution and so Z may be substituted for $1 - Z$:

$$Y = -\frac{1}{\lambda}\ln Z$$

Composition Method

Some probability distributions can be defined in terms of others, a property that is exploited in the method of composition. Then, if sequences having the underlying distributions can be generated, the required "composite" distribution can be constructed. For example, this technique may be used to generate sequences of numbers having a Poisson distribution. The inverse of the distribution function cannot be found; therefore, the inverse transform method cannot be used. For a Poisson arrival process, the number of arrivals within a fixed time interval T has a Poisson distribution. Also, the interarrival distribution is negative exponential, so the two distributions may be combined. To do this, count the number of exponentially distributed interarrival times required to yield a total duration greater than the time interval T, and subtract 1. A collection of many such results constitutes a sequence of numbers having a Poisson distribution.

Accept–Reject Method

Another method, accept–reject, can often be used if the other methods are not suitable. Consider a random variable X having a density $f(t)$, as shown in Figure 15.12, which cannot be generated by other methods. X might be generated by randomly choosing a point lying between the pdf and the t-axis, with a uniform distribution over the entire area. Once the point had been chosen, the corresponding point on the t-axis, with its corresponding value of t, would be selected. But because $f(t)$ cannot be generated by other methods, this

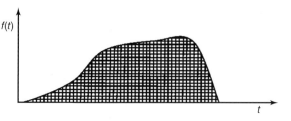

Fig. 15.12 A probability density function f(t), where the area under the curve is divided into a large number of infinitesimally small squares, one of which could be chosen at random to define a value of the random variable.

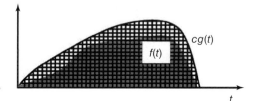

Fig. 15.13 The accept–reject method permits a point under f(t) to be selected randomly even if the corresponding random variable cannot be generated by other methods. To do this, a point is chosen at random under cg(t), and if it lies above f(t), it is rejected and a new point is chosen.

cannot be done, and if it could the procedure would be pointless since it requires the desired result beforehand. However, this does hint at how the problem can be solved.

The solution is to generate a random variable Y having density $g(t)$ so that for all t, $cg(t) \geq f(t)$, where c is a constant (Figure 15.13). It is often convenient to make $g(t)$ a uniform density function. A third variable U must be generated, uniformly distributed between 0 and 1 but independent of Y. Pick a point at random under the graph for $cg(t)$ by generating a value of t from $g(t)$—this specifies the point on the t-axis, and $Ucg(t)$ is the point on the vertical axis, completely defining the point. If the point lies between $f(t)$ and the t-axis (shaded area), then t is chosen as the value generated by $f(t)$. Otherwise (the unshaded area), the process is repeated until a value is generated for t. In other words, if $Ucg(t) \leq f(t)$, then accept the value of t as the next value of X; otherwise reject and start again. c and $g(t)$ must often be chosen carefully to avoid excessive computation.

Generation of Network Traffic

During simulation, traffic must be generated for transmission over the network, either as calls in circuit switching or as packets. This forms an important and crucial part of any communications network simulator. Simulation permits more sophisticated and accurate models of network traffic than mathematical analysis, where the models must be simplified to ensure tractability. For example, when modeling packets in slotted systems analytically, the Bernoulli model is often used, with a fixed and independent probability that each slot contains a packet. For unslotted packet systems, the packet starting times may be defined by a Poisson process, and packet lengths follow a negative exponential distribution.

While these models are simple and analytically convenient, they rarely reflect reality. For circuit-switched voice traffic, on the other hand, Poisson call arrivals with negative exponentially distributed call-holding times represent a good approximation to real telephone traffic, as well as being analytically tractable. More elaborate models exist, for example, capturing bursty behavior, but these are more complex and hence more difficult to incorporate into an analytical model. Indeed, sometimes the analytical model becomes so complex that no closed-form solution exists.

With simulation the same models could be used, but much more complex and accurate models are possible for a variety of traffic types, most significantly packetized data, voice, and video (Adas, 1997; Michiel and Laevens, 1997). Internet data traffic has been shown from measurements to be self-similar, meaning that the traffic possesses the same statistical properties on all time scales, and this concept may be expressed rigorously via mathematics. These mathematical principles, perhaps with information gathered from real networks, can be used to construct traffic generators for use in simulations, which mimic real data traffic.

Caution must be exercised when modeling TCP sources via traffic generators, since TCP adjusts its transmission rate in response to packet loss conditions in the network. Since a traffic generator cannot respond in this way, it yields incorrect simulation results (Arvidsson and Karlsson, 1999). For example, even if traffic traces from a real network were used to drive a simulation program, the traffic that is generated would adjust itself to those packet losses that occurred in the original real network, and not those that were, in fact, taking place in the network being modeled.

Programming Simulations

When programming simulations, an *operation* or *event* refers to a change of state in the system model, such as the arrival of a packet, completion of a telephone call, or failure of a component. Simulations model the real world on a completely different timescale. When 24 hours of the operation of a system are modeled, 24 hours represent the simulation time, but the time to run the corresponding simulation program is the run time or execution time. Usually, the run time is much less than the simulation time, but may not be true for very large networks or systems; it also depends on the hardware being used to run the simulation program. Time flows evenly for the operation of the real system (*real* time), but simulation time may be erratic.

There are two main approaches to writing simulators. A time-based simulation is simpler but relatively inefficient, whereas an event-based simulation (or "discrete-event simulation") is more efficient but more complex, providing more accurate timing.

In a time-based simulation (Figure 15.14), simulation time jumps forward repeatedly by fixed time intervals of length δt, with the implicit assumption that the order of events in a time interval is unimportant. A time variable t is incremented by δt upon each new interval, and the program then decides which events will occur within it. Each type of event is represented by a specific function (or equivalent) within the program, which is invoked if appropriate when a new interval arises and which updates the simulation statistics if necessary. Time-based simulation is very useful when modeling slotted packet systems, since each slot corresponds exactly to a time interval. It is inefficient when there is a low probability of an event occurring in an interval, since many time intervals must be processed without achieving anything. Also, if there are many events per time interval (high event rate), events often get out of order.

In event-based simulation (Figure 15.15), simulated time does not progress in fixed intervals; it simply jumps to the time of the next event. This is more efficient since no time is wasted processing when there are no events, but the main difficulty lies in determining which event occurs next. This is achieved by an event queue, which is a time-ordered list of events, each event being represented by the address of the corresponding function (or equivalent), where each event is labeled with its scheduled time of execution.

The program operates by taking the first event from the event queue—if there are several with the same time label, they may be executed concurrently. As before, each event

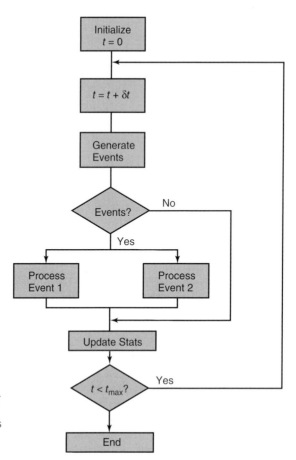

Fig. 15.14 A flow chart for a time-based simulation, where simulation time jumps forward repeatedly by fixed time intervals of δt.

can update the simulation statistics as appropriate. Each event can generate one or more new events; for example, when transmitting a packet over a link the simulator puts an event in the queue representing the arrival at the other end. Some events require other events to be either removed from the queue or rescheduled; failure of a transmission link, say, often implies early completion of all affected calls.

Collection and Verification of Simulation Results

A simulator must gather information that shows how the real system would operate under the given conditions. It may store or print out a file that traces all events—this is useful for debugging and verifying; however, it takes too much processing to obtain useful results. Instead, the desired statistics are usually compiled as the simulation runs. Summary statistics (such as means or standard deviations) are sometimes inaccurate due to arithmetic overflow and limited precision of floating-point numbers. On the other hand, frequency plots (or probability distributions) require large amounts of RAM or disk memory for storage. However, they can be used to calculate (and recalculate if necessary) summary statistics after

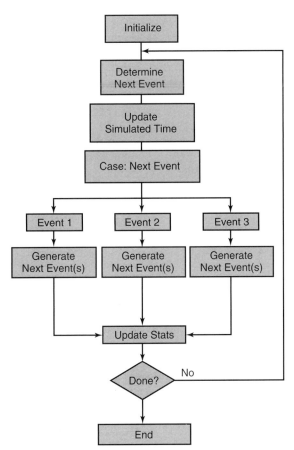

Fig.15.15 A flow chart for an event-based simulation, where, after processing each event, simulation time jumps forward to the next event.

simulation is complete. It is important to choose carefully what statistics are collected; otherwise the simulation may have to be rerun later.

It is important to verify that a simulation program is producing reliable results. However, it is never possible to be absolutely certain that it is correct. It is merely possible to establish a greater or lesser degree of *confidence* in the simulator. One way of doing so is to check as many of the programmer's unconscious assumptions as possible by placing assertions in the code. For example, check that a packet must be present in the server before service can begin, and that queue lengths must not be negative. Assertions can be checked at run time during program development, then switched off (if necessary) for efficiency during program execution. Alternatively, analytical checks can be made by running the simulator under restricted or boundary conditions on a problem that can be solved analytically. It is also possible to run the simulator on a problem for which simulation results from another simulator already exist. In either case, the simulator results (from several runs) can be compared with the known results using the appropriate statistical tests.

Occasionally, simulation results are required that describe the system when starting from cold, for example, after a system restart. More usually, the behavior of the system when running normally is of interest, for example, busy-hour traffic performance. Hence, start-up conditions usually must be removed from the overall run. One way of doing this is to have long runs so that the conditions at start-up have negligible influence, but this is expensive in computer time. Alternatively, the simulator may only record statistics after the system appears to be in equilibrium, or for a fixed proportion of the total run. For example, ignore the first 10 percent of the simulation time. However, there is usually no "scientific" way of telling what this proportion is. Finally, the simulator could artificially set the system to the expected mean state, but this may not be known.

A simulator is written to use pseudo-random number generation, so if it is run again with the same parameters and random seed, it will produce exactly the same results. But having only one such "sample path" of a stochastic process is not representative, and it is necessary to determine what the mean is for the real system. To do this, it is necessary to execute several runs with the same parameters but different random seeds, and then perform a Student's *t*-test on the results to generate confidence intervals for each point. Although the assumption made in the test of an underlying normal distribution may not be correct, it is a reasonable assumption if there are many runs.

Resources

Adas, A. (1997) "Traffic Models in Broadband Networks," *IEEE Communications* (July), pp. 82–89.

Arvidsson A., and Karlsson, P. (1999) "On Traffic Models for TCP/IP," *ITC-16* (June), pp. 457–466.

Besterfield, D. H. (1979) *Quality Control*, Prentice Hall.

Caplen, R. H. (1982) *A Practical Approach to Quality Control*, Business Books.

Chalk, G. O., and Stick, A. W. (1975) *Statistics for the Engineer*, Butterworth.

Cohen, S. S. (1988) *Practical Statistics*, Edward Arnold.

David, H. A. (1981) *Order Statistics*, Wiley.

Dudewicz, E. J., and Mishra, S. N. (1988) *Modern Mathematical Statistics*, Wiley.

Dunn, R. A., and Ramsing, K. D. (1981) *Management Science: A Practical Approach to Decision Making*, Macmillan.

Fitzsimmons, J. A. (1982) *Service Operations Management*, McGraw-Hill.

Grant, E. I., and Leavenworth, R. S. (1980) *Statistical Quality Control*, McGraw-Hill.

Hahn, W. C. (1979) *Modern Statistical Methods*, Butterworth.

Jones, M. E. M. (1988) *Statistics*, Schofield & Sims.

Leemis, Larry H., and Park, Stephen K. (2006) *Discrete-Event Simulation: A First Course*, Prentice Hall.

Mazda, F. F. (1979) *Quantitative Techniques in Business*, Gee & Co.

Michiel, H., and Laevens, J. (1997) "Teletraffic Engineering in a Broad-Band Era," *Proceedings of the IEEE* (December), pp. 2007–2033.

Siegel, A. F. (1988) *Statistics and Data Analysis*, Wiley.

Taylor, W. J., and Watling, T. F. (1985) *The Basic Arts of Management*, Business Books.

16 Queuing Theory

Phil Whiting

Introduction

Some Queuing Problems

Queuing is a ubiquitous feature of electronic systems such as computers and communication networks. Queues will build up, even if the capacity of the system exceeds the load on it. This is an inevitable consequence of the fact that arrivals and services take place in a random fashion. Indeed, the more random the services and the arrivals are, the more queues build up. Thus, it is possible for long queues to develop at even light loads, arising as a consequence of sheer variability in the pattern of arrivals and services.

If the queue length distribution does not depend on time, the queue is said to be in equilibrium. This is only possible if the queue's capacity is not exceeded. In most queuing situations the equilibrium builds up rapidly. Even when the load itself is not constant, the changes in it take place more slowly than in the queue so that the equilibrium is effectively "tracked." This means that equilibrium models can be used to determine all the usual key performance measures: delay, queue lengths, and so on. Based on this, the results here are confined to equilibrium models.

Typical delay performance for a single server queue in equilibrium with a single customer class is shown steadily increasing in the graph depicted in Figure 16.1. The horizontal axis is the occupancy (arrival rate/service rate), which is the proportion of time the customers keep the server busy. The vertical axis is the average delay. As the graph shows, there is an increasing delay penalty for using the system closer and closer to full capacity (occupancy = 1). Also shown is the graph for a deterministic system (constant time between arrivals and constant service time). No queue builds up until the system is at full load. When the occupancy is less than 1, it is said that the queue is stable and equilibrium will then exist, as already mentioned.

In most queuing situations, arrivals and services do not take place in a deterministic fashion, and so it is the graph in Figure 16.1 that is more representative of system perfor-

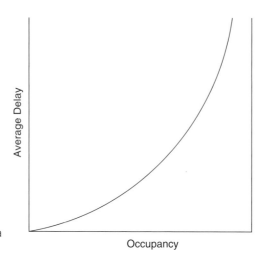

Fig. 16.1 Delay performance of a single server queue.

mance. A fast response is a strict requirement for many electronic systems; thus modeling delays and finding strategies that make efficient use of the resources within these systems becomes an essential part of their design. For example, if the graph represents the performance of a computer controlling a telephone exchange, one way to reduce the average delays would be to schedule the work so as to process shorter jobs, such as call setups and clear downs, first. This procedure "shares out" delay in a sensible fashion. Longer jobs, for example, those associated with background routines, are given lower priority and are held back. Telephone calls are connected as quickly as possible, while the longer jobs take on an increase in delay, which is often small in comparison with the amount of time needed to process them.

It is not always quite so obvious what the capacity of a queuing system is as it was in the example depicted in Figure 16.1, where the queue remains stable provided the arrival rate is less than the service rate. A communication channel that allows a group of users to share it by means of random multiple access protocols is a case in point. Such a channel will be occupied not only with the transmission of users' information but also with reservation periods where the users actually obtain rights to transmit. Some time is used in transmissions, some in reservation, but how much time is spent in each task? This depends on the nature of the protocols themselves.

For example, consider the operation of slotted ALOHA depicted in Figure 16.2. Packets of information are all the same length, and a packet is received successfully if it is the only one to be transmitted in a particular slot. If two or more are transmitted, there is a collision and none are successfully received.

Such packets are retransmitted at random in later slots. But there are many ways to choose a slot for retransmission at random and this choice is critical. It determines how successful the protocol is in avoiding collisions and how many packets can be eventually transmitted. A poor protocol choice may give very little useful throughput or, worse, cause the channel to jam. In this case it turns out that using fixed retransmission probabilities does not work very well. They make no allowance for the variation in the backlog of packets awaiting retransmission. However, adjusting the retransmission probability according to an estimate of this backlog does work adequately. Each packet is retransmitted with high probability when the backlog is estimated to be low and with low probability when the estimated backlog is high.

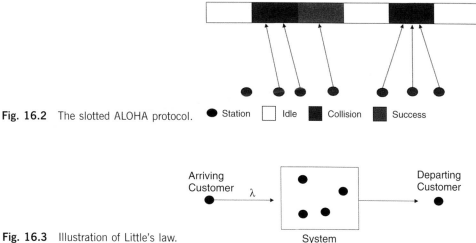

Fig. 16.2 The slotted ALOHA protocol.   ● Station   ☐ Idle   ■ Collision   ■ Success

Fig. 16.3 Illustration of Little's law.

In queuing network problems, there is more than one queue and customers make a series of visits from one queue to the next. In such networks there are interactions between the individual queues within them. For example, the pattern of arrivals at one queue is determined by the departure patterns from the queues that have customers feeding into them (together with the new arrivals at the network). Since these departure patterns themselves depend on the arrivals to those queues, analysis can prove very difficult.

It is interesting to learn, therefore, that for one very wide class of networks it is relatively easy to determine the equilibrium distribution of the customers within those networks. Furthermore, the form of this distribution is particularly striking. It is simply the product of the equilibrium distribution for each queue operating as if it were in isolation!

Such models have been used to analyze the performance of packet-switched networks. In these networks, packets of information are transmitted from source to destination via a series of transmission links. These transmission links are interconnected by switching nodes that queue the packets in storage buffers until it is their turn to be transmitted. (Packet-switched networks are sometimes referred to as store-and-forward networks.)

From these models it is known that the delays that packets experience and, indeed, the throughput of the network itself depend strongly on the routes the packets take through it. But determining the best choice of routes using such models is not easy because the assignment of one particular route or set of routes to one group of packets has an effect on the delays, which other packets experience. However, effective routing algorithms have been proposed based on models such as the ones mentioned before (Gallager, 1977).

Little's Law

There are very few completely general queuing results. However, there is one, which is of considerable utility for the analysis of queues in equilibrium, Little's law. It even applies to systems that can hardly be thought of as queues at all. Suppose that customers arrive at the system (depicted as a black box in Figure 16.3) at a rate λ and that the average sojourn time for each customer is $E[D]$, in equilibrium. Suppose further that the time average number of customers within the system is $E[N]$. Then Little's law gives Equation 16.1.

$$\lambda E[D] = E[N] \qquad\qquad (16.1)$$

Little's law is an accounting identity and can be understood as such. Suppose customers are charged at a rate of \$1 per each second they remain in the system. There are two equivalent ways in which the money can be collected. The first is to collect the appropriate amount from each customer as they leave. Since the average duration within the system is $E[D]$ and the mean departure rate must be λ, the rate at which money is being collected is $\lambda E[D]$. The other way to collect the money is for the customers to pay continuously as they remain in the system. Since the average number of customers within the system is $E[N]$, the long run rate at which money is being collected is also $E[N]$. By altering the rates of payment, one may obtain other similar identities.

Kendall Queuing Notation

As randomness plays such an important part in queuing performance, queues are categorized accordingly. The notation A/B/m/S/P is in common use, as in Table 16.1. Omission of either the storage or the population entry means they are infinite. The most used interarrival and service distributions are given in Table 16.2.

Markovian is equivalent to a Poisson process, if the customer population is infinite and too exponential for the service distribution. The Erlangian distribution is just the sum of k independent, identically distributed exponential variants. Sometimes this is referred to as a server having k exponential service "phases." See Figure 16.4(a), which illustrates E_3. (This distribution is also referred to as the gamma.) The hyperexponential is constructed from the exponential distribution as well, but the customer undergoes a single service phase allocated at random from a number of possible phases; see Figure 16.4(b).

Table 16.1 Notations in Common Use

Item	Symbol
Interarrival distribution	A
Service distribution	B
Number of servers	m
Storage capacity	S
Customer population	P

Table 16.2 Distribution Symbols in Use with Kendall Queuing Notation

Item	Symbol
Deterministic	D
Erlangian with k degrees of freedom	E_k
General	G
Hyperexponential with j degrees of freedom	H_j
Markovian	M

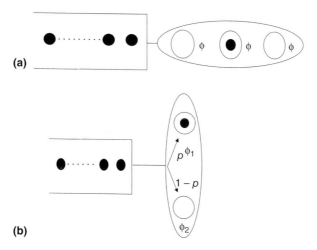

Fig. 16.4 An Erlangian E_3 and a hyperexponential H_2 server: (a) E_3, (b) H_2.

Some examples of queues represented in Kendall queuing notation are:

- *D/G/1:* This is a single server queue with deterministic arrivals and an arbitrary service time.
- *M/M/3/10:* This is a queue with Poisson arrivals, three exponential servers, and room for 10 customers, including those in service.
- *M/M/1/∞/20:* This is a queue with exponential interarrival times with the rate proportional to the number of customers not in queue, a single exponential server, and infinite waiting room. The customer population is 20.

In the case of queuing networks, one may wish to specify the queue types while ignoring the nature of the processes altogether. In such cases it may be written, for example, that there is a network of ./M/1 queues, meaning that there is only a single server at each of the nodes with exponentially distributed service times.

Models

The Memoryless Property and the Poisson Process

The exponential distribution is encountered frequently in queuing analysis. One reason is that the exponential can be used as a building block to construct other distributions, as was shown earlier. Indeed, the distribution of virtually any continuous positive random variable may be approximated using mixtures of sums of iid exponents (Kelly, 1979). However, the following property is a more significant reason for its importance in queuing theory.

The density of the exponential is $\varphi\, e^{-\varphi s}$, and integration of this gives the corresponding distribution function $1 - e^{-\varphi s}$. Suppose X is the random variable drawn from this distribution. Say that X "completes" when the time corresponding to X is reached starting from 0. Suppose t seconds have elapsed and X has not completed. The memoryless property is this: The distribution of remaining time until X completes in no way depends on t and is given by the same exponential distribution as X.

To see that this must be the case, consider the following example. A random job from a class with exponential processing time requirements, with rate parameter φ, has been the

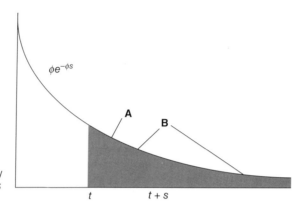

Fig. 16.5 Conditional probability that the job is completed within *s* seconds.

sole task within a computer for *t* seconds. The total processing requirement is *X*. Let us compute the probability that the job will take at most a further *s* seconds to be completed. The proportion of random jobs whose processing times lie in the interval $(t, t + s)$ is given by area A in the graph in Figure 16.5. However, this particular job is one that has taken *t* seconds already and it is the proportion of all such jobs completed in a further *s* seconds that is required. The probability that a job takes at least *t* seconds is given by the tail region B. Thus, the conditional distribution of the remaining time until the job is completed, given that *t* seconds have elapsed, is as in Equation 16.2.

$$\Pr\{X \le t+s \,|\, t < X\} = \frac{\text{Area A}}{\text{Area B}}$$

$$= \frac{\Pr\{t < X \le t+s\}}{\Pr\{t < X\}} \tag{16.2}$$

$$= \frac{e^{-\varphi t} - e^{-\varphi(t+s)}}{e^{-\varphi t}} = 1 - e^{-\varphi s}$$

This is seen to be independent of *t*. The conditional probability that the job is completed in a further *s* seconds is given by the very same exponential distribution as would be used to determine the probability of the job being completed in *s* seconds starting from time 0.

Often it is said that the job is being processed at rate φ, meaning that the probability that the job is completed in the next infinitesimal time interval *dt* is $\varphi \, dt$, given that it has not been completed already.

The Poisson process may be constructed from the exponential. Indeed, it is useful to regard the Poisson process as a sequence of customers arriving at a queue with mutually independent and identical exponential interarrival times. As such, the Poisson process "inherits" the memoryless property from the exponential. This means that the probability that an arrival takes place in any given interval is independent of the history of the queue prior to that interval and in particular of the state of the queue immediately before the arrival. The parameter of the Poisson process is the same rate parameter as the underlying exponential distribution.

In fact, given a sequence of independent random variables, $X_1, X_2, X_3, \ldots$ with common exponential distribution rate parameter λ, the sequence of arrival times is determined by Expression 16.3, starting from time 0.

$$X_1, X_1 + X_2, X_1 + X_2 + X_3, \ldots \tag{16.3}$$

Now that the construction follows, the Poisson distribution can be checked. To do this, recall that the density of the sum of k independent exponential random variables with rate parameter λ is given by Expression 16.4, which is the k-Erlangian distribution with parameter λ (Feller, 1970).

$$\lambda \frac{(\lambda t)^{k-1}}{(k-1)!} e^{-\lambda t} \tag{16.4}$$

Denote the number of arrivals in time interval $(0, t)$ by N_t; then Equation 16.5 may be obtained.

$$\begin{aligned}
\Pr\{N_t = k\} &= \Pr\{X_1 + \ldots + X_k \le t; X_1 + \ldots + X_k + X_{k+1} > t\} \\
&= \int_0^t \lambda \frac{(\lambda s)^{k-1}}{(k-1)!} e^{-\lambda s} e^{-\lambda(t-s)} ds \\
&= \frac{(\lambda t)^k}{k!} e^{-\lambda t}
\end{aligned} \tag{16.5}$$

Markov Processes and Markov Chains

Markov process models underlie a great deal of queuing analysis, and a grasp of these models is essential to understand analytical models of queues. A stochastic process $X(s)$ defined on a countable state space K is said to be Markov if, given a finite sequence of $n > 1$ points (nearly always points in time) $s_1 < s_2 \ldots < s_n$, Equation 16.6 holds.

$$\begin{aligned}
&\Pr\{X(s_n) = j_n \mid X(s_1) = j_1; \ldots; X(s_{n-1}) = j_{n-1}\} \\
&= \Pr\{X(s_n) = j_n \mid X(s_{n-1}) = j_{n-1}\}, j_1, \ldots j_n \varepsilon K
\end{aligned} \tag{16.6}$$

The conditioning event must have positive probability. In words, the probability of finishing up in a particular state, conditional on the process visiting given states at given times, depends only on the most recent of those states stipulated. Thus, to determine the probability of future events, one need only take the current state into account. Notice that the above probability, the transition probability from j_{n-1} to j_n, may also depend on the time but if it depends only on the difference $s_n - s_{n-1}$ the process is called time homogeneous. If any state can be attained from any other state, the process is said to be irreducible.

A discrete time Markov process is called a Markov chain. For simplicity it may be supposed that transitions take place at unit intervals. In Markov chains, it is possible that certain states can be visited at intervals that are constant multiples of some given number $v > 1$. This is illustrated in Figure 16.6, which is for the absolute difference of heads to tails in a series of coin tossings. The even states are visited every other transition, so $v = 2$. Such a chain is called periodic. We shall not consider such chains here. All the chains considered will be aperiodic.

Time-homogeneous Markov chains that are irreducible and aperiodic may have an equilibrium probability distribution $\pi(j)$ given by Equation 16.7, where $p(k, j)$ is given by Equation 16.8.

$$\pi(j) = \sum_{k \varepsilon K} \pi(k) p(k, j) \tag{16.7}$$

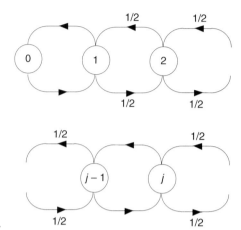

Fig. 16.6   A coin-tossing chain with period 2.

$$p(k, j) = \Pr\{X(s+1) = j \,|\, X(s) = k\} \tag{16.8}$$

If an equilibrium distribution exists, the distribution of the process will always converge to it as in Equation 16.9, and, having this property, it must be unique. $\pi(j)$ can be regarded as the long-run average proportion of time the process spends in state j.

$$\lim_{s \to \infty} \Pr\{X(s) = j \,|\, X(0) = k\} = \pi(j) \tag{16.9}$$

As far as the Markov processes here are concerned, they may be regarded as arising in the following way. The construction is based on the exponential that was discussed in the previous section. In each state j, the rate at which transitions are made to state k is $q(j, k)$. Thus, the probability that the process enters state k in the next interval of length dt is $q(j, k)\,dt$. The probability that a transition out of state j will take place at all in time interval dt is therefore given by Equation 16.10.

$$\sum_{k=j} q(j, k)dt = q(j)dt \tag{16.10}$$

Put another way, each visit to j has a duration that is exponential with rate parameter $q(j)$. The probability that the process then enters state k is given by Equation 16.11.

$$p(j, k) = \frac{q(j, k)}{q(j)} \tag{16.11}$$

The equilibrium distribution, if it exists, satisfies Equation 16.12, which is referred to as the global balance equation. $\pi(j) \, q \, (j, k)$ is called the probability flux out of j into state k in equilibrium.

$$\sum_{k \varepsilon K} \pi(j)q(j, k) = \sum_{k \varepsilon K} \pi(k)q(k, j) \tag{16.12}$$

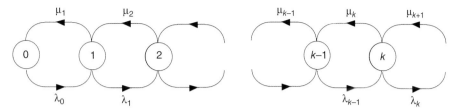

Fig. 16.7 A linear birth–death model.

Birth–Death Models of Queues

The underlying idea of this class of queuing models is to regard arriving customers as births and departures as deaths; thus the queue alters by at most one. The most general form of the linear birth–death process is depicted in Figure 16.7.

The global balance equations can be obtained by equating the probability flux out of each state with the probability flux into it, as already described. (See equation 16.13)

$$\lambda_0 \pi_0 = \mu_1 \pi_1;$$
$$(\lambda_k + \mu_k)\pi_k = \lambda_{k-1}\pi_{k-1} + \mu_{k+1}\pi_{k+1}, k \geq 1 \tag{16.13}$$

However, these equations may be solved by using a simpler set of equations, known as the local (detailed) balance equations, as in Equation 16.14.

$$\lambda_k \pi_k = \mu_{k+1}\pi_{k+1}, k \geq 0 \tag{16.14}$$

From these equations, the equilibrium solution is easily seen to be as in Equation 16.15, and π_0 is found from the fact that the probabilities must sum to 1.

$$\pi_k = \pi_0 \prod_{j=0}^{j=k} \frac{\lambda_j}{\mu_{j+1}}, k \geq 1 \tag{16.15}$$

It can be seen that if a probability distribution that satisfies the detailed balance equations can be found, then it must satisfy the global balance equations. This is so because each global balance equation is the sum of two detailed balance equations, excluding the first and the last, if there is a last. These two are just a single local balance equation. (Apart from the simplification provided, the significance of local balance equations is that they show that the linear birth–death process is reversible in equilibrium. In crude terms, reversible Markov processes cannot be distinguished, statistically, as going forward in time or going backward in time. From this it is clear that a Markov process must be in equilibrium to be reversible.)

By taking different birth and death rates, one can obtain a wide range of queuing models, as for example in Table 16.3. Note that in each case the mean sojourn time, $E[D]$, can be found by an application of Little's law. However, in the case of M/M/K/∞/L, there is only a finite number of customers and, in the case of M/M/1/S, customers are lost because there is only finite storage. In neither case is the arrival rate of customers entering queue λ.

For M/M/K/∞/L, Equation 16.16 may be obtained, where θ is the mean rate at which customers complete a cycle of standing idle, queuing, and service.

Table 16.3 Equilibrium Queue Distribution for Some Birth–Death Models

Queue type	Parameters	Equilibrium queue distribution
M/M/I	$\lambda_k = \lambda,\ k \geq 0$	$Q_k = (1 - \rho)\rho^k,\ k \geq 0$
	$\mu_k = \mu,\ k \geq 1$	$E[Q] = \dfrac{\rho}{(1-\rho)}$
M/M/K/∞/L	$L > K,\ \lambda_k = (L - k)\lambda$	$Q_k = \dbinom{L}{k}\left(\dfrac{\lambda}{\mu}\right)^k Q_0,\ k \leq K$
	$\mu_k = k\mu,\ k \leq K$	
	$\mu_k = K\mu,\ k > K$	$Q_k = \dbinom{L}{k}\left(\dfrac{\lambda}{\mu}\right)^k \dfrac{k!}{K!}K^{K-k}Q_0,\ K < k \leq L$
M/M/1/S	$\lambda_k = \lambda,\ k < S$	$Q_k = \dfrac{(1-\rho)}{(1-\rho^{S+1})}\rho^k,\ k \leq S$
	$\lambda_k = 0,\ k \geq S$	
	$\mu_k = \mu,\ k > 0$	$E[Q] = \dfrac{\rho}{(1-\rho)} - \dfrac{(S+1)\rho^{S+1}}{1-\rho^{S+1}}$
M/M/∞	$\lambda_k = \lambda,$	$Q_k = e^{-\rho}\dfrac{\rho^k}{k!},\ k \geq 0$
	$\mu_k = k\mu$	$E[Q] = \rho$

$$\theta\left(E[D] + \frac{1}{\lambda}\right) = L,\ E[Q] = \theta E[D]; \tag{16.16}$$

Equation 16.16 follows from Little's law applied to all the customers, with $\dfrac{1}{\lambda}$ the mean time a customer stands idle before returning to queue for service. The second part comes from Little's law applied to the queuing and service stages, $\sum_k kQ_k = E[Q]$.

Substituting for θ gives Equations 16.17 and 16.18.

$$E[Q] = \frac{LE[D]}{\left(E[D] + \dfrac{1}{\lambda}\right)} \tag{16.17}$$

$$E[D] = \frac{E[D]}{\lambda(L - E[Q])} \tag{16.18}$$

For M/M/1/S, the mean arrival rate of customers entering queue is given by Expression 16.19.

$$\frac{\lambda(1-\rho^S)}{(1-\rho^{S+1})} \tag{16.19}$$

Again from Little's law we have Equation 16.20.

$$\begin{aligned}
E[D] &= \left[\frac{\rho(1-\rho^{S+1})}{(1-\rho)(1-\rho^S)} - \frac{(S+1)\rho^{S+1}}{(1-\rho^S)}\right]\frac{1}{\lambda} \\
&= \left[\frac{(1-\rho^{S+1})}{(1-\rho)(1-\rho^S)} - \frac{(S+1)\rho^S}{(1-\rho^S)}\right]\frac{1}{\mu}
\end{aligned} \tag{16.20}$$

Transform Methods

Arival and Departure Distributions

It is possible to think of customers' distribution in a queue in three ways:

1. The distribution of the number of customers found in queue by an arriving customer
2. The distribution of the number of customers left in queue by a departing customer
3. The distribution of the number of customers in queue at a particular time

Fortunately, it is the case that, in equilibrium, the first two of these distributions coincide in very general circumstances. This is demonstrated below. It is even possible that all three of these distributions will coincide in equilibrium. This is the case when the arrival process is Poisson.

For the equality of the first two distributions in equilibrium, it is sufficient that arrivals and departures should take place singly and that at least one of the two equilibrium distributions should exist. This latter condition holds for most queues, even non-Markovian ones, provided that the queue is stable. To see this, suppose, for simplicity, that the queue is initially empty. (The argument also works when there is an initial queue of customers, but this minor complication is ignored.)

Figure 16.8 shows the kth departure, leaving $j \leq m$ customers still in the system. This means that customer $k + m + 1$ is yet to arrive and that when he does he will find m or fewer customers in the system. (See Equation 16.21.)

$$\Pr\{d_k \leq m\} \geq \Pr\{a_{m+k+1} \leq m\} \tag{16.21}$$

Now consider the arrival of customer $m + k + 1$ and suppose that she too finds $j \leq m$ customers in the system. This means that the kth departure has already taken place and left behind at most m customers. (See Equation 16.22.)

$$\Pr\{a_{k+m+1} \leq m\} \geq \Pr\{d_k \leq m\} \tag{16.22}$$

Hence, Equation 16.23 can be obtained, and if we suppose that the equilibrium arrival distribution $A(m)$ exists, then Equation 16.24 follows.

$$\Pr\{a_{k+m+1} \leq m\} \geq \Pr\{d_k \leq m\} \tag{16.23}$$

$$
\begin{aligned}
A(m) &= \lim_{k \to \infty} \Pr\{a_{k+m+1} \leq m\} \\
&= \lim_{k \to \infty} \Pr\{d_k \leq m\} = D(m)
\end{aligned}
\tag{16.24}
$$

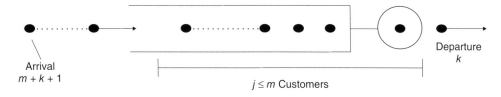

Fig. 16.8 Customers "seen" on arrival and departure.

Similarly, the equilibrium departure distribution can be shown to exist provided that the equilibrium arrival distribution does and the two are once again equal.

Outline of the Method

The approach is based on transforming the Markov equations, which describe the queuing process, and so obtaining algebraic equations for the transforms, and then solving these equations. This latter step usually involves finding zeros of functions. The method can be applied to both equilibrium and transient queue (Takacs, 1962).

The most commonly used transforms are the probability-generating function (*z*-transform) for discrete random variables and the Laplace transform for nonnegative, continuous random variables. If N is a discrete random variable and T is a nonnegative continuous (or mixed) random variable, Equation 16.25 may be obtained, where p_k is given by Equation 16.26 and $H(t)$ is the distribution of T.

$$
\begin{aligned}
H^*(s) = E[e^{-sT}] = \int_0^\infty e^{-st}dH(b) \\
= E[e^{-sT}] = \int_0^\infty e^{-st}dH(t)
\end{aligned}
\tag{16.25}
$$

$$
p_k = \Pr\{N = k\}
\tag{16.26}
$$

A little care is needed in working with the Laplace transforms of queuing variables since mixed distributions can occur (e.g., waiting times); these are a mixture of an atom at the origin (no wait at all) and a continuous density (waiting time conditional on having to wait).

Moments can be obtained from these by taking derivatives at 1 and 0, respectively, as in Equation 16.27. The derivatives give factorial moments in the case of the probability-generating functions and moments about the origin (to within a sign) in the case of the Laplace transform.

$$
\begin{aligned}
\left.\frac{d^n f}{dz^n}\right|_{z=1} = E[N(N-1)\ldots(N-n+1)], \\
\left.\frac{d^k H^*}{ds^k}\right|_{s=0} = (-1)^k E[T^k] = (-1)^k \int_0^\infty t^k dH(t) = (-1)^k h_k
\end{aligned}
\tag{16.27}
$$

The M/G/1 Queue in Equilibrium

In general, the M/G/1 queue is not Markov unless one includes the time to complete the customer in service as well as the number of customers in queue.

However, an analysis can be conducted by examining the queue at special instances. In fact, the number of customers in queue between each departure is a Markov chain (referred to as the embedded Markov chain). This is so because the arrival process is Poisson.

Suppose the arrival rate of customers is v and the service distribution is $H(x)$ with density $h(x)$. The probability p_k that k customers arrive during a customer service is given by Equation 16.28.

$$
p_k = \int_0^\infty e^{-vx} \frac{(vx)^k}{k!} h(x)dx
\tag{16.28}
$$

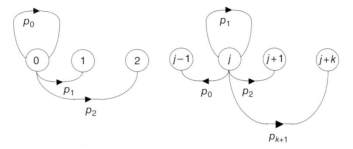

Fig. 16.9 The embedded Markov chain of the M/G/1 queue.

Transforming to get the generating function for the p_k's gives Equation 16.29.

$$\sum_{k=0}^{\infty} p_k z^k = \int_0^{\infty} e^{-\nu x} \frac{(\nu z x)^k}{k!} h(x) dx$$

$$= \int_0^{\infty} e^{-\nu x} \sum_{k=0}^{\infty} \frac{(\nu z x)^k}{k!} h(x) dx \qquad (16.29)$$

$$= \int_0^{\infty} e^{-\nu x + \nu x z} h(x) dx = H^*(\nu - \nu z)$$

The interchange of summation and integral can be justified readily. Thus, the transform of p is obtained through the Laplace transform of H. The embedded Markov chain of arrivals is determined through p but, as can be seen from Figure 16.9, the empty state is a special case. A customer departing and leaving the queue empty marks the start of an idle period.

The simplest way to analyze this chain is to work with the random variables directly. (Alternatively, transform the equilibrium equations of the embedded Markov chain.) Let Q_n denote queue length immediately following the nth departure, A_n the number of arrivals during the nth service; then Equation 16.30 can be obtained, where I_n is 1 to allow for the departure of customer $n - 1$ except if customer $n - 1$ departs before customer n arrives, in which case I_n is 0 because $Q_{n-1} = 0$.

$$Q_{n+1} = Q_n + A_{n+1} - I_{n+1} \qquad (16.30)$$

The probability-generating function for Q_{n+1} is given by Equation 16.31.

$$Q_{n+1}(z) = E[z^{Q_{n+1}}] = E[z^{Q_n - I_{n+1}}] E[z^{A_{n+1}}]$$

$$= \left(\frac{E[z^{Q_n}] - q_{0,n}}{z} \right) H^*(\nu - \nu z) \qquad (16.31)$$

The number of arrivals during each service is independent of the current queue length, allowing this distribution to be factored out at the start of Equation 16.31. By taking limits as $n \to \infty$, an equation for the equilibrium-generating function, Q, for the number of customers left behind by one departing may be found from Equation 16.32.

$$Q(z) = \left(\frac{Q(z) - q_0}{z} + q_0 \right) H^*(\nu - \nu z)$$

$$Q(z) = \frac{H^*(\nu - \nu z)(z - 1) q_0}{z - H^*(\nu - \nu z)} \qquad (16.32)$$

Q exists whenever the queue is stable. However, $-H^{*\prime}(0) = \dfrac{1}{\mu}$ is the mean service duration and so this condition is $\rho = -\upsilon H^{*\prime}(0) < 1$.

Setting $z = 0$ shows that q_0 is the equilibrium probability that a customer departs leaving the queue empty. Dividing by $z - 1$, taking limits as $z \to 1$, and applying L'Hopital's rule determines q_0 as in Equation 16.33 and hence Equation 16.34.

$$1 = Q(1) = \frac{q_0}{\lim\limits_{z \to 1} \dfrac{z - H^*(\nu - \nu z)}{z - 1}} = \frac{q_0}{1 + \nu H^{*\prime}(0)} \tag{16.33}$$

$$q_0 = 1 - \rho, Q(z) = \frac{(1-\rho)H^*(\nu - \nu z)(z-1)}{z - H^*(\nu - \nu z)} \tag{16.34}$$

$Q(z)$, as derived, is the generating function for the number of customers left behind in queue by a departing customer. However, this is then the equilibrium distribution for the number of other customers that the customer finds on arrival.

This follows from the equality of equilibrium customer arrival distributions and customer departure distributions result described earlier. Note also that Q gives the time equilibrium of customers in queue as well because of Poisson arrivals.

Q can be used to obtain the equilibrium waiting time transform. Let S be the total time a customer spends in the system (queuing time plus service time) with equilibrium Laplace transform $T^*(s) = E[c^{-Ss}]$.

By making an analogous argument to the one we made for the distribution of the number of customers that arrive during a service, $Q(z)$ may be found. This is just the number of arrivals during the customers' system time, so that Equation 16.35 may be obtained.

$$Q(z) = T^*(\nu - \nu z) \tag{16.35}$$

However, the service time and waiting time are independent, giving Equation 16.36.

$$W^*(s)H^*(s) = T^*(s) \tag{16.36}$$

Putting these two together and replacing $\upsilon - \upsilon z$ with s gives Equation 16.37 and Equation 16.38.

$$W^*(s)H^*(s) = T^*(s) = \frac{(1-\rho)H^*(s)\left[\left(1 - \dfrac{s}{\nu}\right) - 1\right]}{\left(1 - \dfrac{s}{\nu}\right) - H^*(s)} \tag{16.37}$$

$$W^*(s) = \frac{(1-\rho)s}{\nu H^*(s) + s - \nu} \tag{16.38}$$

This is the equilibrium transform for the waiting time.

The G/M/m Queue in Equilibrium

The analysis is very similar to the preceding but works with the queue immediately after each arrival. The waiting distribution follows immediately from this because the remaining time to complete the service of the customer at the front of the queue is exponential. (See the previously mentioned "memoryless" property of the exponential.) Thus, if $r(z)$ is the generating function of customers in queue ahead of an arrival, the transform of the waiting distribution is given by Equation 16.39 and Equation 16.40.

$$W^*(s) = r\left(\frac{\mu}{\mu + s}\right) \tag{16.39}$$

$$r(z) = \sum_0^\infty R_m z^m \tag{16.40}$$

The form of r is particularly simple for the G/M/m queue. Work the opposite way around from the M/G/1 queue and look at the number of departures between two arrivals. Suppose that the interarrival density is given by $f(x)$. Then assume that there are $n > j$ customers in the queue. The probability that there are j departures before the next arrival is given by Equation 16.41 if $n - j \geq m$ so that all servers are busy during this period.

$$q_{n,j} = \int_0^\infty e^{-m\mu x} \frac{(m\mu x)^j}{j!} f(x)dx \tag{16.41}$$

For R_n, $n \geq m - 1$, the equilibrium probability that an arrival finds n customers ahead of him in the queue is given by Equation 16.42, where the equation includes all the possibilities: The preceding customer may find $n - 1$, n, $n + 1$, ... customers in the queue and then 0, 1, 2, ... corresponding departures.

$$R_n = \sum_{k=0}^\infty R_{k+n-1} \int_0^\infty e^{-m\mu x} \frac{(m\mu x)^k}{k!} f(x)dx, n \geq m \tag{16.42}$$

These equations have the solution given by Equation 16.43 provided Equation 16.44 holds.

$$R_n = C\eta^n, n \geq m - 1 \tag{16.43}$$

$$\eta = F^*(m\mu - m\mu\eta) < 1 \tag{16.44}$$

Thus, η is determined through the Laplace transform of F. A unique root exists if $m\mu$ $\alpha < 1$ where α is the arrival rate. By setting $n = m - 1$, we obtain Equation 16.45. In the special case where $m = 1$, the normalization of Equation 16.46 gives Equation 16.47 and the queue length distribution is geometric.

$$C = R_{m-1}\eta^{l-m} \tag{16.45}$$

$$\sum_0^\infty R_n = 1 \tag{16.46}$$

$$R_0 = 1 - \eta \tag{16.47}$$

To obtain the remaining probabilities, in the case when $m > 1$, R_{m-1}, R_{m-2}, ..., R_0, compute the probability q_{nj} of there being j departures between two customer arrivals when there are n customers initially. This has been done already in the case $n - j \geq m$. Suppose $n \leq m$; then Equation 16.48 may be obtained; if $n > m \geq n - j$, then Equation 16.49 is obtained.

$$q_{n,j} = \int_0^\infty \binom{n}{j}(1 - e^{-\mu x})^j e^{-\mu(n-j)x} f(x)dx \tag{16.48}$$

$$q_{n,j} = \int_0^\infty \binom{m}{n-j} e^{-\mu(n-j)x} \int_0^x \frac{m\mu(m\mu t)^{n-m-1}}{(n-m-1)!} \times (e^{-\mu t} - e^{-\mu x})^{m-n+j} dt \, f(x)dx \tag{16.49}$$

The equilibrium probabilities that there are k customers found in queue on arrival for $k = m - 1, m - 2, \ldots$ are given by a series of equations such as Equation 16.50, Equation 16.51, and so on.

$$R_{m-1} = \sum_{j=0}^\infty q_{m+j-1,j} R_{m+j-2}$$

$$= q_{m-1,0} R_{m-2} + R_{m-1} \sum_{j=1}^\infty \eta^{j-1} q_{m+j-1,j} \tag{16.50}$$

$$R_{m-2} = \sum_{j=0}^\infty q_{m+j-2,j} R_{m+j-3}$$

$$= q_{m-2,0} R_{m-3} + q_{m-1,1} R_{n-2} + R_{m-1} \sum_{j=2}^\infty \eta^{j-2} q_{m+j-2,j} \tag{16.51}$$

These equations can be solved recursively by setting $R_{m-1} = 1$ and then determining R_{m-2}, R_{m-3}, ... in turn. The solution is completed by renormalizing the probabilities to sum to 1, as in Equation 16.52.

$$1 = \sum_0^{m=1} R_j + R_{m-1} \sum_{j=1}^\infty \eta^j = \sum_0^{m-1} R_j + R_{m-1} \frac{\eta}{(1-\eta)} \tag{16.52}$$

Results from the Transforms

By differentiating Equation 16.38 and setting s to 0, the mean waiting time in an M/G/1 queue is obtained. This formula is known as the Pollazcek-Kinchine formula (Equation 16.53).

$$E[W] = \frac{v h_2}{2(1-\rho)} \tag{16.53}$$

Differentiating again, the variance of the equilibrium wait may be determined, as in Equation 16.54.

$$\text{Var}[W] = \{E[W]\}^2 + \frac{v h_3}{3(1-\rho)} \tag{16.54}$$

In fact, by a manipulation, the waiting time transform can be inverted, as in Equation 16.55.

$$W^*(s) = \frac{(1-\rho)}{1 - \rho \dfrac{\left(1 - H^*(s)\right)}{s/\mu}}$$

$$= (1-\rho) \sum_{k=0}^{\infty} \left[\rho \frac{\left(1 - H^*(s)\right)}{s/\mu} \right]^k \qquad (16.55)$$

However, the Laplace transform of the residual service time density $r(x) = \mu(1 - H(x))$ is given by Expression 16.56, so that Equation 16.57 gives the equilibrium waiting time density (Benes, 1956) in an M/G/1 queue.

$$\frac{\left(1 - H^*(s)\right)}{s/\mu} \qquad (16.56)$$

$$w(x) = (1-\rho) \sum_{k=0}^{\infty} \rho^k r^{wk}(x) \qquad (16.57)$$

From which we may deduce that

$$\Pr\{W > x\} = \sum_{n=1}^{\infty} (1-\rho) \rho^n \Pr\left\{ H_1^* + \ldots + H_n^* > x \right\} \qquad (16.57a)$$

where H_k^* represents independent residual service time variables with density $r(x)$. An important case is when the service distribution exhibits "heavy tail behavior" as in regularly varying distributions that satisfy

$$1 - H(x) = x^{-V} L(x), \, x \geq 0 \qquad (16.57b)$$

where $L(x)$ is slowly varying at infinity for all $t > 0$:

$$\lim_{x \to \infty} \frac{L(xt)}{L(x)} = 1 \qquad (16.57c)$$

(e.g., if L is constant or a logarithmic function).

Such distributions arise in the modeling of long-range dependence on the Internet. For more information, see *Self Similar Traffic and Performance Evaluation* edited by K. Park and W. Willinger, which provides an introduction to modeling of such queues in this context.

Since, in the case of heavy tail service times, we have as $x \to \infty$,

$$\Pr\left\{ H_1^* + \ldots + H_n^* > x \right\} \sim n \Pr\left\{ H_1^* > x \right\} \qquad (16.57d)$$

an example of the "single jump" asymptotic for heavy tails. (Here $\sim$ implies that the ratio of the two sides tends to 1 as $x \to \infty$.)

By reversing the order of limits, we thus obtain that

$$\Pr\{W > x\} \sim \frac{\rho}{1-\rho} \Pr\left\{ H_1^* > x \right\} \qquad (16.57e)$$

because $x \rightarrow \infty$ (Equation 16.57e) actually applies to the waiting time of G/G/1 queues with regularly varying behavior. See Chapter 6 and references therein of *Self Similar Traffic and Performance Evaluation* for additional results.

A great deal of caution should be used in the analysis and simulation of such heavy-tailed queues in general. Performance results depend heavily on the details of the queuing system being modeled. In particular, asymptotic results such as Equation 16.57e depend on the service discipline. Nevertheless, for work-conserving systems, Equation 16.57e still holds if we replace W with the equilibrium workload, and it also still holds for a FIFO system if we replace W with the sojourn time (service + waiting time). For additional results, refer to A. P. Zwart's *Queuing Systems with Heavy Tails*.

For the G/M/1 queue, the equilibrium distribution of queue length on arrival (and therefore on departure as well) is geometric, as in Equation. 16.58.

$$\Pr\{Q = k\} = (1 - \eta)\eta^k \tag{16.58}$$

From this it follows immediately that the equilibrium G/M/1 waiting time distribution is exponential, as in Equation 16.59.

$$\Pr\{W \leq x\} = 1 - \eta e^{-\mu(1-\eta)x}, x \geq 0 \tag{16.59}$$

The probability of not waiting at all is given by setting x to be 0, $1 - \eta$. This will not be the probability that the server is idle in general, with an exception for the case when the arrival process is Poisson. In fact, the waiting time distribution of the G/M/m queue is always exponential with a point probability of not having to wait at all, as the reader may verify.

Queuing Networks

With General Customer Routes

The results of this section can be applied to both open and closed queuing networks. In an open queuing network, customers make a sequence of visits to various nodes within the network before departing from the network altogether. In closed networks, on the other hand, the customer population within the network remains fixed with no customers entering the network or departing from it. Examples of these kinds of network with yet more general servers can be found in Kelly (1979).

The results are similar in both cases, but begin by considering the open case. The network is supposed to have J nodes. The server, at each node j, applies service at a rate that depends on the total number of customers present $\varphi_j(n_j)$, taken to be zero if there are no customers present. Each customer within the queue receives a part of this effort according to the service allocation vector $\gamma(k, n_j)$, $k \leq n_j$. The amounts of service a customer requires at each queue are independent of one another and distributed exponentially with unit mean. Some examples of how this vector might be set, along with the service rate, to obtain various kinds of server are as follows:

1. ./M/1 queue with first-come, first-served scheduling, as in Equation 16.60
2. ./M/m queue with first-come, first-served scheduling, as in Equation 16.61 and Equation 16.62
3. Processor sharing, as in Equation 16.63

$$\varphi(n) = \varphi, \gamma(1,n) = 1, \gamma(j,n) = 0; \text{ otherwise,} \tag{16.60}$$

$$\varphi(n) = n\varphi\gamma(j,n) = \frac{1}{n}, \, j \le n \le m \quad \varphi(n) = m\varphi n > m \tag{16.61}$$

$$\gamma(j,n) = \frac{1}{m}, \, j \le m, \gamma(j,n) = 0; \text{ otherwise,} \tag{16.62}$$

$$\varphi(n) = \varphi, \, n \ge 1, \gamma(j,n) = \frac{1}{n}, \, j \le n \tag{16.63}$$

$\varphi > 0$ is the service effort per unit time available from each server. The customers' routes through the network are determined by an ordered set of nodes, which may be regarded as the list of nodes that the customer will visit in the order in which he will visit them. A list is denoted $r = (r(1), \ldots, r(S(r)))$, where $S(r)$ is the number of nodes visited. Repeat visits are permitted.

It is supposed that customers following route r arrive as a Poisson stream with rate v_{r3} independently. Furthermore, it will be supposed that the number of routes is finite, although this is not a necessary assumption.

Apart from specifying in which order the customers of a particular routing class visit a queue, it is also necessary to specify how they enter each queue. This is determined by the random entrance vector δ, which is independent of routing class. Given that there are n_j customers in the queue immediately prior to the entry of the new customer, the position of the new customer is given by $\delta(k, n_j + 1)$. By choice of this vector it is possible to model queues with last-come, first-served scheduling or service in random order.

It remains to specify the state of the network. It is not enough simply to stipulate the number of customers in each queue, since this ignores which classes of customers are in what positions. The state of node j is given by a vector $c_j(n_j)$, where the kth component of c_j is determined by the routing class of the customer in position k together with the stage that he has reached along his route. The state of the network consists of a vector whose components are the individual queue state vectors C. Let $q(C, D)$ be the transition rate from state C to state D.

Network Transitions

There are three kinds of transition within the network: an arrival at the network, a customer moving between queues, and a departure. The new state of the network immediately after the arrival of a customer, following route r, into position k of its initial queue is given by $A_k^r(C)$. The state of the network immediately after the departure of a customer on route r from position k at its last queue is $L_k^r(C)$. The state of the queue immediately after a customer following route r moves from queue i, position e, to queue h, position k is $M_{ie, hk}^r(C)$. Denote by j the first queue on route r and by g the last queue on that route; then these transitions take place at the rates given in Table 16.4.

Table 16.4   Queuing Network Transition Rates

Transition	Rate
Arrivals	$v_r \delta(k, n_j + 1)$
Departures	$\varphi_g(n_g)\gamma(k, n_g)$
Moving between queues	$\gamma(e, n_i)\varphi_i(n_i)\delta(k, n_h + 1)$

The equilibrium queue distribution is of product form, as in Equations 16.64 to 16.66, where G_j is a normalization constant, a_j is the total traffic arriving at node j, and is given by Equation 16.67, and $\alpha_j(r)$ is the number of visits that customers on route r make to node j and (with a little abuse of notation) $v_{r(k)}$ is the arrival rate of the class of the customer who is currently in position k.

$$\pi(C) = \prod_j \pi_j(c_j) \tag{16.64}$$

$$\pi_j(c_j) = G_j \prod_{k=1}^{n_j} \frac{v_r(k)}{\varphi_j(k)} \tag{16.65}$$

$$G_j^{-1} = \sum_{n=0}^{\infty} \frac{a_j^n}{n \prod_{k=1}^{n} \varphi_j(k)} \tag{16.66}$$

$$a_j = \sum_r \alpha_j(r)v_r \tag{16.67}$$

The following result makes it easy to verify that Equation 16.64 is indeed the equilibrium queue distribution.

Theorem
Suppose that numbers $q'(C, D) > 0$ for each pair of states C, D can be found so that Equation 16.68 is obtained.

$$\sum_{D=C} q'(C, D) = q'(C) = q(C) = \sum_{D=C} q(C, D) \tag{16.68}$$

Furthermore, suppose numbers $\pi(C) > 0$ for each state C can be found, satisfying Equation 16.69 for each pair of states C, D. Then π is the equilibrium probability distribution for both Markov processes whose transition rates are determined by $q(C, D)$, $q'(C, D)$, respectively.

$$\pi(C)q(C,D) = \pi(D)q'(D,C) \tag{16.69}$$

It is sufficient to obtain the equilibrium equations from Equation 16.68, as in Equation 16.70, but the left side is given by Equation 16.71.

$$\sum_{D=C} \pi(C)q'(C,D) = \sum_{D=C} \pi(D)q(D,C) \tag{16.70}$$

$$\begin{aligned} \pi(C)q'(C) &= \pi(C)q(C) \\ &= \sum_{D=C} \pi(C)q(C,D) \end{aligned} \tag{16.71}$$

To make use of this result, we must find suitable transition rates $q'(C, D)$. This is done by constructing the time-reversed queuing network that has transition rates given by Table 16.5.

In the reversed network, customers start from their last queue and follow their route backward, and the roles of γ and δ are swapped. Thus the first queue on route r is g, the last is j, and the queue transition is from queue h, position k, to queue i, position e.

Table 16.5   Transition Rates for the Reversed Network

Transition	Rate
Arrivals	$v_r\gamma(k, n_g + 1)$
Departures	$\varphi_j(n_j)\delta(k, n_j)$
Moving between queues	$\gamma(1, n_i + 1)\varphi_j(n_h)\delta(k, n_h)$

Consider the probability flux of arrivals on route r to their first node j, $r(1) = j$. It is easily verified that Equation 16.72 holds, and, by symmetry, a similar relationship holds for departures, as in Equation 16.73.

$$v_r\delta(k, n_j + 1)\pi(C) = \varphi_j(n_j + 1)\delta(k, n_j + 1)\pi(A_k^r(C)) \tag{16.72}$$

$$\varphi_g(n_g + 1)\gamma(k, n_g + 1)\pi(C) = v_r\gamma(k, n_g + 1)\pi(L_k^r(C)) \tag{16.73}$$

Similarly, for transitions between queues from node i, position e, to node h, position k, following route r, Equation 16.74 is also easily verified.

$$\varphi_i(n_i)\gamma(e, n_i)\delta(k, n_h + 1)\pi(C) = \varphi_h(n_h + 1)\delta(k, n_{h+1})\gamma(e, n_i)\pi(M_{ie,hk}^r(C)) \tag{16.74}$$

By adding Equations 16.72 to 16.74 and summing over all possible transitions out of state C, Equation 16.68 may be obtained. The total rate out of state C in both networks is given by Equation 16.75.

$$q(C) = q'(C) = \sum_{j=1}^{J}\varphi_j(n_j) + \sum_r v_r \tag{16.75}$$

Thus the conditions of theorem 1 are verified and Equation 16.64 is indeed the equilibrium distribution of the network.

As described so far, customers have to call at each node in turn along their route. However, it is possible to model customer classes in which each new customer is assigned one from a possible set of routes at random. This is achieved as follows. Let s be a route that can be followed. Suppose the probability that this route is followed is f_s. Each such route s is then offered traffic $v_s = vf_s$ where v is the total traffic offered by the customer class using these random routes.

The approach to closed queuing networks is very similar. To begin with, customers may be regarded as following a route given by a list of nodes in the order in which they are to be visited. However, when a customer departs the final node in its list, it returns to the first and cycles through the network again. For each route there is a population of customers that follows it N_r. Let N denote the population vector giving the number of customers following each route r.

These definitions resemble those for an open network but with no departures or arrivals. A reasonable candidate for the equilibrium is to use the same form as the open network. However, there are no arrivals and so a set of dummy arrival rates are assigned to each route v_r. The precise vales are not important as long as they are positive. The reason for this freedom in choice of arrival rates is mentioned in the following text.

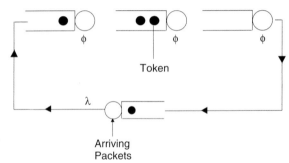

Fig. 16.10 Window flow control over three consecutive transmission links.

Only certain states of the corresponding open network can be attained in the closed network. Thus, to obtain the candidate closed equilibrium distribution, the open distribution is renormalized to the total probability of all the possible states T in the closed queuing network, as in Equation 16.76, where $\pi(C)$ is defined in Equation 16.64 so that the equilibrium probability of being in state C is given by Expression 16.77.

$$G^{-1}(N) = \sum_{C \varepsilon T} \pi(C) \tag{16.76}$$

$$G(N)\pi(C) \tag{16.77}$$

That this is indeed the equilibrium may be verified using the same argument based on theorem 1. Since the system is closed the equilibrium distribution cannot depend on the dummy arrival rates. The normalization constant cancels out the dependence of the equilibrium distribution on them.

As an example, consider a simple model of a series of transmission links, access to which is regulated by window flow control (see Figure 16.10). As depicted, a packet gains access by obtaining one of the window flow control tokens. This token is carried with the packet over the three transmission links before being released. It is supposed to return immediately to be available for other packets to use. As Figure 16.10 shows, this can be represented by a closed cyclic queuing network with the tokens as customers. These are "served" by the arrivals that take place at rate λ and then by the transmission links. Arrivals that find no tokens are lost.

Suppose that there are N tokens and that the service rate at each of the transmission links is φ. The network state is represented by Equation 16.78.

$$n = (n_1, n_2, n_3, n_4), \, n_1 + n_2 + n_3 + n_4 = N \tag{16.78}$$

Using Equation 16.64, the equilibrium distribution is then given by Equation 16.79, where G is given by Equation 16.80.

$$\pi(n) = G^{-1} \frac{1}{\varphi^{n_1}} \frac{1}{\varphi^{n_2}} \frac{1}{\varphi^{n_3}} \frac{1}{\lambda^{n_4}} \tag{16.79}$$

$$G = \sum_{n:n_1+n_2+n_3+n_4=N} \frac{1}{\varphi^{n_1}} \frac{1}{\varphi^{n_2}} \frac{1}{\varphi^{n_3}} \frac{1}{\lambda^{n_4}} \tag{16.80}$$

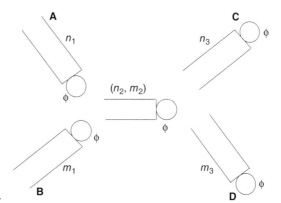

Fig. 16.11 Communication
network with two customer routes.

The probability that a packet is lost because there is no token available is given by Expression 16.81.

$$G^{-1} = \sum_{n:n_4=0} \frac{1}{\varphi^{n_1}} \frac{1}{\varphi^{n_2}} \frac{1}{\varphi^{n_3}}$$

(16.81)

As a second example, consider the communication network shown in Figure 16.11. Customers move from A to C via the central queue or from B to D via the central queue with arrival rates υ and λ, respectively. The service rate at each queue is φ.

Let n_i denote the number of customers going from A to D at each stage i and m_i denote the number at each stage going from B to C; then, again using Equation 16.64, the equilibrium distribution is as in Equation 16.82.

$$\pi(n,m) = G^{-1} \left(\frac{\upsilon}{\varphi}\right)^{n_1} \left(\frac{\lambda}{\varphi}\right)^{m_1} \binom{n_2+m_2}{n_2} \times \left(\frac{\upsilon}{\varphi}\right)^{n_2} \left(\frac{\lambda}{\varphi}\right)^{m_2} \left(\frac{\upsilon}{\varphi}\right)^{n_3} \left(\frac{\lambda}{\varphi}\right)^{m_3}$$

(16.82)

Note the combinatorial term for the center queue. The state reflects only the number of customers of each type at each queue, and the combinatorial term accounts for all possible arrangements, which are all equally likely (Equation 16.83).

$$G^{-1} = \left(1-\frac{\upsilon}{\varphi}\right)^2 \left(1-\frac{\lambda}{\varphi}\right)^2 \left(1-\frac{\upsilon}{\varphi}-\frac{\lambda}{\varphi}\right)$$

(16.83)

Fixed-Point Models for Closed Queuing Networks

In many communication applications, the distributions of system random variables may be known only imperfectly. Consider the distribution of sojourn (queuing plus transmission) time of a packet at a node within a communication network. This varies with the flow of packets through the node. It is possible to have a very good estimate of how the mean sojourn time depends on this quantity but not have more detailed information about the distribution than this.

In these circumstances, detailed modeling of the sojourn time distribution appears inappropriate. As will now be shown, it is possible to deduce a great deal about performance without having to make further assumptions, which is difficult to justify and which may also complicate analysis (Kelly, 1989). These models are capable of considerable generalization, including having priority queues, although this extension will not be discussed here.

The nodes within the network are labeled $1, 2, \ldots, j$. Each customer class r has N_r customers in it. For each such class there is a rule that determines how the customers in the class move around within the network. This rule is based on cycles, with one being completed every time certain sequences of visits are executed.

A simple rule of this type is given if a route is determined by a list of nodes that have to be visited in order. A cycle is completed once all the nodes in the list have been visited. Note that repeat visits to nodes within the list are permitted.

A more general rule is given if it is supposed that routing in each class is determined by an irreducible homogeneous Markov chain. Let π_{jk} denote the probability that the customer visits node k immediately after departing node j. A cycle is completed when a given node is returned to.

As can be seen, the routing rules include those mentioned in the previous section.

Given routing rules such as these, let the mean number of visits to each node per cycle be given by α_{jr}. Finally, denote the mean rate at which class r completes cycles by θ_r and the flow of customers through node j by ρ_j. Conservation of flow gives ρ_j in terms of the θ_r's, as in Equation 16.84.

$$\rho_j = \sum_j \alpha_{jr}\theta_r \tag{16.84}$$

The model is determined once the mean sojourn time at each node is specified. In the simplest cases, this will be a function of mean flow and service capacity only (Equation 16.85).

$$D_j = D_j(\rho_j;\varphi_j) \tag{16.85}$$

This definition allows all the queuing models described in this chapter.

The conservation of flow equation (Equation 16.85), together with Little's law applied to each customer class, gives Equation 16.86, which yields $J + R$ equations in $J + R$ unknowns.

$$N_r = \theta_r \sum_r \alpha_{jr}D_j(\rho_j;\varphi_j) \tag{16.86}$$

If D_j is nonnegative and an increasing function of ρ_j, then the solution is unique. We now demonstrate this. Define H as in Equation 16.87.

$$H = -\sum_j \int_0^{\rho_j} D_j d\rho_j' + \sum_r N_r \log\theta_r \tag{16.87}$$

However, H is strictly concave because the D_j's are continuous and monotonic increasing. Furthermore, any maximum of H must be unique since H is strictly concave. In addition, the maximum must be located at an interior point. This is therefore determined by the zeros of the derivative and must be satisfied at only that point.

H has a derivative as in Equation 16.88.

$$\frac{\partial H}{\partial \theta_r} = -\sum_j \alpha_{jr}D_j + \frac{N_r}{\theta_r} \tag{16.88}$$

It can now be seen that Equation 16.86 must have just one solution since the solution determines the unique maximum of H.

Repeated substitution is one of the most numerically efficient ways of solving the previous sets of equations. One begins with an initial value of the ρ_j's, for example as in Equation 16.89. The θ_r's and ρ_j's may then be solved recursively as in Equations 16.90 and 16.91.

$$\rho_j^{(1)} = 0, \, j = 1, \ldots, J \tag{16.89}$$

$$\theta_r^{(n)} = \frac{N_r}{\sum_r \alpha_{jr} D_j \left(\rho_j^{(n)}; \varphi_j \right)} \tag{16.90}$$

$$\rho_j^{(n+1)} = \sum_r \alpha_{jr} \theta_r^{(n)} \tag{16.91}$$

On many occasions, this recursion will not converge. However, the use of damping usually gets over this problem. With damping, factor γ, $0 < \gamma < 1$, Equation 16.90 can be written as in Equation 16.92.

$$\theta_r^{(n+1)} = \frac{\gamma N_r}{\sum_r \alpha_{jr} D_j \left(\rho_j^{(n)}; \varphi_j \right)} + (1 - \gamma) \theta_r^{(n)} \tag{16.92}$$

On some occasions heavy damping may be needed (i.e., values of γ close to 0).

As an example, consider again the model of a series of transmission links, access to which is regulated by window flow control (see Figure 16.10). Again there are N tokens, which circulate at rate θ. From Figure 16.10 this is determined by the fixed-point Equation 16.93.

$$N = \frac{\theta}{(\lambda - \theta)} + \frac{3\theta}{(\varphi - \theta)} \tag{16.93}$$

The rate of transmitting packets is θ, of course, and, therefore, the proportion of packets that are lost is given by Expression 16.94.

$$1 - \frac{\theta}{\lambda} \tag{16.94}$$

Multiaccess Channels

Queuing analysis has been successfully applied to determine the throughput of a wide range of access protocols used in radio communications and elsewhere. The following provide two examples illustrating a typical approach to the problem of determining the throughput.

ALOHA Models

Consider first asynchronous (unslotted) ALOHA, as shown in Figure 16.12. All packets are the same length, and the transmission of a packet will be successful provided that another packet does not transmit in its vulnerable period. The probability of a successful transmission is therefore given by Equation 16.95, where γ is the total transmission attempt rate.

$$\Pr\{\text{successful transmission}\} = \Pr\{\text{no transmission in vulnerable period}\} = e^{-2\gamma} \tag{16.95}$$

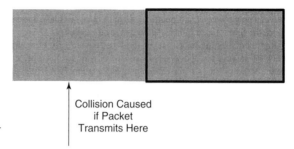

Collision Caused
if Packet
Transmits Here

Fig. 16.12 Vulnerable period for unslotted ALOHA.

The total throughput is the rate of making successful transmissions, as in Equation 16.96.

$$\text{Throughput} = \text{transmission attempt rate} \times \text{probability of success} = \gamma e^{-2\gamma} \tag{16.96}$$

This is maximum for $\gamma = \frac{1}{2}$ so that the throughput cannot exceed $\frac{1}{2}e$.

Note that the total rate of transmissions and retransmissions is being modeled as a Poisson process of rate γ. Even if the process of fresh transmissions is Poisson, the assumption being that the retransmissions are independent, Poisson can only be justified if the retransmissions take place at random over long intervals following the collisions that give rise to them.

The throughput can be increased by adopting a slotted version of this protocol so that packets can only be transmitted at the beginning of each slot; this was depicted in Figure 16.2. In this way, if a collision takes place, the two packets coincide completely, eliminating the possibility of packet loss by partly overlapping packets. The probability of successbecomes $e^{-\gamma}$ and the throughput $\gamma\,e^{-\gamma}$. The maximum throughput becomes $\frac{1}{e}$ at $\gamma = 1$.

The results just given must be treated with caution. The conclusions are correct but only if they are qualified. The reason for this is that the analysis does not take into account the nature of the underlying Markov chain that it purports to analyze. Indeed, if an infinite station model is adopted, the equilibrium can be shown *not* to exist (Kelly, 1985). The slotted ALOHA channel will jam with probability 1.

Even in the case where the station population is finite, the channel may perform inadequately. This is because the system may have bi-stable operation, with one stable point at low traffic and most stations idle, giving high throughput, and a second one where most stations are busy and the throughput is low because of the large number of stations all attempting together. The system may remain in this second state for long periods.

The underlying difficulty is that the protocol does not adapt the number of transmission attempts according to the number of busy stations. The retransmission probability should be high if the number of busy stations is small but low when there are large numbers of busy stations. This problem is overcome by having the stations keep an estimate of the backlog, which is updated following each slot according to whether it was idle or contained a successful transmission or a collision (Gallager and Bertsekas, 1987).

Nonpersistent Carrier Sense Multiple Access

Under Carrier Sense Multiple Access (CSMA), mobiles in a communication network having packets to transmit do so if they "sense" that the radio channel is not being used. We do not go into a description of how this is achieved, but suppose that all mobiles become aware of a packet transmission within τ time units after it was begun. The period τ is called the vulnerable period. The protocol has only two steps:

1. The channel is "sensed." If it is found to be idle, the packet is transmitted immediately.
2. If the channel is "sensed" busy, the packet transmission is rescheduled using the retransmission distribution. At the end of the delay period the protocol moves to step 1.

Figures 16.13(a) and 16.13(b) show successful and unsuccessful transmissions, respectively. The analysis for nonpersistent CSMA is as follows. During the start of a transmission there is a short period during which other transmissions may start before all other mobiles

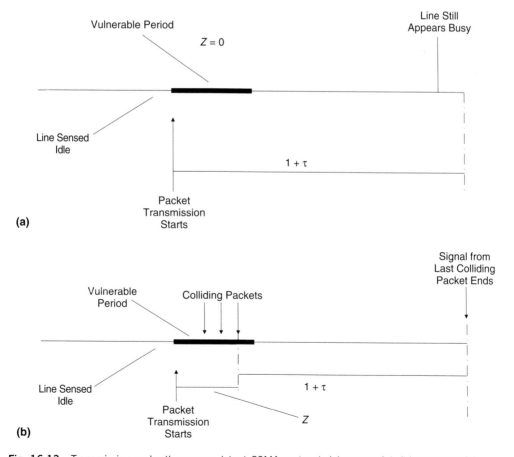

Fig. 16.13 Transmission under the nonpersistent CSMA protocol: (a) successful, (b) unsuccessful.

become aware that the transmission has started. This interval, in units of packet transmission time, is τ, as already mentioned. Making the usual assumption that transmission attempts are taking place at rate λ as a Poisson process, the probability that the packet is transmitted successfully is that there are no further transmission attempts during this vulnerable period, which is given by Expression 16.97.

$$e^{-\tau\lambda} \tag{16.97}$$

Contending packets may arrive at any time during the initial vulnerable period, and it is the last of these arrivals that determines when the contention period ends and the line goes idle. Let Z be a random variable defined as in Equation 16.98.

$$Z = \begin{cases} 0 \text{ no collisions} \\ \text{time between first and last colliding packet} \end{cases} \tag{16.98}$$

Observe that Z must be less than τ. Given that there is at least one colliding packet, the distribution of time between the last packet to arrive and the end of the vulnerable period is the same as the time until the first packet transmission. This observation shows that the distribution of Z is given by Expression 16.99.

$$e^{-\lambda(\tau - z)} \tag{16.99}$$

From Equation 16.99 the mean value of Z is obtained, as in Equation 16.100.

$$E[Z] = \tau - \frac{1}{\lambda} + \frac{e^{-\lambda\tau}}{\lambda} \tag{16.100}$$

Packet transmissions and contention go in cycles. There are only two kinds:

1. Successful transmission, equal to the idle period plus packet transmission time
2. Unsuccessful transmission, equal to the idle period plus the contention period

The mean duration of an idle period is $\frac{1}{\lambda}$.

The definition of Z takes into account both contention and a successful transmission. The mean for this second part of the cycle is given by Expression 16.101.

$$1 + \tau + E[Z] \tag{16.101}$$

The system throughput C is then determined as in Equation 16.102.

$$C = \frac{e^{-\tau\lambda}}{1 + \tau + E[Z] + \dfrac{1}{\lambda}} = \frac{\lambda e^{-\tau\lambda}}{\lambda + 2\lambda\tau + e^{-\tau\lambda}} \tag{16.102}$$

The maximum possible throughput can now be found from examining C over all possible values of λ. Other forms of CSMA protocol are discussed in Kleinrock (1975b).

Note, once again, that these results must be treated cautiously; stability questions arise with CSMA protocols as well.

References

Benes, V. E. (1956) On Queues with Poisson Arrivals, *Annals of Mathematical Statistics* 28, pp. 670–677.

Feller, W. (1958) *An Introduction to Probability Theory and Its Applications*, 1, 3rd ed, Wiley.

———. (1970) *An Introduction to Probability Theory and Its Applications*, 2, Wiley.

Gallager, R. G. (1977) A Minimum Delay Distributed Routing Algorithm, *IEEE Transactions on Communications*, 23, pp. 73–85.

Gallager, R. G., and Bertsekas, D. M. (1987) *Data Networks*, Prentice-Hall.

Kelly, F. P. (1979) *Reversibility and Stochastic Networks*, Wiley.

———. (1985) Stochastic Models of Computer Communication Systems, *Journal of the Royal Statistical Society* Series B, 47, pp. 379–395.

———. (1989) On a Class of Queuing Approximations for Closed Queuing Networks, *Queuing Systems* 4, pp. 69–76.

Kleinrock, L. (1975a) *Queuing Theory*, 1, Wiley Interscience.

———. (1975b) *Queuing Theory*, 2, Wiley Interscience.

Park, K., and Willinger, W., Eds. (2001) *Self Similar Traffic and Performance Evaluation*, Wiley Interscience.

Takacs, L. (1962) *An Introduction to the Theory of Queues*, Oxford University Press.

Zwart, A. P. (2001) *Queueing Systems with Heavy Tails*, Ph.D. thesis, University of Eindhoven.

17 Teletraffic Theory

R. J. Gibbens

Introduction

The subject of teletraffic theory has a long history dating back to the work of the Danish mathematician A. K. Erlang (Brockmeyer et al., 1948), first published in 1917.

Today the subject has grown enormously and is still growing at a rapid pace, with research into the performance modeling of areas as diverse as, for example, dynamic routing strategies, cellular mobile radio systems, and the application of fast packet-switching techniques. Hui (1990) is a good text for some of these developments.

The first section of this chapter studies the basic model of a single link in a circuit-switched telecommunications network. This section also introduces the important control mechanism of trunk reservation.

The next section shows how the single-link models studied in the first section can be incorporated into models of networks of links using the effective link independence approximation to construct the Erlang fixed-point systems of equations.

These equations allow the investigation of the rich behavior of dynamic routing strategies in the setting of symmetric, fully connected networks. This section on network-based results ends with two bounds on the overall best possible behavior of dynamic routing strategies.

The final section presents some of the dynamic routing strategies that have been either implemented or proposed for real networks.

This is an area that has received much attention over the last decade and is one in which teletraffic theory has provided the models and understanding leading to more efficient, flexible, and reliable networks.

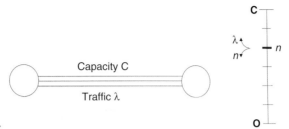

Fig. 17.1   Single-link model.

Single-link Models

Erlang's Loss Formula

The first model to be considered is the classical model for a single link in a circuit-switched network. Suppose that the arrival process of calls offered to a link is a Poisson process of rate λ. Let the holding times of calls be exponentially distributed with a mean of 1 independent of the arrival process. Suppose that the link has a capacity of C circuits. Figure 17.1 shows a single-link model. See Chapter 16 for an introduction to the theory of stochastic processes applied to queuing networks.

This system may be modeled by the stochastic process, n, the number of circuits occupied by the link with n taking values in the range 0 to C inclusive. It can be seen that n is a birth-and-death process since it either increases by one when a new call is accepted on the link or decreases by one when an existing call clears down. The transition rates for the birth-and-death process are therefore given by Equations 17.1 and 17.2.

$$q(n, n+1) = \lambda \quad n = 0, 1, \ldots, C-1 \tag{17.1}$$

$$q(n, n-1) = n \quad n = 1, \ldots, C \tag{17.2}$$

A general result (Kelly, 1979) of birth-and-death processes gives the equilibrium distribution in terms of the birth and death rates, as in Equation 17.3, where π_0 is determined from the normalization condition of Equation 17.4.

$$\pi_n = \pi_0 \prod_{i=1}^{n} \frac{q(i-1, i)}{q(i, i-1)} \tag{17.3}$$

$$\sum_{n=0}^{C} \pi_n = 1 \tag{17.4}$$

Hence, from Equations 17.1 and 17.2, one can obtain Equations 17.5 and 17.6.

$$\pi_n = \pi_0 \prod_{i=1}^{n} \frac{\lambda}{i} = \pi_0 \frac{\lambda^n}{n!} \tag{17.5}$$

$$\pi_0 = \left[\sum_{n=0}^{C} \frac{\lambda^n}{n!} \right]^{-1} \tag{17.6}$$

In summary, the equilibrium distribution for the utilization, n, of a single link is given by Equation 17.7.

$$\pi_n = \frac{\dfrac{\lambda^n}{n!}}{\displaystyle\sum_{i=0}^{C} \dfrac{\lambda^i}{i!}} \quad (n = 0, 1, \ldots, C) \tag{17.7}$$

In particular, the probability, L, that an arriving call finds the link full to capacity is π_C, given by Equation 17.8.

$$L = \pi_C = \frac{\lambda^C}{C!}\left[\sum_{i=0}^{C}\frac{\lambda^i}{i!}\right]^{-1} \tag{17.8}$$

Equation 17.8 is known as Erlang's loss formula and is undoubtedly the most widely used result in teletraffic theory. The standard notation $E(\lambda, C)$ for the quantity given in Equation 17.9 will be used.

$$E(\lambda, C) = \frac{\lambda^C}{C!}\left[\sum_{i=0}^{C}\frac{\lambda^i}{i!}\right]^{-1} \tag{17.9}$$

Later, the question of how best to compute $E(\lambda, C)$ in such a way as to reduce numerical inaccuracies caused by rounding errors will be considered. First, let us look at some of its properties.

In Figure 17.2, the loss probability $L = E(\lambda, C)$ is shown as a function of the traffic intensity $\rho = \lambda/C$ for C held fixed at the values $C = 100, 250, 1000,$ and $10,000\,\text{B}$. This figure demonstrates the important trunking efficiency effect where, for a fixed level of the traffic intensity, ρ, the loss probability, L, decreases with increasing link capacity, C: Larger links are more efficient at carrying calls. Also shown in Figure 17.2 is the limiting case as C increases to infinity, which is given by Equation 17.10.

$$L = \max\left(1 - \frac{1}{\rho}, 0\right) \tag{17.10}$$

Numerical Considerations

Several methods are available for the efficient and accurate numerical evaluation of Erlang's loss formula. For small values of C, perhaps up to several hundreds of circuits, $E(\lambda, C)$ can be computed using the recursive formula of Equation 17.11 and $(\lambda, 0) = 1$.

$$E(\lambda, C) = \frac{\lambda E(\lambda, C-1)}{C + \lambda E(\lambda, C-1)} \tag{17.11}$$

This result can be derived by substituting the expression for $E(\lambda, C)$ given in Equation 17.9 and simplifying the resulting expression.

For larger values of C, which make a recursive computation impractical, a second method is available that avoids the direct calculation of large factorials. After some rearrangement, Equation 17.12 may be obtained.

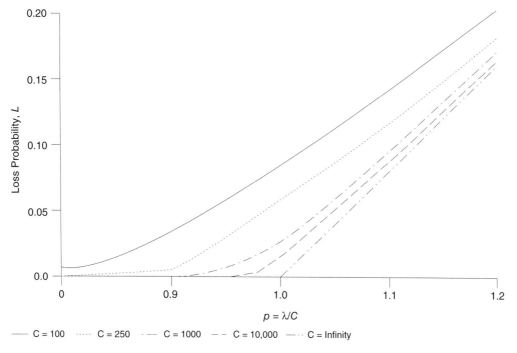

Fig. 17.2 Erlang's blocking formula.

$$E(\lambda, C) = \left[1 + \frac{C}{\lambda} + \frac{C(C-1)}{\lambda^2} + \ldots + \frac{C!}{\lambda^C}\right]^{-1}$$ **(17.12)**

Written in this form, each successive term in the summation is a simple multiple of the previous one and so direct calculation of factorials may be avoided. The final sum is then inverted to give the loss probability.

Multiple Priorities and Trunk Reservation

Consider again a single link with C circuits, but now suppose that there are two independent Poisson arrival streams offered at rates λ_1 and λ_2. Suppose we wish to give priority to the calls at rate λ_1 over those at rate λ_2. A very simple mechanism has been devised to implement these priorities known as trunk reservation. In this mechanism, calls of the low priority stream are only accepted when there are at least $r + 1$ free circuits available. In contrast, the high-priority calls are accepted so long as there is at least 1 free circuit available. The integer value, r, is known as the trunk reservation parameter. An equivalent notation that is sometimes more convenient is to let C^i be the capacity available to calls of priority i or lower so that $C^1 = C$ and $C^2 = C - r$.

Again, a birth-and-death process can be used to model the system. The transition rates are now given by Equations 17.13 and 17.14.

$$q(n, n+1) = \begin{cases} \lambda_1 + \lambda_2 & n = 0, \ldots, C - r - 1 \\ \lambda_1 & n = C - r, \ldots, C \end{cases}$$ **(17.13)**

$$q(n, n-1) = n \quad n = 1, \dots, C \tag{17.14}$$

The equilibrium distribution, using the general result in Equation 17.3, is given, after some rearrangement, by Equation 17.15, where π_0 is determined by Equation 17.16.

$$\frac{\pi_n}{\pi_0} = \begin{cases} \dfrac{(C-r)!}{n!}(\lambda_1 + \lambda_2)^{n-(C-r)} & n = 0, \dots, C-r \\[2ex] \dfrac{(C-r)!}{n!}\lambda_1^{n-(C-r)} & n = C-r+1, \dots, C \end{cases} \tag{17.15}$$

$$\sum_{n=0}^{C} \pi_n = 1 \tag{17.16}$$

The preceding procedure avoiding large factorials in Erlang's loss formula can also be used here to calculate the distribution π_n.

Two criteria for selecting the trunk reservation parameter r will now be considered.

Optimality Criterion for r

Consider the situation given by Equations 17.17 and 17.18.

$$B_1(r) = \pi_C \tag{17.17}$$

$$B_2(r) = \sum_{k=C-r}^{C} \pi_k \tag{17.18}$$

Then $B_1(r)$ and $B_2(r)$ are the blocking probabilities to the high- and low-priority streams respectively when the trunk reservation parameter level is set at the value r.

A criterion for choosing r, which is applicable in much of the work to follow on control in routing strategies in fully connected networks, is to minimize with respect to r the expression of 17.19.

$$\lambda_1 B_1(r) + \frac{1}{2}\lambda_2 B_2(r) \tag{17.19}$$

This would correspond to a valuation by the link that the carrying of a low-priority call is worth just half the carrying of a high-priority call. Here we think of high-priority calls as the directly routed calls and the low-priority calls as the overflow calls using two links.

In Figure 17.3, the value of r minimizing Expression 17.19 is illustrated with λ_1 held fixed in such a way that $E(\lambda, C) = 1\%$ for two values of C. The value of Expression 17.19 is relatively insensitive to r above a low threshold. Thus, while minimizing the value of r can be quite sensitive to the precise values of C, λ_1, and λ_2, close to optimal performance over a wide range of values of C, λ_1 and λ_2 can be obtained by quite a crude choice of the parameter r.

Secondary Criterion for r

A further criterion that may be applied to selecting the trunk reservation parameter r is as follows. The approach begins by first recognizing that the offered traffics in a network are,

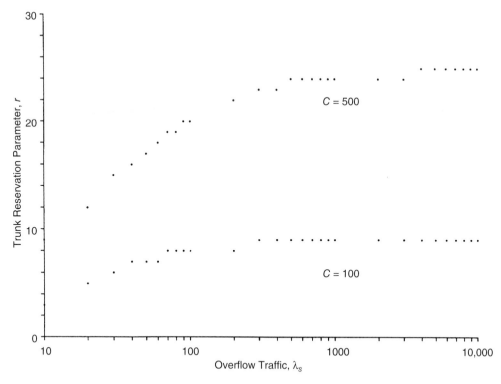

Fig. 17.3 Optimality criterion.

in practice, very uncertain and larger, low-priority overflow traffics may occur. Trunk reservation may then be used to guarantee performance in the worst-case conditions of infinite overflow traffic, where the number of circuits in use on a link is forced to remain in the states $C - r, \ldots, C$, where $r > 0$ is the trunk reservation parameter. Then the transition rates for the birth-and-death process are simply as in Equations 17.20 and 17.21.

$$q(n, n+1) = \lambda \quad n = C - r, \ldots, C - 1 \tag{17.20}$$

$$q(n, n-1) = n \quad n = C - r + 1, \ldots, C \tag{17.21}$$

If $B(\lambda, C, r)$ is the blocking probability for fresh traffic under these circumstances then Equation 17.22 may be obtained.

$$B(\lambda, C, r) = \left[\sum_{n=C-r}^{C} \frac{C(C-1)\ldots(n+1)}{\lambda^{C-n}} \right]^{-1} \tag{17.22}$$

A rationale for choosing the trunk reservation parameter is then to relate λ and C by Equation 17.23, where $B > 0$ is a fixed constant, and to take the trunk reservation parameter, $R(C)$, as in Equation 17.24, for some constant K with $1 < K < 1/B$.

$$E(\lambda, C) = B \tag{17.23}$$

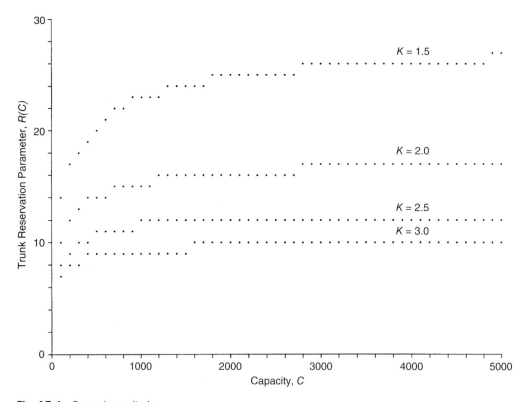

Fig. 17.4 Secondary criterion.

$$R(C) = \min\{r : B(\lambda, C, r) \le KB\} \tag{17.24}$$

Suppose that fresh traffic alone would suffer a blocking probability of B. Then the trunk reservation mechanism with parameter $r = R(C)$ chosen in accordance with this criterion ensures that the blocking probability for fresh traffic under arbitrary conditions of overflow traffic is no worse than K times that without overflow traffic. Figure 17.4 gives some examples of $R(C)$ with several values of the parameter K.

Network Models

Erlang Fixed Point

Models of a single link were described in the previous section. Here, the scope will be broadened to investigate models of networks of links. The simplest, and most commonly used, approach to creating a model of a network of links is to use an approximation that blockings on links within a route are statistically independent. In order to describe this approximation, consider a simple situation of a network of modes fully connected by links, each of capacity C and with independent Poisson arrival streams of traffic offered between the pairs of nodes, each of rate λ.

The fully connected network architecture has been the focus of much attention in recent years in relation to studies of dynamic routing strategies for national trunk networks. The

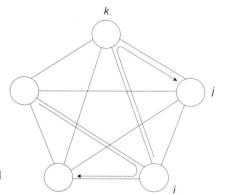

Fig. 17.5 A fully connected network.

final section in this chapter considers in some depth the various strategies that have so far been proposed. In this section, a simple strategy for symmetric, fully connected networks is considered that will suffice to illustrate many of the modeling aspects. In practice, networks are not symmetric and, in particular, traffics may not be well matched to capacity. Nevertheless, one can learn much about the behavior of routing schemes from a consideration of the symmetric case. (For further discussion on dynamic routing strategies, see Gibbens and Kelly, 1990.)

This simple strategy is called random routing and operates as follows: A call arriving at the network from the source destination pair (i, j) is routed along link (i, j) if there is at least one free circuit on this link; otherwise, an intermediate node $k(k \neq \sim i, j)$ (sometimes called a tandem node) is chosen at random and the call is routed along the two-link path between nodes i and j via node k provided that there are at least $r + 1$ free circuits on each of the links (i, k) and (k, j); otherwise, the call is lost. Notice that each link receives direct and overflow traffic. (See Figure 17.5.) Let B_1 and B_2 be the blocking probabilities of a link for direct and overflow traffic respectively. Observe that B_1 and B_2 will differ as a result of the use of the trunk reservation mechanism with parameter r.

Consider a single link in such a network and let n be the number of circuits in use. The process n can be modeled by a birth-and-death process with transition rates as in Equations 17.25 and 17.26, where σ is the rate of overflow traffic.

$$q(n, n+1) = \begin{cases} \lambda + \sigma & n = 0, \ldots, C - r - 1 \\ \lambda & n = C - r, \ldots, C - 1 \end{cases} \tag{17.25}$$

$$q(n, n-1) = n \quad n = 1, \ldots, C \tag{17.26}$$

The independent blocking approximation allows σ to be expressed in terms of B_1 and B_2 by Equation 17.27.

$$\sigma = 2\lambda B_1(1 - B_2) \tag{17.27}$$

The interpretation of Equation 17.27 is that blocking conditions of different links are approximately independent, and that overflow traffic arriving at a link is approximately Poisson. A call blocked with probability B_1 on its direct route will attempt an alternative route through two links, but may be blocked at either of these links. The transition rates correspond to the acceptance by a link of direct traffic and overflow traffic (at rates λ and σ, respectively) in states $\{0, \ldots, C - r - 1\}$ and of direct traffic only (at rate λ) in states $\{C - r, \ldots, C - 1\}$.

If L is the probability that a call offered to the network is lost, then Equation 17.28 may be obtained.

$$1 - L = (1 - B_1) + B_1(1 - B_2)^2 \tag{17.28}$$

Since calls are first offered to the direct links and are accepted with probability $(1 - B_1)$ and with probability B_1, they overflow to a randomly chosen two-link alternative path on each link of which they are independently accepted with probability $(1 - B_2)$.

The task is now to solve for B_1 and B_2. Let $\{\pi_n : n = 0, \ldots, C\}$ be the stationary distribution of the process n, then Equations 17.29 and 17.30 are obtained.

$$B_1 = \pi_C \tag{17.29}$$

$$B_2 = \sum_{n=C-r}^{C} \pi_n \tag{17.30}$$

The distribution $\{\pi_n : n = 0, \ldots, C\}$ is itself a function of B_1 and B_2, and so Equations 17.25 to 17.30 define B_1 and B_2 implicitly as in Equations 17.31 and 17.32 for some functions φ_1 and φ_2.

$$B_1^{l+1} = \varphi_1(B_1^l, B_2^l) \tag{17.31}$$

$$B_2^{l+1} = \varphi_2(B_1^l, B_2^l) \tag{17.32}$$

Choosing initial starting values for B_1^0 and B_2^0, one can use repeated substitution in Equations 17.31 and 17.32 to readily obtain a solution. By the Brouwer fixed-point theorem there exists a solution, and indeed as will be seen later there may be more than one solution.

Now, suppose that a call blocked on its direct route is allowed to attempt not just one but up to M two-link alternatives before it is lost. Again, use trunk reservation to insist that at least $r + 1$ circuits be free on a link before it accepts an alternatively routed call. Then the process n modeling the utilization of a link is again a birth-and-death process with rates as given by Equations 17.25 and 17.26 but now with the overflow rate (σ) given by Equation 17.33 and the loss probability (L) given by Equation 17.34.

$$\sigma = 2\lambda B_1 (1 - B_2)^{-1} \left\{ 1 - \left[1 - (1 - B_2)^2 \right]^M \right\} \tag{17.33}$$

$$L = B_1 \left[1 - (1 - B_2)^2 \right]^M \tag{17.34}$$

These expressions reduce to those in Equations 17.27 and 17.28 with the choice $M = 1$. The repeated substitution method can be used to determine a solution for B_1 and B_2 given some initial starting values.

Figure 17.6 shows solutions to these fixed-point equations when $r = 0$ (i.e., no trunk reservation is applied) with $M = 1, 5$ and a range of values of λ and C. Note that when $r = 0$ we have that $B_1 = B_2 = B$, say. Observe the possibility of multiple solutions for B over a narrow range of the ratio C/N when C is large enough, and that these effects are magnified in the

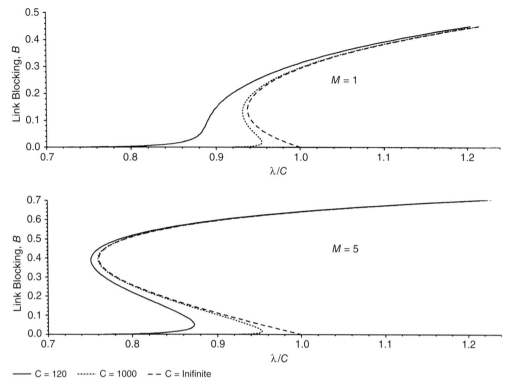

Fig. 17.6 Blocking probability B with $M = 1, 5$.

case of larger M. The upper and lower solutions correspond to stable fixed-point solutions while the middle solution corresponds to an unstable fixed point.

Simulations also exhibit this bistable behavior where the network has two quasi-stable states: a high-blocking state corresponding to many calls being carried on two-link routes and a low-blocking state corresponding to only a few calls carried over two links, with the majority carried on one-link routes.

Trunk reservation has been found to be an ideal way of preventing this bistable behavior in network performance and for providing more efficient use of network resources. In Figure 17.7, the loss probability, L, is shown as a function of r for values $M = 1$ and 5, $\lambda = 110$, and $C = 120$. Observe that a value of r in a broad range will give close to optimal performance.

Now consider the effect of increasing the number of retries M. Figure 17.8 shows the loss probability for varying λ and $C = 120$ and for various values of the retry parameter M. The curves are obtained by the optimal choice of r for each traffic (as in Figure 17.7). The optimal choices of r are also shown in Figure 17.8 together with the limiting case where M tends to infinity.

Implied Costs

Now examine the issue of implied costs (Kelly, 1990) in the context of fixed routing strategies. Suppose that the circuit-switched network consists of links labeled $k = 1, 2, \ldots, K$, with

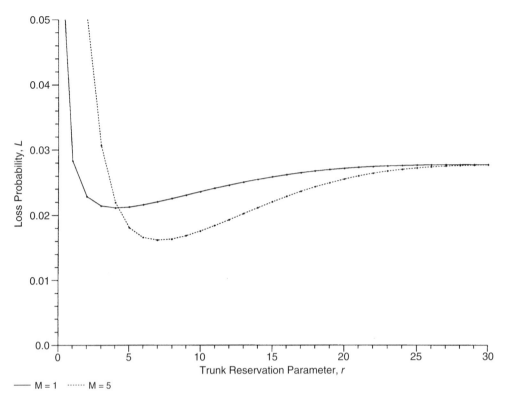

Fig. 17.7 Optimal choice of *r*.

link k comprising C_k circuits. A subset $r \subset \{1, 2, \ldots, K\}$ identifies a route. Calls requesting route r arrive as independent Poisson processes of rate v_r. A call requesting route r is blocked and lost if on any link $k \in r$ there are no free circuits. Otherwise, the call is connected and simultaneously holds one circuit on each link $k \in r$ for the holding period of the call. The call holding period is independent of earlier arrival times and holding periods; holding periods of calls on route r are identically distributed with unit mean. Write R for the set of possible routes.

The Erlang fixed-point equations for this setup are as in Equation 17.35, and the proportion of calls requesting route r that are lost is given by Equation 17.36.

$$B_k = E\left(\sum_{r:k \in r} v_r \prod_{j \in r - \{k\}} (1 - B_j), C_k \right) \tag{17.35}$$

$$L_r = 1 - \prod_{k \in r} (1 - B_k) \tag{17.36}$$

Suppose that a call accepted on route r generates an expected revenue ω_r. The rate of return from the network will be as in Equation 17.37 where λ_r is given by Equation 17.38.

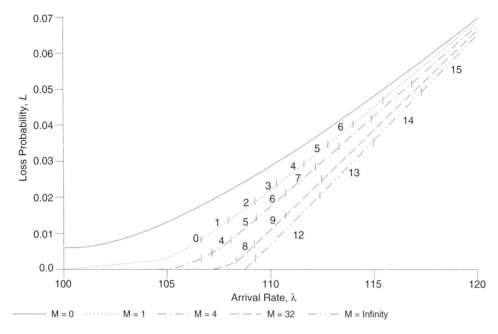

Fig. 17.8 Loss probability with varying number of retries, M.

$$W(v; C) = \sum_{r \in R} \omega_r \lambda_r \tag{17.37}$$

$$\lambda_r = v_r \prod_{k \in r} (1 - B_k) \tag{17.38}$$

Let $c = (c_1, c_2, \ldots, c_k)$ be the unique solution of the system of equations given by Equation 17.39.

$$c_k = \eta_k (1 - B_k)^{-1} \sum_{r:k \in r} \lambda_r \left(\omega_r - \sum_{j \in r - \{k\}} c_j \right) \tag{17.39}$$

The values of η_k and ρ_k are given by Equations 17.40 and 17.41.

$$\eta_k = E(\rho_k, C_k - 1) - E(\rho_k, C_k) \tag{17.40}$$

$$\rho_k = \sum_{r:k \in r} v_r \prod_{j \in r - \{k\}} (1 - B_k) \tag{17.41}$$

Thus, ρ_k is simply the traffic offered to link k under the approximation procedure. It may be proved that Equations 17.42 and 17.43 hold.

$$\frac{d}{dv_r} W(v; C) = (1 - L_r) \left(\omega_r - \sum_{k \in r} c_k \right) \tag{17.42}$$

$$\frac{d}{dC_k} W(v; C) = c_k \tag{17.43}$$

The definition of Erlang's loss formula can be extended to nonintegral values of capacity C by linear interpolation and integer values of C_k define the derivative of $W(v; C)$ with respect to C_k to be the left derivative.

Equation 17.42 shows that the effect of increasing the offered traffic on route r can be assessed from the following rule of thumb: An additional call offered to route r will be accepted with probability $1 - L_r$; if accepted it will earn ω_r directly, but at a cost c_k for each link $k \in r$. The costs c measure the knock-on effects of accepting a call on the other routes in the network. From Equation 17.43 it follows that the costs c also have an interpretation as shadow prices, with c_k measuring the sensitivity of the rate of return to the capacity C_k of link k. Note that Equation 17.39 can be rewritten in the form shown in Equation 17.44 where s_r is given by Equation 17.45.

$$c_k = \eta_k (1 - B_k)^{-1} \sum_{r:k \in r} \lambda_r (c_k + s_r) \tag{17.44}$$

$$s_r = \omega_r - \sum_{j \in r} c_j \tag{17.45}$$

Here s_r is the surplus value of a call on route r.

Applications of Implied Costs

Network Dimensioning

The Equations 17.42 and 17.43 can be used directly as the basis for an efficient hill climbing algorithm to maximize Expression 17.37. Through analogies with deterministic network flow, the shadow price interpretation of Equation 17.43 could be used in algorithms to aid capacity expansion decisions (Key and Whitehead, 1988). The overall approach also has important economic implications for pricing policies and for the apportionment of revenue between different sections of network operation.

Decentralized Adaptive Routing Strategy

Note that the implied cost, c_k, can be written in as shown in Equation 17.46 (where s_r is given by Equation 17.47) in terms of observables such as the carried traffics on routes and links.

$$c_k = \rho_k \eta_k \sum_{r:k \in r} \frac{\text{carried traffic on route } r}{\text{carried traffic through link } k} (c_k + s_r) \tag{17.46}$$

$$s_r = \omega_r - \sum_{k \in r} c_k \tag{17.47}$$

Hence, we may construct over time estimates of the implied costs c_k and surplus values s_r. This might be achieved by using, for example, moving average estimators.

Suppose that the loss probability, L_r, on route r has also been estimated by similar means. A call offered to route r will generate a net expected revenue of $(1 - L_r)s_r$. Should this quantity be negative for any route, then that route should be avoided: More revenue will be lost elsewhere in the network than can be generated by accepting calls on this route. Otherwise, traffic should be shared out between possible routes so as to reflect the net expected revenues $(1 - L_r)s_r$. Routes with higher values of $(1 - L_r)s_r$ than others should receive an increased share of the traffic. Adjustments to routing patterns made on this basis will need to be gradual since the effect of increasing the traffic to a route will be to push up the loss probability of the route and the implied cost c_k along that route, and hence reduce the net expected revenue $(1 - L_r)s_r$.

Network Bounds

This section aims to obtain bounds on the performance of dynamic routing schemes under a fixed stationary pattern of offered traffic. One method makes minimal assumptions concerning the stochastic structure of the system, and within this rather weak framework the bounds are the best possible. Another method considers various ways to improve these bounds for Poisson-offered traffic.

Max-flow Bound

For $(\lambda_{i,j} : i < j)$, let $F(\lambda)$ be the maximum attained in the linear program of Equation 17.48, subject to Equations 17.49 through 17.51.

$$\text{maximize } F(\lambda) = \sum_{i<j} \left(x_{ij} + \sum_{k \neq i,j} x_{ikj} \right) \tag{17.48}$$

$$x_{ij} + \sum_{k \neq i,j} x_{ikj} \leq \lambda_{ij} \quad \forall_i < j \tag{17.49}$$

$$x_{ij} + \sum_{k \neq i,j} (x_{ijk} + x_{jik}) \leq C_{ij} \quad \forall_i < j \tag{17.50}$$

$$x_{ij} \geq 0,\ x_{ijk} \geq 0,\ x_{ikj} = x_{jki} \quad \forall i, j, k \tag{17.51}$$

The interpretation of this problem is as follows. Regard x_{ij} as the direct flow between nodes i and j along the link (i, j), and x_{ikj} as the flow between nodes i and j through the tandem node k. Regard λ_{ij} as the offered traffic between nodes i and j. Then the linear program has as its objective function the total flow through the network, and has as constraints the limits imposed by the levels of offered traffic and the capacities of the links.

It may be shown that the bound of Expression 17.52 holds for the overall network loss probability, L, where Λ is given by Equation 17.53.

$$L \geq 1 - \frac{F(\lambda)}{\Lambda} \tag{17.52}$$

$$\Lambda = \sum_{i<j} \lambda_{ij} \tag{17.53}$$

In this equation, Λ is just the total offered traffic to the network and $F(\lambda)$ is the maximum carried load obtained by solving the above linear program.

Note also that if the arrival streams are in fact deterministic, then the bound can be attained by a dynamic routing strategy that routes traffic according to a solution to the linear program.

Erlang Bound

Suppose now that calls arrive to be connected between nodes i and $j (i, j = 1, 2, \ldots, J)$ as a Poisson process with rates λ_{ij}, and that as the pair $\{i, j\}$ varies it indexes independent Poisson streams. Suppose also that a connected call has a duration that is arbitrarily distributed with unit mean, is independent of earlier arrival times and call durations and of the node pair $\{i, j\}$, and is unknown to the dynamic routing strategy at the time of call arrival. Let L be the overall network loss probability. Then it may be shown that a lower bound on L is provided by l, the optimum attained in the linear program of Expression 17.54 subject to Expressions 17.55 and 17.56.

$$\text{minimize } l = \sum_{i<j} \frac{\lambda_{ij} b_{ij}}{\sum\limits_{i<j} \lambda_{ij}} \tag{17.54}$$

$$\sum_{j\in S, j\notin S} \lambda_{ij} b_{ij} \geq \left(\sum_{i\in S, j\notin S} \lambda_{ij} \right) \times E\left(\sum_{i\in S, j\notin S} \lambda_{ij}, \sum_{i\in S, j\notin S} C_{ij} \right) \quad \forall S \subset \{1, 2, \ldots, J\} \tag{17.55}$$

$$b_{ij} = b_{ji} \geq 0 \quad \forall i < j \tag{17.56}$$

Dynamic Routing Strategies

Advances in the technology of modern telecommunications systems have led to considerable interest in schemes that can dynamically control the routing of calls within a network. The purpose of such dynamic routing schemes is to adjust routing patterns within the network in accordance with varying and uncertain offered traffics, to make better use of spare capacity in the network resulting from dimensioning upgrades or forecasting errors, and to provide extra flexibility and robustness to respond to failures or overloads.

Two approaches in particular have received considerable attention. In the United States, AT&T (see Ash et al., 1981) implemented a scheme called Dynamic Non-hierarchical Routing (DNHR), which uses traffic forecasts for different times of day in a large-scale optimization procedure to predetermine a routing pattern. This pattern may be changed hourly, typically in relation to time zone differences.

In Canada, Bell Northern Research has proposed a scheme called Dynamically Controlled Routing (DCR), based on a central controller that receives information on the current state of all links at intervals of 5 to 10 seconds. This is used by the controller to determine a routing pattern that is then distributed back to the nodes.

Further dynamic routing strategies have been proposed, such as Dynamic Alternative Routing (DAR) for the British Telecom national network (Stacey and Songhurst, 1987). Key and Cope (1990) give a survey of distributed dynamic routing strategies.

Dynamic Non-hierarchical Routing

The DNHR strategy has been in operation throughout the entire AT&T trunk network since 1987 following initial operation within a subnetwork since 1984. The strategy maintains fixed sequences of alternative routes for each source-destination node pair but the sequences are changed during the day to follow changes in the offered traffic patterns. It is designed to take advantage of traffic noncoincidence whereby not all the nodes in the network reach their peak traffic at the same time of day. In international networks, and in trunk networks of geographically large countries, such hourly noncoincidence is due to the presence of multiple time zones.

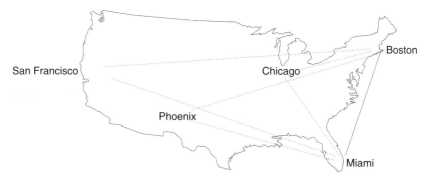

Fig. 17.9 DNHR alternative routes.

Figure 17.9 shows an example of how this might operate in practice. Traffic between Boston and Miami during the morning peak period may experience congestion on the direct route. A sequence of alternative routes for this time of day might be to attempt overflowing via Chicago, then to try via Phoenix, and then finally San Francisco using links that are only lightly congested with direct traffic thanks to the time zone effect, which makes the peak periods for traffic between San Francisco and Boston and between San Francisco and Miami, say, occur a few hours later in the day.

In DNHR, these patterns are calculated in a network management center at intervals of months, with weekly updates using data on traffics and circuit capacities provided by the nodes. This procedure is integrated within a large-scale optimization procedure for updating the link capacities to cope with the growth in forecasted traffic demands. By taking advantage of idle capacity that can be found in the network due to traffic noncoincidence, overall network cost savings can be achieved. In a case study of a 24-node network model, it was reported that with DNHR the trunk cost was reduced by about 15 percent compared with fixed routing strategies. In international networks even greater savings on the order of 25 percent have been reported.

Learning Automata

Learning automata schemes have been applied to telephone routing problems (Narenda and Mars, 1983). They continually offer calls across the available routes $r = 1, 2, \ldots, R$, say, for calls between each source destination node pair according to a probability distribution (p_r: $r = 1, 2, \ldots, R$). This probability distribution is updated over time according to the acceptance or rejection of offered calls. A scheme can thereby reward a route on which a call is successful and punish a route on which a call fails.

For example, the learning automation scheme called $L_{R \in P}$ updates the route selection probabilities as follows. If route i is chosen at time step n and the call is successful, then Equations 17.57 and 17.58 are used.

$$p_i(n+1) = p_i(n) + a[1 - p_i(n)] \tag{17.57}$$

$$p_j(n+1) = (1-a)p_j(n) \quad j \neq i \tag{17.58}$$

However, if the call fails, then Equations 17.59 and 17.60 are used.

$$p_i(n+1) = (1-\epsilon)p_i(n) \tag{17.59}$$

$$p_j(n+1) = (1-\epsilon)p_j(n) + \frac{\epsilon\, p_i(n)}{R-1} \quad j \neq i \tag{17.60}$$

The learning parameter, a, and the penalty parameter, ϵ, are two parameters of the scheme such that $0 < \epsilon < a < 1$ and ϵ is small compared to a. The value a is typically small so that the updating step is gradual. If B_r is the probability that a call is blocked on route r, then it can be shown that the $L_{R\epsilon P}$ scheme tends to approximately equalize blocking probabilities, B_r.

In practice, in a fully connected network it is preferable to first try the direct route between two nodes and then apply the automaton to the choice of two-link overflow routes, rather than to use the automaton to include the single-link direct route.

Dynamic Alternative Routing

Dynamic Alternative Routing (DAR) is a simple, decentralized dynamic routing strategy that was originally designed for the British Telecom trunk network. It consists of between 50 and 60 main switches or nodes that are fully connected. The strategy operates as follows (Gibbens and Kelly, 1990). Calls arriving between a source destination node pair first attempt the direct route and are accepted so long as there is at least one free circuit available. If there are no free circuits on the direct route, then the call attempts a currently nominated two-link alternative route via a tandem node.

The call is attempted on this route with trunk reservation applied against it on both links. If the call is successful, then this tandem node remains the currently nominated tandem node for overflow calls. Otherwise, if the call is rejected by this two-link route, the call is lost and the tandem node for future overflow calls is reselected by choosing at random from among the set of feasible tandem nodes for that source destination node pair.

Note that the tandem node is not reselected if the call is successfully routed on either the direct route or the two-link, alternative route. The term *sticky random routing* has been coined to emphasize this property of the strategy.

The DAR strategy attempts to construct a random search algorithm to find and then efficiently utilize spare capacity available within a network under varying conditions of traffic and capacity mismatches.

The simplicity of DAR permits us to extend the Erlang fixed-point models described above to model DAR's long-run average behavior. Suppose that n_{ij} is the birth-and-death process for the utilization of the link (i, j). Then, as in Equations 17.25 and 17.26, Equations 17.61 and 17.62 may be obtained where C_{ij} and r_{ij} are the (i, j) link capacities and trunk reservation parameters, respectively.

$$q(n_{ij}, n_{ij}+1) = \begin{cases} \lambda_{ij} + \sigma_{ij} & n_{ij} = 0, \dots, C_{ij} - r_{ij} - 1 \\ \lambda_{ij} & n_{ij} = C_{ij} - r_{ij}, \dots, C_{ij} - 1 \end{cases} \tag{17.61}$$

$$q(n_{ij}, n_{ij}-1) = n_{ij} \quad n_{ij} = 1, \dots, C_{ij} \tag{17.62}$$

Now suppose that $p_k(i, j)$ (where $k \neq \sim i, j$) is the long-run proportion of (i, j) overflow traffic offered via node k. Then Equation 17.27 generalizes to the asymmetric case of Equation 17.63, where $B_1(i, j)$ and $B_2(i, j)$ are the (i, j) link-blocking probabilities for fresh and overflow calls, respectively.

$$\sigma_{ij} = \sum_{k \neq i, j} \lambda_{ik} B_1(i, k) p_j(i, k)[1 - B_2(j, k)] + \sum_{k \neq i, j} \lambda_{ij} B_1(j, k) p_i(j, k)[1 - B_2(i, k)] \qquad (17.63)$$

Let $L_k(i, j)$ be the blocking on the two-link alternative route via tandem node k given by Equation 17.64.

$$1 - L_k(i, j) = [1 - B_2(i, k)][1 - B_2(k, j)] \qquad (17.64)$$

In the case of DAR, it may be shown that the route blocking rates are approximately equalized. This leads to Expression 17.65, which does not depend on k. Therefore $p_k(i, j)$ may be determined by the relation of Equation 17.66 in terms of $L_k(i, j)$ and hence $B_2(i, j)$ by Equation 17.64.

$$p_k(i, j) L_k(i, j) \qquad (17.65)$$

$$\sum_{k \neq i, j} p_k(i, j) = 1 \qquad (17.66)$$

The proportions $p_k(i, j)$ may then be used to calculate new estimates for the offered traffics and, hence, for the link blockings $B_1(i, j)$ and $B_2(i, j)$. Using the method of direct repeated substitution has been found to lead to oscillatory behavior in some cases. In practice, this may be readily overcome by damping the iteration in the sense that, if solving the fixed-point Equation 17.67 for some function f, then one should update x by Equation 17.68, where a $(0 < a \leq 1)$ is the damping factor. A value around $a = 0.8$ has been found to be satisfactory in practice.

$$x = f(x) \qquad (17.67)$$

$$x_{n+1} = (1 - a)x_n + af(x_n) \qquad (17.68)$$

DAR Applied to the BT Network

Figure 17.10 shows a typical routing problem similar to those that were faced in the British Telecom national trunk network. A call requires a route from its trunk exchange (labeled as node 1) to its destination local exchange (node 3). There are two trunk exchanges that connect to node 3—namely, nodes labeled 2 and 4. This situation, where a local exchange connects to two trunk exchanges, is often repeated throughout the network and is called "dual-parenting." First, the call attempts the route (1, 2, 3) and if a free circuit cannot be found on each of these two links, it overflows to the route (1, 4, 3) via the alternative destination trunk exchange (node 4). If this second route does not have a free circuit available on each link, then either the call is lost or further, but necessarily longer, routes are attempted.

A situation where a network may implicitly judge it better to lose the call rather than accept it using routes containing many links is when it is subject to a general overload of traffic. Accepting calls over long routes in these circumstances just adds to the levels of congestion felt by all calls.

The "sticky random" or Dynamic Alternative Routing strategy handles this situation by remembering a current tandem trunk exchange (labeled node 5) through which to send calls

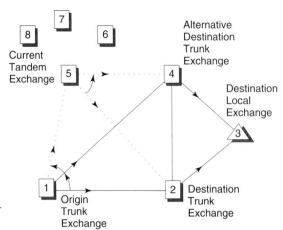

Fig. 17.10 DAR in the BT network.

attempting to overflow to longer routes. First, it attempts the route (1, 5, 2, 3) and then overflows to the alternative destination trunk exchange along route (1, 5, 4, 3). For each of these two routes the trunk reservation is used to give priority to the more directly routed calls in preference to the indirect calls using longer routes. If none of these routes is prepared to accept the call, then the call is at this stage finally rejected from the network.

The sticky random strategy then again comes into play to reset the current tandem trunk exchange by choosing at random from among a predefined set (nodes 5, 6, 7, 8, . . .). In this way, the sticky random strategy is able to shift the patterns of offered traffic to match the presence of spare capacity that it finds to form these longer paths. It should be pointed out that this strategy applies, in parallel, in a decentralized fashion throughout the network. In an example such as the British Telecom national trunk network, there could well be 60 trunk exchanges spread throughout the United Kingdom.

Least Busy Alternative

The least busy alternative (LBA) routing strategy is an example of a state-dependent routing strategy defined for fully connected networks that operates as follows. A call is first attempted on the direct route and is accepted so long as at least one free circuit is available. Otherwise, the strategy looks at the occupancies of all the available two-link alternative routes and attempts to route the call along the least busy such route subject to trunk reservation. Specifically, if m_{ik} and m_{kj} give the number of circuits free on the links (i, k) and (k, j), then the alternative chosen is via tandem node k, which maximizes $\min\{m_{ik}, m_{kj}\}$ over all tandem nodes $k \neq \sim i, j$. Then trunk reservation with parameter r will mean that the chosen alternative is only used if $\min\{m_{ik}, m_{kj}\} > r$.

It is possible to obtain an Erlang fixed-point model for the LBA strategy (Wong and Yum, 1990). We present the model in the simpler case of a symmetric network of N nodes with link capacities C and offered traffics at rate λ. Let n be the birth-and-death process representing the utilization of a link; then the transition rates are as in Equations 17.69 and 17.70.

$$q(i, i+1) = \begin{cases} \lambda + \sigma_i & i = 0, \ldots, C - r - 1 \\ \lambda & i = C - r, \ldots, C \end{cases} \tag{17.69}$$

$$q(i, i-1) = i \quad i = 1, \ldots, C \tag{17.70}$$

In this equation, σ_i is a state-dependent birth rate given in terms of the equilibrium distribution $\{\pi_i : i = 0, \ldots, C\}$ for n as in Equation 17.71, where y_i is given by Equation 17.72 and φ_i by Equation 17.73.

$$\sigma_i = 2N \left[y_i \sum_{j=0}^{i} \pi_j + \sum_{j=i+1}^{C-1} y_j \pi_j \right] \tag{17.71}$$

$$y_i = \frac{\lambda \pi_C}{N} \sum_{j=0}^{N-1} (\Phi_i)^j (\Phi_{i+1})^{N-1-j} \tag{17.72}$$

$$\Phi_i = 1 - \left(\sum_{j=0}^{i-1} \pi_j \right)^2 \tag{17.73}$$

The loss probability, L, is then as in Equation 17.74, where B_1 and B_2 are the link-blocking probabilities for fresh and overflow traffics, respectively, and are given by Equations 17.75 and 17.76.

$$L = B_1 \left[1 - (1 - B_2)^2 \right]^N \tag{17.74}$$

$$B_1 = \pi_C \tag{17.75}$$

$$B_2 = \sum_{i=C-r}^{C} \pi_i \tag{17.76}$$

In the United States, AT&T has recently started implementing a closely related dynamic routing strategy called Real Time Network Routing (RTNR) (see Ash et al., 1991).

References

Ash, G. R., Cardwell, R. H., and Murray, R. P. (1981) Design and optimization of networks with dynamic routing, *Bell. Syst. Tech. J.*, 60(8), pp. 1787–1820.

Ash, G. R., et al. (1991) Real time network routing in a dynamic class of service network. In *Proc. 13th International Teletraffic Congress*, Copenhagen, North-Holland, Amsterdam.

Brockmeyer, E., Halstrom, H. L., and Jensen, A. (1948) *The Life and Works of A. K. Erlang*, Academy of Technical Sciences, Copenhagen.

Gibbens, R. J., and Kelly, F. P. (1990) Dynamic routing in fully connected networks, *IMA J. Math. Control Inform.*, 7, pp. 77–111.

Hui, Y. J. (1990) *Switching and Traffic Theory for Integrated Broadband Networks*, Kluwer, Boston.

Kelly, F. P. (1979) *Reversibility and Stochastic Networks*, Wiley, Chichester.

———. (1990) Routing in circuit-switched networks: optimization, shadow prices, and decentralization, *Adv. in Appl. Probab.*, 20, pp. 112–144.

Key, P. B., and Whitehead, M. J. (1988) Cost-effective use of networks employing Dynamic Alternative Routing. In *Proc. 12th International Teletraffic Congress*, Turin, North-Holland, Amsterdam.

Key, P. B., and Cope, G. A. (1990) Distributed dynamic routing schemes, *IEEE Communications Magazine*, *Advanced Traffic Control Methods for Circuit-Switched Telecommunication Networks*, 28(10), pp. 54–64.

Narenda, K. S., and Mars, P. (1983) The use of learning algorithms in telephone traffic routing—A methodology, *Automatica*, 19(5), pp. 495–502.

Stacey, R. R., and Songhurst, D. J. (1987) Dynamic alternative routing in the British Telecom trunk network, *Int. Switching Symposium*, Phoenix.

Voegtlin, J-C. (1994) The applications of traffic simulation, *Telecommunications*, March.

Wong, E. W. M., and Yum, T. S. (1990) Maximum free circuit routing in circuit-switched networks, *Proc. IEEE Infocom '90*, IEEE Computer Society Press, pp. 934–937.

Part Six

Mathematical Techniques

Telecommunications networks rely heavily on a wide range of mathematical techniques for their design, analysis, and optimization, so it is wholly appropriate to include a description of some of the fundamental methods that are widely used in this first volume of *The Cable and Telecommunication Professionals' Reference*. Some techniques related to circuit- and packet-switched networks were covered in Part 5. Other more fundamental methods are covered here in Part 6.

In telecommunications, received signals are usually combined with noise or interference. Recently developed methods, based on Bayesian techniques, to extract signals from noise are the subject of particle filtering in Chapter 18. Chapter 19 provides a set of trigonometric functions and general formulae. Chapter 20 describes different forms of calculus, including differential calculus, which relates to the slope of a functions graph, and integral calculus, which relates to the area beneath a graph. Mathematical series (Chapter 21), matrices, and determinants (Chapter 22) are also relevant to a wide variety of telecommunications system representations.

In designing or analyzing the performance of a telecommunications system, it is often necessary to consider basic signal properties in both the frequency and time domains. As the two domains are intrinsically related, it is important to be able to transform from one domain to another. The Fourier transform provides a mathematical technique to allow this. Many circuits or systems are linear and time invariant, such as electrical circuits or optical devices. These may be analyzed using Laplace transform methods. Transform methods are described in Chapter 23. Chapter 24 considers the ideas associated with information and entropy and related to probability defined in terms of incomplete knowledge. Mutual and self information is introduced, and these concepts are applied to various models for communications channels.

18 Particle Filtering in Digital Communications

E. Punskaya, A. Doucet, and W. J. Fitzgerald

Introduction

In many engineering applications, one needs to extract the signal from the data corrupted by additive random noise and interference of different kinds in order to recover the unknown quantities of interest. This is a problem of scientific inference, and to carry out consistent reasoning and inference, one may use the Bayesian paradigm.

The main challenge of the Bayesian approach, however, is the fact that quite often it requires multidimensional integration over parameter space, which is generally only possible when the models are linear in the parameters and when one makes a simplifying noise assumption, such as Gaussianity. In more realistic situations, the integrations have to be performed numerically and this requirement has lead to a full-scale investigation of such integration methods.

When the set of data under study does not evolve while the parameters of the system are being estimated, batch simulation-based methods (in particular, Markov chain Monte Carlo [MCMC] methods) have emerged as the tools of choice. There are numerous scenarios, however, where the data arrive sequentially in time and, therefore, require online decision-making responses. Except for several special cases restricted to a narrow linear Gaussian class of models, this problem still presents a major challenge to researchers. A number of approximate filters have been devised for more complicated scenarios. However, until recently there existed no general methodology whenever nonlinearity or non-Gaussianity was involved. The Sequential Monte Carlo (SMC), also known as particle-filtering methods, recently emerged in the field of statistics and engineering and is believed to be a key leading technology.

We describe here a generic SMC method that includes and generalizes many algorithms presented recently in the literature. The idea is to approximate the posterior distribution of

417

interest by swarms of N ($N \gg 1$) weighted points in the sample space, called particles, which evolve randomly in time in correlation with each other and either give birth to offspring particles or die according to their ability to represent the different zones of interest of the state space. The algorithm combines sequential importance sampling, selection procedure, and some variance reduction techniques. We demonstrate the applications of this method to the problems arising in digital communications.

Bayesian Filtering

State-space Model

For the sake of clarity, we restrict ourselves here to the following state-space model, which is of interest in this chapter. The unobserved n_x-dimensional state of interest, $(\mathbf{x}_n; n = 0, 1, 2, \ldots,)$, changes over time according to the *system equation* of the following form:

$$\mathbf{x}_n = \Psi(\mathbf{x}_{n-1}, \mathbf{v}_n) \tag{18.1}$$

and the n_y-dimensional measurements $\mathbf{y}_n$, taken at discrete time points $n = 1, 2, \ldots,$ are given by

$$\mathbf{y}_n = \Omega(\mathbf{x}_n, \varepsilon_n) \tag{18.2}$$

where $\Psi(\cdot)$ is a system transition function, $\Omega(\cdot)$ is a measurement function, and $\mathbf{v}_n$ and ε_n are independent noise vectors with a known distribution. For simplicity, the same notation is used throughout for both random variables and their realization, and we also assume that the analytical forms of the functions $\Psi(\cdot)$, $\Omega(\cdot)$, and the initial probability density of the state $p(\mathbf{x}_0)$ are known.

Bayesian Inference

The Bayesian approach provides an elegant and consistent method of dealing with uncertainty. The Bayesian posterior, $p(\mathbf{x}_{0:n}|\mathbf{y}_{1:n})$, reflects all the information we have about the state of the system $\mathbf{x}_{0:n}$, contained in the measurements $\mathbf{y}_{1:n}$ and the prior $p(\mathbf{x}_{0:n})$, and gives a direct and easily applicable means of combining the two last-mentioned densities (Bayes's theorem)[1]:

$$p(\mathbf{x}_{0:n}|\mathbf{y}_{1:n}) = \frac{p(\mathbf{y}_{1:n}|\mathbf{x}_{0:n})p(\mathbf{x}_{0:n})}{p(\mathbf{y}_{1:n})} \tag{18.3}$$

Taking into account that the observations up to time n are independent given $\mathbf{x}_{0:n}$, the likelihood $p(\mathbf{y}_{1:n}|\mathbf{x}_{0:n})$ in the above equation can be factorized as follows:

$$p(\mathbf{y}_{1:n}|\mathbf{x}_{0:n}) = \prod_{i=1}^{n} p(\mathbf{y}_i|\mathbf{x}_{0:n}) \tag{18.4}$$

---

[1] The sequences $\mathbf{x}_{0:n}$ and $\mathbf{y}_{0:n}$ denote, respectively, the signal and the observations up to time n: $\mathbf{x}_{0:n} = \{\mathbf{x}_0, \ldots, \mathbf{x}_n\}$ and $\mathbf{y}_{1:n} = \{\mathbf{y}_1, \ldots, \mathbf{y}_n\}$.

and, since, conditional on $\mathbf{x}_i$, the measurement $\mathbf{y}_i$, is independent of the states at all other times, it is given by:

$$p(\mathbf{y}_{1:n}|\mathbf{x}_{0:n}) = \prod_{i=1}^{n} p(\mathbf{y}_i|\mathbf{x}_i) \tag{18.5}$$

In addition, as a result of the Markov structure of the system (Equation 18.1), the prior $p(\mathbf{x}_{0:n})$ takes the following form:

$$p(\mathbf{x}_{0:n}) = p(\mathbf{x}_0) \prod_{i=1}^{n} p(\mathbf{x}_i|\mathbf{x}_{i-1}) \tag{18.6}$$

resulting in the posterior probability density being equal to

$$p(\mathbf{x}_{0:n}|\mathbf{y}_{1:n}) = \frac{p(\mathbf{x}_0) \prod_{i=1}^{n} p(\mathbf{y}_i|\mathbf{x}_i) p(\mathbf{x}_i|\mathbf{x}_{i-1})}{p(\mathbf{y}_{1:n})} \tag{18.7}$$

Filtering Objectives

Our objective is to obtain the estimates of the state at time n, conditional upon the measurements up to time $\check{n}$, such as, for example, Minimum Mean Square Estimate (MMSE) of $\mathbf{x}_n$:

$$\hat{\mathbf{x}}_n^{MMSE} = \mathrm{E}_{p(\mathbf{x}_n|\mathbf{y}_{1:\check{n}})}[\mathbf{x}_n] = \int \mathbf{x}_n p(\mathbf{x}_n|\mathbf{y}_{1:\check{n}}) d\mathbf{x}_n \tag{18.8}$$

or Marginal Maximum A Posteriori (MMAP) given by:

$$\hat{\mathbf{x}}_n^{MMAP} = \arg\max_{\mathbf{x}_n} p(\mathbf{x}_n|\mathbf{y}_{1:\check{n}}) \tag{18.9}$$

These can obviously be extended to estimating the functions of the state instead of the state itself:

$$\widehat{\Upsilon(\mathbf{x}_n)}^{MMSE} =$$

$$\mathrm{E}_{p(\mathbf{x}_n|\mathbf{y}_{1:\check{n}})}[\Upsilon(\mathbf{x}_n)] = \int \Upsilon(\mathbf{x}_n) p(\mathbf{x}_n|\mathbf{y}_{1:\check{n}}) d\mathbf{x}_n$$

Calculating $p(\mathbf{x}_n|\mathbf{y}_{1:\check{n}})$ for $\check{n} - n$, and hence the estimates of $\mathbf{x}_n$, given the data up to time n, is the aim of Bayesian *filtering* and a primary subject of this chapter. One might also be interested in the related problems of evaluating $p(\mathbf{x}_{0:n}|\mathbf{y}_{1:\check{n}})$ for $\check{n} > n$ (*smoothing*) or for $\check{n} < n$ (*prediction*).

Sequential Scheme

The probability density of interest $p(\mathbf{x}_n|\mathbf{y}_{1:n})$ can be obtained by marginalization of Equation 18.7; however, the dimension of the integration in this case grows as n increases. This can be avoided by using a sequential scheme.

A recursive formula for the joint probability density can be obtained straightforwardly from Equation 18.7:

$$p(\mathbf{x}_{0:n}|\mathbf{y}_{1:n}) = p(\mathbf{x}_{0:n-1}|\mathbf{y}_{1:n-1})\frac{p(\mathbf{y}_n|\mathbf{x}_n)p(\mathbf{x}_n|\mathbf{x}_{n-1})}{p(\mathbf{y}_n|\mathbf{y}_{1:n-1})} \tag{18.10}$$

with the marginal $p(\mathbf{x}_n|\mathbf{y}_{1:n})$ also satisfying the recursion (Sorenson, 1988):

$$p(\mathbf{x}_n|\mathbf{y}_{1:n-1}) = \int p(\mathbf{x}_n|\mathbf{x}_{n-1})p(\mathbf{x}_{n-1}|\mathbf{y}_{1:n-1})d\mathbf{x}_{n-1} \tag{18.11}$$

$$p(\mathbf{x}_n|\mathbf{y}_{1:n}) = \frac{p(\mathbf{y}_n|\mathbf{x}_n)p(\mathbf{x}_n|\mathbf{y}_{1:n-1})}{p(\mathbf{y}_n|\mathbf{y}_{1:n-1})} \tag{18.12}$$

where

$$p(\mathbf{y}_n|\mathbf{y}_{1:n-1}) = \int p(\mathbf{y}_n|\mathbf{x}_n)p(\mathbf{x}_n|\mathbf{y}_{1:n-1})d\mathbf{x}_n \tag{18.13}$$

Equations 18.11 and 18.12 are called, respectively, *prediction* and *updating*.

The preceding expressions are deceptively simple, however, since the integrations involved are usually intractable. One cannot typically compute the normalizing constant $p(\mathbf{y}_{1:n})$ and the marginals of $p(\mathbf{x}_n|\mathbf{y}_{1:n})$, particularly, $p(\mathbf{x}_n|\mathbf{y}_n)$, except for several special cases when the integration can be performed exactly.

The problem is of great importance, though, which is why a great number of different approaches and filters have been proposed. All of them suffer from quite serious drawbacks and do not provide a general methodology for nonlinear, non-Gaussian filtering. The SMC, also known as the particle filtering approach, described in detail in the next section, shows great promise in providing such a fundamental technology.

Sequential Monte Carlo

The idea to use Monte Carlo integration methods in filters can be traced back to a number of people (Handschin and Mayne, 1969; Akashi and Kumamoto, 1977; Zaritskii et al., 1975). However, it was not until recently, with the increase of computational power, that Monte Carlo-based filters gained much interest in different areas of statistics and engineering. The interest has arisen with the proposal of the so-called *bootstrap* filter, simultaneously developed by Gordon and Kitagawa (Gordon et al., 1993; Kitagawa, 1993). Since then, SMC algorithms, under the names of particle filters, sequential importance resampling (SIR), and condensation trackers, have been applied to a wide range of problems in the fields of engineering, financial data analyses, genetics, medicine, and biology, to name a few. (See Doucet et al., 2001a, for the list of references.)

The Sequential Monte Carlo methods approximate the posterior distribution of interest by swarms of points in the sample space, called particles, which evolve randomly in time according to a simulation-based rule, and either give birth to offspring particles or die according to their ability to represent the different zones of interest of the state space dictated by the observation process and the dynamics of the underlying system. Some similarities can be seen with the grid-based methods; however, particle filters use adaptive stochastic grid approximation (i.e., naturally follow the movement of the state instead of being chosen arbitrarily by the user). The probability density of interest is, thus, represented more accurately, and, in addition, the rate of convergence of the approximation error toward zero is theoretically not sensitive to the size of the state space (Crisan et al., 1999).

This section describes the basic particle-filtering algorithm based on SIR, and points out some of its limitations. The issues concerning some improvements on SIR are discussed in the next section.

Monte Carlo Methods

Monte Carlo methods are commonly used for approximation of intractable integrals and rely on the ability to draw a random sample from the required probability distribution. The idea is to simulate N independent identically distributed (*i.i.d.*) samples $\{\mathbf{x}_{0:n}^{(i)}\}_{i=1}^{N}$ from the distribution of interest, which in our case is the posterior $p(\mathbf{x}_{0:n}|\mathbf{y}_{1:n})$, and use them to obtain an empirical estimate of the distribution:

$$\hat{p}_N(d\mathbf{x}_{0:n}|\mathbf{y}_{1:n}) = \frac{1}{N}\sum_{i=1}^{N}\delta(d\mathbf{x}_{0:n} - \mathbf{x}_{0:n}^{(i)}) \tag{18.14}$$

The function $\delta(d\mathbf{x}_{0:n} - \mathbf{x}_{0:n}^{(i)})$ here denotes the Dirac delta function.

The expected value of $\mathbf{x}_{0:n}$:

$$\mathrm{E}_{p(\mathbf{x}_{0:n}|\mathbf{y}_{1:n})}[\mathbf{x}_{0:n}] = \int \mathbf{x}_{0:n} p(d\mathbf{x}_{0:n}|\mathbf{y}_{1:n}) \tag{18.15}$$

or, indeed, of any function $\Upsilon(\cdot)$ of $\mathbf{x}_{0:n}$:

$$\mathrm{E}_{p(\mathbf{x}_{0:n}|\mathbf{y}_{1:n})}[\Upsilon(\mathbf{x}_{0:n})] = \int \Upsilon(\mathbf{x}_{0:n}) p(d\mathbf{x}_{0:n}|\mathbf{y}_{1:n}) \tag{18.16}$$

can be obtained consequently by approximating the corresponding integrals by the sums:

$$\mathrm{E}_{\hat{p}N(\mathbf{x}_{0:n}|\mathbf{y}_{1:n})}[\mathbf{x}_{0:n}] = \tag{18.17}$$

$$\int \mathbf{x}_{0:n}\hat{p}_N(d\mathbf{x}_{0:n}|\mathbf{y}_{1:n}) = \frac{1}{N}\sum_{i=1}^{N}\mathbf{x}_{0:n}^{(i)} \tag{18.18}$$

$$\mathrm{E}_{\hat{p}N(\mathbf{x}_{0:n}|\mathbf{y}_{1:n})}[\Upsilon(\mathbf{x}_{0:n})] = \tag{18.19}$$

$$\int \Upsilon(\mathbf{x}_{0:n})\hat{p}_N(d\mathbf{x}_{0:n}|\mathbf{y}_{1:n}) = \frac{1}{N}\sum_{i=0}^{N}\Upsilon(\mathbf{x}_{0:n}^{(i)}) \tag{18.20}$$

The estimate (Equation 18.19) is unbiased with the variance proportional to $1/N$ for the finite variance of $\Upsilon(\mathbf{x}_{0:n})$ (see Doucet et al., 2000, for more details), and is easily obtained providing one can sample from $p(\mathbf{x}_{0:n}|\mathbf{y}_{1:n})$. This is usually not the case, however, with $p(\mathbf{x}_{0:n}|\mathbf{y}_{1:n})$ being multivariate, nonstandard, and typically only known up to a normalizing constant. In the next section, we briefly discuss an alternative approach for drawing random samples from such distribution.

Bayesian Importance Sampling

The approach is based on the following remark. Suppose one cannot efficiently sample from $p(\mathbf{x}_{0:n}|\mathbf{y}_{1:n})$ however, there is another arbitrary convenient probability distribution, $\pi(\mathbf{x}_{0:n}|\mathbf{y}_{1:n})$ (such that $p(\mathbf{x}_{0:n}|\mathbf{y}_{1:n}) > 0$ implies $\pi(\mathbf{x}_{0:n}|\mathbf{y}_{1:n}) > 0$), that is easy to sample from. Then the estimate of the function $\Upsilon(\cdot)$ of $\mathbf{x}_{0:n}$ can be represented as

$$E_{p(\mathbf{x}_{0:n}|\mathbf{y}_{1:n})}[\Upsilon(\mathbf{x}_{0:n})] = \int \Upsilon(\mathbf{x}_{0:n}) \frac{p(\mathbf{x}_{0:n}|\mathbf{y}_{1:n})}{\pi(\mathbf{x}_{0:n}|\mathbf{y}_{1:n})} \pi(\mathbf{x}_{0:n}|\mathbf{y}_{1:n}) d\mathbf{x}_{0:n} \qquad (18.21)$$

$$= E_{\pi(\cdot|\mathbf{y}_{1:n})}\left[\Upsilon(\mathbf{x}_{0:n}) \frac{p(\mathbf{x}_{0:n}|\mathbf{y}_{1:n})}{\pi(\mathbf{x}_{0:n}|\mathbf{y}_{1:n})} \right] \qquad (18.22)$$

with $w(\mathbf{x}_{0:n})$ being the *importance weight*:

$$w(\mathbf{x}_{0:n}) = \frac{p(\mathbf{x}_{0:n}|\mathbf{y}_{1:n})}{\pi(\mathbf{x}_{0:n}|\mathbf{y}_{1:n})} \qquad (18.23)$$

Taking into account Bayes's theorem (Equation 18.3), one obtains:

$$E_{p(\mathbf{x}_{0:n}|\mathbf{y}_{1:n})}[\Upsilon(\mathbf{x}_{0:n})] = \int \Upsilon(\mathbf{x}_{0:n}) \frac{p(\mathbf{y}_{1:n}|\mathbf{x}_{0:n})p(\mathbf{x}_{0:n})}{p(\mathbf{y}_{1:n})\pi(\mathbf{x}_{0:n}|\mathbf{y}_{1:n})} \pi(\mathbf{x}_{0:n}|\mathbf{y}_{1:n}) d\mathbf{x}_{0:n} \qquad (18.24)$$

$$= \frac{1}{p(\mathbf{y}_{1:n})} \int \Upsilon(\mathbf{x}_{0:n}) w_n \pi(\mathbf{x}_{0:n}|\mathbf{y}_{1:n}) d\mathbf{x}_{0:n} \qquad (18.25)$$

where w_n is so-called *unnormalized importance weights*:

$$w_n = \frac{p(\mathbf{y}_{1:n}|\mathbf{x}_{0:n})p(\mathbf{x}_{0:n})}{\pi(\mathbf{x}_{0:n}|\mathbf{y}_{1:n})} \qquad (18.26)$$

and $\pi(\mathbf{x}_{0:n}|\mathbf{y}_{1:n})$ is called the *importance distribution*. From Equation 18.25 follow

$$E_{p(\mathbf{x}_{0:n}|\mathbf{y}_{1:n})}[\mathbf{x}_{0:n}] = \frac{1}{p(\mathbf{y}_{1:n})} \int \Upsilon(\mathbf{x}_{0:n}) w_n \pi(\mathbf{x}_{0:n}|\mathbf{y}_{1:n}) d\mathbf{x}_{0:n}$$

$$= \frac{\int \Upsilon(\mathbf{x}_{0:n}) w_n \pi(\mathbf{x}_{0:n}|\mathbf{y}_{1:n}) d\mathbf{x}_{0:n}}{\int p(\mathbf{y}_{1:n}|\mathbf{x}_{0:n})p(\mathbf{x}_{0:n}) \dfrac{\pi(\mathbf{x}_{0:n}|\mathbf{y}_{1:n})}{\pi(\mathbf{x}_{0:n}|\mathbf{y}_{1:n})} d\mathbf{x}_{0:n}} \qquad (18.27)$$

$$= \frac{\int \Upsilon(\mathbf{x}_{0:n}) w_n \pi(\mathbf{x}_{0:n}|\mathbf{y}_{1:n}) d\mathbf{x}_{0:n}}{\int w_n \pi(\mathbf{x}_{0:n}|\mathbf{y}_{1:n}) d\mathbf{x}_{0:n}}$$

$$= \frac{E_{\pi(\cdot|\psi_{1:n})}[\Upsilon(\mathbf{x}_{0:n}) w_n]}{E_{\pi(\cdot|\psi_{1:n})}[w_n]} \qquad (18.28)$$

The estimate of interest can, thus, be approximated by

$$E_{\hat{p}_N(\mathbf{x}_{0:n}|\mathbf{y}_{1:n})}[\Upsilon(\mathbf{x}_{0:n})] = \frac{\frac{1}{N}\sum_{i=1}^{N}\Upsilon(\mathbf{x}_{0:n}^{(i)})w_n^{(i)}}{\frac{1}{N}\sum_{i=0}^{N}w_n^{(i)}} \qquad (18.29)$$

$$= \sum_{i=1}^{N}\Upsilon(\mathbf{x}_{0:n}^{(i)})\tilde{w}_n^{(i)}$$

where $\mathbf{x}_{0:n}^{(i)}$ denotes samples drawn from $\pi(\mathbf{x}_{0:n}|\mathbf{y}_{1:n})$ and $\tilde{w}_n^{(i)}$ is the *normalized importance weights*:

$$\tilde{w}_n^{(i)} = \frac{w_n^{(i)}}{\sum_{i=1}^{N}w_n^{(i)}} \qquad (18.30)$$

This estimate is biased since it involves a ratio of estimates. However, under the assumptions of $\mathbf{x}_{0:n}^{(i)}$ being a set of i.i.d. samples drawn from $\pi(\mathbf{x}_{0:n}|\mathbf{y}_{1:n})$, the support of $\pi(\mathbf{x}_{0:n}|\mathbf{y}_{1:n})$ including the support of $p(\mathbf{x}_{0:n}|\mathbf{y}_{1:n})$, and $E_{p(\mathbf{x}_{0:n}|\mathbf{y}_{1:n})}[\Upsilon(\mathbf{x}_{0:n})]$ existing and being finite, one could obtain a convergence of the empirical distribution $\sum_{i=1}^{N}\tilde{w}_n^{(i)}\delta(d\mathbf{x}_{0:n}-\mathbf{x}_{0:n}^{(i)})$ toward $p(d\mathbf{x}_{0:n}|\mathbf{y}_{1:n})$ in the sense of almost sure convergence of $E_{\hat{p}_N(\mathbf{x}_{0:n}|\mathbf{y}_{1:n})}[\Upsilon(\mathbf{x}_{0:n})]$ toward $E_{p(\mathbf{x}_{0:n}|\mathbf{y}_{1:n})}[\Upsilon(\mathbf{x}_{0:n})]$ as $N \to +\infty$. Under the additional assumptions of the expectations $E_{p(\cdot|\mathbf{y}_{1:n})}[\Upsilon^2(\mathbf{x}_{0:n})w(\mathbf{x}_{0:n})]$ and $E_{p(\cdot|\mathbf{y}_{1:n})}[w(\mathbf{x}_{0:n})]$ existing and being finite, a central limit theorem also holds. (The details are given in Doucet et al., 2000, and Geweke, 1989.)

The preceding means that the algorithm could be interpreted as a simulation-based method for sampling from $p(d\mathbf{x}_{0:n}|\mathbf{y}_{1:n})$, with $p(d\mathbf{x}_{0:n}|\mathbf{y}_{1:n})$ being approximated by point mass estimate:

$$\hat{p}_N(d\mathbf{x}_{0:n}|\mathbf{y}_{1:n}) = \sum_{i=1}^{N}\tilde{w}_n^{(i)}\delta(d\mathbf{x}_{0:n}-\mathbf{x}_{0:n}^{(i)}) \qquad (18.31)$$

The "perfect" simulation case would correspond to

$$\pi(\mathbf{x}_{0:n}|\mathbf{y}_{1:n}) = p(\mathbf{x}_{0:n}|\mathbf{y}_{1:n}) \qquad (18.32)$$

and $\tilde{w}_n^{(i)} = N^{-1}$ for all i.

Sequential Importance Sampling

The method just described is a batch method. In order to obtain the estimate of $p(d\mathbf{x}_{0:n}|\mathbf{y}_{1:n})$ sequentially, one should be able to propagate $\hat{p}_N(d\mathbf{x}_{0:n}|\mathbf{y}_{1:n})$ in time without modifying the past simulated states $\{\mathbf{x}_{0:n}^{(i)}\}_{i=1}^{N}$. In general, the importance function $\pi(d\mathbf{x}_{0:n}|\mathbf{y}_{1:n})$ can be expanded as:

$$\pi(\mathbf{x}_{0:n}|\mathbf{y}_{1:n}) = x(\mathbf{x}_0|\mathbf{y}_{1:n})\prod_{j=1}^{n}\pi(\mathbf{x}_j|\mathbf{x}_{0:j-1}, \mathbf{y}_{1:n}) \qquad (18.33)$$

In order to fulfill the previous condition, though, the adopted distribution should be of the following form:

$$\pi(\mathbf{x}_{0:n}|\mathbf{y}_{1:n}) = \pi(\mathbf{x}_0)\prod_{j=1}^{n}\pi(\mathbf{x}_j|\mathbf{x}_{0:j-1}, \mathbf{y}_{1:j})$$

$$= \pi(\mathbf{x}_{0:n-1}|\mathbf{y}_{1:n-1})\pi(\mathbf{x}_n|\mathbf{x}_{0:n-1}, \mathbf{y}_{1:n}) \qquad (18.34)$$

(i.e., it should admit $\pi(\mathbf{x}_{0:n-1}|\mathbf{y}_{1:n-1})$ as marginal distribution).

A recursive expression for the importance weights can then be derived by substituting Equations 18.34 and 18.5–6 into Equation 18.26:

$$w_n = \frac{p(\mathbf{x}_{0:n-1}|\mathbf{y}_{1:n-1})p(\mathbf{y}_n|\mathbf{x}_n)p(\mathbf{x}_n|\mathbf{x}_{n-1})}{\pi(\mathbf{x}_{0:n-1}|\mathbf{y}_{1:n-1})\pi(\mathbf{x}_n|\mathbf{x}_{0:n-1}, \mathbf{y}_{1:n})} = w_{n-1}\frac{p(\mathbf{y}_n|\mathbf{x}_n)p(\mathbf{x}_n|\mathbf{x}_{n-1})}{\pi(\mathbf{x}_n|\mathbf{x}_{0:n-1}, \mathbf{y}_{1:n})} \qquad (18.35)$$

Although sequential importance sampling poses only one restriction on the importance density, Equation 18.34, with the number of choices otherwise being unlimited, the design of the appropriate proposal function is, in fact, one of the most critical issues in importance sampling algorithms. Poor choice leads to poor approximation and to poor algorithm performance in general. The problem, therefore, has received a lot of interest in the literature, with different importance distributions being advocated by different researchers. Following are descriptions of the most popular choices.

Prior Distribution

This, by all means, is the most popular and most widely used proposal distribution, which is largely due to the simplicity of its implementation (Handschin and Mayne, 1969; Gordon et al., 1993). If one takes

$$\pi(\mathbf{x}_n|\mathbf{x}_{0:n-1}, \mathbf{y}_{1:n}) = p(\mathbf{x}_n|\mathbf{x}_{n-1}) \qquad (18.36)$$

the importance weights could be evaluated straightforwardly:

$$w_n^{(i)} \propto w_{n-1}^{(i)}p(\mathbf{y}_n|\mathbf{x}_n^{(i)}) \qquad (18.37)$$

often resulting in a reduced computational complexity of the designed algorithm. This distribution, however, does not incorporate the information contained in the most recent observations, and, therefore, may be inefficient and especially sensitive to outliers.

Optimal Distribution

This choice of the proposal distribution was introduced in *Automation and Remote Control* (Zaritskii et al., 1975) and later used for a particular case in several other texts, where lists of references can be found (Akashi and Kumamoto, 1977; Liu and Chen, 1995; Doucet et al., 2000). The distribution of the form

$$\pi(\mathbf{x}_n|\mathbf{x}_{n-1}, \mathbf{y}_{1:n}) = p(\mathbf{x}_n|\mathbf{x}_{n-1}, \mathbf{y}_n) \qquad (18.38)$$

is, indeed, optimal in the sense that it minimizes the variance of the importance weights $w(\mathbf{x}_{1:n})$ conditional upon the simulated trajectory $\mathbf{x}_{1:n-1}^{(i)}$ and the observations $\mathbf{y}_{1:n}$ (Doucet et al., 2000), thus limiting the problem of degeneracy of the algorithm. Interestingly, the weights in this case do not depend on the current value of the state, $\mathbf{x}_n^{(i)}$:

$$w_n^{(i)} \propto w_{n-1}^{(i)} \frac{p(\mathbf{y}_n|\mathbf{x}_n^{(i)})p(\mathbf{x}_n^{(i)}|\mathbf{x}_{n-1}^{(i)})}{p(\mathbf{x}_n^{(i)}|\mathbf{x}_{n-1}^{(i)}, \mathbf{y}_n)} \qquad (18.39)$$

$$\propto w_{n-1}^{(i)} p(\mathbf{y}_n|\mathbf{x}_{n-1}^{(i)}) \qquad (18.40)$$

which facilitates parallelization of the simulation of $\{\mathbf{x}_n^{(i)}\}_{i=1}^N$ and evaluation of $w_n^{(i)}$ for $i = 1, \dots, N$. The selection step in this case can be done prior to the sampling step. Unfortunately, sampling from $p(\mathbf{x}_n|\mathbf{x}_{n-1}^{(i)}, \mathbf{y}_n)$, and evaluating $p(\mathbf{y}_n|\mathbf{x}_{n-1}^{(i)})$

$$p(\mathbf{y}_n|\mathbf{x}_{n-1}) = \int p(\mathbf{y}_n|\mathbf{x}_n)p(\mathbf{x}_n|\mathbf{x}_{n-1})d\mathbf{x}_n \qquad (18.41)$$

for many models is impossible, in which case other, suboptimal importance distributions should be designed.

Suboptimal Distributions

Since any probability density fulfilling the condition specified in Equation 18.34 could be used as an importance density, a great number of "clever" suboptimal proposals could be derived. Unfortunately, there is still no general strategy specifying how to perform this procedure, and the design of each suboptimal function should be considered on a case-by-case basis.

Selection Procedure

Degeneracy of the Algorithm

Unfortunately, the algorithm presented earlier has a serious limitation. What happens is that, after running it for a few iterations, one typically finds that one of the normalized importance weights tends to 1, while the remaining weights are negligible. A large computational effort, thus, is directed into updating trajectories that virtually do not contribute to the final estimate.

This occurs due to an increase in the variance of the importance weights over time, which has a negative effect on the accuracy of the algorithm. Indeed, in the ideal case, when we are able to sample directly from the distribution of interest $\pi(\mathbf{x}_{0:n}|\mathbf{y}_{1:n}) = p(\mathbf{x}_{0:n}|\mathbf{y}_{1:n})$, the mean and the variance of the importance weights are equal to 1 and 0, correspondingly:

$$E_{\pi(\cdot|\mathbf{y}_{1:n})}[w(\mathbf{x}_{0:n})] = E_{\pi(\cdot|\mathbf{y}_{1:n})}\left[\frac{p(\mathbf{x}_{0:n}|\mathbf{y}_{1:n})}{\pi(\mathbf{x}_{0:n}|\mathbf{y}_{1:n})}\right] = 1 \qquad (18.42)$$

$$var_{\pi(\cdot|\mathbf{y}_{1:n})}[w(\mathbf{x}_{0:n})] = E_{\pi(\cdot|\mathbf{y}_{1:n})}\left[\left(\frac{p(\mathbf{x}_{0:n}|\mathbf{y}_{1:n})}{\pi(\mathbf{x}_{0:n}|\mathbf{y}_{1:n})} - E_{\pi(\cdot|\mathbf{y}_{1:n})}\left[\frac{p(\mathbf{x}_{0:n}|\mathbf{y}_{1:n})}{\pi(\mathbf{x}_{0:n}|\mathbf{y}_{1:n})}\right]\right)^2\right] = 0 \qquad (18.43)$$

and this is the situation we would like to be as close to as possible. It is, however, proven that the unconditional variance (that is, when the measurements are regarded as random) of the importance weights increases over time (Doucet et al., 2000; Kong et al., 1994), and, therefore, it is impossible to avoid a degeneracy of the algorithm.

Selection Step

One way of limiting the degeneracy would be to choose an appropriate importance function that minimizes the conditional variance of the importance weights. Unfortunately, this is not always possible. Another approach would be to introduce a forgetting factor on the weights (Moral and Salut, 1995). Although this technique does slow down the degeneracy, the problem is not eliminated completely. A solution might be to introduce a selection procedure in the algorithm that discards the low-weighted trajectories and replicates samples with the high-normalized importance weights as in sequential importance resampling (SIR) (Rubin, 1988).

In general, a selection scheme introduces an additional sampling step based on the obtained discrete distribution, which places a probability mass $\tilde{w}_n^{(i)}$ at each of the original points, say $\tilde{\mathbf{x}}_{0:n}^{(i)}$, referred to as particles:

$$\hat{p}_N(d\mathbf{x}_{0:n}|\mathbf{y}_{1:n}) = \sum_{i=1}^{N} \tilde{w}_n^{(i)}\delta\big(d\mathbf{x}_{0:n} - \tilde{\mathbf{x}}_{0:n}^{(i)}\big) \qquad (18.44)$$

At each step, a certain number of offspring (replicas), say N_i, are assigned to each trajectory $\tilde{\mathbf{x}}_{0:n}^{(i)}$ according to its weight $\tilde{w}_n^{(i)}$, with N_i being equal to 0 for particles with negligible weights that should be discarded. These values are chosen so that a total number of samples in the scheme stays the same (i.e., $\Sigma_{i=1}^{N}N_i = N$). The resulting samples $\{\mathbf{x}_{0:n}^{(i)}\}_{i=1}^{N}$ form an approximate sample from $p(d\mathbf{x}_{0:n}|\mathbf{y}_{1:n})$ and the approximating distribution follows as

$$\hat{p}_N(d\mathbf{x}_{0:n}|\mathbf{y}_{1:n}) = \frac{1}{N}\sum_{i=1}^{N}\delta\big(d\mathbf{x}_{0:n} - \mathbf{x}_{0:n}^{(i)}\big) \qquad (18.45)$$

that is, the weights are reset to N^{-1}, and the "surviving" particles $\{\mathbf{x}_{0:n}^{(i)}\}_{i=1}^{N}$ are distributed approximately according to $p(d\mathbf{x}_{0:n}|\mathbf{y}_{1:n})$.

Selection Schemes

A number of different selection schemes have been previously proposed in the literature. These include sequential importance resampling, residual resampling, and stratified sampling, briefly presented in the following text. All these schemes are unbiased (i.e., ensure that $E[N_i] = N\tilde{w}_n^{(i)}$ at each step n, while they differ in terms of $var[N_i]$ and computational load).

The most popular SIR/Multinomial Resampling, based on drawing uniformly from the discrete set $\{\tilde{\mathbf{x}}_{0:n}^{(i)}\}_{i=1}^{N}$ with probabilities $\{\tilde{w}_n^{(i)}\}_{i=1}^{N}$, can be implemented in $O(N)$ operations (Ripley, 1987; Doucet et al., 2000) instead of original $O(N\log N)$ (Gordon et al., 1993). The variance for this algorithm is $N\tilde{w}_n^{(i)}(1 - \tilde{w}_n^{(i)})$. This variance could be reduced by setting $\tilde{N}_i$ equal to the integer part of $N\tilde{w}_n^{(i)}$, and then selecting the remaining $\bar{N} = N - \Sigma_{i=1}^{N}\tilde{N}_i$ samples with the new weights $\tilde{w}_n'^{(i)} = \bar{N}^{-1}(N\tilde{w}_n^{(i)} - \tilde{N}_i)$ using SIR. The result is residual resampling with the lower variance of $\bar{N}\tilde{w}_n'^{(i)}(1 - \tilde{w}_n'^{(i)})$ (Liu and Chen, 1998).

However, one could do even better by employing stratified sampling (Kitagawa, 1996; Carpenter et al., 1999). The idea is to generate N points equally spaced in the interval $[0, 1]$, and to set the number of offspring N_i for each particle to be equal to the number of points lying between the partial sums of weights q_{i-1} and q_i, where $q_i = \Sigma_{j=1}^{i}\tilde{w}_l^{(i)}$. This scheme can still be implemented in $O(N)$ operations and has the minimum variance one can achieve in the class of unbiased schemes (Crisan, 2001), namely,

$$var[N_i] = \left\{ N\tilde{w}_n^{(i)} \right\} \left(1 - \left\{ N\tilde{w}_n^{(i)} \right\} \right)$$

where, for any α, $[\alpha]$ is the integer part of α and $\{\alpha\} \triangleq \alpha - [\alpha]$.

Recent theoretical results (Crisan et al., 1999; Crisan, 2001) suggest that it is not necessary for the selection schemes to be unbiased (i.e., it is possible to have $N_i \neq N\tilde{w}_i^{(i)}$) or to be randomized (Kitagawa, 1994).

Whether to Resample

It has been argued (Liu and Chen, 1995, 1998) that when all the importance weights are nearly equal, it is not beneficial to introduce the selection step in the algorithm, and a measure of degeneracy in a form of the *effective sample size* N_{eff} is proposed. It is suggested that one should resample only if N_{eff} defined as

$$N_{eff} = \frac{1}{\sum_{i=1}^{n} \left(\tilde{w}_n^{(i)} \right)^2} \tag{18.46}$$

is below a fixed threshold. This is an intuitively reasonable result; however, the performance largely depends on the threshold employed. In addition, although resampling increases the variance of the estimate $E_{p(\mathbf{x}_{0:n}|\mathbf{y}_{1:n})}[\Upsilon(\mathbf{x}_{0:n})]$ at time n, it usually leads to a decrease of variance of the future estimates (Liu and Chen, 1995; Doucet et al., 2001b).

Problems with the Selection Procedure

The selection procedure helps to solve the problems associated with the degeneracy of the algorithm. However, it introduces some practical and theoretical limitations. In particular, the statistical independence of the simulated trajectories no longer holds, and the convergence results of the algorithm should be reestablished. (See Berzuini et al., 1997.)

The parallelizability of the algorithm employing the resampling procedure is also limited, and there is loss of diversity due to numerous copies of the same particles in the approximating sample (Gordon et al., 1993; Higuchi, 1995a, 1995b). Despite these limitations, the algorithm is the basis for numerous works, and many strategies have been developed to increase its efficiency.

Particle-filtering Algorithm

The resulting algorithm for the nth step, with $n \geq 1$, $\mathbf{x}_0^{(i)} \sim p(\mathbf{x}_0) w_0 = 1$ for $i = 1, \ldots, N$, is described in the following text.

Sequential Importance Sampling Step
- For $i = 1, \ldots, N$, sample $\tilde{\mathbf{x}}_0^{(i)} \sim \pi(\mathbf{x}_n | \mathbf{x}_{0:n-1}^{(i)}, \mathbf{y}_{1:n})$ and set $\tilde{\mathbf{x}}_{0:n}^{(i)} = (\mathbf{x}_{0:n-1}^{(i)}, \tilde{\mathbf{x}}_n^{(i)})$.
- For $i = 1, \ldots, N$, evaluate the importance weights up to a normalizing constant:

$$w_n^{(i)} \propto w_{n-1}^{(i)} \frac{p\left(\mathbf{y}_n | \tilde{\mathbf{x}}_n^{(i)}\right) p\left(\tilde{\mathbf{x}}_n^{(i)} | \tilde{\mathbf{x}}_{n-1}^{(i)}\right)}{\pi\left(\tilde{\mathbf{x}}_n^{(i)} | \tilde{\mathbf{x}}_{0:n-1}^{(i)}, \mathbf{y}_{1:n}\right)}$$

- For $i = 1, \ldots, N$, normalize the importance weights:

$$\tilde{w}_n^{(i)} = \frac{w_n^{(i)}}{\sum_{j=1}^{N} w_n^{(j)}}$$

Selection Step
- Multiply/discard particles $\{\tilde{\mathbf{x}}_{0:n}^{(i)}\}_{i=1}^{N}$ with respect to high/low normalized importance weights $\tilde{w}_n^{(i)}$ to obtain N particles $\{\mathbf{x}_{0:n}^{(i)}\}_{i=1}^{N}$ with $\tilde{w}_n^{(i)} = \dfrac{1}{N}$.

Numerous algorithms proposed in the literature are, in fact, special cases of this general (and simple) algorithm.

Applications to Digital Communications

The generic particle-filtering algorithm developed in the previous section can be particularly useful for solving a number of problems arising in digital communications. One of these problems is joint symbol and code delay estimation in direct sequence (DS) spread spectrum systems in multipath environments. DS spread spectrum systems are robust to many channel impairments, allow multiuser (CDMA) and low-detectability signal transmission, and, therefore, are widely used in different areas of digital communications. Unlike many other communication systems, however, spread spectrum receivers require additional code synchronization, which can be a rather challenging task under conditions of multipath fading, when severe amplitude and phase variations take place.

The problem of joint delay and multipath estimation has been addressed in literature before (see Iltis, 2001, 1990, for example), and has proved to be difficult due to its inherited nonlinearity. The previously proposed approaches are mainly based on the use of the Extended Kalman Filter (EKF). However, many of them concentrate on the channel parameters and delay estimation only; moreover, in a number of cases, when EKF methods are applied, the estimated parameters are divergent (Iltis, 2001).

In this section, we demonstrate how the channel parameters, code delays, and symbols can be estimated jointly using particle-filtering techniques. We begin with formulating the model specifications and estimation objectives, then consider an application of the generic particle-filtering method for joint symbol/channel coefficients/code delay estimation, and finally present some simulation results.

Problem Statement and Estimation Objectives

Let us begin with the model specification for signal transmission in DS spread spectrum systems in a multipath environment. For simplicity, we address a single-user case with one receiving antenna, and set $K = 1$ and $L = 1$, suppressing the indices k and l hereafter. No channel coding is employed in the system, and the additive channel noise is assumed to be Gaussian. Extension to more complicated scenarios is straightforward.

Transmitted Waveform
Let us denote for any generic sequence κ_i, $\kappa_{i:j} \triangleq (\kappa_i, \kappa_{i+1}, \ldots, \kappa_j)^T$, and let d_n be the nth information symbol and $s_{\text{trans}}(\tau)$ be the corresponding analogue bandpass spread spectrum signal waveform transmitted in the symbol interval of duration T_n:

$$s_{\text{trans}}(\tau) = \text{Re}[s_n(d_{1:n})u(\tau)\exp(j2\pi f_{\text{car}}\tau)], \quad \text{for } (n-1)T_n < \tau \leq nT_n \qquad (18.47)$$

where $s_n(\cdot)$ maps the digital sequence to waveforms and depends on the modulation technique employed, $d_n = r_n$, with $q = n$ since uncoded symbols are considered, f_{car} denotes the carrier frequency, and $u(\tau)$ is a wideband pseudo-noise (PN) waveform defined by

$$u(\tau) = \sum_{h=1}^{H} c_h \eta(\tau - hT_{ch}) \qquad (18.48)$$

Here, $c_{1:H}$ is a spreading code sequence consisting of H chips (with values $\{\pm 1\}$) per symbol, $\eta(\tau - hT_{ch})$ is a rectangular pulse of unit height and duration T_{ch} transmitted at $(h-1)T_{ch} < \tau \leq hT_{ch}$, and T_{ch} is the chip interval satisfying the relation $T_{ch} = T_n/H$.

Channel Model

The signal is passed through a noisy multipath fading channel that induces random amplitude and phase variations. The channel can be represented by a time-varying tapped-delayed line with taps spaced T_s seconds apart, where T_s is the Nyquist sampling rate for the transmitted waveform; $T_s = T_{ch}/2$ due to the PN bandwidth being approximately $1/T_{ch}$. The equivalent discrete-time impulse response of the channel is given by

$$h_{\text{channel},t} = \sum_{g=0}^{G-1} f_t^{(g)} \delta(t-g) \qquad (18.49)$$

where t is a discrete time index, G is the number of paths of the channel, $\mathbf{f}_t^{(g)}$ is the complex-valued, time-varying multipath coefficients (arranged into the vector $\mathbf{f}_t$), and δ denotes the Dirac delta function.

We assume here that the channel coefficients f_t and code delay θ_t propagate according to the first-order autoregressive (AR) model:

$$\mathbf{f}_t = \mathbf{A}\mathbf{f}_{t-1} + \mathbf{B}\mathbf{v}_t, \ \mathbf{v}_t \overset{i.i.d.}{\sim} \mathcal{N}_c(0, \mathbf{I}_G) \qquad (18.50)$$

$$\theta_t = \gamma\theta_{t-1} + \sigma_\theta\vartheta_t, \ \vartheta_t \overset{i.i.d.}{\sim} \mathcal{N}(0, 1) \qquad (18.51)$$

which corresponds to a Rayleigh uncorrelated scattering channel model; here $\mathbf{A} \triangleq diag(\alpha^{(0)} \ldots, \alpha^{(G-1)})$, $\mathbf{B} \triangleq diag(\sigma_f^{(0)}, \ldots, \sigma_f^{(G-1)})$, where $\sigma_f^{(g)}$ is the standard deviation, and $\alpha^{(g)}$ accounts for the Doppler spread. (See Iltis, 1990, and Komninakis et al., 1999, for details and discussion of the use of higher-order AR.) In this work, matrices $\mathbf{A}$, $\mathbf{B}$, and parameters γ and σ_θ are assumed known. Directions on the choice of these parameters are given in Iltis (1990) and Komninakis et al. (1999).

Received Signal

The complex output of the channel sampled at the Nyquist rate (in which case samples $2H(n-1)+1, \ldots, 2Hn$ correspond to the nth symbol transmitted, that is, $d_n \leftrightarrow y_{2H(n-1)+1:2Hn}$) can thus be expressed as

$$y_t = \mathbf{C}(d_{1:n}, \theta_{1:t}) + \sigma\varepsilon_t, \ \varepsilon_t \overset{i.i.d.}{\sim} \mathcal{N}_c(0, 1) \qquad (18.52)$$

where $\mathbf{C}(d_{1:n}, \theta_{1:t}) = \sum_{g=0}^{G-1} \mathbf{f}_t^{(g)} s_{\text{receive}}((t-g)T_s - \theta_t)$ and σ^2 is the noise variance. The noise sequences ϑ_t, ε_t and $\mathbf{v}_t^{(g)}$, $n = 0, \ldots, G - 1$ are assumed mutually independent and independent of the initial states $\mathbf{f}_0 \sim N_c(\hat{\mathbf{f}}_0, \Sigma_{f,0})$, $\theta_0 \sim \mathcal{N}(\hat{\theta}_0, \Sigma_{\theta,0})$. The received waveform $S_{\text{receive}}(\tau)$ is obtained after ideal low-pass filtering of rectangular pulses and is given (see Iltis, 1990) by:

$$s_{\text{receive}}(\tau) = s_n(d_{1:n}) \sum_{h=1}^{H} c_h \frac{1}{\pi} \left[\text{Si}\left(2\pi \frac{\tau - (h-1)T_{ch}}{T_{ch}} \right) - \text{Si}\left(2\pi \frac{\tau - hT_{ch}}{T_{ch}} \right) \right]$$

for $(n-1)T_n < \tau \leq nT_n$

where

$$\text{Si}(\phi) = \int_0^\phi \frac{\sin(\varphi)}{\varphi} d\varphi \tag{18.53}$$

Estimation Objectives

The symbols d_n, which are assumed i.i.d., the channel characteristics f_t, and the code delay θ_t are unknown for $n, t > 0$. Our aim is to obtain sequentially in time an estimate of the joint posterior probability density of these parameters $p(d_{1:n}, \mathbf{f}_{0:2Hn}, \theta_{0:2Hn} | y_{1:Hn})$, and some of its characteristics, such as the MMAP (marginal maximum *a posteriori*) estimates of the symbols

$$\hat{d}_{1:n} = \arg\max_{d_{1:n}} p(d_{1:n} | y_{1:2Hn}) \tag{18.54}$$

and the minimum mean square error (MMSE) estimates of the channel characteristics $E(\mathbf{f}_{0:2Hn} / y_{1:2Hn})$ and the delays $E(\theta_{0:2Hn} | y_{1:2Hn})$. This problem, unfortunately, does not admit any analytical solution and, thus, approximate methods must be employed. One of the methods that has proved to be useful in practice is particle filtering, and in the next section we propose a receiver based on the use of this technique.

Particle-filtering Receiver

Rao-Blackwellization

A generic particle-filtering algorithm developed previously can be directly applied to obtain the estimate of the joint posterior distribution $p(d_{1:n}, \mathbf{f}_{0:2Hn}, \theta_{0:2Hn} | y_{1:2Hn})$. It is beneficial, however, to improve the standard approach by making the most of the structure of the model and applying variance reduction techniques. Indeed, the problem of estimating $p(d_{1:n}, d\mathbf{f}_{0:2Hn}, d\theta_{0:2Hn} | y_{1:2Hn})$ can be reduced to one of sampling from a lower-dimensional posterior $p(d_{1:n}, d\theta_{0:2Hn} | y_{1:2Hn})$ using the so-called Rao-Blackwellization method. If the approximation of $p(d_{1:n}, d\theta_{0:2Hn} | y_{1:2Hn})$ can be obtained, say, via particle filtering:

$$\hat{p}_N(d_{1:n}, d\theta_{0:2Hn} | y_{1:2Hn}) = \sum_{i=1}^{N} \tilde{w}_n^{(i)} \delta(\{d_{1:n}, d\theta_{0:2Hn}\} - \{d_{1:n}^{(i)}, \theta_{0:2nH}^{(i)}\}) \tag{18.55}$$

one can compute the probability density $p(\mathbf{f}_{0:2Hn} | y_{1:2Hn}, d_{1:n}, \theta_{0:2Hn})$ using the Kalman filter associated with Equations 18.50 and 18.52. As a result, the posterior $p(\mathbf{f}_{0:2Hn} / y_{1:2Hn})$ can be approximated by a random mixture of Gaussians:

$$\hat{p}_N(f_{0:2Hn}|y_{1:2Hn}) = \int \theta_{0:2Hn} \sum_{d_{1:n}} p(f_{0:2Hn}|y_{1:2Hn}, d_{1:n}, \theta_{0:2Hn}) \times \hat{p}_N(d_{1:n}, \theta_{0:2Hn}|y_{1:2Hn}) d\theta_{0:2Hn}$$

$$= \sum_{i=1}^{N} \tilde{w}_n^{(i)} p(f_{0:2Hn}|y_{1:2Hn}, d_{1:n}^{(i)}, \theta_{0:2Hn}^{(i)})$$

leading to lower variance of the estimates and, therefore, increased algorithm efficiency (Doucet et al., 2000).

Strictly speaking, we are interested in estimating the information symbols only with the tracking of the channel being naturally incorporated into the proposed algorithm. However, the MMSE (conditional mean) estimates of fading coefficients can, of course, be obtained if necessary as follows:

$$E_{\hat{p}_N}[f_{2H(n-1)+1:2Hn}|y_{1:2Hn}] = \int f_{2H(n-1)+1:2Hn} \hat{p}_N(f_{0:2Hn}|y_{1:2Hn}) df_{0:2Hn}$$

$$= \sum_{i=1}^{N} \tilde{w}_n^{(i)} E[f_{2H(n-1)+1:2Hn}|y_{1:2Hn}, d_{1:n}^{(i)}, \theta_{0:2Hn}^{(i)}]$$

with $E[f_{2H(n-1)+1:2Hn}|y_{1:2Hn}, d_{1:n}^{(i)}, \theta_{0:2Hn}^{(i)}]$ being computed by the Kalman filter, with $2H$ steps required for each symbol transmitted.

Sequential Importance Sampling Step

For the sake of simplicity we assume that the prior is taken to be the importance distribution (although more efficient subsampling schemes can be devised) and

$$\pi(d_n, \theta_n|d_{1:n-1}, \theta_{0:n-1}, \mathbf{y}_{1:n}) = p(d_n)p(\theta_n|\theta_{n-1})$$

$$= p(d_n) \prod_{t=2H(n-1)+1}^{2Hn} p(\theta_t|\theta_{t-1})$$

Then w_n becomes

$$w_n \propto p(\mathbf{y}_n|\mathbf{y}_{1:n-1}, d_{1:n}, \theta_{0:n}) = \prod_{t=2H(n-1)+1}^{2Hn} p(y_t|d_{1:n}, \theta_{0:t}, y_{1:t-1}) \qquad (18.56)$$

and requires evaluation of $2H$ one-step Kalman filter updates for each symbol.

Selection Step

The selection step is done according to a stratified sampling scheme described earlier.

Simulations

The algorithm presented earlier was applied to perform joint symbols/channel coefficients/code delay estimation for DS spread spectrum systems with $H = 15$, $G = 4$. A binary DPSK modulation scheme was employed with the multipath channel response and AR coefficients chosen (Iltis, 1990, channel B).

As it is shown in Figure 18.1, the algorithm employing 100 particles exhibits good bit-error-rate (BER) performance. A tracking error trajectory for 100 information symbols (corresponding to 1500 chips and 3000 channel samples) and an average signal-to-noise ratio (SNR) equal to 10 dB is presented in Figure 18.2. Figure 18.3 also illustrates the mean-square delay error as a function of SNR.

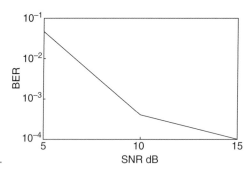

Fig. 18.1 Bit error rate.

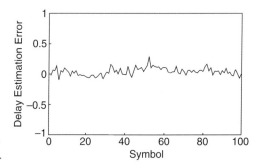

Fig. 18.2 Error in delay
estimation; SNR = 15 dB.

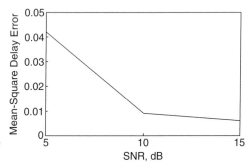

Fig. 18.3 Mean-square delay
error via SNR.

Conclusion

In this chapter, we aimed at developing a general framework for addressing a variety of
problems arising in the field of digital communications, and presented a simple example
as an illustration of this general approach. (The same problem has been treated in more
detail in Punskaya et al., 2004.) In general, a great deal of work has appeared in the literature on the application of particle filtering, and the number is rapidly growing (Chen
et al., 2000; Dong et al., 2003; Iltis, 2003; Huang and Djuric, 2002; Ghirmai et al., 2003; Yang
and Wang, 2002; Punskaya et al., 2001, 2002). Although there is still a lot to be done, one
may only hope that all these efforts will eventually help to increase the ease and breadth of
communicating.

References

Akashi, H., and Kumamoto, H. (1977) Random sampling approach to state estimation in switching environments. *Automatica*, 13, 429–434.

Berzuini, C., Best, N., Gilks, W., and Larizza, C. (1997) Dynamic conditional independence models and Markov chain Monte Carlo methods. *Journal of American Statistical Association*, 92, 1403–1412.

Carpenter, J., Clifford, P., and Fearnhead, P. (1999) An improved particle filter for non-linear problems. *IEE Proceedings—Radar, Sonar and Navigation*, 146, 2–7.

Chen, R., Wang, X., and Liu, J. (2000) Adaptive joint detection and decoding in flat-fading channels via mixture Kalman filtering. *IEEE Transactions on Information Theory*, 46, 2079–2094.

Crisan, D. (2001) Particle filters—A theoretical perspective. In *Sequential Monte Carlo Methods in Practice*, eds. A. Doucet, J. F. G. de Freitas, and N. J. Gordon, Springer-Verlag.

Crisan, D., Moral, P. D., and Lyons, T. (1999) Discrete filtering using branching and interacting particle systems. *Markov Processes and Related Fields*, 5, 293–318.

Dong, B., Wang, X., and Doucet, A. (2003) A new class of soft MIMO demodulation algorithms. *IEEE Transactions on Signal Processing*.

Doucet, A., Godsill, S., and Andrieu, C. (2000) On sequential Monte Carlo sampling methods for Bayesian filtering. *Statistics and Computing*, 10, 197–208.

Doucet, A., de Freitas, J. and Gordon, N., eds. (2001a) *Sequential Monte Carlo Methods in Practice*. Springer-Verlag.

Doucet, A., Gordon, N., and Krishnamurthy, V. (2001b) Particle filters for state estimation of jump Markov linear systems. *IEEE Transactions on Signal Processing*, 49, 613–624.

Geweke, J. (1989) Bayesian inference in econometrics models using Monte Carlo integration. *Econometrica*, 57, 1317–1339.

Ghirmai, T., Bugallo, M., Miguez, J., and Djuric, P. (2003) Joint symbol detection and timing estimation using particle filtering. In *Proceedings ICASSSP 2003*.

Gordon, N., Salmond, D., and Smith, A. (1993) Novel approach to nonlinear/non-Gaussian Bayesian state estimation. *IEE Proceedings-F*, 140, 107–113.

Handschin, J., and Mayne, D. (1969) Monte Carlo techniques to estimate the conditional expectation in multi-stage non-linear filtering. *International Journal of Control*, 9, 547–559.

Higuchi, T. (1995a) Kitagawa Monte Carlo filter from the perspective of genetic algorithm. Research Memorandum, The Institute of Statistical Mathematics, Tokyo.

———. (1995b) Kitagawa Monte Carlo filter using the genetic algorithm operators. Research Memorandum. The Institute of Statistical Mathematics, Tokyo.

Huang, Y., and Djuric, P. (2002) A blind particle filtering detector for for joint channel estimation, tracking and data detection over flat fading channels. In *Proceedings EUSIPCO 2002*.

Iltis, R. (1990) Joint estimation of PN code delay and multipath using the Extended Kalman Filter. *IEEE Transactions on Communications*, 38, 1677–1685.

———. (2001) A DS-CDMA tracking mode receiver with joint channel/delay estimation and MMSE detection. *IEEE Transactions on Communications*, 49, 1770–1779.

———. (2003) A Sequential Monte Carlo Filter for joint linear/nonlinear state estimation with application to DS-CDMA. *IEEE Transactions on Signal Processing*.

Kitagawa, G. (1993) A Monte Carlo filtering and smoothing method for non-Gaussian nonlinear state space models. In *Proceedings of the 2nd US-Japan Joint Seminar on Statistical Time Series Analysis*, Honolulu, pp. 110–131.

———. (1994) The two-filter formula for smoothing and an implementation of the Gaussian-sum smoother. *Annals of the Institute of Statistical Mathematics*, 46, 605–623.

———. (1996) Monte Carlo filter and smoother for non-Gaussian nonlinear state space models. *Journal of Computational and Graphical Statistics*, 5, 1–25.

Komninakis, C., Fragouli, C., Sayed, A., and Wesel, R. D. (1999) Channel estimation and equalization in fading. In *Proceedings of the 33rd Asilomar Conference on Signals, Systems, and Computers*, vol. 2, Pacific Grove, CA, pp. 1159–1163.

Kong, A., Liu, J., and Wong, W. (1994) Sequential imputations and Bayesian missing data problems. *Journal of the American Statistical Association*, 89, 278–288.

Liu, J., and Chen, R. (1995) Blind deconvolution via sequential imputation. *Journal of the American Statistical Association*, 90, 567–576.

————. (1998) Sequential Monte Carlo methods for dynamic systems. *Journal of the American Statistical Association*, 93, 1032–1044.

Moral, P. D., and Salut, G. (1995) Non-linear filtering using Monte Carlo particle methods. *C. R. Academy of Science Paris*, 320, 1147–1152 (in French).

Punskaya, E., Andrieu, C., Doucet, A., and Fitzgerald, W. (2001) Particle filtering for demodulation in fading channels with non-Gaussian additive noise. *IEEE Transactions on Communications*, 49, 579–582.

————, Doucet, A., and Fitzgerald, W. (2002) On the use and misuse of particle filtering in digital communications. In *Proceedings EUSIPCO*.

————. (2004) Particle filtering for joint symbol and code delay estimation in DS spread spectrum systems in multipath environment. *EURASIP Journal on Applied Signal Processing*, 2306–2314.

Ripley, B. (1987) *Stochastic Simulation*, Wiley.

Rubin, D. (1988) Using the SIR algorithm to simulate the posterior distributions. In *Bayesian Statistics 3*, eds. J. M. Bernado, M. H. DeGroot, D. V. Lindley, and A. F. M. Smith, Oxford University Press, pp. 395–402.

Sorenson, H. (1988) Recursive estimation for nonlinear dynamic systems. In *Bayesian Analysis of Time Series and Dynamic Models*, ed. J. C. Spall, Dekker.

Yang, Z., and Wang, X. (2002) Blind detection of OFDM signals in multipath fading channels via Sequential Monte Carlo. *IEEE Transactions on Signal Processing*, 50, 271–280.

Zaritskii, V., Svetnik, V., and Shimelevich, L. (1975) Monte Carlo technique in problems of optimal data processing. *Automation and Remote Control*, 12, 95–103.

19 Trigonometric Functions and General Formulae

J. Barron

Mathematical Signs and Symbols

Sign, symbol	Quantity
$=$	equal to
$\neq$	not equal to
$\equiv$	identically equal to
$\triangle$	corresponds to
$\approx$	approximately equal to
$\rightarrow$	approaches
$\propto$	proportional to
∞	infinity
$<$	smaller than
$>$	larger than
$\leq$	smaller than or equal to
$\geq$	larger than or equal to
$\lvert a \rvert$	magnitude of a
a^n	a raised to the power n
$a^{1/2}$ $\sqrt{a}$	square root of a
$\bar{a}$ $\langle a \rangle$	mean value of a

Continued

Sign, symbol	Quantity
$p!$	factorial p $(1 \times 2 \times 3 \times \ldots \times p)$
$\binom{n}{p}$	binomial coefficient
Σ	Sum
Π	Product
$f(x)$	function f of the variable x
$[f(x)]_a^b$	$f(b) - f(a)$
Δx	delta x = finite increment of x
δx	delta x = variation of x
$\dfrac{df}{dx}$ $f'(x)$	differential coefficient of $f(x)$ with respect to x
$\dfrac{d^n f}{dx^n}$ $f^{(n)}(x)$	differential coefficient of order n of $f(x)$
$\dfrac{\partial f(x, y, \ldots)}{\partial x}$ $\left(\dfrac{\partial f}{\partial x}\right)_{y\ldots}$	partial differential coefficient of $f(x, y, \ldots)$ with respect to x, when y $\ldots$ are held constant
df	total differential of f
$\int f(x)dx$	indefinite integral of $f(x)$ with respect to x
$\int_a^b f(x)\,dx$	definite integral of $f(x)$ from $x = a$ to $x = b$
e	base of natural logarithms
e^x $\exp x$	e raised to the power x
$\log_a x$	logarithm to the base a of x
$\lg x$ $\log x$ $\log_{10} x$	common (Briggsian) logarithm of x
$\lb x$ $\log_2 x$	binary logarithm of x
$\sin x$	sine of x
$\cos x$	cosine of x
$\tan x$	tangent of x
$\cot x$	cotangent of x
$\sec x$	secant of x
$\operatorname{cosec} x$	cosecant of x
$\arcsin x$	inverse sine of x
$\arccos x$	inverse cosine of x
$\arctan x$	inverse tangent of x
$\operatorname{arccot} x$	inverse cotangent of x
$\operatorname{arcsec} x$	inverse secant of x
$\operatorname{arccosec} x$	inverse cosecant of x
$\sinh x$	hyperbolic sine of x
$\cosh x$	hyperbolic cosine of x
$\tanh x$	hyperbolic tangent of x
$\coth x$	hyperbolic cotangent of x

Sign, symbol	Quantity
sech x	hyperbolic secant of x
cosech x	hyperbolic cosecant of x
arsinh x	inverse hyperbolic sine of x
arcosh x	inverse hyperbolic cosine of x
artanh x	inverse hyperbolic tangent of x
arcoth x	inverse hyperbolic cotangent of x
arsech x	inverse hyperbolic secant of x
arcosech x	inverse hyperbolic cosecant of x
i, j	imaginary unity, $i^2 = -1$
Re z	real part of z
Im z	imaginary part of z
arg z	argument of z
z^*	conjugate of z, complex conjugate of z
$\bar{A}$ A' A^t	transpose of matrix A
A^*	complex conjugate matrix of matrix $\boldsymbol{A}$
A^+	Hermitian conjugate matrix of matrix $\boldsymbol{A}$
$\boldsymbol{A}$ a	vector
$|A|$	magnitude of vector
$\boldsymbol{A} . \boldsymbol{B}$	scalar product
$\boldsymbol{A} \times \boldsymbol{B}$ $\boldsymbol{A} \wedge \boldsymbol{B}$	vector product
∇	differential vector operator
$\nabla\varphi$ grad φ	gradient of φ
$V . A$ div $\boldsymbol{A}$	divergence of $\boldsymbol{A}$
$\nabla \times \boldsymbol{A}$ curl $\boldsymbol{A}$	curl of $\boldsymbol{A}$
$\nabla^2\varphi$ $\Delta\varphi$	Laplacian of φ

Trigonometric Formulae

$$\sin^2 A + \cos^2 A = \sin A \operatorname{cosec} A = 1$$

$$\sin A = \frac{\cos A}{\cot A} = \frac{1}{\operatorname{cosec} A} = \left(1 - \cos^2 A\right)^{1/2}$$

$$\cos A = \frac{\sin A}{\tan A} = \frac{1}{\sec A} = \left(1 - \sin^2 A\right)^{1/2}$$

$$\tan A = \frac{\sin A}{\cos A} = \frac{1}{\cot A}$$

$$1 + \tan^2 A = \sec^2 A$$

$$1 + \cot^2 A = \operatorname{cosec}^2 A$$

$$1 - \sin A = \operatorname{coversin} A$$

$$1 - \cos A = \text{versin}\, A$$

$$\tan \tfrac{1}{2}\theta = t; \quad \sin\theta = \frac{2t}{1+t^2}; \quad \cos\theta = \frac{1-t^2}{1+t^2}$$

$$\cot A = \frac{1}{\tan A}$$

$$\sec A = \frac{1}{\cos A}$$

$$\operatorname{cosec} A = \frac{1}{\sin A}$$

$$\cos(A \pm B) = \cos A \cos B \mp \sin A \sin B$$

$$\sin(A \pm B) = \sin A \cos B \pm \cos A \sin B$$

$$\tan(A \pm B) = \frac{\tan A \pm \tan B}{1 \mp \tan A \tan B}$$

$$\cot(A \pm B) = \frac{\cot A \cot B \mp 1}{\cot B \pm \cot A}$$

$$\sin A \pm \sin B = 2 \sin \tfrac{1}{2}(A \pm B) \cos \tfrac{1}{2}(A \mp B)$$

$$\cos A + \cos B = 2 \cos \tfrac{1}{2}(A + B) \cos \tfrac{1}{2}(A - B)$$

$$\cos A - \cos B = 2 \sin \tfrac{1}{2}(A + B) \sin \tfrac{1}{2}(B - A)$$

$$\tan A \pm \tan B = \frac{\sin(A \pm B)}{\cos A \cos B}$$

$$\cot A \pm \cot B = \frac{\sin(B \pm A)}{\sin A \sin B}$$

$$\sin 2A = 2 \sin A \cos A$$

$$\cos 2A = \cos^2 A - \sin^2 A = 2\cos^2 A - 1 = 1 - 2\sin^2 A$$

$$\cos^2 A - \sin^2 B = \cos(A + B)\cos(A - B)$$

$$\tan 2A = \frac{2 \tan A}{1 - \tan^2 A}$$

$$\sin \tfrac{1}{2}A = \left(\frac{1 - \cos A}{2}\right)^{1/2}$$

$$\cos \tfrac{1}{2}A = \pm\left(\frac{1 + \cos A}{2}\right)^{1/2}$$

$$\tan \tfrac{1}{2}A = \left(\frac{\sin A}{1 + \cos A}\right)$$

$$\sin^2 A = \tfrac{1}{2}(1 - \cos 2A)$$

$$\cos^2 A = \tfrac{1}{2}(1 + \cos 2A)$$

$$\tan^2 A = \frac{1 - \cos 2A}{1 + \cos 2A}$$

$$\tan \tfrac{1}{2}(A \pm B) = \frac{\sin A \pm \sin B}{\cos A + \cos B}$$

$$\cot \tfrac{1}{2}(A \pm B) = \frac{\sin A \pm \sin B}{\cos B - \cos A}$$

Approximations for Small Angles

$$\sin \theta = \theta - \frac{\theta^3}{6}$$

$$\cos \theta = 1 - \frac{\theta^2}{2} \quad (\theta \text{ in radians})$$

$$\tan \theta = \theta + \frac{\theta^3}{3}$$

Trigonometric Values

Angle	0°	30°	45°	60°	90°	180°	270°	360°
Radians	0	$\pi/6$	$\pi/4$	$\pi/3$	$\pi/2$	π	$3\pi/2$	2π
Sine	0	$\tfrac{1}{2}$	$\tfrac{1}{2}\sqrt{2}$	$\tfrac{1}{2}\sqrt{3}$	1	0	−1	0
Cosine	1	$\tfrac{1}{2}\sqrt{3}$	$\tfrac{1}{2}\sqrt{2}$	$\tfrac{1}{2}$	0	−1	0	1
Tangent	0	$1/\sqrt{3}$	1	$\sqrt{3}$	∞	0	∞	0

Solution of Triangles

$$\frac{\sin A}{a} = \frac{\sin B}{b} = \frac{\sin C}{c}$$

$$\cos A = \frac{b^2 + c^2 - a^2}{2bc}$$

$$\cos B = \frac{c^2 + a^2 + b^2}{2ca}$$

$$\cos C = \frac{a^2 + b^2 - c^2}{2ab}$$

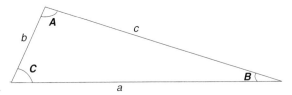

Fig. 19.1   Triangle.

where A, B, C and a, b, c, are shown in Figure 19.1.

If $s = \frac{1}{2}(a + b + c)$:

$$\sin \frac{A}{2} = \sqrt{\frac{(s-b)(s-c)}{bc}}$$

$$\sin \frac{B}{2} = \sqrt{\frac{(s-c)(s-a)}{ca}}$$

$$\sin \frac{C}{2} = \sqrt{\frac{(s-a)(s-b)}{ab}}$$

$$\cos \frac{A}{2} = \sqrt{\frac{s(s-a)}{bc}}$$

$$\cos \frac{B}{2} = \sqrt{\frac{s(s-b)}{ca}}$$

$$\cos \frac{C}{2} = \sqrt{\frac{s(s-c)}{ab}}$$

$$\tan \frac{A}{2} = \sqrt{\frac{(s-b)(s-c)}{s(s-a)}}$$

$$\tan \frac{B}{2} = \sqrt{\frac{(s-c)(s-a)}{s(s-b)}}$$

$$\tan \frac{C}{2} = \sqrt{\frac{(s-a)(s-b)}{s(s-c)}}$$

Spherical Triangle

$$\frac{\sin A}{\sin a} = \frac{\sin B}{\sin b} = \frac{\sin C}{\sin c}$$

$$\cos a = \cos b \cos c + \sin b \sin c \cos A$$

$$\cos b = \cos c \cos a + \sin c \sin a \cos B$$

$$\cos c = \cos a \cos b + \sin a \sin b \cos C$$

where A, B, C and a, b, c are now as in Figure 19.2.

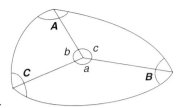

Fig. 19.2 Spherical triangle.

Exponential Form

$$\sin\theta = \frac{e^{i\theta} - e^{-i\theta}}{2i}$$

$$\cos\theta = \frac{e^{i\theta} + e^{-i\theta}}{2}$$

$$e^{i\theta} = \cos\theta + i\sin\theta$$

$$e^{-i\theta} = \cos\theta - i\sin\theta$$

De Moivre's Theorem

$$(\cos A + i\sin A)(\cos B + i\sin B) = \cos(A + B) + i\sin(A + B)$$

Euler's Relation

$$(\cos\theta + i\sin\theta)^n = \cos n\theta + i\sin n\theta = e^{in\theta}$$

Hyperbolic Functions

$$\sinh x = \frac{e^x - e^{-x}}{2}$$

$$\cosh x = \frac{e^x + e^{-x}}{2}$$

$$\tan x = \frac{\sinh x}{\cosh x}$$

Relations between hyperbolic functions can be obtained from the corresponding relations between trigonometric functions by reversing the sign of any term containing the product or implied product of two sines. For example:

$$\cosh^2 A - \sinh^2 A = 1$$

$$\cosh 2A = 2\cosh^2 A - 1 = 1 + 2\sinh^2 A = \cosh^2 A + \sinh^2 A$$

$$\cosh(A \pm B) = \cosh A \cosh B \pm \sinh A \sinh B$$

$$\sinh(A \pm B) = \sinh A \cosh B \pm \cosh A \sinh B$$

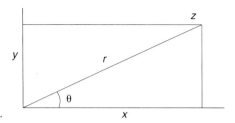

Fig. 19.3 Argand diagram.

$$e^x = \cosh x + \sinh x$$

$$e^{-x} = \cosh x - \sinh x$$

Complex Variables

If $z = x + iy$, where x and y are real variables, z is a complex variable and is a function of x, and y, z may be represented graphically in an Argand diagram (Figure 19.3).

Polar form:

$$z = x + iy = |z|e^{i\theta} = |z|(\cos\theta + i\sin\theta)$$

$$x = r\cos\theta$$

$$y = r\sin\theta$$

where $r = |z|$.

Complex arithmetic:

$$z_1 = x_1 + iy_1$$

$$z_2 = x_2 + iy_2$$

$$z_1 \pm z_2 = (x_1 \pm x_2) + i(y_1 \pm y_2)$$

$$z_1 \cdot z_2 = (x_1 x_2 - y_1 y_2) + i(x_1 y_2 - x_2 y_1)$$

Conjugate:

$$z^* = x - iy$$

$$z \cdot z^* = x^2 + y^2 = |z|^2$$

Function: another complex variable $w = u + iv$ may be related functionally to z by:

$$w = u + iv = f(x + iy) = f(z)$$

which implies:

$$u = u(x, y)$$

$$v = v(x, y)$$

For example:

$$\cosh z = \cosh(x + iy) = \cosh x \cosh iy + \sinh x \sinh iy$$
$$= \cosh x \cos y + i \sinh x \sin y$$

$$u = \cosh x \cos y$$

$$v = \sinh x \sin y$$

Cauchy–Riemann Equations

If $u(x, y)$ and $v(x, y)$ are continuously differentiable with respect to x and y:

$$\frac{\partial u}{\partial x} = \frac{\partial v}{\partial y}$$

$$\frac{\partial u}{\partial y} = -\frac{\partial v}{\partial x}$$

$w = f(z)$ is continuously differentiable with respect to z and its derivative is:

$$f'(z) = \frac{\partial u}{\partial x} + i\frac{\partial v}{\partial x} = \frac{\partial v}{\partial y} - i\frac{\partial u}{\partial y} = \frac{1}{i}\left(\frac{\partial u}{\partial y} + i\frac{\partial v}{\partial y}\right)$$

It is also easy to show that $\nabla^2 u = \nabla^2 v = 0$. Since the transformation from z to w is conformal, the curves $u = $ constant and $v = $ constant intersect each other at right angles, so that one set may be used as equipotentials and the other as field lines in a vector field.

Cauchy's Theorem

If $f(z)$ is analytic everywhere inside a region bounded by C and a is a point within C:

$$f(a) = \frac{1}{2\pi i}\int_C \frac{f(z)}{z - a}dz$$

This formula gives the value of a function at a point in the interior of a closed curve in terms of the values on that curve.

Zeros, Poles, and Residues

If $f(z)$ vanishes at the point z_0, the Taylor series for z in the region of z_0 has its first two terms zero, and perhaps others also: $f(z)$ may then be written:

$$f(z) = (z - z_0)^n g(z)$$

where $g(z_0) \neq 0$. Then $f(z)$ has a zero of order n at z_0. The reciprocal:

$$q(z) = \frac{1}{f(z)} = \frac{h(z)}{(z - z_0)^n}$$

where $h(z) = \dfrac{1}{q(z)} \neq 0$ at z_0, $q(z)$ becomes infinite at $z = z_0$ and is said to have a pole of order n at z_0. $q(z)$ may be expanded in the form (Laurent series):

$$q(z) = c_{-n}(z - z_0)^{-n} + \ldots + c_{-1}(z - z_0)^{-1} + c_0 + \ldots$$

where c_{-1} is the residue of $q(z)$ at $z = z_0$.

From Cauchy's theorem it may be shown that if a function $f(z)$ is analytic throughout a region enclosed by a curve C except at a finite number of poles, the integral of the function around C has a value of $2\pi i$ times the sum of the residues of the function at its poles within C. This fact can be used to evaluate many definite integrals whose indefinite form cannot be found.

Some Standard Forms

$$\int_0^{2\pi} e^{\cos\theta}(\cos n\theta - \sin\theta)d\theta = \frac{2\pi}{n!}$$

$$\int_0^x \frac{x^{a-1}}{1+x}dx = \pi\operatorname{cosec} a\pi$$

$$\int_0^x \frac{x^{a-1}}{1-x}dx = \pi\cot a\pi$$

$$\int_0^\infty \frac{\sin\theta}{\theta}d\theta = \frac{\pi}{2}$$

$$\int_0^\infty \exp(-h^2x^2)dx = \frac{\sqrt{\pi}}{2h}$$

$$\int_0^\infty x\exp(-h^2x^2)dx = \frac{1}{2h^2}$$

$$\int_0^\infty x^2\exp(-h^2x^2)dx = \frac{\sqrt{\pi}}{4h^3}$$

Coordinate Systems

The basic system is the rectangular Cartesian system (x, y, z) to which all other systems are referred. Two other commonly used systems are as follows.

Cylindrical Coordinates

Coordinates of point P are (x, y, z) or (r, θ, z), as in Figure 19.4, where:

$$x = r\cos\theta \quad y = r\sin\theta \quad z = z$$

In these coordinates the volume element is $r\,dr\,d\theta\,dz$.

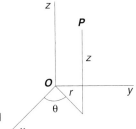

Fig. 19.4 Cylindrical coordinates.

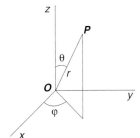

Fig. 19.5 Spherical polar coordinates.

Spherical Polar Coordinates

Coordinates of point P are (x, y, z) or (r, θ, φ), as in Figure 19.5, where:

$$x = r \sin \theta \cos \varphi$$

$$y = r \sin \theta \sin \varphi$$

$$z = r \cos \theta$$

In these coordinates the volume elements are $r^2 \sin \theta \, dr \, d\theta \, d\varphi$.

Transformation of Integrals

$$\iiint f(x, y, z) \, dx \, dy \, dz = \iiint \varphi(u, v, w) |J| \, du \, dv \, dw$$

where:

$$J = \begin{vmatrix} \dfrac{\partial x}{\partial u} & \dfrac{\partial y}{\partial u} & \dfrac{\partial z}{\partial u} \\[2mm] \dfrac{\partial x}{\partial v} & \dfrac{\partial y}{\partial v} & \dfrac{\partial z}{\partial v} \\[2mm] \dfrac{\partial x}{\partial w} & \dfrac{\partial y}{\partial w} & \dfrac{\partial z}{\partial w} \end{vmatrix} = \dfrac{\partial(x, y, z)}{\partial(u, v, w)}$$

is the Jacobian of the transformation of coordinates. For Cartesian to cylindrical coordinates $J = r$, and for Cartesian to spherical polars $J = r^2 \sin \theta$.

Laplace's Equation

The equation satisfied by the scalar potential, at a source-free point in space, from which a vector field may be derived by taking the gradient is Laplace's equation, written as:

$$\nabla^2\varphi = \frac{\partial^2\varphi}{\partial x^2} + \frac{\partial^2\varphi}{\partial y^2} + \frac{\partial^2\varphi}{\partial z^2} = 0$$

In cylindrical coordinates:

$$\nabla^2\varphi = \frac{1}{r}\frac{\partial}{\partial r}\left(r\frac{d\varphi}{\partial r}\right) + \frac{1}{r^2}\frac{\partial^2\varphi}{\partial\theta^2} + \frac{\partial^2\varphi}{\partial z^2}$$

In spherical polars:

$$\nabla^2\varphi = \frac{1}{r^2}\frac{\partial}{\partial r}\left(r^2\frac{\partial\varphi}{\partial r}\right) + \frac{1}{r^2\sin\theta}\frac{\partial\varphi}{\partial\theta} + \frac{1}{r^2\sin^2\theta}\frac{\partial^2\varphi}{\partial\Phi^2}$$

The equation is solved by setting:

$$\varphi = U(u)V(v)W(w)$$

in the appropriate form of the equation, separating the variables, and solving separately for the three functions, where (u, v, w) is the coordinate system in use.

In Cartesian coordinates, typically the functions are trigonometric, hyperbolic, and exponential; in cylincrical coordinates, the function of z is exponential, that of θ is trigonometric and that of r is a Bessel function. In spherical polars, typically the function of r is a power of r, that of φ is trigonometric, and that of θ is a Legendre function of $\cos\theta$.

Solution of Equations

Quadratic Equation

In practical calculations if $b^2 > 4ac$, so that the roots are real and unequal, calculate the root of larger modulus first, using the same sign for both terms in the formula. Then use the fact that $x_1 x_2 = c/a$, where x_1 and x_2 are the roots.

$$ax^2 + bx + c = 0$$

$$x = -\frac{b}{2a} \pm \frac{\sqrt{b^2 - 4ac}}{2a}$$

This avoids the severe cancellation of significant digits that may otherwise occur in calculating the smaller root.

For polynomials other than quadratics, and for other functions, several methods of successive approximation are available.

Bisection Method

By trial, find x_0 and x_1 such that $f(x_0)$ and $f(x_1)$ have opposite signs (see Figure 19.6). Set $x_2 = (x_0 + x_1)/2$ and calculate $f(x_2)$. If $f(x_0)f(x_2)$ is positive, the root lies in the interval (x_1, x_2); if negative, in the interval (x_0, x_2); and if zero, x_2 is the root. Continue if necessary using the new interval.

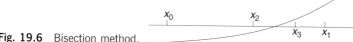

Fig. 19.6   Bisection method.

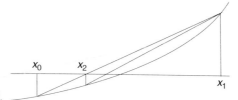

Fig. 19.7   Regula falsi.

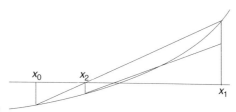

Fig. 19.8   Accelerated method.

Regula Falsi

By trial, find x_0 and x_1 as for the bisection method; these two values define two points $(x_0, f(x_0))$ and $(x_1, f(x_1))$. The straight line joining these two points cuts the x axis at the point (see Figure 19.7):

$$x_2 = \frac{x_0 f(x_1) - x_1 f(x_0)}{f(x_1) - f(x_0)}$$

Evaluate $f(x_2)$ and repeat the process for whichever of the intervals (x_0, x_2) or (x_1, x_2) contains the root. This method can be accelerated by halving at each step the function value at the retained end of the interval, as shown in Figure 19.8.

Fixed-point Iteration

Arrange the equation in the form:

$$x = f(x)$$

Choose an initial value of x by trial, and calculate repetitively:

$$x_{k+1} = f(x_k)$$

This process will not always converge.

Newton's Method

Calculate repetitively (Figure 19.9):

$$x_{k+1} = x_k - f(x_k)/f'(x_k)$$

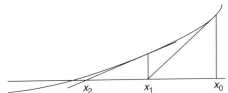

Fig. 19.9   Newton's method.

This method will converge unless: (a) x_k is near a point of inflexion of the function; (b) x_k is near a local minimum; or (c) the root is multiple. If one of these cases arises, most of the trouble can be overcome by checking at each stage that:

$$f(x_{k+1}) < f(x_k)$$

and, if not, halving the preceding value of $|x_{k+1} - x_k|$.

Newton's Method of Complex Roots

Newton's method may also be used to find complex roots. Suppose the function $f(z)$, where $z = x + iy$ and both x and y are real, is expressed in the form $f(x) \doteq X(x, y) + iY(x, y)$. The previous iteration for the real case is then replaced by the two iterations:

$$x_{k+1} = x_k - \frac{\left[X\dfrac{\partial X}{\partial x} + Y\dfrac{\partial Y}{\partial y} \right]_k}{\left[\dfrac{\partial X}{\partial x} \right]_k^2 + \left[\dfrac{\partial Y}{\partial x} \right]_k^2}$$

$$y_{k+1} = y_k - \frac{\left[Y\dfrac{\partial X}{\partial x} - X\dfrac{\partial Y}{\partial x} \right]_k}{\left[\dfrac{\partial X}{\partial x} \right]_k^2 + \left[\dfrac{\partial Y}{\partial x} \right]_k^2}$$

The subscript k indicates that the subscripted quantity is evaluated at the point (x_k, y_k). It is important that neither of the initial values of x and y be zero; otherwise, it may remain so. The real case iteration is quadratically convergent; the complex, linearly.

Method of Least Squares

To obtain the best fit between a straight line $ax + by = 1$ and several points (x_1, y_1), (x_2, y_2), $\dots$, (x_n, y_n) found by observation, the coefficients a and b are to be chosen so that the sum of the squares of the errors:

$$e_i = ax_i + by_i - 1$$

is a minimum. To do this, first write the set of inconsistent equations:

$$ax_1 + by_1 - 1 = 0$$
$$ax_2 + by_2 - 1 = 0$$
$$\vdots$$
$$ax_n + by_n - 1 = 0$$

Multiply each equation by the value of x it contains and add, obtaining:

$$a\sum_{i=1}^{n} x_i^2 + b\sum_{i=1}^{n} x_i y_i - \sum_{i=1}^{n} x_i = 0$$

Similarly, multiply by y and add, obtaining:

$$a\sum_{i=1}^{n} x_i y_i + b\sum_{i=1}^{n} y_i^2 - \sum_{i=1}^{n} y_i = 0$$

Lastly, solve these two equations for a and b, which will be the required values giving the least squares fit. Many modern calculators are programmed to do this calculation.

dB	I_1/I_2 or V_1/V_2	I_2/I_1 or V_2/V_1	P_1/P_2	P_2/P_1
0.1	1.012	0.989	1.023	0.977
0.2	1.023	0.977	1.047	0.955
0.3	1.035	0.966	1.072	0.933
0.4	1.047	0.955	1.096	0.912
0.5	1.059	0.944	1.122	0.891
0.6	1.072	0.933	1.148	0.871
0.7	1.084	0.923	1.175	0.851
0.8	1.096	0.912	1.202	0.832
0.9	1.109	0.902	1.230	0.813
1.0	1.122	0.891	1.259	0.794
1.1	1.135	0.881	1.288	0.776
1.2	1.148	0.871	1.318	0.759
1.3	1.162	0.861	1.349	0.741
1.4	1.175	0.851	1.380	0.724
1.5	1.188	0.841	1.413	0.708
1.6	1.202	0.832	1.445	0.692
1.7	1.216	0.822	1.479	0.676
1.8	1.230	0.813	1.514	0.661
1.9	1.245	0.804	1.549	0.645
2.0	1.259	0.794	1.585	0.631
2.5	1.334	0.750	1.778	0.562
3.0	1.413	0.708	1.995	0.501
3.5	1.496	0.668	2.240	0.447
4.0	1.585	0.631	2.510	0.398
4.5	1.679	0.596	2.820	0.355
5.0	1.778	0.562	3.160	0.316
5.5	1.884	0.531	3.550	0.282
6.0	1.995	0.501	3.980	0.251

Continued

dB	I_1/I_2 or V_1/V_2	I_2/I_1 or V_2/V_1	P_1/P_2	P_2/P_1
6.5	2.110	0.473	4.470	0.224
7.0	2.240	0.447	5.010	0.200
7.5	2.370	0.422	5.620	0.178
8.0	2.510	0.398	6.310	0.158
8.5	2.660	0.376	7.080	0.141
9.0	2.820	0.355	7.940	0.126
9.5	2.980	0.335	8.910	0.112
10.0	3.160	0.316	10.00	0.100
10.5	3.350	0.298	11.20	0.0891
11.0	3.550	0.282	12.60	0.0794
15.0	5.620	0.178	31.60	0.0316
15.5	5.960	0.168	35.50	0.0282
16.0	6.310	0.158	39.80	0.0251
16.5	6.680	0.150	44.70	0.0224
17.0	7.080	0.141	50.10	0.0200
17.5	7.500	0.133	56.20	0.0178
18.0	7.940	0.126	63.10	0.0158
18.5	8.410	0.119	70.80	0.0141
19.0	8.910	0.112	79.40	0.0126
19.5	9.440	0.106	89.10	0.0112
20.0	10.00	0.100	100	0.0100
20.5	10.59	0.0944	112	0.00891
21.0	11.22	0.0891	126	0.00794
21.5	11.88	0.0841	141	0.00708
22.0	12.59	0.0794	158	0.00631
22.5	13.34	0.0750	178	0.00562
23.0	14.13	0.0708	200	0.00501
23.5	14.96	0.0668	224	0.00447
24.0	15.85	0.0631	251	0.00398
24.5	16.79	0.0596	282	0.00355
25.0	17.78	0.0562	316	0.00316
25.5	18.84	0.0531	355	0.00282
26.0	19.95	0.0501	398	0.00251
26.5	21.10	0.0473	447	0.00224
27.0	22.40	0.0447	501	0.00200
27.5	23.70	0.0422	562	0.00178
28.0	25.10	0.0398	631	0.00158
28.5	26.60	0.0376	708	0.00141
29.0	28.20	0.0355	794	0.00126
29.5	29.80	0.0335	891	0.00112
30.0	31.6	0.0316	1000	0.00100

Decibels, Current and Voltage Ratio, and Power Ratio

$$\mathrm{dB} = 10 \log \frac{P_1}{P_2} = 20 \log \frac{V_1}{V_2} = 20 \log \frac{I_1}{I_2}$$

20 Calculus

J. Barron

Derivative

$$f'(x) = \lim_{\partial x \to 0} \frac{f(x + \partial x) - f(x)}{\partial x}$$

If u and v are functions of x:

$$(uv)' = u'v + uv'$$

$$\left(\frac{u}{v}\right)' = \frac{u'v - uv'}{v^2}$$

$$(uv)^{(n)} = u^{(n)}v + v^{(n-1)}v^{(1)} + \ldots + {}^nC_p\, u^{(n-p)}v^{(p)} + \ldots + uv^{(n)}$$

where:

$$^nC_p = \frac{n!}{p!(n-p)!}$$

If $z = f(x)$ and $y = g(z)$, then:

$$\frac{dy}{dx} = \frac{dy}{dz}\frac{dz}{dx}$$

Maxima and Minima

$f(x)$ has a stationary point wherever $f'(x) = 0$: The point is a maximum, minimum, or point of inflexion according to $f''(x)$ being less than, greater than, or equal to zero. $f(x,y)$ has a stationary point whenever $\dfrac{\partial f}{\partial x} = \dfrac{\partial f}{\partial y} = 0$.

Let (a, b) be such a point, and let

$$\frac{\partial^2 f}{\partial x^2} = A \quad \frac{\partial^2 f}{\partial x \partial y} = H \quad \frac{\partial^2 f}{\partial y^2} = B$$

all at that point; then:

- If $H^2 - AB > 0$, $f(x,y)$ has a saddle point at (a, b).
- If $H^2 - AB < 0$ and if $A < 0$, $f(x,y)$ has a maximum at (a,b), but if $A > 0$, $f(x,y)$ has a minimum at (a,b).
- If $H^2 = AB$, higher derivatives need to be considered.

Integral

$$\int_a^b f(x)dx = \lim_{N \to \infty} \sum_{n=0}^{N-1} f\left(a + \frac{n(b-a)}{N}\right)\left(\frac{b-a}{N}\right)$$

$$= \lim_{N \to \infty} \sum_{n=1}^{N} f\left(a + (n-1)\delta x\right)\delta x$$

where $\delta x = (b - a)/N$.

If u and v are functions of x, then $\int uv' dx = uv - \int u'v dx$ (integration by parts).

Derivatives and Integrals

y	$\dfrac{dy}{dx}$	$\displaystyle\int y dx$
x^n	nx^{n-1}	$\dfrac{x^{n+1}}{(n+1)}$
$\dfrac{1}{x}$	$-\dfrac{1}{x^2}$	$\ln(x)$
e^{ax}	ae^{ax}	$\dfrac{e^{ax}}{a}$
$\ln(x)$	$\dfrac{1}{x}$	$x[\ln(x) - 1]$
$\log_a x$	$\dfrac{1}{x}\log_a e$	$x\log_a\left(\dfrac{x}{e}\right)$

y	$\dfrac{dy}{dx}$	$\displaystyle\int y\,dx$
$\sin ax$	$a\cos ax$	$-\dfrac{1}{a}\cos ax$
$\cos ax$	$-a\sin ax$	$\dfrac{1}{a}\sin ax$
$\tan ax$	$a\sec^2 ax$	$-\dfrac{1}{a}\ln(\cos ax)$
$\cot ax$	$-a\operatorname{cosec}^2 ax$	$\dfrac{1}{a}\ln(\sin ax)$
$\sec ax$	$a\tan ax\sec ax$	$\dfrac{1}{a}\ln(\sec ax+\tan ax)$
$\operatorname{cosec} ax$	$-a\cot ax\operatorname{cosec} ax$	$\dfrac{1}{a}\ln(\operatorname{cosec} ax-\cot ax)$
$\arcsin\left(\dfrac{x}{a}\right)$	$\dfrac{1}{\left(a^2-x^2\right)^{1/2}}$	$x\arcsin\left(\dfrac{x}{a}\right)+\left(a^2-x^2\right)^{1/2}$
$\arccos\left(\dfrac{x}{a}\right)$	$\dfrac{-1}{\left(a^2-x^2\right)^{1/2}}$	$x\arccos\left(\dfrac{x}{a}\right)-\left(a^2-x^2\right)^{1/2}$
$\arctan\left(\dfrac{x}{a}\right)$	$\dfrac{a}{\left(a^2+x^2\right)}$	$x\arctan\left(\dfrac{x}{a}\right)-\dfrac{1}{2}a\ln(a^2+x^2)$
$\operatorname{arccot}\left(\dfrac{x}{a}\right)$	$\dfrac{-a}{\left(a^2+x^2\right)}$	$x\operatorname{arccot}\left(\dfrac{x}{a}\right)+\dfrac{1}{2}a\ln(a^2+x^2)$
$\operatorname{arcsec}\left(\dfrac{x}{a}\right)$	$\dfrac{a}{\left(x^2-a^2\right)^{1/2}x}$	$x\operatorname{arcsec}\left(\dfrac{x}{a}\right)-a\ln\!\left[x+(x^2-a^2)^{1/2}\right]$
$\operatorname{arccosec}\left(\dfrac{x}{a}\right)$	$-\dfrac{a}{\left(x^2-a^2\right)^{1/2}x}$	$x\operatorname{arccosec}\left(\dfrac{x}{a}\right)+a\ln\!\left[x+(x^2-a^2)^{1/2}\right]$
$\sinh ax$	$a\cosh ax$	$\dfrac{1}{a}\cosh ax$
$\cosh ax$	$a\sinh ax$	$\dfrac{1}{a}\sinh ax$
$\tanh ax$	$a\operatorname{sech}^2 ax$	$\dfrac{1}{a}\ln(\cosh ax)$
$\coth ax$	$-a\operatorname{cosech}^2 ax$	$\dfrac{1}{a}\ln(\sinh ax)$
$\operatorname{sech} ax$	$-a\tanh ax\operatorname{sech} ax$	$-\dfrac{1}{a}\ln\!\left(\coth\dfrac{ax}{2}\right)$
$\operatorname{cosech} ax$	$-a\coth ax\operatorname{cosech} ax$	$\dfrac{1}{a}\ln\!\left(\tanh\dfrac{ax}{2}\right)$
$\operatorname{arsinh}\left(\dfrac{x}{a}\right)$	$\left(x^2+a^2\right)^{-1/2}$	$x\operatorname{arsinh}\left(\dfrac{x}{a}\right)-\left(x^2+a^2\right)^{1/2}$
$\operatorname{arcosh}\left(\dfrac{x}{a}\right)$	$\left(x^2-a^2\right)^{-1/2}$	$x\operatorname{arcosh}\left(\dfrac{x}{a}\right)-\left(x^2-a^2\right)^{1/2}$

Continued

y	$\dfrac{dy}{dx}$	$\displaystyle\int y\,dx$
$\operatorname{artanh}\left(\dfrac{x}{a}\right)$	$a(a^2-x^2)^{-1}$	$x\operatorname{artanh}\left(\dfrac{x}{a}\right)+\dfrac{1}{2}a\ln(a^2-x^2)$
$\operatorname{arcoth}\left(\dfrac{x}{a}\right)$	$-a(x^2-a^2)^{-1}$	$x\operatorname{arcoth}\left(\dfrac{x}{a}\right)+\dfrac{1}{2}a\ln(x^2-a^2)$
$\operatorname{arsech}\left(\dfrac{x}{a}\right)$	$\dfrac{-a(a^2-x^2)^{-1/2}}{x}$	$x\operatorname{arsech}\left(\dfrac{x}{a}\right)+a\arcsin\left(\dfrac{x}{a}\right)$
$\operatorname{arcosech}\left(\dfrac{x}{a}\right)$	$\dfrac{-a(x^2+a^2)^{-1/2}}{x}$	$x\operatorname{arcosech}\left(\dfrac{x}{a}\right)+a\operatorname{arsinh}\left(\dfrac{x}{a}\right)$
$(x^2\pm a^2)^{1/2}$		$\dfrac{1}{2}x(x^2\pm a^2)^{1/2}\pm\dfrac{1}{2}a^2\operatorname{arsinh}\left(\dfrac{x}{a}\right)$
$(a^2-x^2)^{1/2}$		$\dfrac{1}{2}x(a^2-x^2)^{1/2}+\dfrac{1}{2}a^2\operatorname{arcsinh}\left(\dfrac{x}{a}\right)$
$(x^2\pm a^2)^p x$		$\dfrac{\frac{1}{2}(x^2\pm a^2)^{p+1}}{(p+1)}\quad(p\neq-1)$ $\dfrac{1}{2}\ln(x^2\pm a^2)\,(p=-1)$
$(a^2-x^2)^p x$		$\dfrac{-\frac{1}{2}(a^2-x^2)^{p+1}}{(p+1)}\quad(p\neq-1)$ $-\dfrac{1}{2}\ln(a^2-x^2)\,(p=-1)$
$x(ax^2+b)^p$		$\dfrac{(ax^2+b)^{p+1}}{2a(p+1)}\quad(p\neq-1)$ $\dfrac{\ln(ax^2+b)}{2a}\quad(p=-1)$
$(2ax-x^2)^{-1/2}$		$\arccos\left(\dfrac{a-x}{a}\right)$
$(a^2\sin^2 x+b^2\cos^2 x)^{-1}$		$\dfrac{1}{ab}\arctan\left(\dfrac{a}{b}\tan x\right)$
$(a^2\sin^2 x-b^2\cos^2 x)^{-1}$		$-\dfrac{1}{ab}\operatorname{arctanh}\left(\dfrac{a}{b}\tan x\right)$
$e^{ax}\sin bx$		$e^{ax}\dfrac{a\sin bx-b\cos bx}{a^2+b^2}$
$e^{ax}\cos bx$		$e^{ax}\dfrac{a\cos bx+b\sin bx}{a^2+b^2}$

y	$\int y\,dx$
$\sin mx \sin nx$	$\dfrac{1}{2}\dfrac{\sin(m-n)x}{m-n} - \dfrac{1}{2}\dfrac{\sin(m+n)x}{m+n}$ $(m \neq n)$
	$\dfrac{1}{2}\left(x - \dfrac{\sin 2mx}{2m}\right)$ $(m = n)$
$\sin mx \cos nx$	$-\dfrac{1}{2}\dfrac{\cos(m+n)x}{m+n} - \dfrac{1}{2}\dfrac{\cos(m-n)x}{m-n}$ $(m \neq n)$
	$-\dfrac{1}{2}\left(\dfrac{\cos 2mx}{2m}\right)$ $(m = n)$
$\cos mx \cos nx$	$\dfrac{1}{2}\dfrac{\sin(m+n)x}{m+n} + \dfrac{1}{2}\dfrac{\sin(m-n)x}{m-n}$ $(m \neq n)$
	$\dfrac{1}{2}\left(x + \dfrac{\sin 2mx}{2m}\right)$ $(m = n)$

Standard Substitutions

Integral a function of	Substitute
$a^2 - x^2$	$x = a\sin\theta$ or $x = a\cos\theta$
$a^2 + x^2$	$x = a\tan\theta$ or $x = a\sinh\theta$
$x^2 - a^2$	$x = a\sec\theta$ or $x = a\cosh\theta$

Reduction Formulae

$$\int \sin^m x\,dx = -\frac{1}{m}\sin^{m-1}x\cos x + \frac{m-1}{m}\int \sin^{m-2}x\,dx$$

$$\int \cos^m x\,dx = \frac{1}{m}\cos^{m-1}x\sin x + \frac{m-1}{m}\int \cos^{m-2}x\,dx$$

$$\int \sin^m x\cos^n x\,dx = \frac{\sin^{m+1}x\cos^{n-1}x}{m+n} + \frac{n-1}{m+n}\int \sin^m x\cos^{n-2}x\,dx$$

If the integrand is a rational function of $\sin x$ and/or $\cos x$, substitute $t = \tan\dfrac{1}{2}x$; then:

$$\sin x = \frac{1}{1+t^2}$$

$$\cos x = \frac{1-t^2}{1+t^2}$$

$$dx = \frac{2dt}{1+t^2}$$

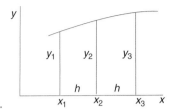

Fig. 20.1 Numerical integration.

Numerical Integration

Trapezoidal Rule

See Figure 20.1.

$$\int_{x_1}^{x_2} y\,dx = \frac{1}{2}h(y_1 - y_2) + O(h^3)$$

Simpson's Rule

See Figure 20.1.

$$\int_{x_1}^{x_3} y\,dx = \frac{h(y_1 + 4y_2 + y_3)}{3} + O(h^5)$$

Change of Variable in Double Integral

$$\iint f(x, y)\,dx\,dy = \iint F(u, v)\,|J|\,du\,dv$$

where:

$$J = \frac{\partial(x, y)}{\partial(u, v)} = \begin{vmatrix} \dfrac{\partial x}{\partial u} & \dfrac{\partial x}{\partial v} \\ \dfrac{\partial y}{\partial u} & \dfrac{\partial y}{\partial v} \end{vmatrix} = \begin{vmatrix} \dfrac{\partial x}{\partial u} & \dfrac{\partial y}{\partial u} \\ \dfrac{\partial x}{\partial v} & \dfrac{\partial y}{\partial v} \end{vmatrix}$$

is the Jacobian of the transformation.

Differential Mean Value Theorem

$$\frac{f(x+h) - f(x)}{h} = f'(x + \theta h) \quad 0 < \theta < 1$$

Integral Mean Value Theorem

$$\int_a^b f(x)g(x)dx = g(a+\theta h)\int_a^b f(x)dx$$
$$h = b - a \quad 0 < \theta < 1$$

Vector Calculus

Let $s(x, y, x)$ be a scalar function of position, and let $v(x, y, x) = iv_x(x, y, z) + jv_y(x, y, z) + kv_z(x, y, z)$ be a vector function of position. Define:

$$\nabla = i\frac{\partial}{\partial x} + j\frac{\partial}{\partial y} + k\frac{\partial}{\partial z}$$

so that:

$$\nabla \cdot \nabla = \nabla^2 = \frac{\partial^2}{\partial x^2} + \frac{\partial^2}{\partial y^2} + \frac{\partial^2}{\partial z^2}$$

then:

$$\operatorname{grad} s = \nabla s = i\frac{\partial s}{\partial x} + j\frac{\partial s}{\partial y} + k\frac{\partial s}{\partial z}$$

$$\operatorname{div} v = \nabla \cdot v = \frac{\partial v_x}{\partial x} + \frac{\partial v_y}{\partial y} + \frac{\partial v_z}{\partial z}$$

$$\operatorname{curl} v = \nabla \times v = i\left(\frac{\partial v_z}{\partial y} - \frac{\partial v_y}{\partial z}\right) + j\left(\frac{\partial v_x}{\partial z} - \frac{\partial v_z}{\partial x}\right) + k\left(\frac{\partial v_y}{\partial x} - \frac{\partial v_x}{\partial y}\right)$$

The following identities are then true:

$\operatorname{div}(sv) = s\operatorname{div} v + (\operatorname{grad} s)\cdot v$
$\operatorname{curl}(sv) = s\operatorname{curl} v + (\operatorname{grad} s) \times v$
$\operatorname{div}(u \times v) = v\cdot\operatorname{curl} u - u\cdot\operatorname{curl} v$
$\operatorname{curl}(u \times v) = u\operatorname{div} v - v\operatorname{div} u + (v\cdot\nabla)u - (u\cdot\nabla)v$
$\operatorname{div}\operatorname{grad} s = \nabla^2 s$
$\operatorname{div}\operatorname{curl} v = 0$
$\operatorname{curl}\operatorname{grad} s = 0$
$\operatorname{curl}\operatorname{curl} v = \operatorname{grad}(\operatorname{div} v) - \nabla^2 v$
where ∇^2 operates on each component of v.
$v \times \operatorname{curl} v + (v\cdot\nabla) v = \operatorname{grad} \frac{1}{2}v^2$
Potentials:

If $\operatorname{curl} v = 0$, $v = \operatorname{grad} \varphi$, where φ is a scalar potential.
If $\operatorname{div} v + 0$, $v = \operatorname{curl} A$, where A is a vector potential.

21 Series and Transforms

J. Barron

Series

Arithmetic Series

Sum of n terms:

$$S_n = a + (a+d) + (a+2d) + \ldots + [a+(n-1)d]$$
$$= \frac{n[2a+(n-1)d]}{2}$$
$$= \frac{n(a+1)}{2}$$

Geometric Series

Sum of n terms:

$$S_n = a + ar + ar^2 + \ldots + ar^{n-1} = \frac{a(1-r^n)}{(1-r)}$$

$$\text{If } (|r| < 1)$$

$$S_\infty = \frac{a}{(1-r)}$$

Binomial Series

$$(1+x)^p = 1 + px + \frac{p(p-1)}{2!}x^2 + \frac{p(p-1)(p-2)}{3!}x^3 + \ldots$$

If p is a positive integer, the series terminates with the term in x^p and is valid for all x; otherwise, the series does not terminate and is valid only for $-1 < x < 1$.

Taylor's Series

Infinite form:

$$f(x+h) = f(x) + hf'(x) + \frac{h^2}{2!}f''(x) + \ldots + \frac{h^n}{n!}f^{(n)}(x) + \ldots$$

Finite form:

$$f(x+h) = f(x) + hf'(x) + \frac{h^2}{2!}f''(x) + \ldots + \frac{h^n}{n!}f^{(n)}(x) + \frac{h^{n+1}}{(n+1)!}f^{(n+1)}(x+\lambda h)$$

where $0 \leq \lambda \leq 1$.

Maclaurin's Series

$$f(x) = f(0) + xf'(0) + \frac{x^2}{2!}f''(0) + \ldots + \frac{x^n}{n!}f^{(n)}(0) + \ldots$$

Neither of these series is necessarily convergent, but both usually are for appropriate ranges of values of h and of x, respectively.

Laurent's Series

If a function $f(z)$ of a complex variable is analytic on and everywhere between two concentric circles center a, then at any point in this region:

$$f(z) = a_0 + a_1(z-a) + \ldots + \frac{b_1}{(z-a)} + \frac{b_2}{(z-a)^2} + \ldots$$

This series is often applicable when Taylor's series is not.

Power Series for Real Variables

The column headed *Math* contains the range of values of the variable x for which the series is convergent in the pure mathematical sense. In some cases, a different range of values is given in the column headed *Comp*, to reduce the rounding errors that arise when computers are used.

	Math	Comp
$e^x = 1 + x + \dfrac{x^2}{2!} + \ldots$	all x	$|x| \leq 1$
$\ln(1+x) = x - \dfrac{x^2}{2} + \dfrac{x^3}{3} - \dfrac{x^4}{4} + \ldots$	$-1 < x \leq 1$	

$$\sin x = x - \frac{x^3}{3!} + \frac{x^5}{5!} - \frac{x^7}{7!} + \ldots \qquad \text{all } x \qquad |x| \le 1$$

$$\cos x = 1 - \frac{x^2}{2!} + \frac{x^4}{4!} - \frac{x^6}{6!} + \ldots \qquad \text{all } x \qquad |x| \le 1$$

$$\tan x = x + \frac{x^3}{3} + \frac{2x^5}{15} - \frac{17x^7}{315} + \ldots \qquad \text{all } x \qquad |x| \le \frac{\pi}{2}$$

$$\arctan x = x - \frac{x^3}{3} + \frac{x^5}{5} - \frac{x^7}{7} + \ldots \qquad \text{all } x \qquad |x| \le 1$$

$$\sinh x = x + \frac{x^3}{3!} + \frac{x^5}{5!} + \frac{x^7}{7!} + \ldots \qquad \text{all } x \qquad |x| \le 1$$

$$\cosh x = 1 + \frac{x^2}{2!} + \frac{x^4}{4!} + \frac{x^6}{6!} + \ldots \qquad \text{all } x \qquad |x| \le 1$$

Integer Series

$$\sum_{n=1}^{N} n = 1 + 2 + 3 + 4 + \ldots + N = \frac{N(N+1)}{2}$$

$$\sum_{n=1}^{N} n^2 = 1^2 + 2^2 + 3^2 + 4^2 + \ldots + N^2 = \frac{N(N+1)(2N+1)}{6}$$

$$\sum_{n=1}^{N} n^3 = 1^3 + 2^3 + 3^3 + 4^3 + \ldots + N^3 = \frac{N^2(N+1)^2}{4}$$

$$\sum_{n=1}^{\infty} \frac{(-1)^{n+1}}{n} = 1 - \frac{1}{2} + \frac{1}{3} - \frac{1}{4} + \ldots = \ln(2) \qquad (\text{see } \ln(1+x))$$

$$\sum_{n=1}^{\infty} \frac{(-1)^{n+1}}{2n-1} = 1 - \frac{1}{3} + \frac{1}{5} - \frac{1}{7} + \ldots = \frac{\pi}{4} \qquad (\text{see } \arctan x)$$

$$\sum_{n=1}^{\infty} \frac{1}{n^2} = 1 + \frac{1}{4} + \frac{1}{9} + \frac{1}{16} + \ldots = \frac{\pi^2}{6}$$

$$\sum_{n=1}^{N} n(n+1)(n+2)\ldots(n+r)$$

$$= 1.\,2.\,3\ldots + 2.\,3.\,4\ldots + 3.\,4.\,5\ldots + N(N+1)(N+2)\ldots(N+r)$$

$$= \frac{N(N+1)(N+2)\ldots(N+r+1)}{r+2}$$

Fourier Series

These expressions for Fourier series are valid for functions having at most a finite number of discontinuities within the period 0 to 2π of the variable of integration.

$$f(\theta) = \frac{1}{2}a_0 + \sum_{n=1}^{\infty} (a_n \cos n\theta + b_n \sin n\theta)$$

with:

$$a_n = \frac{1}{\pi} \int_0^{2\pi} f(\theta) \cos n\theta \, d\theta$$

$$b_n = \frac{1}{\pi} \int_0^{2\pi} f(\theta) \sin n\theta \, d\theta$$

or:

$$f(\theta) = \sum_{n=-\infty}^{\infty} c_n \exp(jn\theta)$$

with:

$$c_n = \frac{1}{2x} \int_0^{2\pi} f(\theta) \exp(-jn\theta) d\theta = \begin{cases} \dfrac{1}{2}(a_n + jb_n) & n < 0 \\ \dfrac{1}{2}(a_n + jb_n) & n > 0 \end{cases}$$

Rectified Sine Wave

$$f(\omega t) = \frac{1}{\pi} + \frac{1}{2} \cos \omega t + \frac{2}{\pi} \sum_{n=1}^{\infty} (-1)^{n+1} \frac{\cos 2n\omega t}{4n^2 - 1}$$

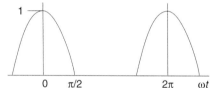

Fig. 21.1 Half wave.

$$f(\omega t) = \frac{\sin(\pi/p)}{\pi/p} + \frac{2p}{\pi} \sin\left(\frac{\pi}{p}\right) \sum_{n=1}^{\infty} (-1)^{n+1} \frac{\cos np\omega t}{p^2 n^2 - 1}$$

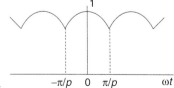

Fig. 21.2 p-phase.

Square Wave

$$f(\omega t) = \frac{4}{\pi} \sum_{n=1}^{\infty} \frac{\sin(2n-1)\omega t}{(2n-1)}$$

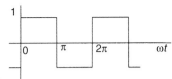

Fig. 21.3   Square wave.

Triangular Wave

$$f(\omega t) = \frac{8}{\pi^2} \sum_{n+1}^{\infty} (-1)^{n+1} \frac{\sin (2n-1)\omega t}{(2n-1)^2}$$

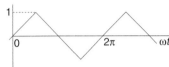

Fig. 21.4   Triangular wave.

Sawtooth Wave

$$f(\omega t) = \frac{2}{\pi} \sum_{n=1}^{\infty} (-1)^{n+1} \frac{\sin n\omega t}{n}$$

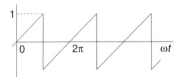

Fig. 21.5   Sawtooth wave.

Pulse Wave

$$f(t) = \frac{\tau}{T} + \frac{2\tau}{T} \sum_{n=1}^{\infty} \frac{\sin (n\omega\tau/T)}{n\pi\tau/T} \cos \left(\frac{2n\pi\tau}{T} \right)$$

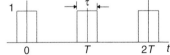

Fig. 21.6   Pulse wave.

Transforms

Fourier Transforms

Among other applications, these are used for converting from the time domain to the frequency domain.

Basic formulae:

$$\int_{-\infty}^{\infty} U(f)\exp{(j2\pi ft)}df = u(t) \leftrightarrow U(f) = \int_{-\infty}^{\infty} u(t)\exp{(-j2\pi ft)}dt$$

Change of sign and complex conjugates:

$$u(-t) \leftrightarrow U(-f) \qquad\qquad u^*(t) \leftrightarrow U^*(-f)$$

Time and frequency shifts (τ and φ constant):

$$u\,(t-\tau) \leftrightarrow U(f)\exp{(-j2\pi f\tau)}\exp{(j2\pi\varphi t)}\,u(t) \leftrightarrow U(f-\varphi)$$

Scaling (T constant):

$$u\left(\frac{t}{T}\right) \leftrightarrow TU(fT)$$

Products and convolutions:

$$u(t)*v(t) \leftrightarrow U(f)\,V(f)$$
$$u(t)v(t) \leftrightarrow U(f)*V(f)$$

Differentiation:

$$u'(t) \leftrightarrow j2\pi fU(f)$$
$$-j2\pi tu(t) \leftrightarrow U'(f)$$
$$\frac{\partial u(t,\alpha)}{\partial\alpha} \leftrightarrow \frac{\partial(U-f,\alpha)}{\partial\alpha}$$

Integration ($U(0) = 0$, a and b real and constants):

$$\int_{-\alpha}^{\tau} u(\tau)d\tau \leftrightarrow \frac{U(f)}{j2\pi f}$$
$$\int_{a}^{b} v(t,\alpha)d\alpha \leftrightarrow \int_{a}^{b} V(f,\alpha)d\alpha$$

Interchange of functions:

$$U(t) \leftrightarrow u(-f)$$

Dirac delta functions:

$$\delta(t) \leftrightarrow 1 \qquad \exp{(j2\pi f_0 t)} \leftrightarrow \delta(f-f_0)$$

Unit length, unit amplitude pulse, centered on $t = 0$:

$$\mathrm{rect}(t) \leftrightarrow \frac{\sin\pi f}{\pi f}$$

Gaussian distribution:

$$\exp(-\pi t^2) \leftrightarrow \exp(-\pi f^2)$$

Repeated and impulse (delta function) sampled waveforms:

$$\sum_{-\infty}^{\infty} u(t - nT) \leftrightarrow \left(\frac{1}{T}\right) U(f) \sum_{-\infty}^{\infty} \delta\left(f - \frac{n}{T}\right)$$

$$u(t) \sum_{-\infty}^{\infty} \delta(t - nT) \leftrightarrow \left(\frac{1}{T}\right) \sum_{-\infty}^{\infty} U\left(f - \frac{n}{T}\right)$$

Parseval's lemma:

$$\int_{-\infty}^{\infty} u(t) v^*(t) dt = \int_{-\infty}^{\infty} U(f) V^*(f) df$$

$$\int_{-\infty}^{\infty} |u(t)|^2 dt = \int_{-\infty}^{\infty} |U(f)|^2 df$$

Laplace Transforms

$$\bar{x}_x = \int_0^{\infty} x(t) \exp(-st) dt$$

Function	Transform	Remarks
$e^{-\alpha t}$	$\dfrac{1}{s + \alpha}$	
$\sin \omega t$	$\dfrac{\omega}{s^2 + \omega^2}$	
$\cos \omega t$	$\dfrac{s}{s^2 + \omega^2}$	
$\sinh \omega t$	$\dfrac{\omega}{s^2 - \omega^2}$	
$\cosh \omega t$	$\dfrac{s}{s^2 - \omega^2}$	
t^n	$\dfrac{n!}{s^{n+1}}$	
1	$\dfrac{1}{s}$	
$H(t - \tau)$	$\dfrac{1}{s} \exp(-s\tau)$	Heaviside step function
$x(t - \tau) H(t - \tau)$	$\exp(-s\tau) X(s)$	Shift in t
$\delta(t - \tau)$	$\exp(-s\tau)$	Dirac delta function

Continued

Function	Transform	Remarks
$\exp(-\alpha t)\,x(t)$	$X(s+\alpha)$	Shift in s
$\exp(-\alpha t)\sin\omega t$	$\dfrac{\omega}{(s+\alpha)^2+\omega^2}$	
$\exp(-\alpha t)\cos\omega t$	$\dfrac{(s+\alpha)}{(s+\alpha)^2+\omega^2}$	
$tx(t)$	$-\dfrac{dX(s)}{ds}$	
$\dfrac{dx(t)}{dt}=x'(t)$	$sX(s)-x(0)$	
$\dfrac{d^2x(t)}{dt^2}=x''(t)$	$x^2X(s)-sx(0)-x'(0)$	
$\dfrac{d^n x(t)}{dx^n}=x^{(n)}(t)$	$s^nX(s)-s^{n-1}x(0)-s^{n-2}x'(0)\ldots$	
	$-sx^{(n-2)}(0)-x^{(n-1)}(0)$	

Convolution integral:

$$\int_0^t x_1(\sigma)x_2(t-\sigma)d\sigma \to X_1(s)X_2(s)$$

22 Matrices and Determinants

J. Barron

Linear Simultaneous Equations

This set of equations:

$$a_{11}x_1 + a_{12}x_2 + \ldots + a_{1n}x_n = b_1$$

$$a_{21}x_1 + a_{22}x_2 + \ldots + a_{2n}x_n = b_2$$

$$\ldots$$

$$a_{n1}x_1 + a_{n2}x_2 + \ldots + a_{nn}x_n = b_n$$

may be written symbolically:

$$Ax = b$$

in which A is the *matrix* of the coefficients a_{ij} and x and b are the column matrices or *vectors* $(x_1 \ldots x_n)$ and $(b_1 \ldots b_n)$. In this case, the matrix A is square $(n \times n)$. The equations can be solved unless two or more of them are not independent, in which case:

$$\det A = [A] = 0$$

and there then exist nonzero solutions x_i only if $b = 0$. If $\det A \neq 0$, there exist nonzero solutions only if $b \neq 0$. When $\det A = 0$, A is *singular*.

Matrix Arithmetic

If A and B are both matrices of m rows and n columns, they are *conformable* and:

$$A \pm B = C \text{ where } C_{ij} = A_{ij} \pm B_{ij}$$

Product

If A is an $m \times n$ matrix and B an $n \times l$ matrix, the product AB is defined by:

$$(AB)_{ij} = \sum_{k=1}^{n} (A)_{ik}(B)_{kj}$$

In this case, if $l \neq m$, the product BA will not exist.

Transpose

The transpose of A is written A' or A^t and is the matrix whose rows are the columns of A. For example:

$$(A^t)_{ij} = (A)_{ji}$$

A square matrix may be equal to its transpose, and it is then said to be *symmetrical*. If the product AB exists, then:

$$(AB)^t = B^t A^t$$

Adjoint

The *adjoint* of a square matrix A is defined as B, where:

$$B_{ij} = A_{ji}$$

and A_{ji} is the cofactor of a_{ji} in det A.

Inverse

If A is nonsingular, the *inverse* A^{-1} is given by:

$$A^{-1} = \frac{\text{adj}\,A}{\det A}$$

$$A^{-1}A = AA^{-1} = I$$

the *unit* matrix.

$$(AB)^{-1} = B^{-1}A^{-1}$$

if both inverses exist. The original equation $Ax = b$ has the solution $x = A^{-1}b$ if the inverse exists.

Orthogonality

A matrix A is orthogonal if $AA^t = I$. If A is the matrix of a coordinate transformation $X = AY$ from variables y_i to variables x_i, then if A is orthogonal, $X^tX = Y^tY$, or:

$$\sum_{i=1}^{n} x_i^2 = \sum_{i=1}^{n} y_i^2$$

Eigenvalues and Eigenvectors

The equation:

$$Ax = \lambda x$$

where A is a square matrix, x a column vector, and λ is a number (in general complex), has at most n solutions (x, λ). The values of λ are *eigenvalues* and those of x are *eigenvectors* of the matrix A. The relation may be written:

$$(A - \lambda I)x = 0$$

so that if $x \neq 0$, the equation $A - \lambda I = 0$ gives the eigenvalues. If A is symmetric and real, the eigenvalues are real. If A is symmetric, the eigenvectors are orthogonal. If A is not symmetric, the eigenvalues are complex and the eigenvectors are not orthogonal.

Coordinate Transformation

Suppose that x and y are two vectors related by the equation:

$$y = Ax$$

when their components are expressed in one orthogonal system, and that a second orthogonal system has unit vectors $u_1, u_2, \ldots, u_n$ expressed in the first system. The components of x and y expressed in the new system will be x' and y', where:

$$x' = U^t x \qquad y' = U^t y$$

and U^t is the orthogonal matrix whose rows are the unit vectors u_1^t, u_2^t, etc. Then:

$$y' = U^t y = U^t Ax = U^t AUx'$$

or

$$y' = A'x'$$

where:

$$A' = U^t A U$$

Matrices A and A' are *congruent*.

Determinants

The determinant:

$$D = \begin{bmatrix} a_{11} & a_{12} & \cdots & a_{1n} \\ a_{21} & a_{22} & \cdots & a_{2n} \\ \vdots & \vdots & \vdots & \vdots \\ a_{n1} & a_{n2} & \cdots & a_{nn} \end{bmatrix}$$

is defined as follows. The first suffix in a_{rs} refers to the row, the second to the column that contains a_{rs}. Denote by M_{rs} the determinant left by deleting the rth row and the sth column from D; then:

$$D = \sum_{k=1}^{n} (-1)^{k+1} a_{1k} M_{1k}$$

gives the value of D in terms of determinants of order $n - 1$, by repeated application, of the determinant in terms of the elements a_{rs}.

Properties of Determinants

If the rows of $|a_{rs}|$ are identical with the columns of $|b_{sr}|$, $a_{rs} = b_{sr}$ and:

$$|a_{rs}| = |b_{sr}|$$

that is, the *transposed* determinant is equal to the original.

If two rows or two columns are interchanged, the numerical value of the determinant is unaltered, but the sign will be changed if the permutation of rows or columns is odd.

If two rows or two columns are identical, the determinant is zero.

If each element of one row or one column is multiplied by k, so is the value of the determinant.

If any row or column is zero, so is the determinant.

If each element of the pth row or column of the determinant of c_{rs} is equal to the sum of the elements of the same row or column in determinants of a_{rs} and b_{rs}, then:

$$|c_{rs}| = |a_{rs}| + |b_{rs}|$$

The addition of any multiple of one row (or column) to another row (or column) does not alter the value of the determinant.

Minor

If row p and column q are deleted from $|a_{rs}|$, the remaining determinant of M_{pq} is called the *minor* of a_{pq}.

Cofactor

The *cofactor* of a_{pq} is the minor of a_{pq} prefixed by the sign that the product $M_{pq} a_{pq}$ would have in the expansion of the determinant and is denoted by A_{pq}:

$$A_{pq} = (-1)^{p+q} M_{pq}$$

A determinant a_{ij} in which $a_{ij} = a_{ji}$ for all i and j is called *symmetric*, while if $a_{ij} = -a_{ji}$ for all i and j, the determinant is *skew-symmetric*. It follows that $a_{ii} = 0$ for all i in a skew-symmetric determinant.

Numerical Solution of Linear Equations

Evaluation of a determinant by direct expansion in terms of elements and cofactors is disastrously slow, and other methods are available, usually programmed on any existing computer system.

Reduction of Determinant or Matrix to Upper Triangular or Diagonal Form

The system of equations may be written:

$$\begin{bmatrix} a_{11} & a_{12} & \cdots & a_{1n} \\ a_{21} & a_{22} & \cdots & a_{2n} \\ \vdots & \vdots & \vdots & \vdots \\ a_{n1} & a_{n2} & \cdots & a_{nn} \end{bmatrix} x_1 \begin{bmatrix} x_1 \\ x_2 \\ \vdots \\ x_n \end{bmatrix} = \begin{bmatrix} b_1 \\ b_2 \\ \vdots \\ b_n \end{bmatrix}$$

The variable x_1 is eliminated from the last $n - 1$ equations by adding a multiple $-a_{i1}/a_{11}$ of the first row to the ith row, obtaining:

$$\begin{bmatrix} a_{11} & a_{12} & \cdots & a_{1n} \\ 0 & a_{22}' & \cdots & a_{2n}' \\ \vdots & \vdots & \vdots & \vdots \\ 0 & a_{n2} & \cdots & a_{nn}'' \end{bmatrix} x_1 \begin{bmatrix} x_1 \\ x_2 \\ \vdots \\ x_n \end{bmatrix} = \begin{bmatrix} b_1 \\ b_2 \\ \vdots \\ b_n'' \end{bmatrix}$$

where primes indicate altered coefficients. This process may be continued by eliminating x_2 from rows 3 to n, and so on.

Alternatively the process may be applied to the system of equations in the form:

$$Ax = Ib$$

where *I* is the unit matrix and the same operations are carried out on *I* the same as on *A*. If the process is continued after reaching the upper triangular form, the matrix *A* can eventually be reduced to diagonal form. Finally, each equation is divided by the corresponding diagonal element of *A*, thus reducing *A* to the unit matrix. The system is now in the form:

$$Ix = Bb$$

and evidently $B = A^{-1}$. The total number of operations required is $O(n^3)$.

23 Fourier Analysis

William J. Fitzgerald

Introduction

This chapter considers Fourier analysis and associated transform methods for both discrete-time and continuous-time signals and systems. Fourier methods are based on using real or complex sinusoids as basis functions, and they allow signals to be represented in terms of sums of sinusoidal components.

Euler and d'Alembert showed that the wave equation (Equation 23.1) could be satisfied by any well-behaved function of the variable $x + ct$ or $x - ct$ (i.e., Equation 23.2).

$$\frac{\partial^2 \Phi}{\partial x^2} = \frac{1}{c^2} \frac{\partial^2 \Phi}{\partial t^2} \tag{23.1}$$

$$\Phi(x, t) = f(x + ct) + g(x - ct) \tag{23.2}$$

Boundary conditions, of course, impose restrictions on the functions f and g, but apart from these restrictions the functions f and g are completely arbitrary. This result started a long controversy, since it had been shown by Daniel Bernoulli (1727) that the solution to the wave equation could also be represented as a superposition of sinusoidal components. If the solution obtained by Bernoulli was as general as that of Euler and d'Alembert, it follows that an arbitrary function could be represented as a superposition of sinusoidal components. Euler had difficulty accepting this, and it was Jean-Baptiste Fourier (1768–1830) whose work *Memoire sur la Chaleur* expounded the method now known as Fourier analysis and who eventually showed the above equivalence.

It is of fundamental importance in both linear systems theory, and signal processing in general, that the input and/or output signals of interest can be represented as linear combinations of simple basis functions, and the Fourier series representation is one such method that allows the signals of interest to be decomposed into harmonic components using a sinusoidal basis. This is clearly of interest if one is considering the spectral composition of signals, and in this area, Fourier analysis and the Fourier transform are used widely.

It should be realized that the Fourier representation is just one of many possible representations, and the choice of the signal representation will be determined by the particular nature of the problem under investigation.

Suppose that we have a function $f(t)$ that we wish to represent on a finite interval (t_1, t_2), where t can obviously represent time, space, or any other dimension relevant to our problem, in terms of a set of basic functions $\psi_1(t)$, $\psi_2(t)$, ..., $\psi_n(t)$. We will assume that these functions are orthogonal on the region of support (t_1, t_2), which is written as Equation 23.3.

$$\int_{t_1}^{t_2} \psi_i(t)\psi_j(t)dt = 0 \tag{23.3}$$

We will use the notation $<\psi_i \mid \psi_j>$ to denote the integral above. The idea of orthogonality, expressed above, is the same as that applied to vectors and vector spaces, and our representation of $f(t)$ in terms of the functions $\psi_i(t)$, $i = 1, 2, ..., n$ is equivalent to representing a vector f in terms of an orthogonal set of vectors that span the space containing f.

We will assume that the representation of $f(t)$ can be written as a linear combination of the basis functions $\psi_i(t)$ as in Equation 23.4.

$$f(t) = \sum_{i=1}^{n} c_i \psi_i(t) \tag{23.4}$$

This will, in general, be in error since it is only a representation of the function $f(t)$, and we will require that the representation should be as close as possible to $f(t)$. Many measures of closeness exist, but a common measure is the mean squared error (MSE). We therefore require that the coefficients c_i, $i = 1, 2, ..., n$ are chosen to minimize its value in Equation 23.5.

$$MSE = \frac{1}{t_2 - t_1} \int_{t_1}^{t_2} \left[f(t) - \sum_{i=1}^{n} c_i \psi_i \right]^2 dt \tag{23.5}$$

This is the mean squared error averaged over the interval (t_1, t_2). Defining the functions in Equations 23.6 and 23.7, the expression for the mean square error above can be written as in Equation 23.8.

$$\alpha_i = \int_{t_1}^{t_2} f(t)\psi_i(t)dt = <f \mid \psi_i> \tag{23.6}$$

$$\beta_i = \int_{t_1}^{t_2} \psi_i^2 dt = <\psi_i \mid \psi_i> \tag{23.7}$$

$$MSE = \frac{1}{t_2 - t_1} \left[\int_{t_1}^{t_2} f(t)^2 dt + c_1^2 \beta_1 + c_2^2 \beta_1 + ... + c_n^2 \beta_n - 2c_1 \alpha_1 - 2c_2 \alpha_2 - ... - 2c_n \alpha_n \right] \tag{23.8}$$

Using the identity of Equation 23.9, the expression for the mean squared error can be written as in Equation 23.10.

$$c_i^2 \beta_i - 2c_i \alpha_i = \left(c_i \sqrt{\beta_i} - \frac{\alpha_i}{\sqrt{\beta_i}} \right)^2 - \frac{\alpha_i^2}{\beta_i} \qquad (23.9)$$

$$\text{MSE} = \frac{1}{t_2 - t_1} \left[\int_{t_1}^{t_2} f(t)^2 \, dt + \sum_{i=1}^{n} \left(c_i \sqrt{\beta_i} - \frac{\alpha_i}{\sqrt{\beta_i}} \right)^2 - \sum_{i=1}^{n} \frac{\alpha_i^2}{\beta_i} \right] \qquad (23.10)$$

From the form of this expression, it is clear that the MSE is always greater than or equal to zero, since it is the sum of squared terms, and that it achieves its least value when the middle term above is zero (i.e., as in Equation 23.11).

$$c_i = \frac{\alpha_i}{\beta_i} = \frac{<f \mid \psi_i>}{<\psi_i \mid \psi_i>} \qquad (23.11)$$

Therefore, the best approximation of an arbitrary signal or function $f(t)$, in the mean squared error sense, over the region of support (t_1, t_2), that can be represented as a linear superposition of orthogonal basis functions, is achieved by choosing the coefficients c_i according to the above expression, which is the normalized projection of the function in the direction of the basis function ψ_i.

As is clear from the preceding, when the coefficients c_i are chosen using the above criterion, the mean squared error is simply the averaged difference between two terms, and this difference can be made as small as one likes by including more and more terms in the summation term. Thus, in the limit the mean squared error is zero when Equation 23.12 is satisfied.

$$\int_{t_1}^{t_2} f(t)^2 \, dt = \sum_{i=1}^{\infty} c_i^2 \int_{t_1}^{t_2} \psi_i^2 \, dt \qquad (23.12)$$

If this relationship holds, then the infinite sum of the weighted basis functions is said to converge in the mean to $f(t)$, an equality known as Parseval's relation, and if it holds for all $f(t)$ of a certain class, then the set $\psi_i(t)$ is said to be complete for that class of functions.

If the basis functions are complex valued functions of a real argument t, orthogonality is defined as in Equation 23.13 where ψ_j^* is the complex conjugate of ψ_j and the generalized Fourier coefficients are given by Equation 23.14.

$$<\psi_i \mid \psi_j^*> = \int_{t_1}^{t_2} \psi_i \psi_j^* \, dt = 1 \text{ or } 0 \qquad (23.13)$$

$$c_i \frac{<f \mid \psi_i>}{<\psi_i \mid \psi_j>} \qquad (23.14)$$

If $<\psi_i \mid \psi_j> = 1$, the set ψ_i is called an orthonormal set of basis functions. For example, the trigonometric system of functions with period 2π forms an orthogonal set over the region $(-\pi, \pi)$, and it can be shown that the following forms an orthonormal set over this region:

$$\frac{1}{\sqrt{2\pi}}, \frac{1}{\sqrt{\pi}} \sin(t), \frac{1}{\sqrt{\pi}} \cos(t), \ldots, \frac{1}{\sqrt{\pi}} \sin(nt), \frac{1}{\sqrt{\pi}} \cos(nt), \ldots$$

Clearly, other orthonormal sets or trigonometric functions can be defined over other intervals (e.g., for the interval $(0, \pi)$), the set

$$\left[\sqrt{\frac{2}{\pi}}\sin(nt)\right]_{n=1}^{\infty}$$

is orthonormal. If ψ_i is an orthonormal set, the generalized Fourier coefficients for the representation of a function f, are given by Equation 23.15. As an example, we can consider the orthonormal trigonometric system of period 2π for which Equations 23.16 through 23.20 are satisfied, for $k = 2, 3, \ldots$:

$$c_n < f | \psi_n > \tag{23.15}$$

$$c_0 = \frac{1}{\sqrt{2\pi}}\int_{-\pi}^{\pi} f(t)dt \tag{23.16}$$

$$c_1 = \frac{1}{\sqrt{\pi}}\int_{-\pi}^{\pi} f(t)\sin(t)dt \tag{23.17}$$

$$c_2 = \frac{1}{\sqrt{\pi}}\int_{-\pi}^{\pi} f(t)\cos(t)dt \tag{23.18}$$

$$c_{2k-1} = \frac{1}{\sqrt{\pi}}\int_{-\pi}^{\pi} f(t)\sin(kt)\,dt \tag{23.19}$$

$$c_{2k} = \frac{1}{\sqrt{\pi}}\int_{-\pi}^{\pi} f(t)\cos(kt)\,dt \tag{23.20}$$

Substituting these coefficients into the general expression for the expansion of $f(t)$ in terms of these orthonormal basis functions, we find that the Fourier series for the signal, or function, $f(t)$ can be expressed as in Equations 23.21 and 23.22, which may be written as in Equation 23.23, where A_0, A_n, and B_n are appropriately defined.

$$f(t) = \left[\frac{1}{\sqrt{2\pi}}\int_{-\pi}^{\pi} f(t)dt\right]\frac{1}{\sqrt{2\pi}}$$
$$+ \sum_{n=1}^{\infty}\left(\left[\frac{1}{\sqrt{\pi}}\int_{-\pi}^{\pi} f(t)\sin(nt)\,dt\right]\frac{1}{\sqrt{\pi}}\sin(nt)\right. \tag{23.21}$$
$$\left.+\left[\frac{1}{\sqrt{\pi}}\int_{-\pi}^{\pi} f(t)\cos(nt)\,dt\right]\frac{1}{\sqrt{\pi}}\cos(nt)\right)$$

$$f(t) = \frac{1}{2}\left[\frac{1}{\pi}\int_{-\pi}^{\pi} f(t)\,dt\right] + \sum_{n=1}^{\infty}\left(\left[\frac{1}{\pi}\int_{-\pi}^{\pi} f(t)\sin(nt)\,dt\right]\sin(nt)\right.$$
$$\left.+\left[\frac{1}{\pi}\int_{-\pi}^{\pi} f(t)\cos(nt)\,dt\right]\cos(nt)\right) \tag{23.22}$$

$$f(t) = \frac{1}{2}A_0 + \sum_{n=1}^{\infty}\left(A_n\cos(nt) + B_n\sin(nt)\right) \tag{23.23}$$

What we have shown is that if a function $f(t)$ can be represented by a Fourier expansion, then the coefficients of the expansion may be calculated by the methods described. However, the question of convergence of the expansion has not yet been addressed. The necessary and sufficient conditions for the convergence of a Fourier expansion are well known, but it is useful to state a sufficient condition, known as the Dirichlet condition, which states that if $f(t)$ is bounded and of period T and if $f(t)$ has at most a finite number of maxima and minima in one period and a finite number of discontinuities, then the Fourier series for $f(t)$ converges to $f(t)$ at all points where $f(t)$ is continuous, and converges to the average of the right-hand and left-hand limits of $f(t)$ at each point where $f(t)$ is discontinuous.

In many applications, such as the design of filters, it may be necessary to use only a finite number of terms of the Fourier series to approximate a function $f(t)$ over $(0, T_p)$, and it is therefore of much interest to inquire what effect the truncation of the series has (Banks, 1990). The error incurred obviously decreases as one takes into account more and more terms when the function $f(t)$ is continuous. However, in the neighborhood of discontinuities, ripples occur, the magnitude of which remains roughly the same even as more and more terms are included in the Fourier expansion. These ripples are referred to as Gibbs's oscillations.

Generalized Fourier Expansion

As an example of using the generalized Fourier expansion, let us consider representing the function shown in Figure 23.1 in terms of the Legendre polynomials, which form an orthogonal set on the interval $(-1, 1)$. The corresponding normalized basis functions are given by Equations 23.24 through 23.27, where the $P_{n(t)}$ are the Legendre polynomials, which may be generated by the expression of Equation 23.28 or by using the recurrence relation of Equation 23.29.

$$\Phi_0(t) = \frac{1}{\sqrt{2}} \tag{23.24}$$

$$\Phi_1(t) = t\sqrt{\frac{3}{2}} \tag{23.25}$$

$$\Phi_2(t) = \sqrt{\frac{5}{2}}\left(\frac{3}{2}t^2 - \frac{1}{2}\right) \tag{23.26}$$

$$\Phi_n(t) = \left(\frac{2n+1}{2}\right)^{1/2} P_n(t) \tag{23.27}$$

$$P_n(t) = \frac{1}{2^n n!}\frac{d^n}{dt^n}(t^2-1)^n \tag{23.28}$$

$$nP_n(t) = (2n-1)tP_{n-1}(t) - (n-1)P_{n-2}(t) \tag{23.29}$$

The generalized Fourier coefficient is given by Equation 23.30, where $n = 0, 1, 2 \ldots$:

$$c_n = <f|\Phi_n> \ = \int_{-1}^{1} f(t)\Phi_n(t)dt \tag{23.30}$$

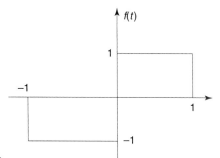

Fig. 23.1   The function $f(t)$.

Using the expressions for Φ_n given before, we obtain Equations 23.31 through 23.34.

$$c_0 = \int_{-1}^{1} \frac{f(t)}{\sqrt{2}} dt = 0 \qquad (23.31)$$

$$c_1 = \int_{-1}^{1} \sqrt{\frac{3}{2}}\, t f(t)\, dt = \sqrt{\frac{3}{2}} \qquad (23.32)$$

$$c_2 = \int_{-1}^{1} \sqrt{\frac{5}{2}} \left(\frac{3}{2} t^2 - \frac{1}{2} \right) f(t) dt = 0 \qquad (23.33)$$

$$c_3 = \int_{-1}^{1} \sqrt{\frac{7}{2}} \left(\frac{5}{2} t^3 - \frac{3}{2} t \right) f(t) dt = \sqrt{\frac{7}{16}} \qquad (23.34)$$

In general, the coefficient c_n is zero when n is even, for this particular example. Hence the function $f(t)$ in Figure 23.1 can be represented in terms of the Legendre polynomials, as in Equation 23.35.

$$\begin{aligned}
f(t) &= \sum_{k=0}^{\infty} c_k \Phi_k(t) \\
&= \frac{3}{2} t + \frac{7}{4\sqrt{2}} \left(\frac{5}{2} t^3 - \frac{3}{2} t \right) + \ldots c_n \Phi_n(t) + \ldots
\end{aligned} \qquad (23.35)$$

We have shown that an arbitrary function, $f(t)$, can be expressed in terms of a super-position of orthogonal basis functions, and we have derived expressions for the generalized Fourier coefficients in terms of the projection of the given function onto the orthogonal basis functions.

If a signal, $x(t)$, repeats itself exactly every T_p seconds, then $x(t)$ may be represented as a linear combination of harmonically related complex exponentials of the form given in Equation 23.36, where the fundamental frequency is given by Equation 23.37.

$$x(t) = \sum_{k=-\infty}^{\infty} c_k\, e^{j 2\pi k f_0 t} \qquad (23.36)$$

$$f_0 = \frac{1}{T_p} \qquad (23.37)$$

Hence one can regard the exponential signals:

$$e^{j2\pi k f_0 t} \quad k = 0, 1, 2, \ldots$$

as building blocks for periodic signals of various forms constructed by choosing the fundamental frequency and the coefficients c_k.

A periodic signal $x(t)$ with period T_p may be represented by a Fourier series where f_0 is selected to be the reciprocal of T_p. The Fourier coefficients c_k are obtained by multiplication of the Fourier representation by the complex exponential $e^{-j2\pi l f_0 t}$, where l is an integer, followed by integration over a single period, either 0 to T_p or more generally t_0 to $t_0 + T_p$, where t_0 is arbitrary (Equation 23.38).

$$
\begin{aligned}
&\int_{t_0}^{t_0+T_p} x(t) e^{-j2\pi l f_0 t} dt \\
&= \int_{t_0}^{t_0+T_p} e^{-j2\pi l f_0 t} \left(\sum_{k=-\infty}^{\infty} c_k e^{j2\pi k f_0 t} \right) dt \\
&= \sum_{k=-\infty}^{\infty} c_k \int_{t_0}^{t_0+T_p} e^{j2\pi f_0 (k-l)t} dt
\end{aligned}
\tag{23.38}
$$

The integral on the r.h.s. is identically zero if $k \neq 1$ and is equal to T_p if $k = 1$, and therefore Equation 23.39 can be obtained.

$$c_k = \frac{1}{T_p} \int_{t_0}^{t_0+T_p} x(t) e^{-j2 p i k f_0 t} dt \tag{23.39}$$

It is now of interest to consider the average power, P, of a periodic signal $x(t)$, which is given by Equation 23.40.

$$P = \frac{1}{T_p} \int |x(t)|^2 dt = \frac{1}{T_p} \int x(t) x^*(t) dt \tag{23.40}$$

Using the Fourier representation of $x(t)$ and the expression found for the Fourier coefficients, we can write Equation 23.41.

$$
\begin{aligned}
P &= \frac{1}{T_p} \int x(t) \left(\sum_{k=-\infty}^{\infty} c_k^* e^{-2j\pi k f_0 t} \right) dt \\
&= \sum_{k=-\infty}^{\infty} c_k^* \left(\frac{1}{T_p} \int x(t) e^{-2j\pi k f_0 t} \right) dt \\
&= \sum_{k=-\infty}^{\infty} c_k^* c_k = \sum_{k=-\infty}^{\infty} |c_k|^2
\end{aligned}
\tag{23.41}
$$

This relationship between the average power and the square of the Fourier coefficients is called Parseval's relation, as mentioned earlier in the context of function representation. The kth harmonic component of the signal has a power $|c_k|^2$, and hence the total average power in a periodic signal is just the sum of the average powers in all the harmonics. If $|c_k|^2$

is plotted as a function of kf_0, the resulting function is called the power density spectrum of the periodic signal $x(t)$. Since the power in a periodic signal exists only at discrete values of frequencies (i.e., Equation 23.42 is satisfied), the signal is said to have a line spectrum, and the frequency spacing between two adjacent spectral lines is equal to the reciprocal of the fundamental period T_p.

$$f = 0, \quad \pm f_0, \quad \pm 2f_0, \ldots \tag{23.42}$$

The Fourier coefficients c_k are in general complex and may be represented as in Equation 23.43, and hence an alternative to the power spectrum representation may be obtained by plotting the magnitude spectrum $|c_k|$ and phase spectrum Θ_k as a function of frequency.

$$c_k = |c_k| e^{j\Theta_k} \tag{23.43}$$

Phase information is therefore lost in using the power spectral density.

For a real valued periodic signal, the Fourier coefficients c_k can easily be shown to satisfy Equation 23.44, which implies that the power spectrum, and hence the magnitude spectrum, are symmetric functions of frequency. However, the phase spectrum is an odd function.

$$c_{-k} = c_k^* \tag{23.44}$$

As an example of this analysis, the Fourier series and power spectral density of the rectangular pulse train shown in Figure 23.2 will be given, and since the signal $x(t)$ is even, we will select the integration interval from $\dfrac{-T_p}{2}$ to $\dfrac{T_p}{2}$, as in Equations 23.45 and 23.46, for $k = \pm 1, = \pm 2, \ldots$, and the power spectral density is obtained by squaring these quantities.

$$c_0 = \frac{1}{T_p} \int_{-\tau/2}^{\tau/2} x(t)\, dt = \frac{A\tau}{T_p} \tag{23.45}$$

$$
\begin{aligned}
c_k &= \frac{1}{T_p} \int_{-\tau/2}^{\tau/2} A e^{-j2\pi k f_0 t}\, dt \\
&= \frac{A}{\pi f_0 k T_p} \frac{e^{j\pi k f_0 \tau} - e^{-j\pi k f_0 \tau}}{2j} = \frac{A\tau}{T_p} \frac{\sin(\pi k f_0 \tau)}{\pi k f_0 \tau}
\end{aligned}
\tag{23.46}
$$

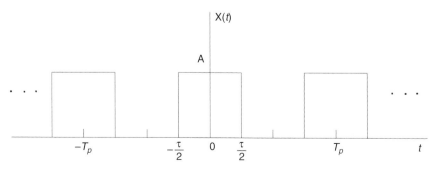

Fig. 23.2 Continuous time periodic train of rectangular pulses.

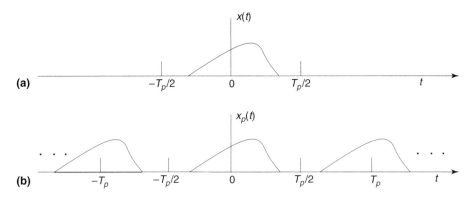

Fig. 23.3   An aperiodic signal and a periodic signal constructed by repeating the periodic signal.

It is now important to introduce the concept of aperiodic signals and transform methods before dealing with discrete data.

Fourier Transforms

Consider an aperiodic signal $x(t)$ that is of finite duration, as in Figure 23.3. A periodic signal $x_p(t)$ can be created from $x(t)$ by translation by fixed amounts, T_p, as shown. This new periodic signal, $x_p(t)$, approaches $x(t)$ in the limit as T_p approaches infinity, and the spectrum of $x_p(t)$ should be obtainable from the spectrum of $x(t)$ in the same limit.

As before, we can obtain Equations 23.47 and 23.48, where $f_0 = 1/T_p$.

$$x_p(t) = \sum_{k=-\infty}^{\infty} c_k\, e^{j2\pi k f_0 t} \tag{23.47}$$

$$c_k = \frac{1}{T_p} \int_{-T_p/2}^{T_p/2} x_p(t) e^{-j2\pi k f_0 t}\, dt \tag{23.48}$$

Between the limits $\pm\dfrac{T_p}{2}$, $x_p(t)$ can be replaced by $x(t)$, and since $x(t)$ is zero outside this range of integration, Equation 23.49 may be obtained.

$$c_k = \frac{1}{T_p} \int_{-\infty}^{\infty} x(t) e^{-j2\pi k f_0 t}\, dt \tag{23.49}$$

A new function, $X(f)$, can now be defined as the Fourier transform of $x(t)$ as in Equation 23.50, and the Fourier coefficient c_k can be written as in Equation 23.51. These equations are samples of $X(f)$ taken at multiples of f_0 and also scaled by the factor f_0 ($= 1/T_p$).

$$X(f) = \int_{-\infty}^{\infty} x(t) e^{-j2\pi f t}\, dt \tag{23.50}$$

$$c_k = \frac{1}{T_p} X(k f_0) \tag{23.51}$$

Therefore, Equation 23.52 can be obtained.

$$x_p(t) = \frac{1}{T_p} \sum_{k=-\infty}^{\infty} \left(\frac{k}{T_p} \right) e^{j2\pi kt/T_p} \tag{23.52}$$

As described previously, we are required to take the limit as T_p approaches infinity, and it is therefore convenient to define a frequency differential $\delta f = 1/T_p$ such that in the limit δf approaches zero. We may therefore write Equation 23.53, and taking the limit as $\delta f \to 0$ we obtain Equations 23.54 and 23.55.

$$x_p(t) = \sum_{k=-\infty}^{\infty} X(k\delta f) e^{j2\pi kt\delta f} \delta f \tag{23.53}$$

$$\lim_{T_p \to \infty} x_p(t) = x(t) = \lim_{\delta f \to 0} \sum_{k=-\infty}^{\infty} X(k\delta f) e^{j2\pi ft} df \tag{23.54}$$

$$x(t) = \int_{-\infty}^{\infty} X(f) e^{j2\pi ft} df \tag{23.55}$$

These equations show that $x(t)$ and $X(f)$ form a so-called Fourier transform pair, and the Fourier transform of $x(t)$ exists if the signal has a finite energy (i.e., as in Equation 23.56).

$$\int_{-\infty}^{\infty} |x(t)|^2 dt < \infty \tag{23.56}$$

Earlier we defined the average power of a periodic signal. It is now possible similarly to define the energy of an aperiodic signal as in Equations 23.57 and 23.58.

$$E = \int_{-\infty}^{\infty} |x(t)|^2 dt = \int_{-\infty}^{\infty} x(t)x^*(t) dt$$
$$= \int_{-\infty}^{\infty} x(t) dt \left(\int_{-\infty}^{\infty} X^*(f) e^{-j2\pi ft} df \right) \tag{23.57}$$

$$E = \int_{-\infty}^{\infty} X^*(f) df \left(\int_{-\infty}^{\infty} x(t) e^{-j2\pi ft} dt \right)$$
$$= \int_{-\infty}^{\infty} |X(f)|^2 df \tag{23.58}$$

Hence the energy of an aperiodic signal can be written as in Equation 23.59, which is also known as Parseval's relation (Wax, 1954).

$$E = \int_{-\infty}^{\infty} |x(t)|^2 dt = \int_{-\infty}^{\infty} |X(f)|^2 df \tag{23.59}$$

Discrete Sequences

In order for a digital computer to manipulate a signal, the signal must be sampled at a chosen sampling rate, $1/T_s$, giving rise to a set of numbers called a sequence. If the continuous signal is $x(t)$, the sampled sequence is represented by $x(nT_s)$, where n is an integer, and the independent variable, t, can represent time or a spatial coordinate, for example. The analysis can, of course, be extended into higher dimensions.

In order to move between the continuous domain and the discrete domain, the idea of a sampling function must be introduced, and from the definition of the Dirac delta function $(\delta(t))$ as in Equation 23.60, it is clear that this function meets the requirement.

$$\int_{-\infty}^{\infty} x(t)\delta(t-\tau)dt = x(\tau) \tag{23.60}$$

Consider an analogue signal $x(t)$ that is continuous in both time and amplitude, and assume that $x(t)$ has infinite duration but finite energy. Let the sample values of the signal $x(t)$ at times $t = 0, \pm T_s, \pm 2T_s, \ldots$, be denoted by the series $x(nT_s)$, $n = 0, \pm 1, \pm 2, \ldots$, where T_s is the sampling period and $f_s = 1/T_s$ is the sampling rate.

The discrete time signal $x_d(t)$ that is obtained by sampling the continuous signal $x(t)$ can then be written as in Equation 23.61, where $\delta(t - nT_s)$ is a Dirac delta function located at time $t = nT_s$ and each delta function is weighted by the corresponding sample value of the input signal $x(t)$.

$$x_d(t) = \sum_{n=-\infty}^{\infty} x(nT_s)\delta(t-nT_s) \tag{23.61}$$

For the case of discrete data, exactly analogous transform methods may be applied as for continuous signals, and we must now discuss the implementation issues for such methods.

The Discrete Fourier Transform

The discrete Fourier transform (DFT) is used extensively in digital signal processing (Burrus and Parks, 1985), and is used routinely for the detection and estimation of periodic signals. The DFT of a discrete-time signal $x(n)$ is defined as in Equation 23.62, where $k = 0, 1, \ldots,$ $N - 1$, and $W_N^{nk} = e^{j2\pi nk/N}$ are the basis functions of the DFT.

$$X(k) = \frac{1}{N}\sum_{n=0}^{N-1} x(n)W_N^{nk} \tag{23.62}$$

These functions are sometimes known as "twiddle factors." The basis functions are periodic and define points on the unit circle in the complex plane. Figure 23.4 illustrates the cyclic property of the basis functions for an eight-point DFT, and the basis functions are equally spaced around the unit circle at frequency increments of F/N, where F is the sampling rate of the input signal sequence. In this figure the cyclic character of the twiddle factors are illustrated as follows:

$$W_8^0 = W_8^8 = W_8^{16} = W_8^{24} = \ldots$$

$$W_8^1 = W_8^9 = W_8^{17} = W_8^{25} = \ldots$$

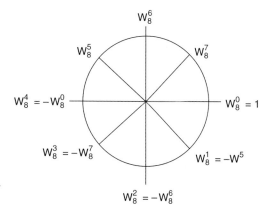

Fig. 23.4 Cyclic properties of the basis functions for an eight-point DFT.

$$W_8^2 = W_8^{10} = W_8^{18} = W_8^{26} = \ldots$$
$$\vdots \quad\quad \vdots \quad\quad \vdots \quad\quad \vdots$$
$$W_8^7 = W_8^{15} = W_8^{23} = W_8^{31} = \ldots$$

The set of frequency samples that define the spectrum $X(k)$ are given on a frequency axis whose discrete frequency locations are given by Equation 23.63, where $k = 0, 1, \ldots, N-1$.

$$f_k = k\frac{F}{N} \tag{23.63}$$

The frequency resolution of the DFT is equal to the frequency increment F/N and is referred to as the bin spacing of the DFT outputs. The frequency response of any DFT bin output is determined by applying a complex exponential input signal and evaluating the DFT bin output response as the frequency is varied.

Consider an input signal given by Equation 23.64; then the DFT of $x(n)$ can be expressed as a function of the arbitrary frequency variable f by Equation 23.65.

$$x(n) = e^{j2\pi fn/F} \tag{23.64}$$

$$X(k) = \frac{1}{N}\sum_{n=0}^{N-1} x(n)W_N^{nk} = \frac{1}{N}\sum_{n=0}^{N-1} e^{j2\pi fn/F}W_N^{nk} \tag{23.65}$$

This summation can be evaluated using the geometric series summation to give Equation 23.66.

$$X(k) = \frac{1}{N}\frac{1 - e^{-j2\pi(k/N - f/F)N}}{1 - e^{-j2\pi(k/N - f/F)}} \tag{23.66}$$

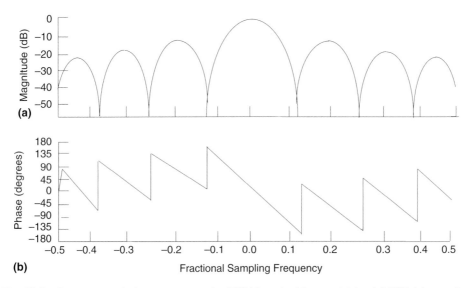

Fig. 23.5   Frequency and phase response of a DFT bin output for an eight-point FFT: (a) magnitude; (b) phase angle.

Defining Ω by Equation 23.67, the DFT of $x(n)$ can be written as in Equation 23.68.

$$\Omega = 2\pi\left(\frac{k}{N} - \frac{f}{F}\right) \tag{23.67}$$

$$X(k) = e^{-j\Omega(N-1)/2} \frac{\sin\dfrac{\Omega N}{2}}{N\sin\dfrac{\Omega}{2}} \tag{23.68}$$

The first term in this expression is the phase of the response, and the ratio of sines is the amplitude response. If k is an integer, then $X(k) = 1$ for these values of k and zero elsewhere. If k is not an integer, then none of the DFT values are zero. This is called spectral leakage. Hence, a unit impulse at the kth frequency location is only obtained when Equation 23.69 is satisfied, where k is an integer.

$$f_k = \frac{kF}{N} \tag{23.69}$$

Figure 23.5 shows the frequency response of a DFT bin output for an eight-point DFT, and the high sidelobe levels are a consequence of truncation.

The Inverse Discrete Fourier Transform

Given the discrete Fourier transform $X(k)$ of the input sequence $x(n)$, the inverse discrete Fourier transform (IDFT) of $X(k)$ is the original time sequence and is given by Equation 23.70.

$$x(n) = \sum_{k=0}^{N-1} X(k)W_n^{-nk} \tag{23.70}$$

The proof is easily obtained by substitution and interchanging the order of summation. The form of the IDFT is the same as the DFT apart from the factor $\frac{1}{N}$ and the sign change in the exponent of the basis functions. Consequently the DFT algorithm can be used to compute these transforms in either direction. It should be noted that the DFT is equal to the z-transform of a sequence, $x(n)$, evaluated at equally spaced inputs on the unit circle in the z-plane (see later).

The Fast Fourier Transform

Special analysis and many other applications often require DFTs to be performed on data sequences in real time and on contiguous sets of input samples. If the input sequence has N samples, the computation of the DFT requires N^2 complex multiplies and $N^2 - N$ complex additions.

The FFT is a fast algorithm for the efficient implementation of the DFT where the number of time samples of the input signal N is transformed into N frequency points, and the required number of arithmetic operations is reduced to the order of $\frac{N}{2}\log_2(N)$.

Several approaches can be used to develop the FFT algorithm.

Starting with the DFT expression, consider factorizing it into two DFTs of length $N/2$ by splitting the input samples into even and odd samples, as in Equations 23.71 and 23.72, where $k = 0, 1, 2, \ldots, N - 1$.

$$X(k) = \frac{1}{N}\sum_{n=0}^{N-1} x(n)W_N^{nk} \tag{23.71}$$

$$X(k) = \frac{1}{N}\sum_{m=0}^{N/2-1} x(2m)W_N^{2mk} + \frac{1}{N}\sum_{m=0}^{N/2-1} x(2m+1)W_N^{(2m+1)k} \tag{23.72}$$

x_1 and x_2 represent the normalized even and odd sample components, as in Equations 23.73 and 23.74, for $m = 0, 1, 2, \ldots (N/2 - 1)$.

$$x_1(m) = \frac{x(2m)}{N} \tag{23.73}$$

$$x_2(m) = \frac{x(2m+1)}{N} \tag{23.74}$$

Therefore, Equation 23.72 may be written as in Equation 23.75, where Equation 23.76 holds and each of the summation terms is reduced to an $N/2$-point DFT.

$$X(k) = \sum_{m=0}^{N/2-1} x_1(m)W_{N/2}^{mk} + W_N^k \sum_{m=0}^{N/2-1} x(m)W_{N/2}^{mk} \tag{23.75}$$

$$W_N^{2n} = W_{N/2}^n \tag{23.76}$$

The general form of the algorithm may be written as in Equation 23.77 and 23.78, where Equations 23.79 and 23.80 hold.

$$X(k) = X_1(k) + W_N^k X_2(k) \tag{23.77}$$

$$X\left(k + \frac{N}{2}\right) = X_1(k) + W_N^{k+N/2} X_2(k) \tag{23.78}$$
$$= X_1(k) - W_N^k X_2(k)$$

$$W_N^{k+N/2} = -W_N^k \tag{23.79}$$

$$W_{N/2}^{m(k+N/2)} = W_{N/2}^{mk} \tag{23.80}$$

Since the DFT output is periodic, Equations 23.81 and 23.82 can be obtained and the form of the algorithm given above is referred to as the decimation in time FFT butterfly. An example for a sixteen-point FFT is shown in Figures 23.6 and 23.7.

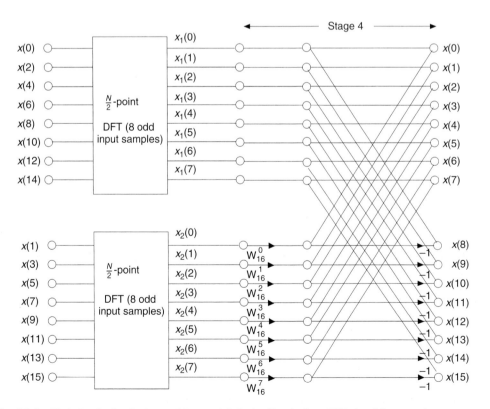

Fig. 23.6   First step in developing a sixteen-point decimation in time FFT signal flow graph.

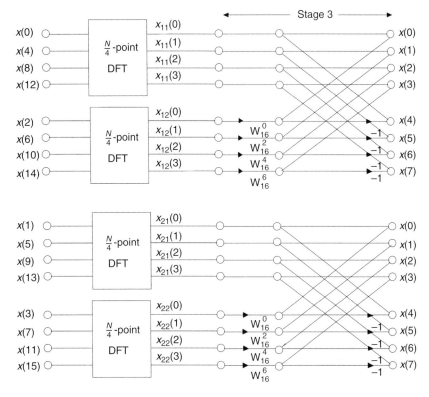

Fig. 23.7 Second step in developing a sixteen-point decimation in time FFT signal flow graph.

$$X_1(k) = X_1\left(k + \frac{N}{2}\right) \tag{23.81}$$

$$X_2(k) = X_2\left(k + \frac{N}{2}\right) \tag{23.82}$$

The decomposition process is repeated until two-point DFTs are generated. Each decomposition is called a stage, and the total number of stages is given by Equation 23.83.

$$M(N) = \log_2 N \tag{23.83}$$

Thus, a sixteen-point DFT requires four stages, as shown in Figure 23.8. The algorithm is now reapplied to each of the $N/2$ sample DFTs.

Assuming that $N/2$ is even, the same process can be carried out on each of the $N/2$-point DFTs to further reduce the computation.

If $N = 2^M$ then the whole process can be repeated M times to reduce the computation to that of evaluating N single-point DFTs.

Linear Time-invariant Digital Systems

The theory of discrete-time, linear, time-invariant systems forms the basis for digital signal processing, and a discrete-time system performs an operation on the input signal according

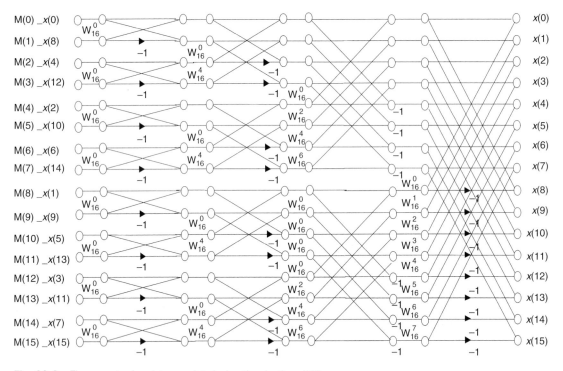

Fig. 23.8 Flow graph of a sixteen-point decimation in time FFT.

to a defined criterion to produce a modified output signal. The input signal $x(n)$ is the system excitation, and $y(n)$ is the response of the system to the excitation, as shown in Figure 23.9 (Kamen, 1990).

The transformation operation can be represented by the operator R. Linear and time-invariant systems can be completely characterized by their impulse response, $h(n)$, which is defined by Equation 23.84.

$$h(n) = R[\delta(n)] \tag{23.84}$$

Once the impulse response is determined, the output of the system for any given input signal is obtained by convolving the input with the impulse response of the system, as in Equation 23.85. A system is linear if and only if the system's response to the sum of two signals, each multiplied by arbitrary scalar constants, is equal to the sum of the system's responses to the two signals taken separately, as in Equation 23.86.

$$y(n) = R[x(n)] = \sum_{k=-\infty}^{\infty} x(k)h(n-k) \tag{23.85}$$

$$y(n) = R[a_1 x_1(n) + a_2 x_2(n)] = a_1 R[x_1(n)] + a_2 R[x_2(n)] \tag{23.86}$$

A system is time invariant if the response to a shifted version of the input is identical to a shifted version of the response based on the unshifted input, and a system operator R is time invariant if Equation 23.87 holds for all values of m.

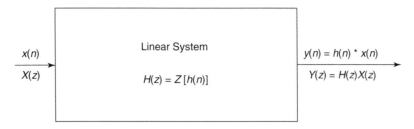

Fig. 23.9 Transfer function relationship for linear systems.

$$R[x(n-m)] = z^{-m}R[x(n)] \qquad (23.87)$$

The operator z^{-m} represents a signal delay of m samples.

The z-transform provides a very powerful method for the analysis of linear, time-invariant discrete systems, which can be represented in terms of difference equations, via the operator R, relating the input signal samples to the output signal samples.

In the following, we will define the properties of the z-transform and then we will apply the methods to the analysis of discrete-time linear time-invariant systems. The z-transform of any data sequence $x(n)$ is defined as in Equation 23.88.

$$X(z) = \sum_{k=-\infty}^{\infty} x_k z^{-k} \qquad (23.88)$$

In this definition, z is a continuous complex variable and $X(z)$ is referred to as the "two-sided z-transform," since both positive and negative values of the index k are allowed.

As an example of a z-transform, let us consider $x(n)$ to be samples of an exponential function such that $x_k = 0$ for $k < 0$ and $e^{-\alpha k}$ for $k \geq 0$ and $\alpha > 0$. The z-transform of this sampled signal is given by Equation 23.89, which is a simple rational function of α and z. Since the signal is zero for $k < 0$, we have a one-sided transform in this case. Table 23.1 shows some common sampled sequences and their z-transforms.

$$X(z) = \sum_{k=0}^{\infty} e^{-\alpha k} z^{-k} = \frac{z}{z - e^{-\alpha}} \qquad (23.89)$$

In linear systems analysis, the concept of a transfer function is fundamental and is defined simply as the transform of the output of the system divided by the transform of the input. The Laplace transform is an important tool that is used in continuous systems theory; for a continuous signal $x(t)$, this transform is defined as in Equation 23.90.

$$X(s) = \int_0^{\infty} x(t)e^{-st}dt \qquad (23.90)$$

If, as before, we sample this continuous signal $x(t)$, we obtain the sampled signal $x_s(t)$ given by Equation 23.91, with $x(t) = 0$ for $t < 0$.

Table 23.1   Commonly Used *z*-transform Pairs

x(n)	X(z)	Region of convergence
$\delta(n)$	1	All z
$u(n)$	$\dfrac{z}{z-1}$	$|z| > 1$
$a^n u(n)$	$\dfrac{z}{z-a}$	$|z| > |a|$
$-a^n u(-n-1)$	$\dfrac{z}{z-a}$	$|z| < |a|$
$na^n u(n)$	$\dfrac{az}{(z-a)^2}$	$|z| > |a|$
$-na^n u(-n)$	$\dfrac{az}{(z-a)^2}$	$|z| < |a|$
$\dfrac{(n+k-1)!}{N!(k-1)} a^n u(n)$	$\dfrac{z^k}{(z-a)^k}$	$|z| > |a|$
$a^{|n|}$	$\dfrac{a^2-1}{a}\dfrac{z}{(z-a)(z-1/a)}$	$|a| < |z| < |a|^{-1}$
$\sin(n\omega)u(n)$	$\dfrac{z\sin\omega}{z^2-2z\cos\omega+1}$	$|z| > 1$
$\cos(n\omega)u(n)$	$\dfrac{z(z-\cos\omega)}{z^2-2z\cos\omega+1}$	$|z| > 1$
$a^n\sin(n\omega)u(n)$	$\dfrac{az\sin\omega}{z^2-2az\cos\omega+a^2}$	$|z| > |a|$

$$x_s(t) = \sum_{n=0}^{\infty} x(t)\delta(t-nT) \tag{23.91}$$

The Laplace transform of the sampled signal $x_s(t)$ is then as in Equation 23.96.

$$\begin{aligned}X_s(s) &= \int_0^\infty \sum_{n=0}^{\infty} x(t)\delta(t-nT)e^{-st}dt \\ &= \sum_{n=0}^{\infty} x(nT)e^{-snT}\end{aligned} \tag{23.92}$$

Making the substitution gives Equation 23.93.

$$z = e^{sT} \tag{23.93}$$

Using the definition of Equation 23.94 gives Equation 23.95 which is the *z*-transform of the sampled sequence $x(nT)$ and is also the Laplace transform of the sampled signal $x_s(t)$.

$$X(z) = X_s(s) \tag{23.94}$$

$$X(z) = \sum_{n=0}^{\infty} x(nT)z^{-n} \tag{23.95}$$

Before deriving an expression for the transfer function and frequency response of a general linear system, a few of the properties of the z-transforms need to be explained.

Linearity

The z-transform of the sum of two sequences multiplied by arbitrary constants is the sum of the z-transforms of the individual sequences, as in Equation 23.96, where Z represents the z-transform operator and $X(z)$ and $Y(z)$ are the z-transforms of the sequences $x(n)$ and $y(n)$, respectively.

$$Z[ax(n) + by(n)] = aX(z) + bY(z) \tag{23.96}$$

Delay Property

Consider a sampled sequence that has been delayed by p samples, and define the delayed sequence to be $x'(nT)$, as in Equation 23.97.

$$x'(nT) = x(nT - pT) \tag{23.97}$$

The z-transform of this delayed sequence is then as in Equation 23.98.

$$X'(z) = \sum_{n=0}^{\infty} x'(n)z^{-n}$$
$$= \sum_{n=0}^{p-1} x'(n)z^{-n} + \sum_{n=p}^{\infty} x'(n)z^{-n} \tag{23.98}$$

This may be written in terms of the undelayed sequence, and with a redefinition of the summation index in the second term, we obtain Equation 23.99.

$$X'(z) = \sum_{n=0}^{p-1} x(n-p)z^{-n} + \sum_{n=0}^{\infty} x(n)^{-(n+p)} \tag{23.99}$$

Therefore, the z-transform of a delayed sequence is given by Equation 23.100.

$$X'(z) = \sum_{n=0}^{p-1} x(n-p)z^{-n} + z^{-p}X(z) \tag{23.100}$$

The first term represents a contribution from any initial conditions, and if $x(n) = 0$ for $n < 0$, the z-transform of a delayed sequence is just $z^{-p}X(z)$, where p is the delay. This is, of course, analogous to the shift theorem for the Laplace transform.

Convolution Summation Property

From the linear systems theory point of view, this represents one of the most valuable properties. Consider the definition of the convolution of an input sequence $x(n)$, with the impulse response $h(k)$ of a linear system, to yield an output $y(n)$, as in Equation 23.101.

$$y(n) = \sum_{k=-\infty}^{\infty} h(k)x(n-k) \qquad \text{(23.101)}$$

Taking the double-sided z-transform of this equation gives Equation 23.102.

$$Y(z) = \sum_{n=-\infty}^{\infty} \left[\sum_{k=-\infty}^{\infty} h(k)x(n-k) \right] z^{-n} \qquad \text{(23.102)}$$

Again, changing the order of summation, and defining $p = n - k$, we obtain Equation 23.103.

$$Y(z) = \sum_{k=-\infty}^{\infty} h(k)z^{-k} \sum_{p=-\infty}^{\infty} x(p)z^{-p} \qquad \text{(23.103)}$$

This can therefore be written as the product of the z-transforms of $h(k)$ and $x(n)$ as in Equation 23.104.

$$Y(z) = H(z)X(z) \qquad \text{(23.104)}$$

Other important properties of the z-transform are shown in Table 23.2. In both signal processing applications and linear systems theory, linear recursive operators are used extensively, examples in both areas being the design of recursive digital filters and feedback systems in control, respectively. The general input-output relationship for such a linear recursive system can be written in the form shown in Equation 23.105.

$$\sum_{k=0}^{p} a_k y(n-k) = \sum_{k=0}^{q} b_k x(n-k) \qquad \text{(23.105)}$$

Taking the z-transform of both sides of the expression and interchanging the order of summation, we obtain Equation 23.106.

$$\sum_{k=0}^{p} a_k \left[\sum_{n=-\infty}^{\infty} y(n-k)z^{-n} \right] = \sum_{k=0}^{q} b_k \left[\sum_{n=-\infty}^{\infty} x(n-k)z^{-n} \right] \qquad \text{(23.106)}$$

Using the shift theorem, this expression can be written as in Equation 23.107.

$$\left[\sum_{k=0}^{p} a_k z^{-k} \right] Y(z) = \left[\sum_{k=0}^{q} b_k z^{-k} \right] X(z) \qquad \text{(23.107)}$$

This is used to define the transfer function, $H(z)$, of the system, which may be written as in Equation 23.108, where $H(z)$ is given by Equation 23.109.

Table 23.2 Properties of the z-transform

Property	Time series	z-transform	Region of convergence
Linearity	$x(n)$	$X(z)$	$r_{cx} < |z| < r_{ax}$
	$y(n)$	$Y(z)$	$r_{cy} < |z| < r_{ay}$
	$ax(n) + by(n)$	$aX(z) + bY(z)$	At least $\max(r_{cx}, r_{cy})$ $< |z| < \min(r_{ax}, r_{ay})$
Time shift	$x(n - m)$	$z^{-m}X(z)$	$r_{cy} < |z| < r_{ax}$
Convolution	$\sum_{k=-\infty}^{\infty} x(k)y(n-k)$	$X(z)Y(z)$	At least $\max(r_{cx}, r_{cy})$ $< |z| < \min(r_{ax}, r_{ay})$
Exponential multiplication	$a_n x(n)$	$X(a^{-1}z)$	$|a| \, r_{cx} < |z| < |a| \, r_{ax}$
Time multiplication	$nx(n)$	$-z\dfrac{dX(z)}{dz}$	$r_{cx} < |z| < r_{ax}$
Product	$x(n)y(n)$	$\dfrac{1}{2\pi j}\oint_C X(w)Y\left(\dfrac{z}{w}\right)w^{-1}dw$	$r_{xc}r_{cy} < |z| < r_{ax}r_{ay}$
Correlation	$\sum_{k=-\infty}^{\infty} x(k)y(n+k)$	$X(z^{-1})Y(z)$	$\max(r_{ax}^{-1}, r_{cy}) < |z|$ $< \min(r_{cx}^{-1}, r_{ay})$
Time transpose	$x(-n)$	$X(z^{-1})$	$r_a^{-1} < |z| < r_c^{-1}$

$$Y(z) = H(z)X(z) \qquad (23.108)$$

$$H(z) = \frac{b_0 + b_1 z^{-1} + \ldots + b_q z^{-q}}{a_0 + a_1 z^{-1} + \ldots + a_p z^{-p}} \qquad (23.109)$$

From the transfer function of a system, the frequency response is found by evaluating the transfer function at $z = e^{j\omega}$, where the angular frequency $\omega = 2\pi f$.

Factorizing the preceding transfer function as in Equation 23.110, where the zeros and poles, z_i and p_i, may be complex numbers, and evaluating this function at $z = e^{j\omega}$, we obtain the associated frequency response as in Equations 23.111 and 23.112.

$$H(z) = \frac{b(z - z_1)(z - z_2)\ldots(z - z_q)}{(z - p_1)(z - p_2)\ldots(z - p_p)} \qquad (23.110)$$

$$H(e^{j\omega}) \frac{b(e^{j\omega} - z_1)(e^{j\omega} - z_2)\ldots(e^{j\omega} - z_q)}{(e^{j\omega} - p_1)(e^{j\omega} - p_2)\ldots(e^{j\omega} - p_p)} \qquad (23.111)$$

$$H(e^{j\omega}) = \frac{b\alpha_1(\omega)\alpha_2(\omega)\ldots\alpha_q(\omega)}{\beta_1(\omega)\beta_2(\omega)\ldots\beta_p(\omega)} \qquad (23.112)$$

The complex functions $\alpha_i(\omega)$ and $\beta_i(\omega)$ are given by Equations 23.113 and 23.114.

$$\alpha_i(\omega) = e^{j\omega} - z_i \qquad (23.113)$$

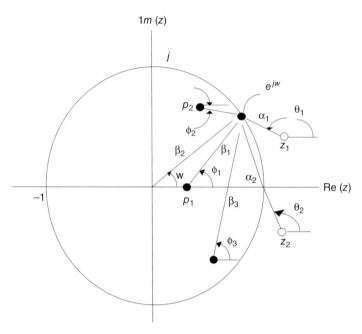

Fig. 23.10   Geometrical construction of the frequency response for a linear system with three poles and two zeros.

$$\beta_i(\omega) = e^{j\omega} - p_i \tag{23.114}$$

A vector interpretation of the frequency response for a linear zeros system with three poles and two zeros is shown in Figure 23.10 and Equations 23.115 and 23.116.

$$|H(e^{j\omega})| = \frac{|b||\alpha_1| * |\alpha_2|}{|\beta_1| * |\beta_2| * |\beta_3|} \tag{23.115}$$

$$\text{angle}\,\{H(e^{j\omega})\} = \Theta_1 + \Theta_2 - \Phi_1 - \Phi_2 - \Phi_3 + \text{angle}\,\{b\} \tag{23.116}$$

The Inverse z-transform

In the section describing the Fourier transform, it was straightforward to relate $x(n)$ to $X(k)$ through the definition of the integral transform, and vice versa. In this section we will consider methods of obtaining the inverse z-transform, that is, $x(nT)$ given $X(z)$.

The formal method is derived using Cauchy's integral theorem. From the definition of the z-transform, multiplying both sides by z^{m-1} and integrating around a closed contour in the z-plane gives Equation 23.117.

$$\oint_C X(z)z^{m-1}\,dz = \oint_C \sum_{n=0}^{\infty} (nT)z^{m-n-1}dz \tag{23.117}$$

If the path of integration encloses the origin, then from Cauchy's integral theorem, the right-hand side is zero except when $m = n$, in which case it is $2\pi j$. Therefore, Equation 23.118 can be obtained, which may be evaluated using the residue theorem.

$$x(nT) = \frac{1}{2\pi j} \oint_C X(z) z^{n-1} dz \qquad (23.118)$$

The remaining methods involve partial fraction expansions and straightforward inversion by division.

Data Truncation

As mentioned before, the Fourier transform of any signal satisfies the periodic relationship, as in Equation 23.119.

$$X(e^{j(\omega+2\pi)}) = X(e^{j\omega}) \qquad (23.119)$$

Due to this periodicity, one can consider the frequency range $-\pi < \omega < \pi$ to be the fundamental set of frequencies.

As seen before, the Fourier transform is of fundamental importance in characterizing the spectral content of signals, be they in the time domain, spatial domain, and so on. However, the Fourier transform requires the entire set of the signal samples, and the question therefore arises concerning the applicability of transform techniques when the data are of finite support. The procedure of filling in the missing data with zeros is a very common method of trying to account for the situation.

As before, the discrete Fourier transform of a finite data set can be written as in Equation 23.120.

$$X_N(e^{j\omega}) = \sum_{n=0}^{N-1} x(n) e^{-j\omega n} \qquad (23.120)$$

Replacing $x(n)$ by its inverse Fourier transform gives Equation 23.121.

$$X_N(e^{j\omega}) = \sum_{n=0}^{N-1} \left[\frac{1}{2\pi} \int_{-\pi}^{\pi} X(e^{jv}) e^{jv} dv \right] e^{-j\omega n} \qquad (23.121)$$

Interchanging the order of the integral and the summation, and using the definition of Equation 23.122, gives Equation 23.123, where * means convolution of the nontruncated Fourier transform $X(e^{j\omega})$ with the rectangular window transform $W_r(e^{j\omega})$.

$$W_r(e^{j\omega}) = \sum_{n=0}^{N-1} e^{-j\omega n}$$

$$= e^{-j\omega(N-1)/2} \frac{\sin(\omega N/2)}{\sin(\omega/2)} \qquad (23.122)$$

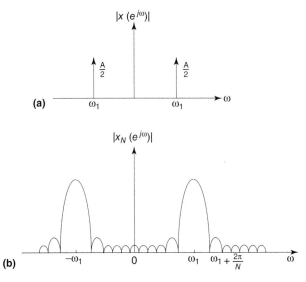

Fig. 23.11 Magnitude of the Fourier transform associated with: (a) an untruncated sinusoid; (b) a truncated sinusoidal time series.

$$X_n(e^{j\omega}) = \frac{1}{2\pi}\int_{-\pi}^{\pi} X(e^{jv})\, W_r(e^{j(\omega-v)})\, dv = \frac{1}{2\pi} X(e^{j\omega}) * W_r(e^{j\omega}) \qquad (23.123)$$

When only a finite amount of data is present, the effect of trying to estimate the underlying Fourier transform manifests in terms of the convolution of the underlying transform with the transform of a rectangular window. This means that in the time domain, the data have been "multiplied" by a window that is unity over the duration of the data and zero outside. As the amount of data increases, the Dirichlet kernel, W_r, tends toward a Dirac delta function. The Dirichlet kernel is plotted as a function of frequency, ω, in Figure 23.11, and it can be seen to consist of many lobes centered at frequencies 0, and $\dfrac{k\pi}{N}$ for $k = 3, 5, 7, \dots$. The main lobe has a width of $\dfrac{4\pi}{N}$ and an amplitude of N, and the side lobes each have a width of $\dfrac{2\pi}{N}$.

To determine the effects of using only a finite amount of data, consider the example where the data consist of a pure sinusoid, as in Equation 23.124.

$$x(n) = A\cos(n\omega_1) \qquad (23.124)$$

The corresponding Fourier transform is given by Equation 23.125 for $-\pi < \omega \le \pi$.

$$X(e^{j\omega}) = \pi A\left[\delta(\omega - \omega_1) + \delta(\omega + \omega_1)\right] \qquad (23.125)$$

The Fourier transform of the truncated sequence is therefore given by Equation 23.126.

$$X_N\left(e^{j\omega}\right) = Ae^{-j(\omega-\omega_1)(N-1)/2} \frac{\sin\left[\dfrac{(\omega-\omega_1)N}{2}\right]}{\sin\left[\dfrac{(\omega-\omega_1)}{2}\right]}$$

$$+ Ae^{j(\omega+\omega_1)(N-1)/2} \frac{\sin\left[\dfrac{(\omega+\omega_1)N}{2}\right]}{\sin\left[\dfrac{(\omega+\omega_1)}{2}\right]}$$

(23.126)

Thus, the effect of using only a small data set is to smear out the original delta function spectrum. This effect is commonly known as spectral leakage.

In attempts to reduce spectral leakage, a variety of so-called data windows have been introduced in order to "smooth out" the ripples introduced by the rectangular window. A few of the window types are triangular, Hanning (or raised cosine), Hamming, Blackman, and so on. For a good introduction to windowing, see Harris (1978). The effects of various data windows are shown in Table 23.3.

Clearly there is a trade-off between side lobe levels and width of the main peak, so caution is needed in using data windows depending on the application at hand. One should also say that there is a school of thought that windowing the data by anything other than a rectangular window distorts the original data, and depending on the application (spectral analysis, for example), model-based signal processing may be more appropriate than just calculating a window-based FFT; again, this will depend on the precise nature of the application (Jaynes, 1987).

Table 23.3 Commonly Used Data Windows

Window	$w(n)$, $0 \leq n \leq (N-1)$	Main lobe width (rad)	$20\log\dfrac{Main\ lobe\ amplitude}{Largest\ side\ lobe\ amplitude}$
Rectangular	1	$\dfrac{4\pi}{N}$	$-13\,\mathrm{dB}$
Bartlett	$\dfrac{2n}{N-1} \quad 0 \leq n \leq \dfrac{N-1}{2}$ $2 - \dfrac{2n}{N-1} \quad \dfrac{N-1}{2} < n \leq N-1$	$\dfrac{8\pi}{N}$	$-27\,\mathrm{dB}$
Hanning	$0.5\left[1-\cos\left(\dfrac{2\pi n}{N}\right)\right]$	$\dfrac{8\pi}{N}$	$-32\,\mathrm{dB}$
Hamming	$0.54 - 0.46\cos\left(\dfrac{2\pi n}{N}\right)$	$\dfrac{8\pi}{N}$	$-43\,\mathrm{dB}$
Blackman	$0.42 - 0.5\cos\left(\dfrac{2\pi n}{N}\right) + 0.8\cos\left(\dfrac{4\pi n}{N}\right)$	$\dfrac{12\pi}{N}$	$-58\,\mathrm{dB}$

Conclusions

In this chapter, we have introduced the concept of orthogonal basis functions and the methods of obtaining the generalized Fourier coefficients that can be used to expand a particular signal of interest in terms of "elementary" basis functions—for example, sines and cosines.

Consideration was given to the representation of continuous signals and functions, and the analysis then considered the case of discrete data sequences.

The Fourier transform was introduced and applied to aperiodic signals and the relationship between the power in a signal and the expansion coefficients was given. Ideas concerning spectral analysis were introduced.

Linear systems theory was briefly touched on, the z-transform was introduced, and the concepts of transfer function and frequency response were discussed.

Fast implementations of the discrete Fourier transform, using the FFT, were discussed, and the ideas of spectral leakage and windowing were mentioned.

There exists a vast literature concerned with function expansion and representation, some references for which are given below.

The subject of transform methods is similarly vast, and it must be appreciated that the Fourier and z-transforms are members of a transform family including radon transforms, Wigner transforms, wavelet transforms, and Hartley transforms (Bracewell, 1986).

References

Banks, S. (1990) *Signal Processing, Image Processing and Pattern Recognition*, Prentice Hall.

Bracewell, R. N. (1986) *The Hartley Transform*, Oxford University Press.

Burrus, C. S., and Parks, T. W. (1985) *DFI/FFT and Convolution Algorithms*, Wiley Interscience.

Harris, F. J. (1978) *Proc. 66, Number 1 On the use of windows for harmonic analysis with the discrete Fourier transform*, IEEE.

Jaynes, E. T. (1987) Bayesian spectrum and Chirp analysis. In *Maximum Entropy and Bayesian Spectral Analysis and Estimation Problems*, Kluwer.

Kamen, E. W. (1990) *Introduction to Signals and Systems*, Maxwell Macmillan.

Walker, J. S. (1988) *Fourier Analysis*, Oxford University Press.

Wax, N. (ed.) (1954) *Noise and Stochastic Processes*, Dover.

24 Information Theory

William J. Fitzgerald

Introduction

In this chapter, the ideas associated with information and entropy are considered and related to probability defined in terms of incomplete knowledge. Mutual and self information are introduced, and these concepts are applied to various models for communications channels.

The concepts of information and related quantity entropy date back to the early days of thermodynamics. The concept of entropy arose in considerations connected with the theory of heat, and its relationship to information was not at first realized. However, in retrospect it is not surprising that the analytical treatment of information should have its origin in this area of physical science, for the subject of thermodynamics is primarily concerned with the determination of the laws that govern the conversion of mechanical and other forms of energy into heat energy. Heat is characterized by its disorder, and ultimately by the irreversibility of certain processes that involve heat transfer.

The degree of organization of a system can also be interpreted as a measure of the quantity of information incorporated in it. This idea was developed by Szilard in 1929, after which time the notion of entropy was another term for information, and the physical measure of entropy was also a measure of the quantity of information or degree of organization of the corresponding physical system. More precisely, the difference in the quantity of information between two states of a physical system is equal to the negative of the corresponding difference in entropy of the two states.

In nature it is found that self-contained systems change from more highly organized structures to less organized ones (i.e., from states of higher information content to states of lower information content).

It is important to note that in the preceding context, a physical system corresponds to many different thermodynamic systems, and entropy is not a property simply of the system but of the experiments undertaken on the system. For example, one normally controls a set of variables for the physical system, and one measures the entropy for that set. A solid with

N atoms has approximately 6*N* degrees of freedom, of which only a few (e.g., pressure, temperature, magnetic field) are usually specified to obtain the entropy. By increasing this set, one obtains a sequence of entropy values, each corresponding to the chosen constraints.

According to Jaynes (1978), the entropy of a thermodynamic system is a measure of the degree of ignorance of a person whose sole knowledge about its microstate consists of the values of the macroscopic quantities (e.g., pressure and temperature, which define its thermodynamic state), and it is a completely objective quantity in the sense that it can be measured in the laboratory.

A very good source of reference concerning both the thermodynamic and the information processing approach to entropy and information can be found in the book *Maxwell's Demon, Entropy, Information, Computing*, edited by Leff and Rex (1990).

Before we concentrate on information theory proper, it will be worth laying a few foundations concerning the role of probability theory in information, and in trying to use probability theory in information theory, one immediately sees that there exist many different views as to what probability theory is, a sample of the various views being:

1. Kolmogorov: the theory of additive measures
2. Jeffreys: the theory of rational belief
3. Fisher: the theory of frequencies in random experiments
4. de Morgan: the calculus of inductive reasoning
5. Laplace: common sense reduced to calculation
6. Bernoulli: the art of conjecture
7. von Mises: the exact science of mass phenomena and repetitive events

For more than one hundred years, controversy has raged between those holding these and other views. However, these views just reflect the particular problems that were being addressed by the authors, and one should take the position that the views listed are valid and useful in different contexts but that some are more general than others, and it is this general approach that will now be discussed before we apply it to information theory.

According to the view put forward by Jeffreys, probability expresses a state of knowledge about any system under investigation and, as such, applies to a very wide range of problems. This approach to probability theory incorporates many of the just mentioned interpretations.

It should not be considered wrong to adopt a narrow view of probabilities in terms of random experiments and frequencies of occurrence, but there is certainly nothing to be lost in adopting a broader view in terms of states of knowledge, and if the problem requires the concept of relative frequencies and so forth, this should emerge out of the broader formulation.

Consider propositions *EA* and *EB*. Using Boolean algebra, we may construct new propositions from *A* and *B* by conjunction, disjunction, and negation, as in Equations 24.1 through 24.3.

$$AB = \text{Both } A \text{ and } B \text{ are true} \tag{24.1}$$

$$A + B = \text{At least one of the propositions is true} \tag{24.2}$$

$$\tilde{A} = A \text{ is false} \tag{24.3}$$

The conditional probability that *A* is true given that *B* is true is a real number between 0 and 1, and is represented by the symbol $P(A \mid B)$. Therefore, $p(A + B \mid CD)$ is the probability that at least one of the propositions (or hypotheses) A, B is true, given that both C and D are true, and so on.

In this formulation of probability as a state of knowledge, or a representation of incomplete knowledge, all probabilities are conditional on some information, and there does not exist an absolute probability.

The rules for conducting scientific inference follow from the simple laws of probability, as in Equations 24.4 and 24.5.

$$P(AB \mid I) = p(A \mid I)p(B \mid AI) = p(B \mid I)p(A \mid BI) \tag{24.4}$$

$$p(A \mid B) + p(\tilde{A} \mid B) = 1 \tag{24.5}$$

If $p(B \mid I) \neq 0$, then Bayes's theorem follows, as in Equation 24.6, where I stands for any prior information.

$$p(A \mid BI) = \frac{p(A \mid I)p(B \mid AI)}{p(B \mid I)} \tag{24.6}$$

Using Bayes's theorem, it is possible to incorporate chains of evidence into one's reasoning, the posterior probability becoming the prior for the next iteration as new data become available. This method of inference goes back to Laplace in the eighteenth century, but the mathematics necessary to show that Bayesian inference is the only logical way to proceed was given in 1946 by R. T. Cox. However, the formalism does not show us how to assign the initial probabilities. There are several methods that may be used to assist in the initial assignment of probabilities, the first one to consider being symmetry. For example, if one considers coins or dice that are unbiased, it would seem reasonable to express this state of knowledge by assigning equal probabilities to the allowed states, and this allows us to make the best predictions we can from our state of incomplete knowledge. This is referred to as Laplace's law of insufficient reason.

Boltzmann in 1877 wished to determine how gas molecules distribute themselves in a conservative field such as gravity. He divided the 6-d "phase space," 3 spatial dimensions and 3 momentum dimensions, into equal cells, with N molecules present in the ith cell. The number of ways this distribution can be realized is as in Equation 24.7.

$$W = \frac{N!}{N_1! N_2! \ldots N_n!} \tag{24.7}$$

The prior knowledge that Boltzmann was able to use was that the total energy and the total number of molecules in the gas were constant, as in Equations 24.8 and 24.9.

$$N = \sum_i N_i = \text{constant} \tag{24.8}$$

$$E = \sum_i N_i E_i = \text{constant} \tag{24.9}$$

Boltzmann considered that the most probable distribution for this case was the one that maximizes W subject to the constraints just given.

If the total number of gas molecules is large, then Striling's approximation gives Equation 24.10, which may be seen to be related to the Shannon entropy.

$$\frac{1}{N}\log W = -\sum_i \left(\frac{N_i}{N}\right)\log\left(\frac{N_i}{N}\right) \tag{24.10}$$

Thus, the most likely distribution is obtained by maximizing entropy subject to the constraints of the problem. This methodology has, over the years, been applied to a vast range of problems with remarkable success.

Information theory says that a random variable, X, which has an associated probability density function, $p(x)$, has an entropy given by Equation 24.11, which means that H bits are sufficient to describe X on the average.

$$H = -\sum p(x)\log p(x) \tag{24.11}$$

Kolmogorov similarly described algorithmic complexity, $K(x)$, to be the shortest binary program that describes X. According to Kolmogorov, information theory must precede probability theory and should not be based on it, and his contribution to information theory was followed by a direct development of his ideas in algorithmic complexity. In particular, he was interested in finding any determinism in random events, and his work on turbulence must be seen in this light as an attempt to find deterministic order in chaotic processes, and indeed turbulence was one of the key phenomena that motivated the resurgence of interest in nonlinear dynamical systems and the range of chaotic phenomena and strange attractors.

It is interesting that many of the ideas borrowed from information theory are used to classify many of the chaotic signals, such as heart beats, brain waves, chemical reactions, lasers, flames, and dimensions, entropies, and Lyapunov exponents are used routinely.

Information Capacity of a Store

Figure 24.1(a) shows a system with two possible states. If the position of the points is unknown, a priori, and we learn that the point is in the left-hand box, say, we gain

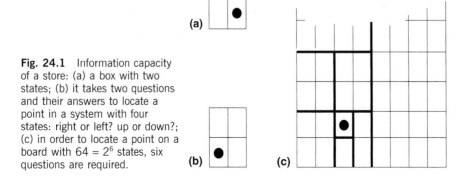

Fig. 24.1   Information capacity of a store: (a) a box with two states; (b) it takes two questions and their answers to locate a point in a system with four states: right or left? up or down?; (c) in order to locate a point on a board with $64 = 2^6$ states, six questions are required.

(a)

(b)

(c)

information amounting to 1 bit. If we obtain this information, we save one question in order to locate the point. Hence the maximum information content of a system with two states is 1 bit.

For a box with two states, one needs two questions in order to locate the point (such as, is the point to the right or left and up or down?) and the maximum amount of information is 2 bits. This can, of course, be written as the logarithm to the base 2 of the number of allowable states. In general, the maximum information content associated with a system having N states is given by Equation 24.12.

$$I = \log_2 N \qquad (24.12)$$

We can also consider the outcome of statistical events and calculate the average gain of information associated with such events. As an example, consider coin tosses such that the outcome, heads or tails, occurs with equal probability of $1/2$.

The information gain acquired by learning that the outcome of an experiment is heads, say, is given by Equation 24.13.

$$I = -\left(\frac{1}{2} \log_2 \frac{1}{2} + \frac{1}{2} \log_2 \frac{1}{2} \right) = 1 \qquad (24.13)$$

In general, therefore, the information gain associated with many states having probabilities associated with each state is given by Equation 24.14.

$$I = -\sum_i p_i \log_2 p_i \qquad (24.14)$$

Information and Thermodynamics

Boltzmann realized that entropy was a measure of disorder of the configuration of the system being considered. This viewpoint was extended by Szilard in 1929, who identified entropy with information, and the measure of information with the negative of the measure of entropy. This laid the foundations of information theory.

This viewpoint came about due to a paradox conceived by Maxwell concerning two chambers separated by a common partition that could be removed to allow objects (gas molecules) to move freely between the two chambers. Normal experience tells us that if one of the chambers initially has gas present and the partition is withdrawn, then gas will rapidly diffuse to fill both chambers. The reverse never occurs.

Maxwell proposed his famous "demon" whose purpose was to open a small frictionless door present in the partition, and when due to thermal motion a particle heads toward the door, the demon opens the door to let the particle through the partition, in one direction only, and as time passes, one chamber gets filled up with particles at the expense of the other, thus seeming to violate the second law of thermodynamics. This paradox was explained by Szilard, who pointed out that in order for the demon to perform, it had to be very well informed about the position and velocity of the particle as it approached the door. Szilard's argument is very well given in the book by Leff and Rex (1990), and it has led to the realization of the connection between entropy and information.

Entropy of Finite Schemes

Consider a complete system of events $A_1, A_2, \ldots, A_n$ such that one and only one event may happen at a given trial (e.g., tossing a coin), for which $n = 2$, and the events are mutually exclusive.

Every finite scheme describes a state of uncertainty, and we only know the probability of possible outcomes (i.e., a one-in-six chance of a particular face of a die being uppermost—for a fair die).

If for one scheme the particular probabilities are different, then there will be different degrees of uncertainty associated with states having different probabilities. It is therefore essential to derive a measure of uncertainty of a particular scheme, and the quantity given in Equation 24.15 is a very suitable measure.

$$H(p_1, p_2, \ldots p_n) = -\sum_{i=1}^{n} p_i \log p_i \tag{24.15}$$

The logarithm is taken to be an arbitrary base, and $p_i \log p_i = 0$ if $P_i = 0$. The quantity H is called the entropy of the finite scheme.

Around the time Szilard was considering Maxwell's demon and the connection between entropy and information, the whole of physics was undergoing vast changes due to the development of quantum theory. Of particular interest is the work of J. von Neumann, who in 1932 showed that in the process of making measurements, the observer must be taken into account, and this gathered together ideas of entropy, information, and irreversibility.

It was R. V. Hartley (1928), of Bell Telephone Laboratories, who gave the logarithmic definition of information from a communications viewpoint, and it was C. Shannon who further refined the ideas in 1949.

Suppose that X is a random variable or random vector that can be described by its probability density function $p(x)$, where x takes values in a certain alphabet, A_x. Shannon (1949) defined the entropy of X by Equation 24.16.

$$H = -\sum_{x \in A_x} p_x(x) \log p_x(x) \tag{24.16}$$

Similarly, given two discrete random variables X and Y, one can define various entropies $H(X, Y)$, $H(X)$, $H(Y)$ as well as the conditional entropy, as in Equation 24.17 (see the next section).

$$H(X \mid Y) = H(X, Y) - H(Y) \tag{24.17}$$

It should also be mentioned that the statistician R. A. Fisher in 1922 defined the information content of a statistical distribution, and this Fisher information still plays a fundamental role in statistical estimation and parameter estimation in general, and allows one to put bounds on the accuracy with which one can estimate parameters.

Mutual and Self Information

Let X and Y be two discrete variables with possible outcomes: x_i, $i = 1, 2, \ldots, n$, and y_i, $i = 1, 2, \ldots, m$, respectively. If an outcome $Y = y_j$ is observed, and we wish to quantitatively determine the amount of information that this observation provides about the event $X = x_i$, $i = 1, 2, \ldots, n$, then the appropriate measure of information has to be selected. If X and Y are statistically independent, then the occurrence of $Y = y_j$ provides no information about the occurrence of $X = x_i$. If, however, the random variables are fully dependent on one another, then the outcome of one determines the outcome of the other. An obvious measure that satisfies these "boundary conditions" is that of the logarithm of the ratio of the conditional probability $P(x_i \mid y_i)$ divided by $P(x_i)$, and this defines the mutual information $I(x_i, y_j)$ as between x_i and y_j, as in Equation 24.18.

$$I(x_i, y_i) = \log \frac{P(x_i \mid y_i)}{P(x_i)} \tag{24.18}$$

When the random variables are statistically independent, Equation 24.19 and therefore Equation 24.20 are obtained.

$$P(x_i \mid y_i) = P(x_i) \tag{24.19}$$

$$I(x_i, y_j) = 0 \tag{24.20}$$

However, when the occurrence of event $Y = y_j$ uniquely determines the occurrence of event $X = x_i$, the conditional probability above is unity. Therefore, Equation 24.21 is obtained, which is called the self information of the event $X = x_i$.

$$I(x_i, y_j) = \log \frac{1}{P(x_i)} = -\log P(x_i) \tag{24.21}$$

It should be noted that an event that occurs with high probability conveys less information than a low-probability event.

Consider the expression given for the mutual information; since we can write Equation 24.22, then the mutual information obeys the symmetry relation given in Equation 24.23 by virtue of Bayes's theorem.

$$\frac{P(x_i \mid y_j)}{P(x_i)} = \frac{P(x_i \mid y_j)P(y_j)}{P(x_i)P(y_j)} = \frac{P(x_i, y_j)}{P(x_i)P(y_j)} = \frac{P(y_j \mid x_i)}{P(y_j)} \tag{24.22}$$

$$I(x_i, y_j) = I(y_j, x_i) \tag{24.23}$$

The average value of the mutual information defined before is given by Equation 24.24.

$$I(X, Y) = \sum_{i=1}^{n} \sum_{j=1}^{m} P(x_i, y_j) I(x_i, y_j) = \sum_{i=1}^{n} \sum_{j=1}^{m} P(x_i, y_j) \log \frac{P(x_i, y_j)}{P(x_i)P(y_j)} \tag{24.24}$$

Likewise, the average value of the self information, $H(X)$, is defined by Equation 24.25.

$$H(X) = \sum_{i=1}^{n} P(x_i)I(x_i) = -\sum_{i=1}^{n} P(x_i)\log P(x_i) \qquad \textbf{(24.25)}$$

For any given communication source transmitting an "alphabet" of symbols, the entropy of the source is a maximum when the output symbols are equally probable, as is clear since, if the symbols are equally probable, Equations 24.26 and 24.27 and in general Equation 24.28 are obtained.

$$P(x_i) = \frac{1}{n} \quad \text{for all } i \qquad \textbf{(24.26)}$$

$$H(X) = -\sum_{i=1}^{n} \frac{1}{n} \log \frac{1}{n} = \log n \qquad \textbf{(24.27)}$$

$$H(X) \le \log n \qquad \textbf{(24.28)}$$

It is very straightforward to show that Equation 24.29 follows with equality if and only if X and Y are independent, and that Equation 24.30 is true where $H(X \mid Y)$ is a conditional uncertainty, and so on.

$$H(X, Y) \le H(X) + H(Y) \qquad \textbf{(24.29)}$$

$$H(X, Y) = H(X) + H(Y \mid X) = H(Y) + H(X \mid Y) \qquad \textbf{(24.30)}$$

These definitions of mutual information, self information, and entropy for discrete random variables may be easily extended to the case of continuous random variables, and if the joint probability density functions and the marginal probability density function are $p(x, y)$, $p(x)$, and $p(y)$, respectively, then the average mutual information between X and Y (or the cross entropy) is given by Equation 24.31, and the entropy of X in the continuous case is given by Equation 24.32, where $m(x)$ is an appropriate measure function (Jaynes, 1989).

$$I(X, Y) = \int_{-\infty}^{\infty} \int_{-\infty}^{\infty} p(x)p(y \mid x) \log \frac{p(y \mid x)P(x)}{p(x)p(y)} dx\, dy \qquad \textbf{(24.31)}$$

$$H(X) = -\int_{-\infty}^{\infty} p(x) \log \frac{p(x)}{m(x)} dx \qquad \textbf{(24.32)}$$

Discrete Memoryless Channels

The discrete memoryless channel serves as a statistical model with an input X and an output Y, which is a noisy version of X, and both X and Y are random variables.

The channel is said to be *discrete* when both the alphabets from which X and Y are selected have finite size, not necessarily the same size, and the channel is said to be *memory-less* when the current output symbol depends only on the current input symbol and not on any previous ones (c., Markov models and ARMA processes, and so on).

In terms of the elements of the input and output alphabets, one can define a set of transition probabilities for all j and k, as in Equation 24.33, and this is just the conditional probability that the channel output is $Y = y_k$ given that the channel input is $X = x_j$.

$$p(y_k \mid x_j) = P(Y = y_k \mid X = x_j) \tag{24.33}$$

Due to transmission errors, when $k = j$, the transition probability represents the conditional probability of correct reception, and when $k \neq j$ it the conditional probability of error. Just as for Markov processes, one can define a transition probability matrix as follows.

$$P = \begin{bmatrix} p(y_0 \mid x_0) & p(y_1 \mid x_0) & \cdots & p(y_{K-1} \mid x_0) \\ p(y_0 \mid x_1) & p(y_1 \mid x_1) & \cdots & p(y_{K-1} \mid x_1) \\ \vdots & \vdots & \vdots & \vdots \\ p(y_0 \mid x_{J-1}) & p(y_1 \mid x_{J-1}) & \cdots & p(y_{K-1} \mid x_{J-1}) \end{bmatrix}$$

This is a $J \times K$ matrix, called the channel matrix. The sum of elements along any row is given by Equation 24.34 for all j.

$$\sum_{k=0}^{K-1} p(y_k \mid x_j) = 1 \tag{24.34}$$

If the inputs to a discrete, memoryless channel are selected according to the probability distribution $p(x_j)$, $j = 0, 1, \ldots, J - 1$, then the event $X = x_j$ occurs with probability given by Equation 24.35.

$$p(x_j) = P(X = x_j) \tag{24.35}$$

The joint probability distribution of the random variables X and Y is given by Equation 24.36, and the marginal distribution of the output variable Y is obtained by integrating out, or averaging out, the dependence on x_j for $(j = 0, 1, \ldots, J - 1)$, as in Equation 24.37 for $k = 0, 1, \ldots, K - 1$.

$$\begin{aligned} p(x_j, y_k) &= P(X = x_j, Y = y_k) \\ &= P(Y = y_k \mid X + x_j)P(X = x_j) = p(y_k \mid x_j)p(x_j) \end{aligned} \tag{24.36}$$

$$p(y_k) = \sum_{J=0}^{J=1} p(y_k \mid x_j)p(x_j) \tag{24.37}$$

Equation 24.37 is very important, since if we are given the input a priori probabilities $p(x_j)$ and the channel matrix, then we may obtain the probabilities of the various output symbols from the previous equation.

Channel Capacity

Regardless of whatever means, or channels, are used for transmission, there is a maximum rate of transmission, called the "capacity" of the channel, which is determined by the intrinsic properties of the channel and is independent of the content of the transmitted information and the way it is encoded. This is measured in bits per second. As an example, in order to transmit a color television picture, a channel with a capacity of about 200 million bits per second is required.

For a discrete memoryless channel with transition probabilities $p(y_k \mid x_j)$ as before, the average mutual information between the output and the input is given by Equation 24.38.

$$I(X, Y) = \sum_{j=0}^{J-1} \sum_{k=0}^{K-1} p(x_j, y_k) \log \frac{p(y_k \mid x_j)}{p(y_k)} \tag{24.38}$$

Also as before, Equations 24.39 and 24.40 may be obtained.

$$p(x_j, y_k) = p(y_k \mid x_j) p(x_j) \tag{24.39}$$

$$p(y_k) = \sum_{j=0}^{J-1} p(y_k \mid x_j) p(x_j) \tag{24.40}$$

From the expression for the average mutual information, it is seen therefore that the average mutual information depends on the channel characteristics expressed in terms of the elements of the channel matrix and also on the input probability distribution, which is clearly independent of the channel. Therefore, by changing the input probability distribution, the average mutual information will change, and we can define the "channel capacity" in terms of the maximum average mutual information with respect to the input probability distribution.

We can therefore write the channel capacity as in Equation 24.41.

$$C = \max I(X, Y) \tag{24.41}$$

The calculation of C is a constrained optimization, since the constraints of Equations 24.42 and 24.43 have to apply to $p(x_j)$.

$$p(x_j) = 0 \tag{24.42}$$

$$\sum_{j=1}^{J-1} p(x_j) = 1 \tag{24.43}$$

This may be compared to the method employed for the maximum entropy analysis of images and inverse problems (Jaynes, 1989; Skilling, 1988).

The channel capacity is an extremely important quantity, since it is possible to transmit information through a channel at any rate less than the channel capacity with an arbitrary small probability of error; completely reliable transmission is not possible if the information processed is greater than the channel capacity. However, in general the calculation of the channel capacity is a difficult problem.

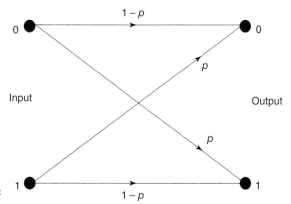

Fig. 24.2 Binary symmetric channel.

Before calculating the channel capacity for some simple channel models, it is useful to introduce certain classes of channel, as follows:

1. A channel is lossless if $H(X \mid Y) = 0$ for all input distributions, which means that the input is determined by the output, and hence no transmission errors can occur.
2. A channel is deterministic if $p(y_j \mid x_i) = 1$ or 0 for all i, j, which means that Y is determined by X, and hence $H(Y \mid X) = 0$ for all input distributions.
3. A channel is noiseless if it is lossless and deterministic.
4. A channel is useless or zero capacity if $I(X \mid Y) = 0$ for all input distributions.
5. A channel is symmetric if each row of the transition matrix contains the same set of numbers, and if each column contains the same set of numbers (different in general for the row set).

It is a consequence of the definition of the symmetric channel that $H(Y \mid X)$ is independent of the input distribution $p(x)$, and depends only on the channel probabilities $p(y_j \mid x_i)$.

The best-known model for a symmetric channel is the binary symmetric channel shown in Figure 24.2. As an example of the calculation of the channel capacity, a closed-form expression for a symmetric channel will be given.

Let us consider a symmetric channel with input alphabet $x_1, \ldots, x_M$ and output alphabet $y_1, \ldots, y_L$, and a channel matrix with row probabilities $p_1', \ldots, p_L'$ and column probabilities $q_1', \ldots, q_M'$.

Since $H(Y \mid X)$ does not depend on the input distribution, the problem of maximizing the information reduces to the problem of maximizing the output uncertainty $H(Y)$, as in Equation 24.44.

$$I(X \mid Y) = H(Y) - H(Y \mid X) \qquad \textbf{(24.44)}$$

It is known that Equation 24.45 holds with equality if and only if all values of Y are equally likely.

$$H(Y) \leq \log L \qquad \textbf{(24.45)}$$

Therefore, if an input distribution can be found for which all values of Y have the same probability, then that input distribution would maximize $I(X \mid Y)$. The uniform distribution does this, since Equation 24.46 holds, and for a uniform distribution Equation 24.47 holds.

$$p(y_j) = \sum_{i=1}^{M} p(x_i)p(y_j \mid x_i) = \frac{1}{M} \sum_{i=1}^{M} p(y_j \mid x_i) \qquad (24.46)$$

$$p(x_i) = \frac{1}{M} \quad \text{for all } i \qquad (24.47)$$

However, the term $\sum_{i=1}^{M} p(y_j \mid x_i)$ is the sum of the entries in the jth column of the channel matrix, and since the channel is symmetric, Equation 24.48 is obtained, which is independent of j, and hence $p(y_j)$ does not depend on j; equivalently all values of Y have the same probability.

$$\sum_{i=1}^{M} p(y_j \mid x_i) = \sum_{k=1}^{M} q_k' \qquad (24.48)$$

The maximum, given by Equation 24.49, is therefore attainable and the channel capacity can be written as in Equation 24.50 since Equation 24.51 is also true.

$$H(Y) = \log L \qquad (24.49)$$

$$C_{sym} = \log L + \sum_{j=1}^{L} p_j' \log p_j' \qquad (24.50)$$

$$H(Y \mid X) = -\sum_{j=1}^{L} p_j' \log p_j' \qquad (24.51)$$

For the binary symmetric channel shown in Figure 24.2, the channel capacity can be written as in Equation 24.52, and the variation with β is shown in Figure 24.3.

$$C_{BSC} = \log 2 + \beta \log \beta + (1-\beta) \log (1-\beta) = 1 - H(\beta, 1-\beta) \qquad (24.52)$$

Source Encoding

An important problem in communications is how to efficiently represent data generated by a discrete source. This is called source encoding. For a source encoder to be efficient, one needs knowledge of the statistics of the source. For example, if some source symbols are known to be more probable than others, then this feature may be exploited by assigning short code words to frequent source symbols and vice versa (e.g., the Morse code).

The average length of a code word, $\bar{L}$ can be written as in Equation 24.53.

$$\bar{L} = \sum_{k=1}^{K} p_k l_k \qquad (24.53)$$

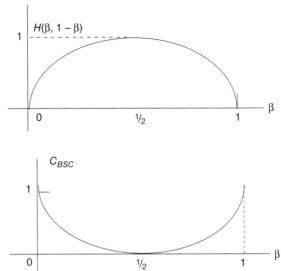

Fig. 24.3 Capacity of a binary symmetric channel.

This represents the average number of bits per source symbol used in the source encoding process.

If L_{min} represents the minimum possible value of $\overline{L}$, the coding efficiency of the source encoder is defined as in Equation 24.54 and the source encoder is said to be efficient when η approaches unity.

$$\eta = \frac{L_{min}}{\overline{L}} \tag{24.54}$$

One question of importance is how to determine the minimum-length code. The answer to this is given in Shannon's first theorem, also called the source coding theorem. This states that given a discrete, memoryless source of entropy $H(X)$, then the average code-word length $\overline{L}$ for any source encoding is bounded as in Equation 24.55.

$$\overline{L} \geq H(X) \tag{24.55}$$

Therefore, the entropy $H(X)$ represents a fundamental limit on the average number of bits per source symbol necessary to represent a discrete memoryless source. Thus Equation 24.56 can be obtained and the efficiency of the source encoder may be written in terms of the entropy, as in Equation 24.57.

$$L_{min} = H(X) \tag{24.56}$$

$$\eta = \frac{H(X)}{\overline{L}} \tag{24.57}$$

A theorem that is complementary to Shannon's first theorem and applies to a channel in which the noise is Gaussian is known as the Shannon–Hartley theorem, which states that the channel capacity of a white, band-limited Gaussian channel is given by Equation 24.58, where B is the channel bandwidth, S the signal power, and N is the total noise within the channel bandwidth.

$$C = B \log_2\left(1 + \frac{S}{N}\right) \quad \text{bits/s} \tag{24.58}$$

This theorem, although restricted to Gaussian channels, is of fundamental importance, since many channels are approximately Gaussian, and the results obtained by assuming a Gaussian channel often provide a lower bound on the performance of a system operating over a non-Gaussian channel. The Shannon-Hartley theorem indicates that a noiseless Gaussian channel $\left(\frac{S}{N} = \infty\right)$ has an infinite capacity. However, when noise is present in the channel, the capacity does not approach infinity as the bandwidth is increased, since the noise power increases as the bandwidth increases, and the channel capacity reaches a finite upper limit with increasing bandwidth if the signal power remains constant.

This limit is easily calculated using the fact that the total noise within the channel bandwidth is given by Equation 24.59, where $\frac{\eta}{2}$ is the two-sided, power-spectral density. The Shannon–Hartley theorem may now be written as in Equations 24.60 and 24.61.

$$N = \eta B \tag{24.59}$$

$$C = B \log_2\left(1 + \frac{S}{\eta B}\right) = \left(\frac{S}{\eta}\right)\left(\frac{\eta B}{S}\right) \log_2\left(1 + \frac{S}{\eta B}\right) \tag{24.60}$$

$$C = \frac{S}{\eta} \log_2\left(1 + \frac{S}{\eta B}\right)^{\eta B/S} \tag{24.61}$$

Using Equation 24.62 gives Equation 24.63.

$$x = \frac{S}{\eta B} \tag{24.62}$$

$$\lim_{B \to \infty} C = \frac{S}{\eta} \log_2 e = 1.44 \frac{S}{\eta} \tag{24.63}$$

The Shannon–Hartley theorem allows for a trade-off between bandwidth and signal-to-noise ratio in the following sense. If, for example, the signal-to-noise ratio $S/N = 7$, and $B = 4\,\text{kHz}$, we obtain $C = 12 \times 10^3$ bits/s. If the signal-to-noise ratio is increased to 15, and the bandwidth increased to $3\,\text{kHz}$, the channel capacity remains the same.

We have seen that in the case of a noiseless channel, the capacity is infinite, and no matter how restricted the bandwidth, it is always possible to receive a signal without error, as predicted by the Shannon limit. The question of how in practice one achieves this performance is complex, one method being the use of orthogonal signals.

Let us now consider the problem of the reliable transmission of messages through a noisy communications channel. We therefore need the best decoding scheme to enable us to obtain the input sequence after seeing the received symbols.

Assume the input alphabet $x_1, \ldots, x_M$, the output alphabet $y_1, \ldots, y_L$, and the channel matrix $p(y_j \mid x_i)$ as before, and consider the special case where a sequence of symbols, chosen at random according to a known distribution, $p(x)$, is transmitted through the channel. For this given distribution $p(x)$, we wish to construct a decision scheme that minimizes the overall probability of error. Such a scheme is called an ideal observer.

Consider a sequence $x = (\alpha_1, \ldots, \alpha_n)$ chosen in accordance with the distribution $p(x)$. The probability that the output sequence $\beta_1, \ldots, \beta_n$ is produced is given by Equation 24.64, which may also be written as Equation 24.65.

$$p(\beta_1, \ldots, \beta_n) = \sum_{\alpha_1, \ldots, \alpha_n} p(\alpha_1, \ldots, \alpha_n) \times p(\beta_1, \ldots, \beta_n \mid \alpha_1, \ldots, \alpha_n) \tag{24.64}$$

$$p(\beta_1, \ldots, \beta_n) = \sum_{\alpha_1, \ldots, \beta_n} p(\alpha_1, \ldots, \alpha_n) \times p(\beta_1 \mid \alpha_1(p(\beta_2 \mid \alpha_2) \ldots p(\beta_n \mid a_n))) \tag{24.65}$$

A decoder scheme may be defined as a function that assigns to each output sequence $(\beta_1, \ldots, \beta_n)$ an input sequence $(\alpha_1, \ldots, \alpha_n)$ and the ideal observer is the scheme that minimizes the overall probability of error for the given input distribution, and this is found by maximizing the conditional probability, as given by Equation 24.66.

$$p(\alpha_1, \ldots, \alpha_n \mid \beta_1, \ldots, \beta_n) = \frac{p(\alpha_1, \ldots, \alpha_n) \prod_{k=1}^{n} p(\beta_k \mid \alpha_k)}{p(\beta_1, \ldots, \beta_n)} \tag{24.66}$$

When all the inputs are equally likely, then Equation 24.67 may be obtained and for a fixed y, maximizing $p(x_i \mid y)$ is equivalent to maximizing the inverse probability $p(y \mid x_i)$. This is referred to as the maximum likelihood decision scheme.

$$p(x_i \mid y) = \frac{p(x_i)p(y \mid x_i)}{p(y)} = \frac{1}{Mp(y)} p(y \mid x_i) \tag{24.67}$$

Discrete Channels with Memory

Previously, we described channels that have no memory, that is, channels in which the occurrence of errors during a particular symbol interval does not influence the occurrence of errors during succeeding symbol intervals. However, in many realistic channels, errors do not occur as independent random events, but tend to occur in bursts. These channels are said to have memory. Examples of channels with memory would be telephone channels that suffer from switching transients, microwave radio links that suffer from fading, and so on.

It is useful to give two simple models for discrete channels with memory, before going onto ideas associated with capacity and coding for these channels. The first model is from Blackwell (1961), and considers "trapdoors," as shown in Figure 24.4. Initially a ball labeled with 0 or 1 is placed in each of the two slots. One of the trapdoors is then opened, with each door having the same probability of being opened. The ball then falls through the open door, and the door then closes behind it. The empty compartment then has another ball placed in it, and the whole process starts over again.

This model defines a channel whose inputs correspond to the balls placed in the empty compartment and whose outputs correspond to the balls that fall through the trapdoors. If

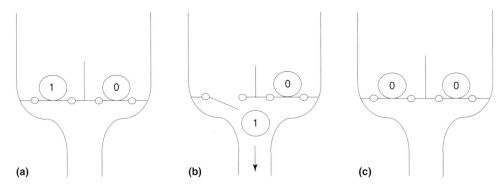

Fig. 24.4 Trapdoor channel.

$$
\text{Input} = 0 \quad
\begin{array}{c c}
& \begin{array}{cccc} S_{00} & S_{10} & S_{01} & S_{11} \end{array} \\
\begin{array}{c} S_{00} \\ S_{10} \\ S_{01} \\ S_{11} \end{array} &
\left[\begin{array}{cccc}
1 & 0 & 0 & 0 \\
1 & 0 & 0 & 0 \\
0 & \frac{1}{2} & 0 & \frac{1}{2} \\
0 & \frac{1}{2} & 0 & \frac{1}{2}
\end{array} \right]
\end{array} = M_0
$$

$$
\text{Input} = 1 \quad
\begin{array}{c c}
& \begin{array}{cccc} S_{00} & S_{10} & S_{01} & S_{11} \end{array} \\
\begin{array}{c} S_{00} \\ S_{10} \\ S_{01} \\ S_{11} \end{array} &
\left[\begin{array}{cccc}
\frac{1}{2} & 0 & \frac{1}{2} & 0 \\
\frac{1}{2} & 0 & \frac{1}{2} & 0 \\
0 & 0 & 0 & 1 \\
0 & 0 & 0 & 1
\end{array} \right]
\end{array} = M_1
$$

Fig. 24.5 Channel matrices for the trapdoor channel.

the symbol b_i corresponds to the condition in which a ball labeled i remains in an occupied slot, and time is started after one of the doors is opened, four states, s_{ij}, for $i_j = 0$ or 1 may be defined as follows. The channel is in state s_{ij} at time $t = n$ if the condition b_j holds at time $t = n$ and condition b_i holds at time $t = n - 1$. An input k therefore corresponds to the placing of a ball labeled k in the unoccupied slot, and the opening of the trapdoor then determines the corresponding output and the next state.

If at time $t = n$ the channel is in states s_{10} and an input 1 is applied, then one ball labeled 0 and one ball labeled 1 rest over the trapdoors. With a probability of $1/2$, the ball 1 falls through, leaving the ball 0 in the occupied slot. The channel then moves to state s_{00} and emits an output of 1. With probability $1/2$ the 0 ball falls through, sending the channel into the state s_{01}, and an output 0 is emitted. The behavior of this channel may thus be described by two matrices, M_0 and M_1. These are called the channel matrices, whose components are the state transition probabilities under the inputs 0 and 1, respectively, and a function that associates an output with each input-state pair, as seen in Figure 24.5.

Another model that is successful in characterizing error bursts and impulsive noise in channels is the so-called Gilbert model, where the channel is modeled as a discrete memoryless binary symmetric channel for which the probability of error is a time-variant parameter. The changes in probability of error are modeled as a Markov process, as shown in

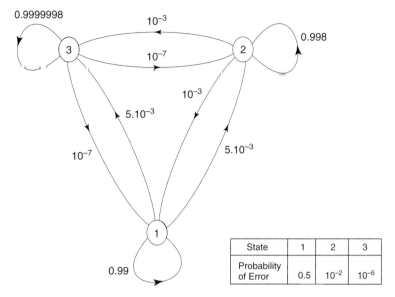

State	1	2	3
Probability of Error	0.5	10^{-2}	10^{-6}

Fig. 24.6 A three-state Gilbert model.

Figure 24.6. The error-generating mechanism in the channel occupies one of the three states. When the channel is in state 2, the probability of error during a bit interval is 10^{-2} and the channel stays in this state during the succeeding bit interval with a probability of 0.998. The channel may make a transition to state 1, which has a bit error probability of 0.5, and since the system stays in this state with a probability of 0.99, errors tend to occur in groups.

The models developed for the case of a discrete channel with memory are those for which the channel has a finite number of internal states, for which the present state of the channel represents a summary of its history. The application of an input will result in a transition to another state, with the production of an output. The resulting source channel matrix then determines a finite Markov chain with states (a_i, s_k), where a_i is the state of the source and s_k is the state of the channel.

If the chain associated with the source channel matrix has steady state probabilities, then given the pair (a_i, s_k) one can determine the corresponding input $f_1(a_i, s_k) = f(a_i)$ and output $g_1(a_i, s_k) = g(f(a_i), s_k)$. Thus the chain associated with the source channel matrix determines a stationary sequence of input-output pairs (X_n, Y_n) for $n = 1, 2, \ldots$ An input, output, and joint uncertainty may therefore be defined as in Equations 24.68 through 24.70.

$$H(X) = \lim_{n \to \infty} H(X_n \mid X_1, \ldots X_{n-1}) \tag{24.68}$$

$$H(Y) = \lim_{n \to \infty} H(Y_n \mid Y_1, \ldots Y_{n-1}) \tag{24.69}$$

$$H(X, Y) = \lim_{n \to \infty} H[(X_n, Y_n) \mid (X_1, Y_1), \ldots (X_{n-1}, Y_{n-1})] \tag{24.70}$$

The information conveyed about the process X by the process Y is given by Equation 24.71.

$$I[X \mid Y] = H(X) + H(Y) - H(X, Y) = I[Y \mid X] \tag{24.71}$$

The capacity of the given channel is defined as the least upper bound of $I\,[X \mid Y]$, taken over all regular Markov sources, and it can be proved that it is possible to transmit information at any rate less than the capacity with an arbitrarily small probability of error.

The preceding discussion says nothing about how one calculates the channel capacity of a given finite-state channel. This is an extremely complex task, and details may be found in Proakis (1987).

It is possible to model a channel such that the model combines features of the discrete memoryless channel with those of the discrete channel with memory, and this defines a compound channel.

Continuous Channels

Communications channels may be continuous in two senses. In the first sense, we may allow the input and output alphabets to contain an infinite number of elements, but we require that the transmitted material be in the form of a discrete sequence of symbols. This type of channel is called a time-discrete, amplitude-continuous channel.

For the second type of continuous channel, we allow the transmission of information to be continuous in time.

Consider an analogue source that emits a message waveform $x(t)$ that is a sample function of a stochastic process $X(t)$. If $X(t)$ is a stationary stochastic process with autocorrelation function $\Phi_{xx}(\tau)$ and power spectral density function $\Phi_{xx}(f)$, and if $X(t)$ is a band-limited process such that Equation 24.72 is satisfied, then $X(t)$ has representation given by the sampling theorem, as in Equation 24.73.

$$\Phi_{xx}(f) = 0 \quad \text{for } |f| > W \tag{24.72}$$

$$X(t) = \sum_{n=-\infty}^{\infty} X\left(\frac{n}{2W}\right) \frac{\sin 2\pi W\left(t - \dfrac{n}{2W}\right)}{2\pi W\left(t - \dfrac{n}{2W}\right)} \tag{24.73}$$

This means that the band-limited signal can be represented by a sequence of samples, as in Equation 24.74, sampled at the rate given by Equation 24.75, the so-called Nyquist rate. These samples may then be encoded using various techniques to give a digital representation of the analogue signal.

$$X_n = X\left(\frac{n}{2W}\right) \tag{24.74}$$

$$f_s = 2W \qquad \text{samples per second} \tag{24.75}$$

One such technique is Pulse Code Modulation (PCM), the essential operations being sampling, quantizing, and encoding, usually performed in the same circuit, called an analogue-to-digital converter (ADC). It should be noted that PCM is not modulation in the conventional sense of the word, since modulation usually refers to the variation of some characteristic of the carrier wave in accordance with an information-bearing signal. The only part of PCM that is similar to this definition is sampling. The subsequent use of quantization, which is basic to PCM, introduces a signal distortion that has no counterpart in conventional modulation.

Let $x(t)$ denote a sample function emitted by a source, and let x_n represent the samples taken at a sampling rate greater than the Nyquist rate, $2W$, where W is the highest frequency present in the signal. In PCM, each sample is quantized to one of 2^b amplitude levels, where b is the number of binary digits used to represent each sample. The rate from the source is therefore bf_s bits per second, where f_s is the sampling frequency.

The quantization may be represented as in Equation 24.76, where $\tilde{x}_n$ represents the quantized value of x_n and q_n represents the quantization error, which is treated as additive noise.

$$\tilde{x}_n = x_n + q_n \tag{24.76}$$

If a uniform quantizer is used, the quantization noise is well represented by a uniform probability density function, and the mean square value of the quantization error can be shown to be as in Equation 24.77, where $\Delta = 2^{-b}$ is the step size of the quantizer.

$$E(q^2) = \frac{\Delta^2}{12} = \frac{2^{-2b}}{12} \tag{24.77}$$

Measured in decibels, the mean square value of the noise is as in Equation 24.78.

$$10\log\frac{\Delta^2}{12} = -6b - 10.8\,\mathrm{dB} \tag{24.78}$$

For example, a 7-bit quantizer gives a quantization noise power of $-52.8\,\mathrm{dB}$.

A uniform quantizer provides the same spacing between successive levels throughout the entire dynamic range of the signal. In practice this can cause problems since for many signals—speech, for example—small signal amplitudes occur more frequently than large signal amplitudes, and a better approach would have more closely spaced levels at low signal amplitudes, and vice versa. The desired form of nonuniform quantization can be achieved by using a compressor followed by a uniform quantizer. By cascading this combination with an expander complementary to the compressor, the original signal samples are restored to their correct values apart for quantization errors. This combination of compressor and expander is called a compander.

There are basically two types of compander in current use, μ-law and A-law. The transfer characteristic of the compressor is represented by a memoryless, nonlinearity $c(x)$, where x is the sample value of a random variable X denoting the compressor input. In the μ-law compander, $c(x)$, is continuous, approximating a linear dependence on x for low input levels and a logarithmic dependence for high input levels, as in Equation 24.79, for $0 \le \frac{|x|}{x_{\max}} \le 1$:

$$\frac{c(|x|)}{x_{\max}} = \frac{\ln\left(1 + \frac{\mu|x|}{x_{\max}}\right)}{\ln(1+\mu)} \tag{24.79}$$

The special case of uniform quantization corresponds to the case where $\mu = 0$. The μ-law is used for PCM telephone systems in the United States, Canada, and Japan.

In A-law companding, the compressor characteristic is piecewise, made up of a linear segment for low-level inputs and a logarithmic segment for high-level inputs. This type of compander is used for the PCM telephone systems in Europe.

In PCM, each sample of the waveform is encoded independently of all the other samples. However, most source signals sampled at or above the Nyquist rate exhibit significant correlation between successive samples (i.e., the average change in amplitude between successive samples is relatively small). An encoding scheme that exploits the redundancy in the samples will result in a lower bit rate for the source output, and one such system called Differential Pulse Code Modulation (DPCM) does this by encoding the differences between successive samples rather than the samples themselves.

A subclass of DPCM is delta modulation, in which the code word is only one binary digit in length. Delta modulators have the advantage over more conventional PCM systems of having a particularly simple architecture, which forms the basis of "single-chip" coder-decoder (CODEC) integrated circuits. Because delta modulation is a 1-digit code, sampling rate and output digit rate are the same. However, in order for delta modulation to achieve performance comparable to that of conventional n-digit PCM systems, the sampling rate must be increased substantially over that required for a PCM system.

Most real signal sources are nonstationary in nature, which means that the variance and the autocorrelation function of the source output vary with time. PCM and DPCM encoders are designed on the basis that the source output is stationary, and the performance of these encoders can be improved by using adaptive methods. One such method is the use of an adaptive quantizer.

PCM, DPCM, and the adaptive counterparts are source encoding techniques that attempt to represent the output waveforms. Consequently, these methods are known as waveform encoding techniques. In contrast to these methods, linear predictive coding models the source as a linear system, which when excited by an appropriate input signal gives rise to an observed source output. Instead of transmitting the samples of the source waveform to the receiver, the model parameters of the linear system are transmitted together with the appropriate excitation signal.

References

Cox, R. T. (1946) *Am. J. Phys.* 17, 1.

Gull, S. (1988) Developments in maximum entropy data analysis. In *Maximum Entropy and Bayesian Methods* (ed. J. Skilling). Kluwer Academic Press.

Hartley, R. V. L. (1928) Transmission of information. *Bell Systems Technical Journal*, July.

Jaynes, E. T. (1978) Where do we stand on maximum entropy. In *The Maximum Entropy Formulism* (ed. R. D. Levine and M. Tribus). M.I.T. Press.

———. (1989) *Papers on Probability, Statistics and Statistical Physics* (ed. R. D. Rosenkranz). Kluwer Academic Press.

Leff, H. S., and Rex. A. F. (ed.) (1990) *Maxwell's Demon, Entropy, Information, Computing.* Adam Hilger Press.

Proakis, J. (1987) *Digital Communications.* McGraw-Hill.

Shannon, C. (1949) *The Mathematical Theory of Communication*, University of Illinois Press.

Skilling, J. (ed.) (1988) *Maximum Entropy and Bayesian Methods.* Kluwer Academic Press.

List of Acronyms

1G	First-generation cellular system	ANI	Application to Network Interface
2.5G	Two-and-a-half-generation cellular system	ANSI	American National Standards Institute
2G	Second-generation cellular system	AP	Application Process
3G	Third-generation cellular system	AR	Autoregressive model
3GPP	3rd Generation Partnership Project	ARCEP	L'autorité de régulation des communications électroniques et des postes
3GPP2	3rd Generation Partnership Project 2	ARIB	Association of Radio Industries and Businesses
4G	Fourth-generation cellular system	ARP	Address Resolution Protocol
π/4 DQPSK	π/4 Differential Quaternary Phase Shift Keying	ARQ	Acknowledge Receipt
		AS	Autonomous System; Application Server
A/D	Analogue-to-digital conversion	ASCII	American Standard Code for Information Interchange
AAA	Authentication, Authorization, and Accounting	ASE	Application Service Element
ABE	Aggregated Business Entities	ASK	Amplitude Shift Keying
ACC	Analogue Control Channel	ASN	Abstract Syntax Notation
ACCH	Associated Control Channel	ATM	Asynchronous Transfer Mode
ACL	Asynchronous Connectionless Link	AUC	Authentication Center
ACSE	Association Control Service Element	BCCH	Broadcast Control Channel
		BCH	Broadcast Channel
ADSL	Asymmetric Digital Subscriber Line	BER	Bit Error Rate
		BGCF	Breakout Gateway Control Function
ADT	Antenna Diversity Technology		
AFH	Adaptive Frequency Hopping	BGP	Border Gateway Protocol
AGCH	Access Grant Control Channel	BHCA	Busy hour call attempts
AH	Authentication Header	BIB	Backward Indicator Bit
Air interface	Implies the RF channel	Bluetooth	Wikipedia, the free encyclopedia1.htm—see this for a list of competing technologies, links in several languages
ALL-IP	All IP networks		
ALOHA	Protocol that became the basis for the Ethernet		
AMA	Active Member Address	BOOTP	BOOT Protocol
AMPS	Advanced Mobile Phone Service/System	BPSK	Binary Phase Shift Key

BS	Base Station	CSMA/CD	Carrier Sense Multiple Access with Collision Detection
BSC	Base Station Controller		
BSN	Backward Sequence Number	CT	Centres du transit
BSS	Base station subsystem	CU	Coordinator Unit
BT	Bandwidth-Time product; British Telecom		
		D-AMPS	Digital-Advanced Mobile Phone Service/System
BTS	Base transceiver station		
		DAR	Dynamic Alternative Routing
C/N	Carrier to Noise ratio	DARPA	Defense Advanced Research Projects Agency
CAVE	Cellular Authentication, Voice Privacy and Encryption		
		dBi	Antenna gain with respect to an isotropic radiator
CC	Connection Counter		
CCH	Control Channel	dBr	Decibels relative to maximum
CCITT	Comite Consultatif Internationale de Telegraphie et Telephonie (now ITU-T)		
		DC	Direct current
		DCC	Digital Control Channel
CCS	Common-channel Signaling	DCCE	Digital Call Center Exchange
CDMA	Code Division Multiple Access	DCR	Dynamically Controlled Routing
CDMA-2000	Code Division Multiple Access 2000		
		DCS	Digital Cellular System
CDMAOne	Code Division Multiple Access system One	DDI	Direct dialing in
		DECT	Digital Enhanced Cordless Telecommunications
CDPD	Cellular Digital Packet Data		
CFP	Contention-Free Period	DF	Don't fragment
CIDR	Classless Interdomain Routing	DFT	Discrete Fourier Transform
CLASS	Customer Line Access Signaling System	DHCP	Dynamic Host Configuration Protocol
CLI	Calling Line Identity	DiffServ	Differentiated Services
CLNP	Connectionless Network Protocol	DMSU	Digital Main Switching Unit
		DNHR	Dynamic NonHierarchical Routing
CM	Connection Management		
CMEA	Cellular Message Encryption Algorithm	DNS	Domain Name System
		DoJ	Department of Justice
CMIP	Common Management Information Protocol	DPCM	Differential Pulse Code Modulation
		DPSK	Differential Phase Shift Key
CMIS	Common Management Information Service	DQPSK	Differential Quaternary Phase Shift Keying
CNET	Cellular NET	DS	Direct Sequence
ComReg	Commission for Communications Regulation	DSCP	Differentiated Services Code Point
CORBA	Common Object Request Broker Architecture	DSL	Digital Subscriber Line/Loop
		DSLAM	Digital Subscriber Line Access Multiplexer
COS	Class of Service		
CRC	Cyclic Redundancy Check	DSSS	Direct Sequence Spread Spectrum
CRM	Customer Relationship Management		
		DTC	Digital Traffic Channel
CSCF	Call Session Control Function	DTMF	Dual Tone Multifrequency
CSD	Circuit-Switched Data	DV	Distance Vector
CSMA	Carrier Sense Multiple Access protocol	DVCC	Digital Verification Color Code
CSMA/CA	Carrier Sense Multiple Access with Collision Avoidance	DXC	Digital Cross Connect

EAC	Enquiry Access Code	FHSS	Frequency-Hopped Spread Spectrum
EC	European Commission		
ECM	Error Correcting fax Machine	FIB	Forward Indicator Bit
ECTA	European Competitive Telecommunications Association	FISU	Fill in signaling units
		FM	Frequency Modulation
		FMC	Fixed Mobile Convergence
EDGE	Enhanced Data GSM Evolution	FSK	Frequency Shift Keying
		FSN	Forward Sequence Number
EDR	Enhanced Data Rate, in Bluetooth	FTP	File Transfer Protocol
		FTTH	Fibre to the Home
EGP	Exterior Gateway Protocol		
EIGRP	Enhanced Interior Gateway Routing Protocol	GDMO	Guidelines for Definitions of Managed Objects
EIR	Equipment Identity Register	GERAN	GSM/EDGE/Radio Access Network
EIRP	Effective Isotropic Radiated Power		
		GFSK	Gaussian Frequency Shift Keying
EKF	Extended Kalman Filter		
ELSR	Edge Label Switched Router	GII	Global Information Infrastructure
ENUM	Electronic numbering		
ERL	Echo Return Loss	GMSK	Gaussian Minimum Shift Keying
eSCO	extended Synchronous Connection Oriented packets		
		GOS	Grade of Service
ETA	Enhanced Temporal Algorithm	GPRS	General Packet Radio Service
ETACS	Extended Total Access Communications System	GSC	Group Switching Center
		GSM	Global System for Mobile communication
ETDMA	Extened TDMA		
ETNO	European Telecommunications Network Operators Association	GTT	Global Title Translation
		H2.5G	High 2.5 Generation
eTOM	enhanced Telecoms Operations Map	HIPERLAN	High Performance Radio LAN
		HLR	Home Location Register
ETSI	European Telecommunications Standards Institute	HSCSD	High-Speed, Circuit-Switched Data
EU	European Union	HSDPA	High-Speed Downlink Packet Access
EVDO	Evolution Data Only/Evolution Data Optimized system		
		HSS	Home Subscriber Server
EVDOV	Evolution Data and Voice	HSUPA	High-Speed Uplink Packet Access
FACCH	Fast Access Control Channel	HTML	Hypertext Markup Language
FCC	Federal Communications Commission	HTTP	Hypertext Transfer Protocol
FCCCH	Frequency Correction Control Channel	ICANN	Internet Corporation for Assigned Names and Numbers
FCS	Frame Check Sequence		
FDD	Frequency Division Duplex	ICE	Interactive Connectivity Establishment
FDDI	Fibre Distributed Data Interface		
FDMA	Frequency Division Multiple Access	ICMP	Internet Control Message Protocol
FEC	Forward Error Correction	ID	Identity
FFSK	Fast Frequency Shift Keying	IDFT	Inverse Discrete Fourier Transform
FFT	Fast Fourier Transform		
FH/TDMA	Frequency Hopped TDMA	IDL	Interface Definition Language

IDN	Integrated Digital Network	ITU	International Telecommunications Union
IEC	International Electrotechnical Commission	ITU-R	ITU Radio communication sector
IEEE	Institute of Electronic and Electrical Engineers	ITU-T	International Telecommunications Union– Telecommunication standardization sector
IEEE 802.11	IEEE LAN standard		
IEEE 802.3	IEEE Ethernet standard		
IEEE 802.5	IEEE Token Ring standard		
IEEE SA	Institute of Electronic and Electrical Engineers Standards Association	IWF	Interworking Function
		IXC	Inter Exchange Carrier
IETF	Internet Engineering Task Force	J2EE	Java 2 Enterprise Edition
IGMP	Internet Group Message Protocol	L2.5G	Low 2.5 Generation
IGRP	Interior Gateway Routing Protocol	LAI	Location Area Information
		LAN	Local Area Network
IHL	IP Header Length	LAPDM	Link Access Protocol in the D Channel (modified)
ILEC	Incumbent Local Exchange Carrier	LATA	Local Access and Transport Area
IMEI	International Mobile Equipment Identity	LBA	Least busy alternative
IMS	IP Multimedia Subsystem	LFN	Long Fat Network
IMSI	International Mobile Subscriber Identity	LLU	Local Loop Unbundling
		LSD	Link State Database
IMT	International Mobile Telecommunications 2000	LSR	Label Switched Router
IMTS	Improved Mobile Telephone Service	MAC	Medium Access Layer; Medium Access Control Protocol
IN	Intelligent Network	MAP	Mobile Application Part
IntServ	Integrated Services	MATS-E	Mobile Automatic Telephone System
INTUG	International Telecommunications Users Group	MCI	Malicious Call Identification
		MCMC	Markov chain Monte Carlo
IP	Internet Protocol	MCS	Mobile Cellular System
IPR	Intellectual Property Rights	MCU	Multipoint Control Unit
IR	Infrared	META	MAC Enhanced Temporal Algorithm
IRC	Interference Rejection Combining	MF	More fragments
IRG	Independent Regulators Group	MGC	Media Gateway Controller
IS	Interface Specification	MGCF	Media Gateway Control Function
ISDN	Integrated Services Digital Network	MGCP	Media Gateway Control Protocol
ISDN	International Subscriber Dialing		
IS-IS	Intermediate System to Intermediate System	MGW	Media Gateway
		MIME	Multipurpose Internet Mail Extension
ISM	Industrial Scientific and Medical frequency band	MM	Mobility Management
ISN	Initial Sequence Number	MMAP	Marginal maximum a posteriori
ISO	International Organization for Standardization	MMI	Man–Machine Interface
		MMS	Multimedia Message Service
ISP	Internet Service Provider	MMSE	Minimum Mean Square Estimate
ISUP	ISDN User Part		

MOS	Mean Opinion Score	OSPF	Open Shortest Path First
MPLS	Multiprotocol Label Switching	OSS	Operational Support System
MRC	Maximum Ratio Combining		
MRFC	Multimedia Resource Function Controller	PABX	Private Automatic Branch Exchange
MRFP	Multimedia Resource Function Processor	PAX	Private Automatic Exchange
		PC	Personal Computer
MS	Mobile Station	PCG	Project Coordination Group
MSC	Mobile Switching Center	PCH	Paging Control Channel
MSE	Mean squared error	PCI	Protocol Control Information
MSISDN	Mobile station ISDN	PCM	Pulse Code Modulation
MSRN	Mobile Subscriber Roaming Number	PCMIA	Personal Computer Memory Card International Association
MTBF	Mean time between failure		
MTP	Message Transfer Protocol	PCS	Personal Communication System
MTU	Maximum Transmission Unit	P-CSCF	Proxy-Call/Session Control Function
NADC	North American digital cellular	PDA	Personal Digital Assistant
N-AMPS	Narrowband Advanced Mobile Phone System	PDU	Protocol Data Unit
		PESQ	Perceptual Evaluation of Speech Quality
NARUC	National Association of Regulatory Utility Commissioners	PHB	Per hop behavior
		PHY	Physical Layer
NAT	Network Address Translation	Piconet	Up to 7 wireless-connected Bluetooth devices
NDC	National Destination Code	PIN	Personal Identification Number
NGN	Next Generation Network		
NGOSS	New Generation Operations Systems and Software	PLC	Packet Loss Concealment
NIC	Network Interface Card	PLMN	Public Land Mobile Network
NMT	Nordic Mobile Telephone System	PMA	Parked Member Address
		PMR	Private Mobile Radio
NNI	Network Node Interface	PN	Pseudo-noise
NPL	National Physical Laboratory	PON	Passive Optical Network
NRA	National Regulatory Authority	POP	Point of Presence
NTIA	National Telecommunications and Information Administration	POP3	Post Office Protocol Version 3
		POTS	Plain ordinary telephone service
NTP	Network Time Protocol	PPP	Point to Point Protocol
		PSK	Phase Shift Keying
OASIS	Organization for the Advancement of Structured Information Standards	PSTN	Public Switched Telephone Network
		PTDF	Proportional Traffic Distribution Facility
OFCOM	Office of Communications	PTO	Public Telecommunications Operator
OFDM	Orthogonal Frequency Division Multiplexing		
OFDMA	Orthogonal Frequency Division Multiple Access	PUC	Public Utility Commission
OLR	Overall Loudness Rating	QAM	Quadrature Amplitude Modulation
OMA	Open Mobile Alliance		
OSA	Open Service Access	QoS	Quality of Service
OSI	Open Systems Interconnection	QPSK	Quadrature Phase Shift Keying

RACH	Random Access Control Channel
RAM	Random Access Memory
RAN	Radio Access Network
RAND	Random Number
RARP	Reverse Address Resolution Protocol
RAS	Registration, Admission and Status
RBOC	Regional Bell Operating Companies
RCU	Remote Concentrator Unit
RIP	Routing Information Protocol
RPE-LPC	Regular Pulse Excited Linear Predictive Coder
RR	Radio Resource
RSVP	Resource Reservation Protocol
RTCP	Realtime Transport Control Protocol
RTNR	Real Time Network Routing
RTP	Realtime Transport Protocol
RTT	Radio Transmission; Round Trip Time
SA	System Aspects
SA1	System Aspects (services)
SA2	System Aspects (architecture)
SACCH	Slow Access Control Channel
SACK	Selective Acknowledgement
SAP	Service Access Point
SAT	Supervisor Audio Tone
SCC	Standards Council of Canada
SCCP	Signaling Connection Control Part
SCH	Synchronization Control Channel
SCO	Synchronous Connection-Oriented link
SCP	Service Control Point
S-CSCF	Serving Call Session Control Function
SDCCH	Standalone Dedicated Control Channel
SDH	Synchronous Digital Hierarchy
SDP	Session Description Protocol
SDU	Service Data Unit
SFD	Start of Frame Delimiter
SID	Shared Information and Data framework
SIG	Special Interest Group
SIM	Subscriber Identity Module
SIP	Session Initiation Protocol

SIPS	Secure version of SIP
SIR	Sequential Importance Resampling
SLEE	Service Logic Execution Environment
SLIC	Subscriber's Line Interface Card
SLIP	Serial Line Internet Protocol
SMC	Sequential Monte Carlo
SMP	Significant Market Player
SMS	Short Message Service
SMTP	Simple Mail Transfer Protocol
SNMP	Simple Network Management Protocol
SNR	Signal-to-noise radio
SOA	Service Oriented Architecture
SOAP	Simple Object Access Protocol
SoLSA	Support of Localized Service Area
SONET	Switched Optical Network
SP	Service Provider
SPC	Stored Program Control
SRES	Signed Response
SS	Spread Spectrum
SS7	Signaling System No 7
SSP	Service Switching Point
STD	Subscriber Trunk Dialing
STM	Synchronous Transport Module
STP	Signal Transfer Point
STS	Space-Time-Space switching
STUN	Simple Traversal of User datagram protocol through Network address translators
SU	Signaling Unit
SYNC	Synchronization Channel
TACS	Total Access Communications System
TBTT	Target Beacon Transfer Time
TCAP	Transaction Capabilities Applications Part
TCH	Traffic Channel
TCM	Trellis Coded Modulation
TCP	Transmission Control Protocol
TDM	Time Division Multiplex
TDMA	Time Division Multiple Access
TELR	Talker Echo Loudness Rating
TISPAN	Telecommunications and Internet Converged Services and Protocols for Advanced Networking

TMF	Telecommunications Management framework; TeleManagement Forum		URL	Uniform Resource Locator
			USO	Universal Service Obligation
TMN	Telecommunication Management Network		UWB	Ultra-Wideband
TMSI	Temporary Mobile Subscriber Identity		VAD	Voice Activity Detector
			VBS	Voice Broadcast Service
TNA	Technology Neutral Architecture		VDSL	Very high bit rate Digital Subscriber Loop
ToS	Time of Service		VF	Voice Frequency
TS	Time slot		VGCS	Voice Group Call Service
TST	Time-Space-Time		VLR	Visitor Location Register
TTA	Telecommunications Technology Association		VoIP	Voice over Internet Protocol
			VPN	Virtual Private Network
TTL	Time to Live field		WAN	Wide Area Network
TU	Traffic Unit		WAP	Wireless Application Protocol
TUP	Telephone User Part		W-CDMA	Wideband Code Division Multiple Access
TURN	Traversal Using Relay NAT		Wi-Fi	Wireless Fidelity
UDDI	Universal Description, Discovery, and Integration language		WiMax	World Interoperability for Microwave Access
			WLAN	Wireless Local Area Network
UDP	User Datagram Protocol		WPAN	Wireless Personal Area Network
UE	User Equipment		WSDL	Web Service Definition Language
UML	Unified Modeling Language			
UMTS	Universal Mobile Telephone Service		xDSL	Any type of Digital Subscriber Line
UNI	User Network Interface			
URI	Uniform Resource Identifier		XML	Extensible Markup Language

Index

A

A-law companders, 517–518
Absolute errors, 339
ACC (Analogue Control Channel), 295
Accept-reject random number method, 359–360
Access channels in CDMA, 296
Access Code element in Bluetooth packets, 323
Access Directive, 17–18
Access issues and techniques
cellular systems, 250–254
interconnections, 26–28
Access network frequency plans, 31
Accumulation of errors, 340
ACK bits, 144–145
Acknowledgment packets, 160
ACL (Asynchronous Connectionless links, 322
A/D. *See* Analogue-to-digital conversion
Adaptive Frequency Hopping (AFH), 329–330
Adaptive routing strategy, 406–407
Addition errors, 340
Address Resolution Protocol (ARP), 140
Addresses, IP. *See* IP addresses
Adjoints of square matrices, 467
Administrative information in SIM, 282
Administrative law, 14–15
ADSL (Asymmetric Digital Subscriber Loop), 29
Advanced Mobile Phone System (AMPS), 274–275
Advanced subscriber features, 108
AFH (Adaptive Frequency Hopping), 329–330
ALL-IP networks, 245

Allocative efficiency, 12
ALOHA networks, 366–367, 389–390
Alternative routing, 83–84
Amplitude Shift Keying (ASK), 234
AMPS (Advanced Mobile Phone System), 274–275
Analogue Control Channel (ACC), 295
Analogue systems
cellular, 270
circuit-switched telephone networks, 72
vs. digital transmissions, 97–98
Analogue-to-digital (A/D) conversion
continuous channels, 516
digital circuit switching, 98
ANI interface, 194
ANSI/TIA/EIA-627 standard, 297–298
Antennas in Bluetooth, 317–318
Application layer
OSI, 46
TCP/IP, 126, 166–168
Application Server (AS) in IMS, 195, 212
Approximation
small angles, 439
in statistical analysis, 338–340
AR (Autoregressive) model, 429
Area-border routers, 162–163
Areas in routing hierarchy, 155, 161
Argand diagrams, 442
Arithmetic for matrices, 467–468
Arithmetic mean, 340
Arithmetic series, 458
ARP (Address Resolution Protocol), 140
Arrival distributions for queues, 375–376

AS (Application Server) in IMS, 195, 212
AS (Autonomous System), 155, 161–164
ASK (Amplitude Shift Keying), 234
Associated mode in SS7, 111
Assurance process in eTOM, 218
Asymmetric Digital Subscriber Loop (ADSL), 29
Asymmetric regulation, 15
Asynchronous ALOHA networks, 389–390
Asynchronous Connectionless links (ACL), 322
Asynchronous Digital Subscriber Line (ADSL) modems, 235–236
Asynchronous Transfer Mode (ATM) networks, 200–201
Attenuation in telephone networks, 56
Authentication Center (AUC) in GSM, 280
Authorization Directive, 17
Automatic alternative routing, 83–84
Autonomous System (AS), 155, 161–164
Autoregressive (AR) model, 429
Average power in generalized Fourier expansion, 478
Averages
dispersion from, 341–343
measurements, 340–341

B

Back-haul connections, 30
Backbones in routing hierarchy, 162–164
Backward Indicator Bit (BIB), 116–117
Backward Sequence Number (BSN), 116